Materials Development and Processing for Biomedical Applications

Materials Development and Processing for Biomedical Applications

Edited by

Savaş Kaya, Sasikumar Yesudass,
Srinivasan Arthanari, Sivakumar Bose,
Goncagül Serdaroğlu

CRC Press
Taylor & Francis Group
Boca Raton London

CRC Press is an imprint of the
Taylor & Francis Group, an **informa** business

First edition published 2022
by CRC Press
6000 Broken Sound Parkway NW, Suite 300, Boca Raton, FL 33487–2742

and by CRC Press
2 Park Square, Milton Park, Abingdon, Oxon, OX14 4RN

CRC Press is an imprint of Taylor & Francis Group, LLC

Library of Congress Cataloging-in-Publication Data
Names: Kaya, Savaş, editor. | Yesudass, Sasikumar, editor. | Arthanari, Srinivasan, editor. | Bose, Sivakumar, editor. | Serdaroğlu, Goncagül, editor.
Title: Materials development and processing for biomedical applications / edited by Savaş Kaya, Sasikumar Yesudass, Srinivasan Arthanari, Sivakumar Bose, Goncagül Serdaroğlu.
Description: First edition. | Boca Raton : CRC Press, 2022. | Includes bibliographical references and index.
Identifiers: LCCN 2021047080 | ISBN 9781032002880 (hardback) | ISBN 9781032002903 (paperback) | ISBN 9781003173533 (ebook)
Subjects: LCSH: Biomedical materials.
Classification: LCC R857.M3 M384 2022 | DDC 610.28—dc23/eng/20220124
LC record available at https://lccn.loc.gov/2021047080

ISBN: 978-1-032-00288-0 (hbk)
ISBN: 978-1-032-00290-3 (pbk)
ISBN: 978-1-003-17353-3 (ebk)

DOI: 10.1201/9781003173533

Typeset in Times
by Apex CoVantage, LLC

Contents

Preface

It is our immense pleasure to introduce the book titled *Materials Development and Processing for Biomedical Applications*. This book is mainly focused on various methods of manufacturing, surface modifications, and advancements in biomedical applications for all kinds of readers. In particular, the fundamental aspects are discussed for a better understanding of the processing of various biomedical materials such as metals, ceramics, polymers, composites, etc. Besides, advancements in various fields of biomedical applications are emphasized. The book is basically focusing on five different aspects such as materials properties, development, processing, surface coatings, and future perspectives of advanced biomedical device fabrications.

The development and applications of various metallic and non-metallic materials are given importance in recent days and advancements in the development of new materials are highly appreciated. Biomedical materials from the micro-nano scale possess various properties which will significantly affect the resulting properties. The introduction to biomaterial and the properties of various biomedical materials are important to understand for the further development of biomedical materials with better biocompatibility and without any adverse effects. Authors have elaborated the various properties of biomedical materials at the beginning in several chapters which will enrich the fundamental knowledge of the readers. Furthermore, synthesis of nano materials, the properties of degradable and non-degradable (permanent) implant materials are discussed extensively. Following the materials properties part, the development of degradable and polymeric materials is discussed. Interestingly, Mg alloys have similar mechanical properties close to the natural bone; however, their chemical reactivity is much faster and a faster degradation rate limits them for implant applications. Ultra-pure Mg possesses a very low corrosion rate (~0.1 mm/y), yet mechanical properties are poor, and alloying and processing are necessary to enhance the properties. The degradable metallic implant materials, particularly of Mg-based alloys developments, processing through severe plastic deformation (SPD) processes are discussed for manufacturing degradable stent materials which shows the improvement in the required properties. Besides, the development of 2D materials and applications of 3D printing technology to fabricate polymeric materials for biomedical applications are discussed.

Laser processing is one of the precise techniques used for the fabrication of micro/nano functional surfaces and is attractive for biomedical applications. Surface laser texturing to fabricate the micro/nano surfaces and their biomedical applications are discussed. Laser powder bed fusion (LPBF) to fabricate Ti-based alloys are discussed in two parts. Part 1 discusses fabrication, the process details, and the influencing parameters of LPBF. Further, Part 2 discusses the characteristic properties of as-built and post-treated titanium alloys. These chapters comprehensively discuss the prospects of LPBF, and it is one of the promising topics for the fabrication of Ti-based alloys for biomedical applications and readers can get the benefits from these chapters. Online monitoring during laser processing is one of the interesting topics to understand and optimizing the processing condition are discussed. In particular, collaborative monitoring and artificial intelligence of optical signals, photoacoustic signals, image signals, temperature signals, etc., during high potential laser surgeries (such as orthopedic surgery, eye surgery, and so on) will be beneficial for the precise control of the process. It is one of the demanding technologies for the modern biomedical industries. Furthermore, laser processing and ablation as one of the rapid processes has also been focused on fabricating the nanoparticles for various biomedical applications. In this context, the laser-assisted production of calcium phosphate nanoparticles from marine origin has also been discussed, which will give the idea to the readers to expand the laser processing for nanoparticle synthesis for the applications in various scales for energy and environmental applications.

Surface treatments and modification of implant materials are beneficial to alter the surface-related properties such as surface energies, chemical composition, corrosion resistance, mechanical properties, and biocompatibility. Furthermore, surface treatments and subsurface modification in certain cases are advantageous for the implant to introduce porous structures without altering the bulk properties of implants. Surface treatments such as surface mechanical attrition treatments (SMAT), chemical conversion treatments/coatings, anodic oxidation, plasma electrolytic oxidation, polymeric coatings, ceramic coatings, composite coatings, vacuum deposition, plasma coatings, electrospinning, etc. are some of the techniques commonly used for metallic implants. The introduction to these techniques, processing conditions, and properties are emphasized by several authors and discussed in detail. The SMAT is a mechanical treatment used to refine the microstructure at the surface and subsurface levels to enhance the corrosion resistance. Anodic oxidation of titanium alloys results in the formation of ordered nanotubes and enhances the biocompatibility of implants. Further, control of process parameters such as anodizing environment, condition, and post-treatments result in porous to ordered surfaces and altered properties. A detailed discussion about the influencing parameters and results properties are explained. Electrospinning is one of the evolving techniques to coat polymeric and composite materials directly over implant surfaces; one of the studies on electrospinning of polymers on degradable implants is discussed for biomedical applications. A wide range of surface treatments covered in this book will be helpful for readers to understand the importance of surface treatments and their future perspectives.

At the end of this book, the chapters discuss the advanced techniques such as flexible electronics, biosensors, microfluidic devices, chips, etc. for advanced health care applications. An overview of the key advances in wearable, flexible, smart biosensors and their potential in sensitivity and reliability is discussed. In particular, materials/components used for advanced health care diagnostic devices, signal measurements, and exercise-based wearable devices, ocular wearable, internet-of-things-based biosensors for the biomedical field are explained. Various types of organs-on-a-chip (OOc), also called micro physiological systems, used in biomedical applications are also explained. Overall, the chapters present in the book are comprised of a wide range of topics for the benefit of the readers working in the area of biomedical applications. Therefore, the editors strongly believe that the resources given in this book from various authors will be helpful for researchers from basic to advanced levels.

Thank you.

Editorial Team

Editors

Savaş Kaya is Associate Professor of Inorganic Chemistry at Sivas Cumhuriyet University, Health Services Vocational School, Department of Pharmacy, Sivas/Turkey. He earned a doctorate degree in 2017 in the field of Theoretical Inorganic Chemistry. He does research in Theoretical Chemistry, Computational Chemistry, Materials Science, Corrosion Science, Physical Inorganic Chemistry and Coordination Chemistry. Savaş Kaya has published more than 180 papers in international journals indexed SCI and SCI expanded with h-index = 26. He is the editor of the book *Conceptual Density Functional Theory and Its Applications in the Chemical Domain*. He is the author of ten book chapters. Recently, he introduced the Kaya chemical reactivity approach and the Kaya combined reactivity descriptor and proposed some electronic structure principles.

Dr. Sasikumar Yesudass is currently working as Post-Doctoral Researcher in the School of Materials Science and Engineering, Tianjin University of Technology, China since 2018. He is presently working on the surface coatings of Mg alloys with a focus on improving bioactivity and is interested in advanced orthopedic implant devices. His major research areas include the electrochemical behavior of magnesium and titanium alloys for bio-implant applications and the corrosion inhibition studies of steels. Prior to joining Tianjin University in China, he received the CAPES Post-Doctoral fellowship award from the Brazilian government, at Central Federal Technical University, Rio de Janeiro, Brazil, and worked there 2017–2018. Besides, he has also received the National Research Fellow (NRF) Innovation Post Doctoral Research Fellowship award through the DST/National Research Foundation, at North West University, South Africa during 2014–2016. After his doctoral research, he worked as Assistant Professor in the Department of Physics, RRASE College of Engineering, Chennai, India 2012–2013. Dr. Sasikumar received his doctoral research (PhD) in surface modification and electrochemical behavior of titanium alloys for biomedical application' from the Department of Chemistry, College of Engineering, Anna University, India in the year 2012. He has published overall 22 peer-reviewed international publications in well-reputed SCI journals with an h-index of 12, written two book chapters and participated in various national and international conferences. In addition, he has delivered keynote lectures and invited talks at various international conferences. Besides, he has received research-oriented fellowships, like 'ICMR-Senior Research Fellowship' (2008) and 'CSIR-Diamond Jubilee Research Internship' (2006).

Dr. Srinivasan Arthanari has been a Researcher in the Chungnam National University, Daejeon, the Republic of Korea since 2021. His major research areas include the development of light metals, laser processing, surface treatments and corrosion studies. Previously, he received the Chinese post-doctoral fellowship at the School of Mechanical Engineering and Automation, Hefei Innovation Research Institute of Beihang University, Beijing, China in the year 2019–2021. He has also received Brain Korea (BK) fellowship and worked as a post-doctoral researcher at the School of Materials Science and Engineering, Seoul National University, South Korea during 2016–2018. Dr. Srinivasan earned his PhD in chemical surface treatment of Mg alloys for biomedical applications from the Department of Chemistry, College of Engineering Guindy, Anna University, India in 2015. Further, he has worked as Junior Research Fellow, and Senior Research Fellow in the Department of Science and Technology-Science and Engineering Research Board (DST-SERB), an Indian government-funded project during 2011–2013. As a recognition of his doctoral work, he earned the best PhD thesis award from the National Corrosion Council of India (NCCI) in 2016 and also received the Young Scientist Award-Runner Up I in 2015. He has published more than 35 peer-reviewed international publications, written book chapters, has been granted patents and has participated in various national/international projects. Dr. Srinivasan has also served as co-guest editor for a special issue in *MDPI-Crystals* journal, delivered webinars and keynote lectures in various international conferences and as a guest of honor in the national conference.

Dr. Sivakumar Bose has been a Postdoctoral Researcher in the Department of Biomedical Engineering, Pukyong National University, South Korea, since 2021. His present research work is focused on synthesis of nanomaterials for the bacterial disinfection (wound healing) applications using photo-induced methods. Previously he received a 'Chinese postdoctoral fellowship' in the School of Materials Science and Engineering, Hunan University, China during 2019–2020. Prior to joining Hunan University in China, he served as a National Postdoctoral Fellow in the Department of Metallurgical and Materials Engineering, Indian Institute of Technology Madras, Chennai, India during 2017–2019. He completed his doctoral research in engineering (2017) at CSIR-National Metallurgical Laboratory, Jamshedpur, India under AcSIR University, India. His PhD thesis work is mainly focused on the 'Fabrication of boride coatings of titanium and Ti-6Al-4V alloy for the improvement of tribological properties'. Overall, his past and current research areas have expanded into various fields such as surface engineering, corrosion, bio-implants, synthesis of nanomaterials and bacterial disinfection. Dr. Sivakumar has published overall research articles in well-reputed SCI journals like (*Mat. Sci. & Eng. C*, *Applied Surf. Sci.*, *Nanoscale*, *Tribology International*, etc). He has presented and participated in many national/international conferences. Besides, he has received various research fellowships, such as 'National Postdoctoral Fellowship' (2019), 'CSIR-Senior Research Fellowship' (2013) and 'CSIR-Diamond Jubilee Research Intern' (2008). In addition, he has also delivered various keynote lectures to different institutions for the benefit of students and research scholars.

Dr. Goncagül Serdaroğlu is an Associate Professor at Sivas Cumhuriyet University (Math. and Sci. Edu. Department), Sivas, Turkey. Her main area of research is chemical reactivity behavior of the pharmaceutical important molecules by using the computational tools. Also, she is experienced with the computational prediction of the molecular spectroscopic properties (IR, NMR, UV) in addition to the electronic-related properties NBO (Natural Bond Orbital), FMO (Frontier Molecular Orbital) and NLO (nonlinear optic) of the molecular systems. Her master work was focused on the Statistical thermodynamics in calculation of the entropy and heat capacity and completed it from Sivas Cumhuriyet University, Department of Physicochemistry (Theoretical Chemistry) in 2003. Then, she earned a PhD degree from the same university in 2008. She has also worked as a post-doctoral fellow with Prof. Joseph Vincent Ortiz, Department of Chemistry and Biochemistry, Auburn University, Alabama, USA, during 2013–2014. She has also visited on scholarship, Instituto de Quimica Medica (CSIC), Madrid, Spain to work on the non-covalency effect and NMR chemical shifts properties of the organic-based compounds. Recently, she visited the University of L'Aquila, Italy for part of their research group work activities. So far, she has published over 60 papers in the Web of Science Core collections and has attended various international conferences which have mainly focused on Computational & Theoretical Chemistry and General Chemistry.

Contributors

Abdul Bakrudeen Ali Ahmed
Department of Biochemistry
PRIST Deemed University
Thanjavur 613 403, India

P. Agilan
Department of Chemistry
CEG Campus, Anna University
Chennai 600 025, India

D. Ajith
Department of Mechanical Engineering
SRM Easwari Engineering College
Chennai 600 089, India

Somasundaram Ambiga
Department of Biochemistry
PRIST Deemed University
Thanjavur 613 403, India

Nacchiappan Annamalai
School of Biosciences
University of Nottingham
Sutton Bonington, United Kingdom

K. Aravind
Department of Prosthodontics
SRM Dental College, Ramapuram
SRMIST, Chennai, India

Srinivasan Arthanari
Department of Mechanical & Materials
    Engineering Education
Chungnam National University (CNU), 34134,
    99 Daehak-ro, Yuseong-gu
Daejeon, Republic of Korea

Arul Kashmir Arulraj
Centre for Materials for Electronics
    Technology (C-MET)
Under the Ministry of Electronics and
    Information Technology (MeitY)
Government of India, Kerala, India

K.G. Ashok
Department of Mechanical Engineering
SRM Easwari Engineering College
Chennai 600 089, India

Aida Badaoui
Materials Engineering, Applied Mechanics and
    Construction Department
University of Vigo
Vigo, Spain

T. Balusamy
CSIR—National Metallurgical Laboratory
Madras Centre, CSIR Complex, Taramani
Chennai, India
and Department of Applied Chemistry
Sri Venkateswara College of Engineering
Sriperumbudur, India
and Department of Analytical Chemistry
University of Madras, Guindy Campus
Chennai, India

Sivakumar Bose
Department of Biomedical Engineering
Pukyong National University
Busan, South Korea

Mohamed Boutinguiza Larosi
LaserON Research Group Research Center
    in Technologies, Energy and Industrial
    Processes (CINTECX)
University of Vigo, Spain
Vigo, Spain

Hua Chai
College of Materials Science and Engineering
Taiyuan University of Technology
Taiyuan, China

Silpa Cheriyan
Department of Sciences
Amrita School of Engineering, Coimbatore
    Amrita Vishwa
Vidyapeetham, India

Rafael Comesaña Piñeiro
LaserON Research Group Research Center
    in Technologies, Energy and Industrial
    Processes (CINTECX)
University of Vigo, Spain
Vigo, Spain
and Materials Engineering, Applied Mechanics
    and Construction Department

University of Vigo, Spain
Vigo, Spain

Sevgi Durna Daştan
Department of Biology, Faculty of Science
Sivas Cumhuriyet University
Sivas, 58140, Turkey

Taner Daştan
Department of Medical Services and Techniques
Yıldızeli Vocational School, Sivas Cumhuriyet
    University
Sivas Yildizeli, Turkey

Jesús del Val García
LaserON Research Group Research Center
    in Technologies, Energy and Industrial
    Processes (CINTECX)
University of Vigo, Spain
Vigo, Spain

P. Dhivya
Department of Chemistry
Nirmala College for Women
Coimbatore, India

G. S. Mary Fabiola
Department of Zoology
Nirmala College for Women
Coimbatore, India

Mónica Fernández-Arias
LaserON Research Group Research Center
    in Technologies, Energy and Industrial
    Processes (CINTECX)
University of Vigo, Spain
Vigo, Spain

Bruno Gago
LaserON Research Group Research Center
    in Technologies, Energy and Industrial
    Processes (CINTECX)
University of Vigo, Spain
Vigo, Spain
and
Department of Plastic and Reconstructive Surgery
University Hospital Complex of Vigo
Vigo, Spain

Jingwen He
State Key Laboratory of Space-Ground
    Integrated Information Technology

Beijing Institute of Satellite Information
    Engineering
Beijing, China

Guoqing Hu
Key Laboratory of the Ministry of Education
    for Optoelectronic Measurement
    Technology and Instrument
Beijing Information Science & Technology
    University
Beijing, China
and Beijing Laboratory of Optical Fiber
    Sensing and System
Beijing Information Science & Technology
    University
Beijing, China

M. Anto Simon Joseph
Department of Biotechnology
Sri Krishna Arts and Science College
Coimbatore, India

M. Kalaiyarasan
Department of Chemistry
CEG Campus, Anna University
Chennai 600 025, India

G. Keerthiga
Indian Institute of Technology-Bombay
    (IITB)
Monash Research Academy
Powai Mumbai, India

Ashish Kumar
Department of Chemistry, Faculty of
    Technology and Science
Lovely Professional University
Phagwara, Punjab, India

Manoharan Arun Kumar
Department of Electrical, Electronics &
    Communication Engineering, School of
    Technology
Gandhi Institute of Technology and
    Management (GITAM) University
Bengaluru 561203, India

Ramu Arun Kumar
Department of Biotechnology
PRIST Deemed University
Thanjavur 613 403, India

Lazarus Vijune Lawrence
Department of Biochemistry
PRIST Deemed University
Thanjavur 613 403, India

Pengbin Lu
College of Materials Science and Engineering
Taiyuan University of Technology
Taiyuan, China

Fernando Lusquiños Rodríguez
LaserON Research Group Research Center
in Technologies, Energy and Industrial
Processes
(CINTECX)
University of Vigo, Spain
Vigo, Spain
and
Galicia Sur Health Research Institute
(IIS Galicia Sur)
SERGAS-UVIGO, Spain

Karthega Mani
Department of Sciences
Amrita School of Engineering
Coimbatore Amrita Vishwa Vidyapeetham, India

L. Mohan
Department of Mechanical Engineering
Toyohashi University of Technology
Toyohashi, Japan

T.S.N. Sankara Narayanan
CSIR—National Metallurgical Laboratory,
Madras Centre, CSIR Complex
Taramani, Chennai, India
and Department of Analytical Chemistry
University of Madras, Guindy Campus
Chennai, India

Xiaohuan Pan
College of Materials Science and Engineering
Taiyuan University of Technology
Taiyuan, China

Arjun Pandian
Department of Biotechnology
PRIST Deemed University
Thanjavur 613 403, India

Raja Suja Pandian
Department of Biochemistry

Annai College for Arts and Science
(Affiliated with Bharathidasan University)
Kumbakonam 612 503, India

Hyung Wook Park
Department of Mechanical Engineering
Ulsan National Institute of Science and
Technology UNIST-gil 50
Eonyang-eup Ulju-gun, Ulsan, Republic
of Korea

Juan Pou Saracho
LaserON Research Group Research
Center in Technologies, Energy
and Industrial Processes Galicia
Sur Health Research Institute
(IIS Galicia Sur)
SERGAS-UVIGO, Spain

Pablo Pou
LaserON Research Group Research Center
in Technologies, Energy and Industrial
Processes (CINTECX)
University of Vigo
Vigo, Spain

N. Rajendran
Department of Chemistry
CEG Campus, Anna University
Chennai 600 025, India

Raju Ramasubbu
Department of Biology,
The Gandhigram Rural Institute—Deemed
University Gandhigram
Tamil Nadu 624 302, India

B.E. Amita Rani
Surface Engineering Division
CSIR—National Aerospace Laboratories
Bangalore, India

K. Ravichandran
Department of Analytical Chemistry
University of Madras, Guindy Campus
Chennai, India

Bhavana Rikhari
Surface Engineering Division
CSIR—National Aerospace Laboratories
Bangalore, India

Antonio Riveiro Rodríguez
LaserON Research Group Research Center
    in Technologies, Energy and Industrial
    Processes (CINTECX)
University of Vigo, Spain

Ananda Babu Sairam
Department of Applied Chemistry
Sri Venkateswara College of Engineering
Sriperumbudur, India

K. Saranya
Department of Chemistry
CEG Campus, Anna University
Chennai 600 025, India

Lakshmanan Saravanan
Department of Micro and Nanoelectronics
Saveetha School of Engineering
Saveetha Institute of Medical and Technical
    Sciences (SIMATS)
Thandalam, 602105, Tamil Nadu, India

Mahalingam Shanmugam
Inorganic Chemistry Laboratory
Department of Chemistry
Chung Ang University
Seoul, Korea

V.S. Simi
Department of Chemistry
Anna University
Chennai, India

Nagarajan Srinivasan
Department of Chemistry
Manonmaniam Sundaranar University
Tirunelveli, India

Abhinay Thakur
Department of Chemistry, Faculty of
    Technology and Science
Lovely Professional University
Phagwara, Punjab, India

Maurizio Vedani
Department of Mechanical Engineering
Politecnico di Milano
Milan, Italy

Lifei Wang
College of Materials Science and Engineering
Taiyuan University of Technology, Taiyuan,
    China
and Key Laboratory of Interface Science
    and Engineering in Advanced Materials,
    Ministry of Education
Taiyuan University of Technology
Taiyuan, China

Xuan Wang
ALIT Life Science Co. Limited
Shanghai, China

Liangliang Xue
College of Materials Science and
    Engineering
Taiyuan University of Technology
Taiyuan, China

Jie Yang
Department of Physics and Electronic
    Information Technology
Yunnan Normal University
Kunming, China

Sasikumar Yesudass
School of Materials Science and Engineering
Tianjin University of Technology
Tianjin 300384, China

Qiang Zhang
Key Laboratory of Interface Science and
    Engineering in Advanced Materials,
    Ministry of Education
Taiyuan University of Technology
Taiyuan, China

Feng Zhao
Department of Environmental Science and
    Engineering
Hunan University
Hefei, Anhui, China

1 Nanomaterials and Its Application as Biomedical Materials

G.S. Mary Fabiola, P. Dhivya and M. Anto Simon Joseph

CONTENTS

1.1 INTRODUCTION

Not all things that are big, are always beautiful. Tiny things make wonders. Nano—a unit prefix—meaning "one billionth"—an element of 10^{-9} or 0.000000001. The term "*nano*" has its root deep from the Greek term "*nanos*" or Latin "*nanus*", meaning "dwarf". Nanomaterials are exploited to designate the fabrication of materials stretching in size from 1–100 nm. Progressive and prospective research over a couple of decades has facilitated the development of hybrid materials via integrated design. The conceptualization of nanotechnology attracted more attention during the 1990's. However, the word nanotechnology was rediscovered and publicized in a lecture delivered by physics Nobel laureate Richard P. Feymann titled "There's Plenty of Room at the Bottom" on the eve of a

DOI: 10.1201/9781003173533-1

gathering of the American Physical Society at Caltech on 29th December 1952. Nanotechnology has reformed the era of science and engineering over a span of two decades since the beginning of the 21st century. The tailor-made nanomaterials have initiated researchers to discover, design, develop, and manipulate the sole properties of constituents on nano scale. The fabrication of nanomaterials has rendered a tremendous contribution to material science. Nanomaterials possess certain unique physiochemical characteristics which make them unique and account for their vivid applications in comparison with the corresponding bulk material.

The incorporation of engineered materials in science and technology has quintessentially replaced the traditional metals to reach new horizons in therapeutics. The implementation of nanoparticles in the recapitulation of technologies is attributable to the nanoscale size of the reinforcing phase and the fact that the surface to volume ratio is expressively higher than conventional materials.

Nanomaterials are vibrantly used in medicine and therapeutics for the profound recognition of strategic biological molecules, specific and benign imaging of ailing tissues, and new forms of therapeutics. In the recent past, numerous nanoparticle-based therapeutic and diagnostic mediators have been established for the treatment of numerous diseases. The exploitation of nanoparticles in medicine provides exceptional choice to alter some essential properties of therapeutic carriers which include their solubility, diffusivity, bio-distribution, release characteristics, and immunogenicity. Accurate nanoparticle engineering has generated longer flow half-lives, enhanced bioavailability, and lower toxicity.

1.2 PROPERTIES OF NANOMATERIALS

Nanomaterials possess remarkable properties that make them distinct from their bulk counterparts as depicted in Figure 1.1. Materials when reduced to nano scale reveal properties exclusively diverse from their bulk material.

For instance, copper in its bulk state is opaque, whereas nano copper is transparent. Similarly, aluminum is stable even at a higher temperature, but nano-aluminum is combustible. Thus materials in their nano scale have significant changes in properties which are entirely different from the micro- and macroscopic materials. The distinctive size, shape, and structure of the nanomaterials justify the reactivity, sensitivity, mechanical, magnetic, optical, and thermal properties of nanomaterials. This accounts for the unique and ubiquitous properties of the nanomaterials, thereby inviting intense scientific research providing a prospective outlook into the future.

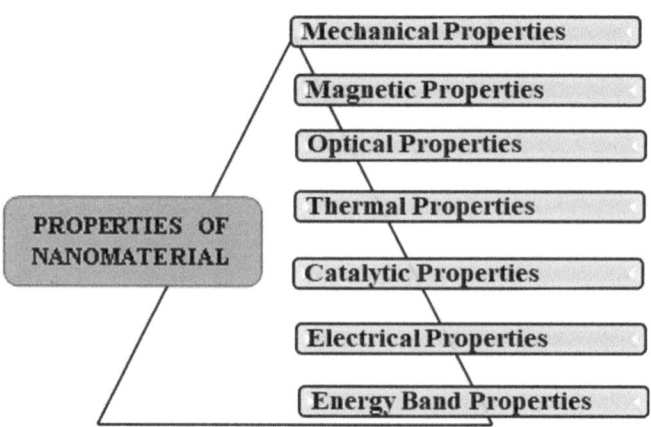

FIGURE 1.1   Materialistic properties of nanomaterials.

1.2.1 Mechanical Properties

Mechanical properties of metals are often associated with mechanical characteristics of metal which include strength, toughness, hardness, brittleness, plasticity, elasticity, rigidity, malleability, and ductility. The traditional inorganic metals are brittle, hard, and rigid but lack plasticity and elasticity. Alternatively, organic materials are flexible but are not rigid, brittle, and strong. These disadvantages are overcome by the nanomaterials which possess high surface area, volume, and quantum effects when compared to micro- and macroscopic materials. The influence of the selection of nanomaterials, the process of fabrication, grain size, and structure of the grain boundary has a noteworthy effect on the mechanical properties of nanomaterials. In comparison with the bulk, nanomaterials refine the grain size and form inter/intragranular structure, improving the grain boundary and thereby enhancing the mechanical properties of nanostructured materials. The flexural strength of nano-Al_2O_3 ceramics is comparatively stronger when compared with micro-scale monolithic alumina ceramics (Teng et al. 2007).

1.2.2 Magnetic Properties

Nanomaterials have properties entirely different from bulk material. The reduction in size or dimension of the nanomaterial introduces quantum confinement by reducing the symmetry of the system. The total energy of a ferromagnetic material is the summation of exchange energy, anisotropic energy, demagnetization energy, and energy due to the applied magnetic field. As in the case of nanomaterials, interaction among the exchange energy, anisotropic energy, and demagnetization energy is more pronounced. When the dimensions of the grain size become smaller, exchange forces dominate because of strong coupling and cause all the neighboring atoms to align in a particular spin. From the alignment, there exists a diameter called critical diameter that can be calculated. When the size of the particle more reduced than the critical diameter, magnetization becomes unstable and loses magnetization and the ferromagnetic material becomes superparamagnetic. The shrinkage in particle size increases the saturated magnetization and the reversal of magnetization becomes insignificant in nanomaterials. The structural and magnetic features of biocompatible Fe_3O_4 magnetic nanoparticles are used in labeling units of biomedical applications.

1.2.3 Optical Properties

The surface morphology has a great effect on the optical and semiconducting properties of nanomaterials. The general optical properties of materials include reflection, refraction, transmission, absorption, and emission. The origin of the color of materials is caused by the surface plasmons. The surface plasmon is a natural phenomenon of oscillation of an electron at the junction of the material. The optical properties are largely dependent on the electronic structure—nanosphere in particular, which in turn varies with the morphology as it depends on the surface atoms. When the sphere is small in comparison with the wavelength of the incident light, and the frequency is quite close to that of the surface plasmon, then the surface plasmon absorbs more energy. Because of reduced dimensionality, the drift of electrons is restricted in nano scale when related to its bulk counterpart. The smaller the magnitude of the particle, the wider will be the optical band gap and shorter will be the wavelength and hence will be blue shifted. For example, spherical gold nanomaterial of 25-nm diameter appears in the region of green whereas gold nanomaterials of 100 nm appear orange. The factors that govern the size-dependent optical properties in nanomaterials are augmented energy level spacing—quantum effect and Surface Plasmon Resonance. When the dimension of the nanomaterial is confined to the nanometer range, with its characteristic wavelength approaching either closer to or less than the de Broglie wavelength of the respective charge carriers, which may be electrons or holes or the wavelength of the light, the periphery of the crystal gets ruptured in case of crystalline solids, whereas the atomic density of amorphous solids changes in the nanometer range.

1.2.4 THERMAL PROPERTIES

The thermal properties involve the transfer of heat in nanomaterials. The thermal properties of materials in nanoscale dimensions largely depend on the surface properties, classical or quantum size, interfacial structure, which is usually insignificant in bulk materials. The thermal properties could be accounted for the conduction of electrons as well as phonons which causes lattice vibrations. In nanostructured materials, the size of the nanomaterial becomes comparable to the mean free path of phonons through phonon scattering, phonon confinement, and quantization effects of phonon. Nanoscale thermal management suffers a slow progression because of the difficulties experimental setup and a controlled thermal transportation feature in the nanoscale dimension.

1.2.5 CATALYTIC PROPERTIES

Nanomaterials have been used as a catalyst in enormous chemical reactions. Nanocatalyst stands as a perfect boundary between the homogeneous and heterogeneous catalyst. Nanocatalyst embarks supreme efficacy in terms of its high activity, selectivity, sensitivity, efficiency, and stability. The catalytic activity of nanomaterials is confined to the structural, quantum size and electronic effects. It is a well-known fact that a decrease in the dimension of the particles increases the surface area thereby enabling more and more reactant molecules to get adsorbed on the nanomaterials, eventually resulting in enhanced catalytic activity. The existence of a greater quantity of surface atoms creates more active sites for the adsorption of reactant molecules. These phenomena cause a greater dissociation of the binding energy of the reactants and a pronounced catalytic activity of the nanomaterials resulting in the formation of the products.

1.2.6 ELECTRICAL PROPERTIES

The electrical conductivity of nanomaterials can be expressed in terms of conductivity or resistivity. The electrical conductivity is justified by the band structure of solids. Unlike nanomaterials, the conductivity of bulk material is independent of measurements like diameter, area of cross-section, twist of conductivity, etc. When a material is condensed to nano size, the electron is restricted for movement to a confined particular dimension and there is an increase in surface scattering. Hence the electrical conductivity drops down with reduced dimensions. However, the electrical conductivity may be altered due to the establishment of a well-ordered microstructure when the particle size is diminished to the nm range. The nanomaterials are associated parallel to the axis, which contributes to the conduction through the tunneling effect. The smaller the diameter, the better alignment of nanomaterials, which results in higher electrical conductivity. The conductivity of a multi-walled carbon nanotube is very much different from the single-walled carbon nanotube of the same dimensions.

1.2.7 ENERGY BAND CONDUCTION PROPERTIES

Semiconductor nanomaterials often refer to a variety of compounds of group II–VI, III–V, or IV–VI of the periodic table into which these elements are formed. For example, silicon and germanium occupy group IV, gallium and indium constitute group III–V, while those compounds of zinc and cadmium form II–VI semiconductors. The semiconducting properties are associated with the electronic structure and as discussed in the previous section, electrical property. The precise surface area and surface-to-volume ratio considerably increase as the size of the material decreases. Factors such as size, shape, and surface characteristics can be changed to control their properties for exclusive applications of interest. A decrease in the particle size towards the nanometer range causes an escalation in the bandwidth between the valence band and conduction band. In other words, the confinement of an electron restricts the transition of an electron from the highest occupied molecular orbital to the lowest unoccupied molecular orbital. This eventually results in the blue shift of the absorbed light, which is towards the high energy region. The transfer of electrons and holes in semiconductor nanomaterials is predominantly directed by the distinguished quantum confinement, and the means of transport of phonons and photons are essentially varied by the size and geometry of the materials.

1.3 CLASSIFICATION OF NANOMATERIALS

1.3.1 CLASSIFICATION BASED ON SPATIAL DIMENSION

According to Richard W. Siegel (Siegel 1993), based on the spatial dimensions, nanomaterials are classified into zero dimensional (0D) nanomaterials, one dimensional (1D) nanomaterials, two dimensional (2D) nanomaterials, and three dimensional (3D) nanomaterials.

1.3.1.1 Zero Dimensional (0D) Nanomaterials

When all the three dimensions are confined to one particular point and thus the movement of electron is restricted in all the three x,y,z directions, it is called zero dimensional. Example: nano dots, nanoparticles, quantum dots.

1.3.1.2 One Dimensional (1D) Nanomaterials

In case of one dimensional nanomaterials, two dimensions are reduced to nm range; only one dimension remains large and the electron is permitted to move in this one dimension. Example: nanowires, nanotubes, nano rods.

1.3.1.3 Two Dimensional (2D) Nanomaterials

Two dimensional nanomaterials have one dimension restrained to nm range and the electron is permitted to move freely in the remaining two dimensions which remains large. Example: nanowells, nanofilms, nanocoatings, nanolayers, nanoflakes, nanoplatelets.

1.3.1.4 Three Dimensional (3D) Nanomaterials

Three dimensional nanomaterials have no confinement in the nm range and the electron is permitted to move in all the x,y,z directions. They possess an arbitrary dimension above 100 nm. Example: bulk nanomaterials, nanopowders.

1.3.2 CLASSIFICATION BASED ON COMPOSITION

On the basis of composition, nanomaterials can be classified as organic nanomaterials, inorganic nanomaterials, and hybrid nanomaterials.

1.3.2.1 Organic Nanomaterials

Organic nanomaterials entail carbon-based nanomaterials taking different forms of spheres, cylinders, ellipsoids, or tubes. There exist non-covalent interactions like hydrogen bonding, pi staking, and electrostatic interactions. Organic nanomaterials are assembled upon either natural or synthetic organic molecules. Example: fullerenes, carbon nanotubes (CNTs), single-walled and multi-walled carbon nanotubes (SWCNTs and MWCNTs) and electrospun nanofibers.

1.3.2.2 Inorganic Nanomaterials

Inorganic nanomaterials include metals based, oxides of metal based, and quantum dots—nanomaterial metalloids in nano dimensions. They may extract the form of the oxide or hydroxide or phosphate or sulphide or chalcogenide of the metal. Example: gold nanoparticles, zinc oxide nanoparticles, mesoporous silica nanoparticles.

1.3.2.3 Hybrid Nanomaterials

Hybrid nanomaterials are composite nanomaterials and are a combination of organic-organic nanomaterial, organic-inorganic nanomaterial and inorganic-inorganic nanomaterials. They are unique chemical conjugates of organic and inorganic nanoparticles; as a result, multifunctional hybrid materials are obtained, which possess immense applications. Example: lipid-polymer hybrid nanoparticle, hybrid silica particles.

1.4 BIOMEDICAL APPLICATION OF NANOMATERIALS

The rise of inventive and fabricated nanomaterials is at the forefront in emergent areas of biomedical applications. The exploitation of rationally designed nano biomaterials in clinical applications has surpassed traditional therapeutic modalities. Nanomaterials have invoked engrossment among researchers because of their unique physicochemical properties, biocompatibility, and desired functionalization modification, size- and shape-dependent optical and magnetic properties. Nanomaterials scale well in biomedical applications which include diagnosis, targeted drug delivery, prostheses, and implants. A multifunctional framework centered around metal, carbon, polymer, biological moieties, and lipid-based nanomaterials is associated with biomedical imaging, diagnostics, and/or therapeutics and serves as a synergistic combinational platform in biomedical applications.

1.4.1 Biomedical Imaging

Early detection and diagnosis offer a quintessential role in biomedical studies. Fluorescence imaging is vividly used for the characterization of novel drugs or of new formulations of prevailing drugs exclusively at the preclinical level. Integrating photo luminescent imaging analytics and molecular probes over the preceding years have been used in imaging of cell/tissue in diagnostics. Photo luminescent imaging constitutes the preliminary stage in the drug development process for the translation phase from *in vitro* assays to preclinical systems, as well as to evaluate their ADME. Functional imaging will provide a detailed picture of the local and real time biological activity of the drug. The technique involves the radioactive labeling of the fluorescent material which can be applied in the early progress of a drug and transform the applicability, *in vitro* on cells, then *in vivo* in small animals and finally rendered into the clinic with restrictions. It is not appropriate to use this method for small drugs due to the large size of the fluorophores. But this can be applied for a nanoparticle, on the basis of the assumption that the labeling should not intensely modify the corresponding property of the nanoparticles or that the fluorophore could be a fragment of the final formulation.

Prominent advances of recent times include the application of optical nanoprobes, such as persistent luminescent nanoparticles (PLNPs) which are developed to replace the usage of long-lasting near infrared (NIR) luminescence capability. The added advantage to the usage of determined luminescence nanoparticles is optical imaging without constant excitation and autofluorescence. The most common and extensively used nanoparticles in biomedical imaging and cancer therapy include nanoparticles, nano rods, nanospheres, nano shells, and nano stars. Nanoparticles aid as drug carriers, imaging contrast agents, photothermal agents, photoacoustic agents, and radiation dose enhancers. Nanoparticles continue to be potential candidates in biomedical imaging for the major imaging techniques Magnetic Resonance Imaging (MRI), Positron Emission Tomography (PET), Single Photon Emission Computed Tomography (SPECT), Magnetic Particle Imaging (MPI), Nuclear Medicine, Ultrasound (US) imaging, Computed Tomography (CT), and Optical Imaging in particular. The breakthrough in the advances of NPs includes the use of iron oxide NPs (Pellico et al. 2017), the design of radio isotope chelator free particles for PET (Dash et al. 2019), and the development of fluorescent NPs such as carbon dots and up-converting nanoparticles (Siddique et al. 2020).

1.4.2 TARGETED DRUG DELIVERY AND CONTROLLED DRUG RELEASE

Drug delivery often refers to the design, construction, engineering technologies and transport of a particular therapeutic compound to attain the desired therapeutic effect. Drugs can be directed into several routes into the human body which may be buccal, oral, pulmonary, transdermal, ocular, sublingual, vaginal, and anal. These conventional approaches of drug delivery involve the transportation of the drug through the blood to the target of interest. The major setback of the traditional method is the damage caused to the normal cells. Hence, research is intensified in seeking a selective and targeted drug delivery where the drug is being delivered to the target without affecting the healthy cell. Another aspect of concern is the advance of biodegradable nanoparticles as drug delivery devices (Idrees et al. 2020). Various morphologies involving large surface-area-to-volume ratios such as nanoparticles, nanospheres, nano-encapsules are used as drug delivery systems. The small size of the nanoparticles penetrate through the smaller capillaries, being effortlessly taken up by the cells, and the biodegradability of the nanomaterial allows efficient and controlled drug accumulation and controlled drug release to the target site over a duration of time. Nanomaterials also defend the captured drug from gastrointestinal interferences. The formulations of targeted drug delivery of nanoparticles and controlled drug release involve drugs to be dissolved, entrapped, adsorbed, attached, and encapsulated into the nanomaterial matrix (Yetisgin et al. 2020).

Each nanomaterial has its own characteristic way of targeted drug delivery. For instance, nano capsules are a vesicular system with the drug bounded by a polymer membrane, whereas nanospheres are matrix types of structures where the drug is physically and uniformly spread. As in the case of nanospheres belonging to the matrix type of system, the drugs are adsorbed at their surface, entrapped or dissolved within the particle. Drug delivery systems are in general polymeric and nano sized and the different forms of drug delivery systems include nanoparticles, ceramic nanoparticles, micelles, polymeric micelles, dendrimers, and liposomes. Polymeric nanoparticles such as PLGA (poly(lactide-co-glycide)), PLA (polylactic acid) is largely used for the drug delivery of estradiol. The degradation of the polymer can be modified by changing the block of the copolymer composition and the molecular weight of the polymer and thus the release of the encapsulated therapeutic agent from the polymeric nanoparticle can be transformed from days to months (Masood 2016).

1.4.3 TISSUE ENGINEERING

Tissue engineering deals with the art of creating, restoring, replacing, and maintaining tissues and organs using biological substituents, which have a very close resemblance to the body's native tissues/organs. It is a connecting discipline of integrated biology, engineering, material science, and medicine. Traditional bone substituents in the biomedical industry include bioceramics like alumina, zirconia, hydroxyapatite, tricalcium phosphates, owing to their low density, biocompatibility, chemical stability, and high wear resistance (Eliaz et al. 2017). The implementation of nanotechnology in tissue engineering has evolved as evolutionary and revolutionary changes. The most common feature of tissue engineering is the fabrication of a three dimensional porous scaffold which serves as a substrate and support for tissue growth and directs the cells to grow in the correct anatomical shape, thereby possessing biocompatibility to avoid inhibition of cell growth. Nanotechnology fabricates biomaterials of nanometer size like nanofibers, nanopatterns, and controlled-release nanoparticles to mimic native tissues/organs engineering.

Tissue engineering involves functionality-dependent design and construction of nanostructures. For instance, the design and construction of neural tissue require electrical conductivity, while bone and cartilage tissues necessitate enhanced mechanical properties (Achachelouei et al. 2019). The fabrication of scaffolding material that reiterates the cellular environment on a nano scale has raised great interest in recent years. Carbon nanotubes are vividly used in tissue engineering as they are chemically stable, conduct electricity, and mechanically strong to be employed as scaffolds. Filamentous carbon nanotubes possess a structural alignment that is analogous to the extracellular environment which supports surrounding cells. Thus the carbon nanotubes may have the capability to kindle cell function in the same way as the extracellular matrix (Huang 2020). The biocompatibility tests of carbon nanotubes in suspension and carbon nanotubes confined in a structure exposed that the loose carbon nanotubes suspended in cell culture were found to decline in cell viability. The cells that are directly attached to carbon nanotube-containing structures created cell growth and demonstrated excellent biocompatibility of carbon nanotubes with living cells.

1.4.4 ARTIFICIAL IMPLANTS

The influx of nanoparticles in medicinal devices is an upcoming field. Devices or materials that are positioned within the body superficially are called artificial implants. They are intended to convey suppositories, monitor body functions, and deliver sustenance to organs and tissues. Earlier implants were made of skin, bone, or other body tissues. Later, metals, ceramics, polymers and their composites were designed and largely employed to support, enhance, or to even replace a fraction. Artificial implants can be used permanently as in the case of stents or hip implants, while chemotherapy ports or screws to repair broken bones are temporary and are removed after healing. But the safety and potential side effect remain a question, though the artificial implants possess good dimensional tolerance, high fracture toughness, good fatigue resistance, comparable strength, and modulus close to the bone, high wear resistance of tissue, biocompatibility, high purity, and reproducibility.

The interaction between the artificial implants and cells has facilitated nanotech research towards the frame of nanomaterials towards artificial implants. The norm of nano-engineered quantum dots and magnetic nanoparticles for stem cell tracking and the enrichment of material properties with carbon nanotubes and graphene are in progress (Zhao et al. 2020). The biocompatibility of the nanomaterials in artificial implants includes promoting biological tissue for implant integration, promoting cell adhesion, providing pathways for vascularization, non-carcinogenesis, non-pyrogenicity, non-toxicity, and non-allergic response (Velu et al. 2020). The ability to undergo sterilization, autoclave and dry heating, ethylene oxide gas, and radiation account for the sterilizability. The primary functionality of nanomaterials in artificial implants is the entrenchment of modulus of elasticity for the stiffness of the material, ultimate tensile strength to withstand a load and dimensional accuracy on an economical fabrication process. The ease of molding, extrusion process, machinability, and ability for fiber forming elucidate the usage of nanomaterials in manufacturing artificial implants. Nanostructures provide antibacterial properties to prevent implants against postoperative infections proposed for bone and implants. Nano-sized silver particles widely aid in the exploration of suitable size, shape as well as a novel method of surface modifications such as SDP technology for orthopedic implants (Qing et al. 2018).

1.4.5 GENE THERAPY

Despite of the blooming advances in the field of medicine, cancer remains with a high mortality rate, especially in developing and underdeveloped countries. One of the leading causes for such a high mortality rate is the limitation of actual treatments based on drugs and radiation. These confines include lack of specificity, reduced drug bioavailability, drug rapid blood clearance, poor drug solubility, patient resistance, and disease relapse. Traditionally used chemotherapeutics which include cisplatin or taxol have been favored over other therapies due to the selective killing of cancer cells preferentially by inhibiting replication or inducing apoptosis. Certain chemotherapeutics produce adverse effects as well. Chemotherapeutics with anthracyclines and cyclophosphamide cores cause serious side effects in patients, killing healthy cells and tissues like bone marrow, epithelial cells, and hair follicles. Hence, the development of alternate and more efficient treatments that may offer fewer side effects in comparison to the actual therapies remains a challenge to researchers.

Novel technologies for cancer treatment have been employed in the recent past based on the research and application of nanotechnology and molecular biology. The targeted level treatment remains the limelight of treatment in the recent past and near future. The advance of molecular biology permits the manipulation of nucleic acid in the management of numerous genetic diseases like cancer. Gene therapy comprises the transmission of genetic material into a target cell nucleus for healing concerns with comparatively negligible side effects. This genetic material could be DNA or RNA, the complete gene sequence, gene segments, or an oligonucleotide.

With the development of genomic technologies, nanoparticles owing to their greater penetrating power play a critical part in incorporating all the desirable characteristics of modification into a single gene delivery system (Rodrigues et al. 2020). Lipid and polymer-based gene delivery vectors are paved to be sophisticated delivery systems in gene therapy. Polymeric nanoparticles are largely employed for gene therapy and protein delivery.

1.4.6 PHOTODYNAMIC THERAPY

Photodynamic therapy is emerging to be a remedial modality for early detection and localized cancers. The three key components in photodynamic therapy include: photosensitizer, light, and molecular oxygen. The photosensitizer is administered either by intravenous injection or by local application depending on the part of the body to be treated. Once the drug is absorbed by the pathologic tissue, light is exposed. The photosensitizer gets activated by light and forms Reactive Oxygen Species (ROS), which in turn kill cancer directly. A major issue faced by prolonged photodynamic therapy is the increased selective accumulation of the photosensitizers within the tumor thereby leading to a lower effective dose of the drug. To improve the efficacy of photodynamic therapy, efforts were laid to bind the photosensitizer itself by ligands such as monoclonal antibodies or low-density lipoprotein (LDL) or via carrier system such as liposomes and micelles (Gibot et al. 2020).

Nanoparticles are proving to be an emerging paradigm in photodynamic therapy. The foremost lead application of nanoparticles in photodynamic therapy is large surface area with a varied functional group for modified biochemical processes, large surface volume, controlled release of drugs, and easy transportation of hydrophobic drugs in blood, high permeability and retention effect. Nanoparticles encompassing inorganic oxide, metallic, ceramic and biodegradable polymer nanomaterials have successfully been in use in photodynamic therapy in the recent past (Chen et al. 2020). Nanoparticles used in photodynamic therapy can be broadly classified into active and passive nanomaterials depending on the mechanism of activation of photosensitizer nanoparticles. The role of the mechanism of active nanoparticles can be sub-classified as activation of photosensitized nanoparticles, up-conversion nanoparticles, and self-lighting nanoparticles. In the case of active nanoparticles, materials for photosensitizers like CdSe/CdS/ZnS cause indirect excitation of photosensitizers through a Fluorescence Resonance Energy Transfer (FRET) mechanism from the nanoparticle to the photosensitizer. Fullerene aids in the transfer of energy from incident light directly to surrounding oxygen.

Based on material composition, passive nanoparticles can be classified as biodegradable and non-biodegradable nanoparticles. Biodegradable nanoparticles include alginate, chitosan, cyclodextrin, albumin, PLA, PLGA, wherein the drug is delivered by micelles, dendrimers, liposomes, or polymeric nanoparticles, ensures the controlled release of the encapsulated photosensitizer through biodegradation. Fabrication of non-biodegradable nanomaterials includes polyacrylamide in which the two-photon dye is encapsulated by microemulsion, silica which assists in the absorption of photosensitizer by covalently bonding through a porous shell, gold nanoparticles act as pure carriers, and magnetic iron oxide nanoparticles in which a drug is carried directly or co-encapsulated in a micelle or polymeric nanoparticle. The added advantage of magnetic iron oxide nanoparticle is the achievement of target delivery by an external magnetic field (Yang et al. 2019).

1.4.7 SONODYNAMIC THERAPY

Sonodynamic therapy has been considered as a safe alternate to the conventional as SDT uses ultrasound at relatively low intensities (ranging from 0.5 to 4 W/cm^2) when thermal or mechanical effects cannot be induced to living cells (Wan et al. 2016). Porphyrin-based molecules or Xanthene dyes are employed in the sonodynamic therapy, as the same were earlier used in photodynamic therapy (Buck et al. 2017). These molecules present a Reactive Oxygen Species (ROS)–mediated cytotoxic effect when stimulated by ultrasound. The major disadvantage of most of the sonosensitizing agents is they are strongly hydrophobic and aggregate easily in the physiological environment, thereby decreasing the efficacy and producing a retarding effect in the pharmacokinetic behavior. These molecules would indeed be toxic and show low selectivity towards tissues.

The evolution of nanoparticle-mediated sonodynamic therapy has made a major stride forward in overcoming the challenges faced by deleterious side effects caused by chemotherapy and radiotherapy (Canavese et al. 2018). Nanomaterials may serve as nano sensitizers or active carriers of sonosensitizers. Titanium dioxide nanoparticles are the most widely used nano sensitizers in SDT. Because of their semiconducting property, they are employed as photosensitizers in photodynamic therapy as well to obtain ROS. The therapeutic enhancements obtained with TiO$_2$ NPs, an enriched and favored binding and internalization of NPs toward cancer cells and functionalization with targeting molecules make nanoparticles auspicious for the advance towards targeted therapy (Kim et al. 2020). The sonodynamic therapy (SDT) of semiconductor metal oxide nanoparticles, for example TiO$_2$ and ZnO$_2$, can provide a therapeutic platform in the future (Bogdan et al. 2017). As termed, the significant role played by the NPs in inducing the cytotoxic effects, as initiators of the SDT process, is tremendous.

1.4.8 CRYOSURGERY

Cryosurgery is a unique technique when extreme cold derived from liquid nitrogen or argon gas is used in surgery to terminate unusual or damaged tissues. Cryosurgery is otherwise called freezing therapy, cryotherapy, or cryoablation and has been increasingly used due to the controlled annihilation of tumor tissue. But a major setback of cryosurgery is when the gases undergo deficit or inappropriate freezing as it fails to destroy the target tumor tissues, and the probability of regenesis of tumor is high and the rate of treatment often a failure. Another major drawback is that the surrounding healthy tissues/cells may suffer from serious injury due to the extreme coldness. Hence, a new strategy of inculcating nanomaterials—nano cryosurgery is invoked in biomedical applications to overcome the freezing efficiency of the traditional cryosurgical procedure.

The primary protocol of nano cryosurgery is to carry a functional suspension of nanoparticles into the target tissues, which then helps as adjuvant or drug carrier either to maximize the freezing heat transfer process, standardize freezing scale, alter ice-ball formation orientation, or avert the surrounding healthy tissues from being frozen (Hou et al. 2018). Furthermore, the introduction of nanoparticles in the course of cryosurgery with potential challenges and future prospects aid in the better imaging of the edge of a tumor as well as the margin of the ice ball. The nano cryosurgery is anticipated to move horizons emerging frontline of nano-biomedical engineering. Typical nanoparticles (NPs), which are nontoxic, biodegradable, and possess excellent thermal properties with a few side effects, produce an accelerated and enlargement of ice-ball formation and enhance cryoinjury, thereby promoting the generation of ice nuclei. The applicability of magnetic nanoparticles with high thermal conductivity and good biological compatibility improves nucleation with increased kinetic and thermodynamic parameters. Polymeric NPs change the morphology of ice crystals and improve thermal conductivity (Stewart et al. 2020). TNF α—conjugated Au NPs causes contraction of the tumor without systematic toxicity and destroys tumor cells within the ice ball efficiently with minimal side effects (Hou et al. 2018). Nanoparticle-encapsulated doxorubicin (nDOX) achieves nearly complete eradication of the cancer stem-like cells (CSCs) with fewer side effects and enhanced targeting.

1.4.9 Magnetic Hypothermia

Over the era, cancer is still considered a deadly disease and most forms of human cancer are not curable. The reasons may be multifactorial. But the chief and primary limiting factor remains in the lack of understanding of the mechanism by which the tumor grows and the therapeutic intervention. The most common and principal types of cancer therapies include chemotherapy, radiation therapy, and surgery. Magnetic hyperthermia is an additional modality to cancer therapy but is yet to be considered as a standard-of-care therapy.

The term hyperthermia involves the mild elevation of temperature to induce the death of cancer cells and to enhance the effect of radiotherapy and chemotherapy. Heat treatment is used as a chief aspect to destroy cancer cells. The principle behind magnetic hyperthermia involves the magnetic nanoparticles being activated by an alternating magnetic field being reconnoitered by targeted heating of the tumors. The basis of heat generation under alternating magnetic fields both for *in vivo* and *in vitro* studies for biomedical applications has been a subject of intense research (Chang et al. 2018). The efficacy of nanomaterials for magnetic imaging-guided hyperthermia, thermal cancer therapy, magnetically actuated drug delivery, and biofilm eradication is research of the recent past. A gradual increase in the temperature around the cancerous region to 40–43 °C induces significant cancer cell death in addition to the cytotoxic effect of radiotherapy and chemotherapy. The magnetic nanoparticles act as intermediaries for cancer therapy. A steady increase in the temperature of the cells above 40 °C produces pronounced effects in the membrane and the interior of the cells. Multifunctionalized hybrid nanocomposites involving the combination of magnetic nanoparticles with materials like graphene oxide (GO), photoactive materials, mesoporous nanoparticles, and polymeric nanoparticles—polymer matrix—embedded with active nanomaterials have been widely investigated (Kim et al. 2018).

1.4.10 Antimicrobial and Wound Healing

The largest organ in the human body—skin—covers and integuments the entire body and provides protection against pathogens, toxins, and trauma and receives sensory stimuli from the external environment. The rupture of the skin leads to a wound, and the wound is often associated with infections. The healing of a wound is the extremely synchronized progression of restoring damaged tissue encompassing four sequential, yet overlying biological stages: hemostasis, inflammation, proliferation, and remodeling. A disconcertion in the previously stated wound-healing phases due to both external and internal factors may prolong the wound-healing stage and may lead to a disappointing outcome, causing a chronic wound status. The colonization of pathogens over the wound retards the healing process and infection control remains crucially important. In the emerging scenario of biomedical applications, nanomaterial-based wound-healing tactics have emerged as an effective tool against bacterial infections for their cell specificity, which was not earlier attainable with conventional wound-dressing materials or present therapies. The constructive use of metal and alloy nanoparticles have minimal concomitant and enhanced curative activity as compared to its ionic counterpart, which is well documented by *in vivo* excision wound-healing activity of silver (Ag), gold (Au), and Ag/Au alloy nanoparticles. It was evident that Ag NPs and Ag/Au NPs actively inhibited the growth of gram-negative bacterial pathogens and opportunistic *Candida spp.* (Shanmugasundaram et al. 2017). Nanomaterials with their large surface-to-area volume ratio, stability, and tunable properties are designed as drug delivery vehicles or the drug may itself be formulated to the nanoscale. These physicochemical properties enable the nanomaterial to penetrate through the layer of skin and interact efficiently at the wounded site with a continuous and controlled release of therapeutics. Nano-based approaches for wound-healing applications include micelles, polymeric nanocomposites, dendrimers, nanoemulsions, liposomes, cyclodextrins, lipid nanoparticles, magnetic nanoparticles, silica nanoparticles, nanographene oxide scaffolds, and metal nanoparticles. The applicability of nanomaterials as core antibacterial agents and as vehicles for the transportation to wound-healing therapeutic agents will be explored in the future.

1.5 CONCLUSIONS

This chapter summarizes the critical role of nanomaterials in biomedical applications. The portfolio set forward by the nano effects anticipate the evolution of stemming growth towards the advancements in biomedicine with engineered structures with novel functionalities. The distinctive properties of nanomaterials such as size, shape, chemical composition, surface structure and charge, biocompatibility, aggregation and agglomeration, and solubility, can prominently stimulate the interactions with biomolecules and cells, and can be exploited in a multifaceted spectrum of biomedical utilities ranging from drug delivery and biosensors to nanorobots. The revolutionary innovations—programmable and precise delivery of nanomaterials—impart a positive impact rendered by the biomedical applications, minimizing the adverse effects of traditional therapeutics and practices on human health and the environment. The key concerns and encounters in nanotechnology-based approaches provide a futuristic scope and vision of inculcating nanomaterials in biomedical applications.

TABLE 1.1
Nanodrug Carries Approved in Recent Past in Clinical Trials [17]

Drug Name	Delivery Material	Condition	Therapeutic Delivered	Clinical trials. Gov. Identifier	Status
Genexol PM	Amphilic diblock copolymer forming micelle	Non-small cell lung cancer	Paclitaxel	NCT01023347	Completed
Docetaxel-PNP	Polymeric nanomaterials (active nanocomponents loaded/entrapped in polymeric core)	Advanced solid malignancies	Docetaxel	NCT01103791	Completed
CYT-6091	Au NP	Unspecified adult solid tumor	TNF	NCT00356980	Completed
Kogenate FS	PEG-liposome	Hemophilia A	Recombinant factor VIII	NCT00629837	Completed
Long circulating liposomal prednisolone disodium phosphate	Liposome	Rheumatoid arthritis	Prednisolone	NCT00241982	Completed
LE-DT	Liposome	Pancreatic cancer	Doxetaxel	NCT01186731	Completed
Cisplatin and Liposomal Doxorubicin	Liposome	Advanced cancer	Cisplatin and Doxorubicin	NCT00507962	Completed
Liposomal doxorubicin and bevacizumab	Liposome	Kaposi's sarcoma	Doxorubicin and bevacizumab	NCT00923936	Completed
AP5346	Drug polymer conjugate	Head and neck cancer	AP5346 and Oxaliplatin	NCT00415298	Status Unknown

1.6 CHALLENGES AND FUTURE SCOPE OF NANOMATERIALS IN BIOMEDICAL APPLICATIONS

The convergence of science and technology has provided a quintessential hope of developing nano-structured materials in the field of medicine. The widespread opportunities of nanomaterials in therapeutics have gained the attention of researchers from multidimensional aspects ranging from medical practitioners to health experts working in government, industries, and academia. Recent clinical trials of nanomaterials in therapeutics are enumerated in Table 1.1. However, the biocompatibility of nanomaterials is a major concern because of adverse effects extending from cytotoxicity to hypersensitivity. Hence, prior to human exposure, all nanomaterials are imperiled to toxicological studies to meet the regulatory standards. In addition to designing, identifying, and validating a nanomaterial down the lane of the pipeline of drug designing, the toxicological analysis, route of exposure, coating material and sterility of the nanomaterial, economic viability have to be considered with utmost care. The futuristic scope of nanomaterials in the bio-medicinal field involves a perfect blend of nanotechnology and Computer-Aided Drug Designing (CADD). Nanorobots skilled in intruding biological system to identify cancer cells is the recent lead of nanomedicine. The potential of applying nanomaterials in this pandemic situation to arrive at more effective vaccines against COVID 19 is always a subject of major concern.

REFERENCES

Achachelouei, M., Knopf-Marques, H. et al. 2019. Use of nanoparticles in tissue engineering and regenerative medicine. *Front. Bioeng. Biotechnol.* 7: 113. www.frontiersin.org/articles/10.3389/fbioe.2019.00113/full

Bogdan, J., Plawinska-Czarnak, J., Zarzynska, J. 2017. Nanoparticles of Titanium and Zinc oxides as novel agents in tumor treatment: A review. *Nanoscale Res. Lett.* 12: 225. https://nanoscalereslett.springeropen.com/articles/10.1186/s11671-017-2007-y

Buck, S.T.G., Bettanin, F., Orestes, E. et al. 2017. Photodynamic efficiency of xanthene dyes and their phototoxicity against a carcinoma cell line: A computational and experimental study. *Journal of Chemistry.* https://doi.org/10.1155/2017/7365263

Canavese, G., Ancona, A., Racca, L. et al. 2018. Nanoparticle-assisted ultrasound: A special focus on Sonodynamic therapy against cancer. *Chemical Engineering Journal* 340: 155–172. https://doi.org/10.1016/j.cej.2018.01.060

Chang, D., Lim, M., Goos, J.A.C.M. 2018. Biologically targeted Magnetic Hyperthermia: Potential and limitations. *Front. Pharm.* https://doi.org/10.3389/fphar.2018.00831

Chen, J., Fan, T., Xie, Z. et al. 2020. Advances in nanomaterials for photodynamic therapy applications: Status and challenges. *Biomaterials.* 237: 119827. https://doi.org/10.1016/j.biomaterials.2020.119827

Dash, A., Chakravarty, R. 2019. Radionuclide generators: The prospects of availing PET radiotracers to meet current clinical needs and future research demands. *Am. J. Nucl. Med. Mol. Imaging.* 9(1): 30–66. www.ncbi.nlm.nih.gov/pmc/articles/PMC6420712/

Eliaz, N., Metoki, N. 2017. Calcium phosphate bioceramics: A review of their history, structure, properties, coating technologies and biomedical applications. *Materials (Basel).* 10(4): 334. www.mdpi.com/1996-1944/10/4/334

Gibot, L., Demazeau, M., Pimienta, V. et al. 2020. Role of polymer micelles in the delivery of photodynamic therapy agent to liposomes and cells. *Cancers (Basel)* 12(2): 384–406. www.ncbi.nlm.nih.gov/pmc/articles/PMC7072360/pdf/cancers-12-00384.pdf

Hou, Y., Sun, Z., Rao, W. et al. 2018. Nanoparticle- mediated cryosurgery for tumor therapy. *Nanomed: Nanotechnol. Biol. Med.* 14(2). https://doi. 10.1016/j.nano.2017.11.018

Huang, B. 2020. Carbon nanotubes and their polymeric composites: The applications in tissue engineering. *Biomanufacturing Reviews* 5(3): 1–26. https://link.springer.com/content/pdf/10.1007/s40898-020-00009-x.pdf

Idrees, H., Zaidi, S.Z.J., Sabir, A. et al. 2020. A review of biodegradable natural polymer based nanoparticles for drug delivery applications. *Nanomaterials* 10: 1970. www.ncbi.nlm.nih.gov/pmc/articles/PMC7600772/pdf/nanomaterials-10-01970.pdf

Kim, D., Shin, K., Kwon, S. et al. 2018. Synthesis and biomedical applications of multifunctional nanoparticles. *Adv. Mater.* 30: 1802309. https://doi.org/10.1002/adma.201802309

Kim, S., Lm, S., Park, E.-Y. et al. 2020. Drug-loaded titanium dioxide nanoparticle coated with tumor targeting polymer as a Sonodynamic Chemotherapeutic agent for anticancer therapy. *Nanomedicine: Nanotechnology, Biology and Medicine* https://doi.org/10.1016/j.nano.2019.102110

Masood, F. 2016. Polymeric nanoparticles for targeted drug delivery system for cancer therapy. *Materials Science and Engineering: C* 60(1): 569–578. https://pubmed.ncbi.nlm.nih.gov/26706565/

Pellico, J., Llop, J., Fernandez-Barahona, I. et al. 2017. Iron oxide nanoradiomaterials: Combining nanoscale properties with radioisotopes for enhanced molecular imaging. *Contrast Media & Molecular Imaging.* https://doi.org/10.1155/2017/1549580

Qing, Y., Cheng, L., Li, R. et al. 2018. Potential antibacterial mechanism of silver nanoparticles and the optimization of orthopedic implants by advanced modification technologies. *Int. J. Nanomed.* 13: 3311–3327.

Rodrigues, R.C., Rivas-Garcia, L., Baptista, P.V. et al. 2020. Gene therapy in Cancer treatment: Why go nano? *Pharmaceutics* 12(3): 233–267. www.ncbi.nlm.nih.gov/pmc/articles/PMC7150812/

Shanmugasundaram, T., Radhakrishnan, M., Gopikrishnan, V., Kadirvelu, K., Balagurunathan, R. 2017. In vitro antimicrobial and in vivo wound healing effect of actinobacterially synthesized nanoparticles of silver, gold and their alloys. *RSC Adv* 7: 51729–51743. https://pubs.rsc.org/en/content/articlelanding/2017/ra/c7ra08483h#!divAbstract

Siddique, S., Chow, J.C.L. 2020. Application of nanomaterials in biomedical imaging and cancer therapy. *Nanomaterials* 10: 1700–1740. www.mdpi.com/2079-4991/10/9/1700/htm

Siegel, R.W. 1993. Nanostructured materials: Mind over matter. *Nanostructured Materials* 3(1–6): 1–18. https://doi.org/10.1016/0965-9773(93)90058-J

Stewart, S., Arminan, et al. 2020. Perspective: Nanoparticle-mediated delivery of cryoprotectants for cryopreservation. *Cryoletters* 41(6): 308–316. www.cryoletters.org/perspective-41-6-308-316-stewart.pdf

Teng, X., Liu, H., Huang, C. 2007. Effect of Al2O3 particle size on the mechanical properties of alumina-based ceramics, *Mater. Sci. Eng. A* 452–453: 545–551. https://doi.org/10.1016/j.msea.2006.10.073

Velu, R., Calais, T., Jayakumar, A. et al. 2020. A comprehensive review on bio-nanomaterials for medical implants by additive manufacturing technique. *Materials (Basel)* 13(1): 92–115. https://pubmed.ncbi.nlm.nih.gov/31878040/

Wan, G.-Y., Liu, Y., Chen, B.-W. et al. 2016. Recent advances of sonodynamic therapy in cancer treatment. *Cancer Biol. Med.* 13(3): 325–338.

Yang, Z., Sun, Z., Ren, Y. et al. 2019. Advances in nanomaterials for use in photothermal and photodynamic therapeutics. *Molecular Medicine Reports* 20(1): 5–15. www.spandidos-publications.com/10.3892/mmr.2019.10218

Yetisgin, A.A., Cetinel, S., Zuvin, M. et al. 2020. Therapeutic nanoparticles and their targeted delivery applications. *Molecules* 25: 2193–2224. www.mdpi.com/1420-3049/25/9/2193

Zhao, C., Song, X., Liu, Y. et al. 2020. Synthesis of graphene quantum dots and their applications in drug delivery. *J. Nanobiotechnol.* 18: 142–174. https://jnanobiotechnology.biomedcentral.com/articles/10.1186/s12951-020-00698-z

2 An Introduction to Properties of Biomedical Materials

Bhavana Rikhari and B.E. Amita Rani

CONTENTS

2.1 INTRODUCTION OF BIOMATERIALS

Biomaterials are used for replacement of damaged or diseased parts of the body. They are usually produced in low volumes as they are customized according to an individual's specific requirement. Implants are used for various applications such as medical/clinical, prosthetics, drug delivery, and biosensors. Requisite for research in biomaterials has taken a quantum leap due to significant expansion in different areas of science to regenerate tissues/organs and various bioactive materials. Injury, degeneration, and illness require appropriate treatment in the human body. This includes

DOI: 10.1201/9781003173533-2

the replacement of different parts of the body, which comprises hips, knees, elbows, hearts, skin, kidneys, etc. In 1985, Boretos proposed the definition of biomaterial:

> Any substance (Biological or synthetic but other than a drug) which can be introduced in a body as a part or whole for therapeutic use (treat, augment, repair or replace any tissue, organ or function of the body) for any period.
>
> (Boretos et al., 1985)

It is also defined as "A non-viable material subjected to interact with the human system" (Sridhar et al., 2003). Clemson University explained biomaterial as a "Systemically and pharmacologically inert substance designed for implantation within or incorporation with living systems by function in intimate contact with living tissue" (Parida et al., 2012). Increase in demand for implants in the markets is expected to reach more than US$116 billion in upcoming years, due to the increase in trauma injuries and road accidents (Chin et al., 2020).

2.2   HISTORY OF BIOMATERIALS

Biomaterials were already in existence since the Egyptian era around 4000 years ago, as linen thread to cover the wounds, sutures in the Neolithic era, gold dentistry, eyeglasses, ivory, and wooden feet by the Chinese and Romani. The concept of artificial eyes, ears, nose, and teeth was adopted by the Egyptians (Pattanayak et al., 2005). Later, these experiments enabled scientists to understand the incompatibility of the human body with artificial materials. During World War II, Sir Harold Ridley discovered that embedded pieces of cockpit canopies were tolerable in the eyes, leading to the use of methyl methacrylate for intraocular lenses to date! In 1929, "Vitallium" a new metal alloy was fabricated for dental purpose (Pal, 2015). The early 1960's witnessed the success of orthopaedic implants and artificial heart valves. Presently, wide varieties of implants are easily accessible in the market for advancement in the quality of human life.

2.3   EVOLUTION OF BIOMATERIALS

Based on the history of biomaterials, implants are categorized into first, second, and third generations (Hench and Polak, 2002). First-generation biomaterials include gold filling, PMMA dental prostheses, gold, ivory, wooden teeth, steel, etc. with access to use in daily life. Titanium (Ti), alumina, stainless steel, partially stabilized zirconia, etc. comprise bio-inert materials, which do not interact with their external environment. Second-generation biomaterials (Ti and its alloy, bioactive glass, calcium phosphate, hydroxyapatite, glass ceramics) tend to merge with the neighbouring tissues. They are bioactive and enhance biological responses by forming an active ionic layer with proximal tissues. The third generation includes bioengineered materials (resorbing bone, repairing cement, tissue-engineered implants), which can regenerate the functional tissue. They stimulate and regenerate specific responses at the tissue level and are bio-resorbable. Here, the material dissolves initially and is later replaced by advanced tissues, thus enhancing the interfacial stability (Tavares et al., 2013; Mano et al., 2007).

Bioactive materials have been furthered as follows (Hench, 1994). Bioactivity occurring at both the intercellular and extracellular interfaces of the implant is called osteoproductive. Extracellular response taking place at the interface of the implant material is called osteoconductive, for example synthetic hydroxyapatite (HAp).

2.4   CHARACTERISTICS OF BIOMATERIALS

A perfect implant needs to have some desirable mechanical properties, non-toxicity, biocompatibility, osseointegration, and corrosion resistance.

2.4.1 Mechanical Properties

The material can be selected based on its mechanical strength and ability to sustain the continuous cyclic load for the desired organ/tissue where it has to be placed in the human system to avoid revision surgery. Moghaddam et al. (2016) report that insufficient strength between bone and the implant material leads to mechanical incompatibility. For instance, biomaterials with flexibility and toughness are required for heart valves, while for orthopaedic applications, Young's modulus, elongation tensile strength, etc. are desired. Human bones exhibit modulus from 5–30 GPa (Niinomi and Nakai, 2011). When bones transfer the stress to the proximal bone, the implant experiences higher stiffness and leads to loosening and failure due to stress shielding.

2.4.2 Non-Toxicity

The implanted material should be biocompatible and not induce toxicity expressed as causing tumour, allergy, inflammation, etc. (Raghavendra et al., 2015).

2.4.3 Biocompatibility

Biocompatibility is a measure of the tendency of the biomaterial to function appropriately without exhibiting adverse effects such as leaching of its constituent material inside the body (Ratner, 2015). The host environment also influences implant behaviour. Therefore, biological testing for a biomaterial is necessary. Ratner (2004) described the various tests of material to confirm the compatibility (Table 2.1).

Our *in-vitro* studies reveal that Ti confirmed the biocompatibility concept of metal (Rikhari et al., 2020). The studies support the viability of MG-63 osteoblast-like cells, evaluated by MTT assay for orthopaedic applications and represented in Figure 2.1.

TABLE 2.1
Tests to Verify the Biocompatibility of the Material

	***In-Vitro* Test**
Cytotoxicity	Elution or extract test, Agar or Agarose overly test, Direct contact test
Haemocompatibility	Haemolysis assay, clotting and complement activation
Hypersensitivity	Lymphocyte transformation test, Leukocyte migration inhibition test
	***In-Vivo* Test**
Short-Term Implantation Tests	Subcutaneous, fibrosis, inflammation, and intramuscular and intraperitoneal implantation tests to evaluate general tissue necrosis
Long-Term Functional Tests	Functionality of device and histopathological evaluation is performed
	Device or compositionally identical prototypes are implanted in appropriate animal models to replicate/stimulate intended end use in humans
Carcinogenicity Tests	Determination of chemical/compound that may be released from biomaterials elicit sensitization reactions—result of immunological medicating reaction results in swelling and redness
Irritation	Ocular, Mucosal and Skin
Sensitization	Occluded patch test, Open epicutaneous test
Genotoxicity Tests	Alteration in DNA, chromosomal structures, other DNA, or damage in gene results in permanent inheritable changes in cell functions
Functional Tests	Evaluation of soft tissues, greater complexity, also specific *in-vivo* physical assessment test (carcinogenicity, genotoxicity, toxicity, etc.)

Source: Ratner (2004).

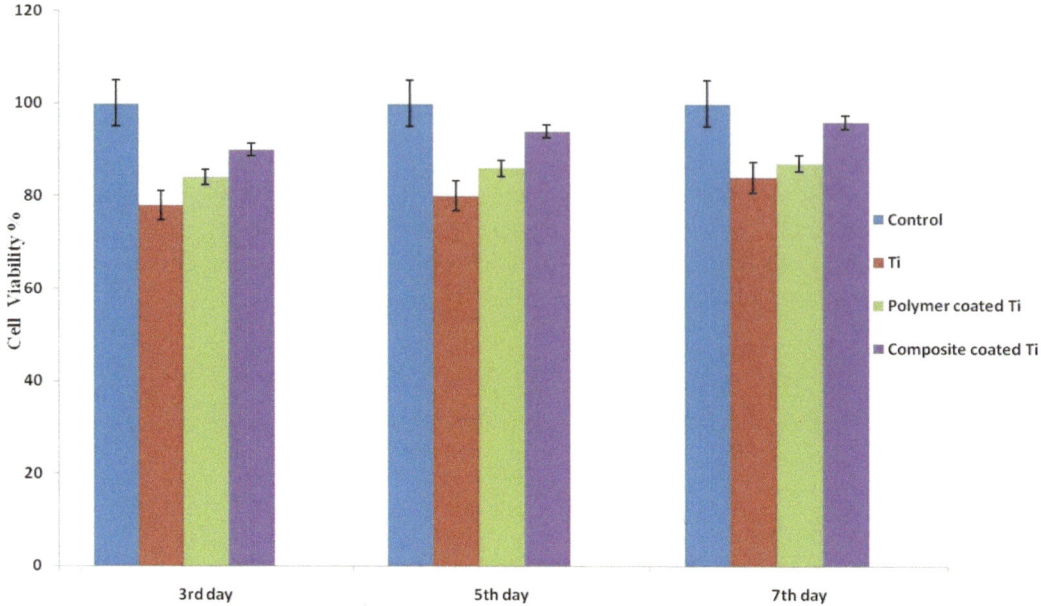

FIGURE 2.1 Analysis of cell viability by MTT assay for control, composite coated, polymer coated and uncoated Ti.

Source: Rikhari et al. (2020).

To assess the biocompatibility, human bone cells are cultured on Ti metal with composite and polymer coating, indicated that the cell viability and growth was more for composite/polymer coating compared to Ti without coating. The composite coating promotes denseness of bone cells compared to without polymer coating on Ti (Figure 2.2). A flattened and polygonal type structure with elongated filopodia of bone cells was witnessed. Composite coating showed elongated morphology of bone cells with spindle shape as compared to Ti with and without polymer coating.

Liu et al. (2014) and Li et al. (2014) also endorse *in-vitro* biocompatibility of composite coated material by enhanced osteogenic differentiation for pre-osteoblastic cell and fibroblast cells respectively. The compatibility of composite coated implants depends on various aspects such as chemical composition, interaction with water and intermolecular bonding between the two components (Vandrovcova and Bacakova, 2011).

2.4.4 OSSEOINTEGRATION

It is the integration between the surface of the living bone and implants without disturbing the soft tissue. The implant surface plays a significant role to amalgamate with the neighbour bone. Material topography is crucial to the successful integration of the bio-implant. The growth of bone cells over the implant is a complex process that includes micrometric interaction, which could interfere in the integration and cause implant failure (Viceconti et al., 2000). Post implant, the body exhibits biological, physical, and chemical changes in the host system.

2.4.5 CORROSION RESISTANT

All metals/alloys are susceptible to corrosion. It is the deterioration of a metal (electrochemical attack) in an aqueous solution leading to the release of metal ions, pH alterations, and oxygen level reduction, leading to implant failure. The reduction in material strength and life, the defence

Uncoated Ti   Polymer Coated Ti   Composite Coated Ti

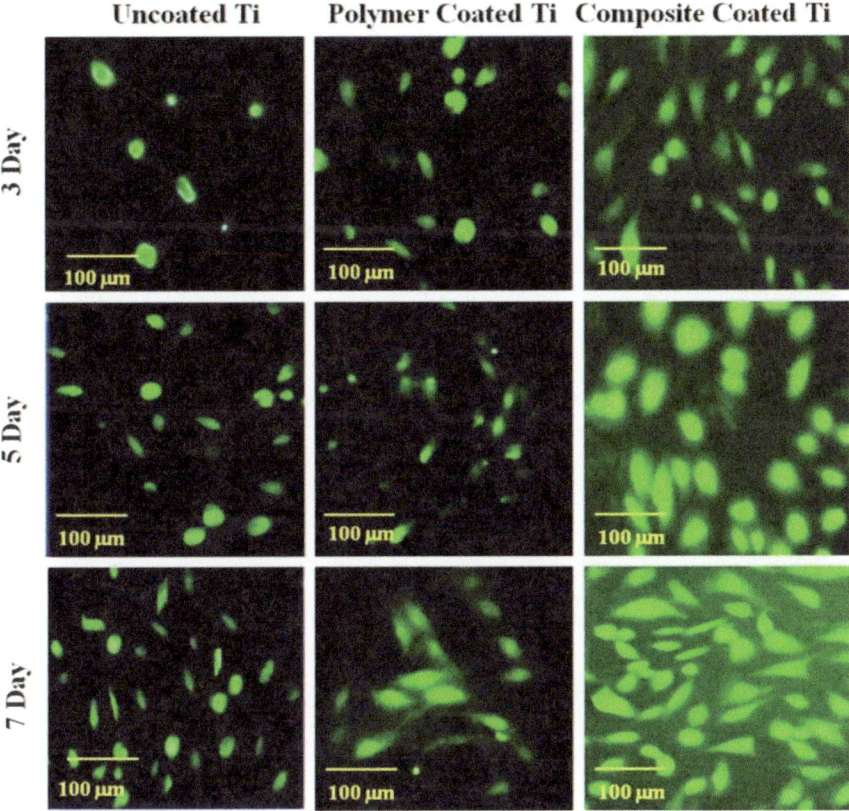

FIGURE 2.2   Confocal images of MG-63 cells on composite coated, polymer coated and uncoated Ti.

Source: Rikhari et al. (2020).

mechanism of the body impels mechanical collapse of the implants, increased infections, and eventual implant failure. Therefore, it is essential for the implant material to be non-reactive and tolerable in the body environment.

Metallic implants are more susceptible to localized corrosion by aggressive ions (Kruger, 1979; Eliaz, 2019; Kamachimudali et al., 2003). Several studies concluded that 80% failure of implants are because of corrosion. Bio-implants encounter two critical environments within the body. One, the hostile body fluids, leading to corrosion, and the other is load.

The types of corrosion that are expected based on the bio-implant materials are categorized and discussed next.

2.4.5.1   Crevice Corrosion

It is generally related to structure and seen beneath the screw head or intersection of two components. Stainless steel (316L) is most vulnerable to corrosion as compared to other materials (Bates, 1973). In the bone plate, the head underside screw area and the countersunk are examples of where crevice corrosion can occur. However, it can be abated with appropriate architecture and retailoring of the material.

2.4.5.2   Galvanic Corrosion

Whenever two unlike metals come under physical interaction in the occurrence of physiological body fluid (ionic electrolyte), a potential difference is generated between the metals. The contact between the stainless steel cerclage wire with Co and Ti alloy femoral stem or titanium screw with stainless steel plate leads to bimetallic corrosion (Jacobs et al., 1998).

2.4.5.3   Fatigue Corrosion

Fatigue corrosion in implants is determined by cyclic loading and high stress in corrosive media. Loading can be any compression, tension, bending, and torsion depending on the type of activity performed. The mechanical degradation gets further accelerated by the formation of wear debris which gives rise to fatigue wear. Disruption of the oxide layer, the inability of material to re-passivate, and immediate exposure of some area of the metal in the physiological environment leads to corrosion due to electrochemical reactions and also by cyclic loading (Cattis and Husain, 1982).

2.4.5.4   Fretting Corrosion

This is generally observed when two materials with different mechanical properties undergo friction due to slight movement or continuous rubbing of like different surfaces such as plates of bone and heads of the screw in prosthetic devices or at countersunk in plates and at hip nails (Syrett and Wing, 1978). The effect of fretting corrosion depends not so much on the presence of corrosive media but wear debris and metal ion concentration present in the surrounding tissue (Pellier et al., 2011). Hip joints, dental implants, knee implants, plates, and screws are examples of where fretting corrosion can take place.

2.4.5.5   Pitting Corrosion

It's often found in chloride ion rich media (Sivakumar et al., 1994) or in the presence of dissimilar materials. It's generally confined to a very small area or point in the metal and characterized by the presence of cavities or small holes in the metals. Of all the types of corrosion, pitting is the most damaging, leading to metallic implant failures.

In stainless steel fracture, pitting corrosion is observed in the fixation hardware under the screw heads (Black, 1988).

2.4.5.6   Stress Corrosion Cracking

When implants are subject to continuous activity of loading in the presence of a corrosive environment, this can lead to stress corrosion cracking. With regular heart function, a cardiovascular stent can experience cyclic loading. Even an orthopaedic implant undergoes stress corrosion due to body movement and walking.

2.5   CLASSIFICATION OF BIOMATERIALS BASED ON THE FUNCTION AND THEIR APPLICATIONS

Based on their functions and applications the biomaterials are classified into metals, ceramics, polymers, composites, biodegradable materials, nanomaterials and are represented in Figure 2.3.

2.5.1   Metals

To set a fracture in bone, metal plates have been used since 1895. During the primitive period, implants suffered mainly from deterioration and inadequate strength issues (Hermawan, 2012). Soon after, stainless steel was identified for use, as it has improved corrosion resistance properties compared to other metals. Thereafter, greater development and clinical research on metallic implants was accomplished. Based on the medical applications, various types of metals are utilized. The essential criteria for choosing metal implants are biocompatibility, anti-corrosion behaviour, and cost. Stainless steel (316L SS) and titanium are widely used as implants. Table 2.2 depicts the different metals for biomedical application (Hermawan et al., 2011).

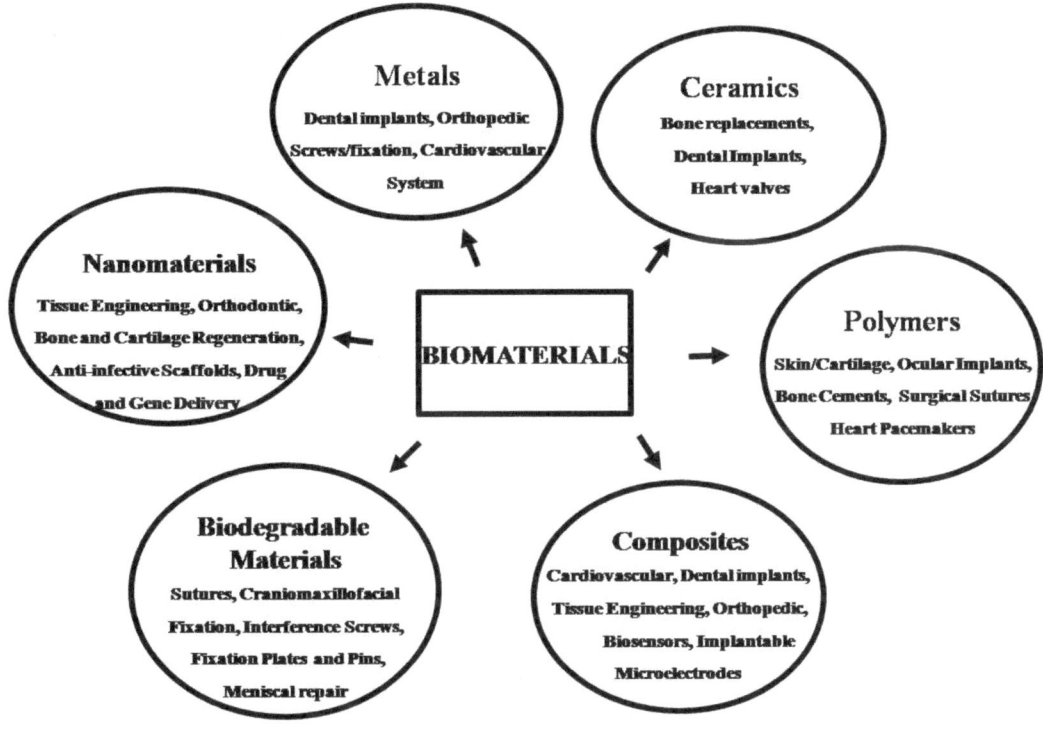

FIGURE 2.3 Classification of biomaterials based on their applications.

TABLE 2.2
Metals/Alloys Used for Biomedical Applications

Types of Metal and Alloys	Example of Implants	Area
316L SS, Ti, Ti6Al4V, CoCr, Ti6Al7Nb	Bone fixation (plates, screws), artificial joints	Orthopaedics
316L SS, Ti, Ti6Al4V	Stents, artificial valves	Cardiovascular system
316 SS, TiMo, Au	Orthodontic wire filling	Dentistry
316L SS, Ti6Al4V	Plates and screws	Craniofacial surgery

Currently, various medical-based companies are dependent on metals/alloys. For orthopaedic, load-bearing purposes, screws and pins for hip joints, knee prostheses, maxillofacial surgery, and dental purposes, metals are the most successful implants (Poinern et al., 2013; Sridhar et al., 2003). It exhibits high strength and toughness compared to other materials, though it undergoes corrosion in an aggressive environment. Generally, three groups of metals and their alloys are dominant for bio-implants: stainless steel, Co-Cr alloys, and titanium.

2.5.1.1   316L SS

Surgical grade stainless steel 316L is used for temporary and permanent implants. Its alloying elements, nickel, molybdenum, chromium, provide some desirable properties for implants (Mikrolegiranih, 2011). However, it was later understood that metal-based alloys (i.e., Co-CrMo, 316L SS) cause some adverse effects on the immune system of the host body. Therefore, it was decided from the overall experiences that even stainless steel will not fulfil the criteria of the best implant material. Crevice and pitting corrosion is more prominent, which can cause premature failure of steel implants and require a second surgery.

2.5.1.2   Cobalt-Chromium Alloys

These alloys are well known for their mechanical characteristics and biocompatibility. They contain cobalt and chromium as alloying elements. These alloys are highly corrosion resistant due to the Cr passive layer formation, similar to stainless steel. Co-Cr-W-Ni and Co-Ni-Cr-Mo-Ti alloys are present in implant material. They are prone to corrosion in different simulated body fluids and leach out Co in the body. When these ions are released in the body and come in contact with an organ or tissue, they can cause metallosis (i.e., implant failure leads to tissue/organ death).

2.5.1.3   Ti and Its Alloys

Globally, Ti and its alloys are most successful and the preferred implant materials for orthopaedic and oral aesthetics due to their lightweight, high mechanical strength, anti-corrosive behaviour, and biocompatibility. Ti metals form a thin oxide (TiO), dioxide (TiO_2), and Ti_2O_3 suboxides (Fekry and Ameer, 2011) film, influenced by many factors such as the thickness of the material, environmental conditions, and they offer excellent corrosion resistance properties. The breakdown of oxide film due to corrosion affects the implant's surrounding tissue (Tamilselvi et al., 2007; Geetha et al., 2009). Ti alloys (Ti6Al4V, Ti-13Nb-13Zr) have excellent tensile strength and corrosion resistance. Our potentiodynamic polarization studies revealed decrease in current density for composite coated Ti after immersion in 168 h of Simulated Body Fluid (SBF) solution in Figure 2.4 and represents the corrosion protection ability of Ti metal (Rikhari et al., 2016). Titanium alloyed with Ni is called Nitinol. They are applicable in dental restorations, cardiovascular stents, guide wires, etc. (Beline et al., 2016). Ti metals usually have their inherent stable oxide (TiO_2) film on their surface, which is highly adherent. Ti metals enhance the bone cells to differentiation at implant/tissue interface (Rautray et al., 2010).

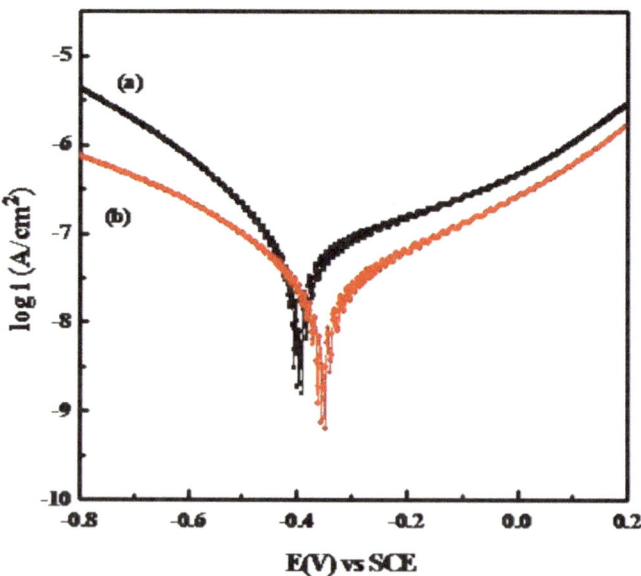

FIGURE 2.4   Potentiodynamic polarization results for (a) polymer coated Ti (b) composite coated Ti immersion in SBF solution.

HAp deposition was observed over composite coated Ti in physiological body fluid solution at different periods (Rikhari et al., 2018). The formation of apatite over metal for the different durations can be observed from SEM images as represented in Figure 2.5(a–c) which exhibits the bone-forming ability of Ti metal. Figure 2.5d shows the EDAX results and explains that composite coated Ti immersed for 168 h in SBF medium exhibited calcium to phosphorus ratio equal to 1.67, which supports the presences of HAp on the surface of the metal. This also assists in the growth and development of bone cells.

2.5.2 Ceramics

Since ancient times ceramics have been used in diversified areas and contributed to various technological growth. Owing to the innate brittleness, hardness, and stiffness, ceramics have constrained their application. Advancement in ceramic processing has helped favourable interaction of the implants with human tissues, due to enhanced mechanical, biological, and physical properties of the implanted biomaterial. These modified ceramics utilized for human applications are termed as bio-ceramics. Bio-ceramics include alumina, silica, titanium dioxide, zirconia, glass ceramics, bioactive glass, calcium phosphate, etc., Ceramics can be utilized for various applications. They are classified as follows.

2.5.2.1 Based on Origin

Bio-ceramics can be produced naturally and synthetically. Ceramics obtained naturally and originating from a dead or living organism such as mollusc shells, bone, teeth are called natural bio-ceramics (Brundavanam et al., 2017). Artificial bio-ceramics include zirconia, alumina, bio-glass, hydroxyapatite (HAp) and could be prepared by sol-gel, co-precipitation, and hydrothermal methods.

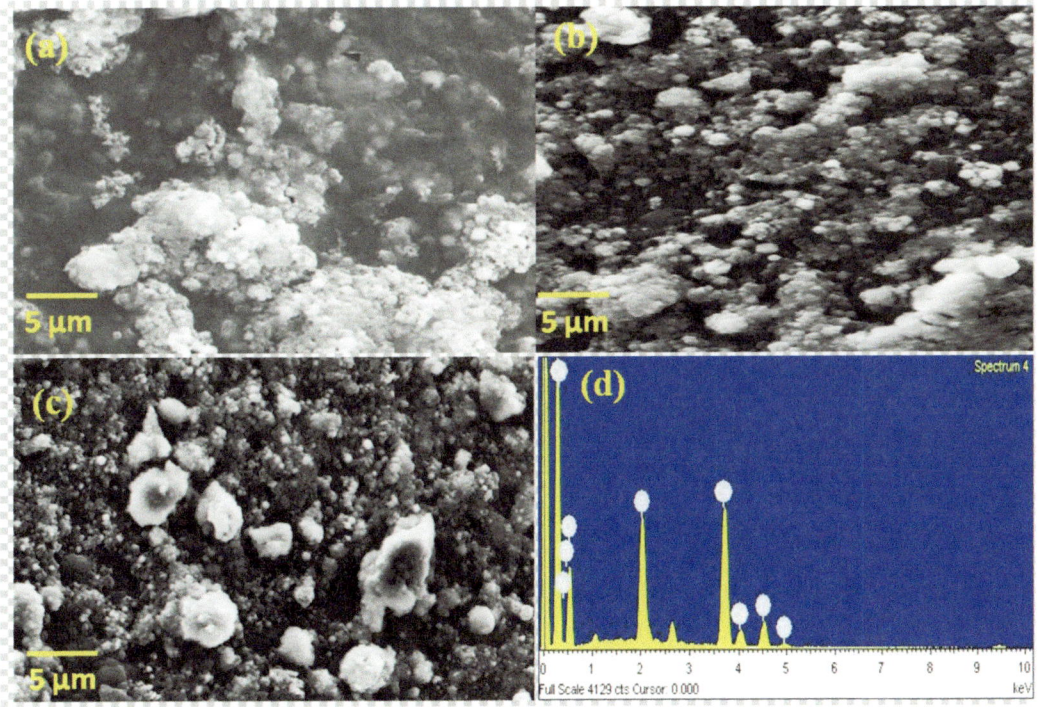

FIGURE 2.5   SEM images of composite coated Ti immersed for (a) 72 h, (b) 120 h, (c) 168 h, (d) EDAX at 168 h.
Source: Rikhari et al. (2018).

HAp is a calcium-phosphate based material that can be synthesized and obtained naturally. HAp is used to replace bone material. It manifests biocompatibility with bone tissue. The main source for natural HAp is fishbone, sheep bone, goat bone (Gul et al., 2020).

2.5.2.2   Based on Tissue—Response

In the 1960s, bio-ceramics were used as an alternative to the metallic material for the enhancement of biocompatibility. Bio-ceramics are further categorized into bio-inert ceramic and bioactive ceramics (Juhasz and Best, 2012).

Bio-ceramics does not evoke any tissue or organ reaction while interacting with the human system (e.g., alumina, titania and zirconia are known as bio-inert ceramics), while bioactive ceramics influence specific tissue/organ reactions (e.g., calcium phosphate materials and bioactive glasses).

Bioactive ceramics are further categorized into resorbable and non-resorbable ceramics based on their solubility. HAp is resorbable and used as bone filling material over the defected bone. Tricalcium phosphates (TCPs) are porous in nature and mainly utilized as bone fillers (Pina and Ferreira, 2012). Non-resorbable ceramics consist of alumina or zirconia-based metal oxides used in oral health and fillers.

2.5.2.3   Based on Crystallinity

Bio-ceramics represent lattice structures formed by the crystallization process. These bio-ceramics appear as crystalline structures (De Aza et al., 2005), which is when crystalline material exhibits regular arrangement of atoms/ions/molecules. HAp is an example of crystalline bio-ceramics which display hexagonal packing with tetrahedral and octahedral voids. Whereas bio-ceramics without any lattice structure are called amorphous bio-ceramics, such as bioactive glasses.

2.5.2.4   Based on Composition

Alumina is the oxide of aluminium, a pure and fine white powder that contains trace amounts of magnesium, silicon, iron, and calcium oxides. It exhibits excellent strength and stability to pressure and heat. Due to the exceptional hardness of alumina, it is used for the load-bearing implant (Chevalier and Gremillard, 2009). Zirconia exhibits high mechanical strength and toughness compared to other ceramics (Patel et al., 2012). Encapsulation by a thin fibre over it shows interfacial instability. Implanted bioactive zirconia establishes a bond with the bone tissue (Piconi and Maccauro, 1999). The apatite deposit on exchanging the ions between the implant and the tissue contributes to orthopaedic applications (Abd El-Ghany and Sherief, 2016).

2.5.3   POLYMERS

Polymers are large molecules, made of repeated small units known as the monomer and exhibit chemical and inter-molecular bonding and are extensively used for scaffolds development in bio-engineering and medical applications. In polymers, this bonding mainly depends on the difference in chain length and energy. The strong chemical bonds hold atoms in the chain together. The difference in structure and bonding of the monomer leads to variation in the mechanical strength and stability of different polymers.

Based on physical and chemical properties, which includes mechanical behaviour, thermal characteristics, and stereochemistry, polymers are categorized into (a) natural and synthetic polymers, (b) organic and inorganic polymers for bio-implant applications.

Natural polymers are promising materials for medical applications (Choueka et al., 1996). Collagen is a natural biodegradable polymer that is opted for ocular, skin, corneal, cartilage, tendon/ligament engineering (Vroman and Tighzert, 2009). Other biodegradable polymer based scaffolds such as polylactic acid and polyglycolic acid are used for orthopaedic purposes. Some of the examples of polymers opted for biomedical applications (Davis, 2003) are given in Table 2.3. Polyethylene

---

TABLE 2.3
Polymers Used for Biomedical Applications (Davis, 2003).

Polymers	Applications
Polylactic acid, polyglycolic acid, nylon	Sutures
Ultra high molecular weight polyethylene	Knee, hip, shoulder joints
Silicone	Finger joints
Acetal, polyethylene, polyurethane	Heart pacemaker
Polyester, polytetrafluoroethylene, PVC	Blood vessels
Polydimethyl siloxane, polyurethane, PVC	Facial prostheses
Polymethyl methacrylate	Bone cement

High-Density (PEHD) and Polyethylene Low-Density (PELD) and Ultra High Molecular Weight Polythene are used for surgery. Polymer is utilized for different parts of the body, namely artificial heart valves, blood vessels, the liver, kidneys, and for pacemakers and joint replacements.

The merits of polymer implants are (a) ease of manufacture and processing, (b) achievement of desired properties (physical/mechanical), and (c) cost. However, the main issue with polymer is its degradation due to chemical and biochemical reactions that show adverse effects on implant material.

2.5.4 COMPOSITES

Composite biomaterials are composed of two or more components with considerably distinct physical, chemical, and mechanical characteristics combined to give a new material superior characteristics different from either of the components (Dorozhkin, 2012). They are desirable due to their superior mechanical properties and load-bearing ability. Fibre-reinforced materials are employed for femoral components of hip joints, dental applications (restoration material and dental filling), ligaments, tendons, bones, etc. Fibre-reinforced composites are successful materials for implant purposes. Carbon-reinforced polymer composites have tremendous applications in joint replacements, prosthetic limbs, and bone repair due to their lower elastic modulus. They have also been opted for vascular graft surgery polyurethane with gelatin or collagen. Composites obtained by the amalgamation of bioactive and bio-inert materials enhance bioactivity and mechanical strength of implants. Apart from many advantages, composite materials also suffer from many failures due to inadequate strength and high degradation under aggressive environment and stress (Kamachimudali et al., 2003).

2.5.5 BIODEGRADABLE MATERIALS

Implants are generally fabricated to be degradable or resorbed inside the human body instead of temporarily lodged, removable materials. Biodegradable materials degrade gradually and support curing. Metals/alloys like magnesium and iron are the first choices over polymers due to their mechanical strength compared to polymers (Chen et al., 2014; Schinhammer et al., 2010) in addition to being biodegradable. It is important that the degraded material is not toxic to the host. However, some natural polymers like collagen, chitosan, and starch have low mechanical strength, but high physiological activity and thus may be used as biodegradable materials. Porous metal alloy implants (Mg and iron) degrade faster compared to non-porous alloys. The porosity helps mimic natural human-like implants.

Synthetic degradable polymers are tailored by their application and desirable properties. The rate of degradation is dependent on the organization of the monomers in the polymer chain (He et al., 2000) and whether they are crystalline/amorphous. Temperature is crucial for regulating the degradability of the polymers. They become stretchy on transition temperature (Lodge and McLeish, 2000) and biodegradable above body temperature. Polylactic acid (PLA), Polylactic acid-co-glycolic acid (PLGA), and polyglycolic acid (PGA) are some examples of commonly used biodegradable polymers (Hollinger and Battistone, 1985; Pezzin et al., 2003). A synthetic polymer exhibits high crystallinity, low solubility, and a high rate of degradation. After implantation due to the high degradation rate, it leads to reduction in its mechanical property (Sung et al., 2004). Even though PCL rate of degradation is very high still, it is most desirable for implants due to its malleability. PPD is used in form of foams and coating types of products for internal fracture fixation. Biodegradable polymers are related to tissue engineering, where tissue repair or remodelling is the main focus whereas long-term stability is not required. They also find application in biodegradable stents, where mechanical strength is not desirable.

Metallic materials show high mechanical properties compared to polymer-based implants. Extensive experimental studies on magnesium alloys have indicated that Mg-based materials are non-toxic for the human system. Many magnesium-based materials and alloys exhibit high mechanical strength with biocompatibility. But the rate of corrosion by hydrogen evolution test was very high (Pezzin et al., 2003) due to its negative effects in the healing process (Vojtěch et al., 2011). Zn-based materials are noble, corrosion-resistant, and have good mechanical properties compared to magnesium alloys (Kubasek and Vojtěch, 2012). Zinc alloys preparations are not very expensive.

2.5.6 NANOMATERIALS

Nanotechnology comprises of materials within 1–100 nm size range, in which completely new physical/chemical properties are analyzed by particles compared to their bulk parts (Kumar et al., 2020).

Since major biological structures are nanoscale ranges, it helps nanomaterials perform efficiently with implants. The nanomaterial assists in the progression of a novel way of treatment with repairing and substitution of tissues. Nanomaterials have a huge surface area to volume proportion, which promotes interactions with surrounding structures. Nanomaterials have important biomedical applications in prostheses, implants, drug delivery systems due to their size.

Nanomaterials can exist in various forms of polymers, metals, or ceramics, etc. Several characteristics of materials are able to be enhanced at the nano level and modified to particular purposes.

Generally, integration of material with host tissue is very hard. It gives rise to infectious reactions and ultimately implant failure takes place. Nano biomaterials tend to replicate the extracellular matrix of bone that assists in cell growth. Biological and morphological characteristics of nanomaterial have been chosen for the development of scaffolds for implant usage.

Nanomaterials are categorized into organic and inorganic nanomaterials (Liu et al., 2016). Organic nanomaterials possess different properties (i.e., anti-toxic, anti-carcinogenic, and biocompatible). Chitosan, collagen, gelatin, liposomes, lipids, and carbon nanotubes are examples of organic nanomaterials. Nanomaterials not containing carbon constituents and employed for replacement/restoration of a body organ are classified as inorganic nanomaterials. Bio-ceramics/bio-glasses are major inorganic materials used for implants. A few examples of inorganic nanomaterials are silver nanoparticles, gold nanostructures, Ti nanostructures, bioactive glasses, nano silica, etc. Amidst various nanostructures, gold and titania are available in different nanotube forms. These nanoparticles have the tendency to regulate nanostructure size/dimensions including physical characteristics.

Various studies indicate that osteoblast cells have better adhesion to nanomaterials. Deposition of nano calcium coating on the metal, compared to conventional materials exhibited good osteoblast cells growth with improved adhesion (Shrivastava and Dash, 2009). This nano-coating helped in better bonding for the propagation of healthy bone tissues (Park and Webster, 2005). The formation of even layers of nanocrystalline structure on artificial implants helps to prevent complications of wear or disintegration of the bio-implant. The natural bone is made up of 70% mineral hydroxyapatite and 30% collagen. Nanoscale HAp grain size is ideal for biocompatibility purposes.

2.6 CLASSIFICATION OF BIOMATERIALS BASED ON THE NUMBER OF CELLULAR COMPONENTS

Implants based on the numeral cellular constituent are classified into biological, biologized, and biofunctional materials.

2.6.1 BIOLOGICAL MATERIALS

These are designed from living cells, protein, and degradable scaffolds/hydrogels which prearrange themselves in the 3D form essential in fabricating the organ by means of bioprinting.

2.6.2 BIOLOGIZED MATERIALS

Biologized implants are generally made of stainless steel, titanium, gold, ceramic, nitinol, etc. They are inert, non-biodegradable and are a mixture of permanent biomaterials and cellular components. They provide mechanical stability.

2.6.3 BIOFUNCTIONAL MATERIALS

As the name suggests the surface is functionalized by surface treatment for enhanced interaction with the biological environment. Hydroxyapatite, glass ceramics are some of the examples of this class of implants.

2.7 APPLICATIONS OF BIOMATERIALS

The human body comprises Comprises of various tissues and cells. These tissues and cells have a certain life span but may be damaged due to chronic diseases, accidents, and serious diseases like cancer or tumour. To retain the ideal performance of living tissues, it becomes mandatory to replace the damaged tissues with synthetic biomaterial. Biomaterials are of a replacement nature for damaged/destructed hard and soft tissues. Some of the applications are described next (Ratner, 2004).

2.7.1 ORTHOPAEDIC APPLICATIONS

Globally, biomaterials are widely used as orthopaedic implants. It is a significant accomplishment in the medicine domain. Osteoarthritis and rheumatoid arthritis majorly affect synovial joints or freely movable joints like in the shoulder, elbow, hip, and ankle. With prostheses, it has become possible to treat many patients. Artificial hips and artificial knees are biomaterials used for orthopaedics.

2.7.2 OPHTHALMOLOGIC APPLICATIONS

Eye tissues are delicate and can get easily infected, which may lead to low eyesight or blindness. Externally, low eyesight can be easily corrected by spectacles whereas contact lenses, composed of implant material, provide contact with eye tissues to provide the same results. Artificial corneas/endothelium and intraocular implants are used for glaucoma.

2.7.3 CARDIOVASCULAR APPLICATIONS

Cardiovascular problems associated with arteries and cardiac valve failures are easily treatable with implants. There are several biomaterial devices available and significantly used for cardiovascular treatments like bypasses, vascular grafts, pacemakers, stents, and even complete artificial hearts.

2.7.4 DENTAL APPLICATIONS

The mouth is prone to bacterial infections in the gums. Tooth dissolution, cavities, plaque, and tartar may result in tooth decay. Several dental implants are used for single or multiple teeth replacement. The endosteal root and tooth gums are the biomaterials for dental applications.

2.7.5 WOUND DRESSING APPLICATIONS

As far as external medical treatment is concerned, wound dressing is a surgical procedure to stop exudates and blood oozing, thus facilitating wound healing. Cellulosic or hydrogels materials come in wound dressing material.

2.7.6 OTHER APPLICATIONS

Almost prominent biomaterial applications have already been discussed and elaborated but there are few other applications like organ implantation which includes artificial skin, hemodialyzers, lungs and artificial pancreases, breast implants, and many more.

2.8 CONCLUSION

The evolution of biomaterials has created a paradigm shift in reconstructive surgery with the hope of advancement in quality of life. The biomaterials are used depending on the body part requiring repair. This chapter discusses the classification, interaction, and application of biomaterial. Our studies revealed that Ti and its amalgams are the most demanding and successful implant materials globally in biomedical applications due to their inherent properties for strength and biocompatibility.

2.9 FUTURE DIRECTIONS

Currently, several biomedical implants are easily accessible due to enormous demand in the biomedical industry. However, there is a need for biomaterials with a combination of mechanical and biocompatible features with enhanced biofunctionality. An extended investigation is required to obtain a desirable implant for the future.

REFERENCES

Abd El-Ghany, O. S. and Sherief, A. H. 2016. Zirconia based ceramics, some clinical and biological aspects. *Future Dental Journal*, 2:55–64.

Bates, J. F. 1973. Cathodic protection to prevent crevice corrosion of stainless steels in halide media. *Corrosion*, 29:28–32.

Beline, T., Marques, I. D., Matos, A. O., Ogawa, E. S., Ricomini-Filho, A. P., Rangel, E. C., Da Cruz, N. C., Sukotjo, C., Mathew, M. T., Landers, R. and Consani, R. L. 2016. Production of a biofunctional titanium surface using plasma electrolytic oxidation and glow-discharge plasma for biomedical applications. *Biointerphases*, 11:11013–11018.

Black, J. 1988. Corrosion and degradation. In *Orthopaedic biomaterials in research and practice*, 235–266. Churchill, Livingstone, New York.

Boretos, J. W., Eden, M. and Fung, Y. C. 1985. Contemporary biomaterials: Material and host response, clinical applications, new technology and legal aspects. *ASME. The Journal of Biomechanical Engineering,* February, 107(1):87.

Brundavanam, R. K., Fawcett, D. and Poinern, G. E. J. 2017. Synthesis of a bone like composite material derived from waste pearl oyster shells for potential bone tissue bioengineering applications. *International Journal of Research in Medical Sciences*, 5:2454–2461.

Cattis, R. A. and Husain, Z. 1982. Corrosion-fatigue initiation processes in a maraging steel. *Metals Technology*, 9:104–108.

Chen, Y., Xu, Z., Smith, C. and Sankar, J. 2014. Recent advances on the development of magnesium alloys for biodegradable implants. *Actabiomaterialia*, 10:4561–4573.

Chevalier, J. and Gremillard, L. 2009. Ceramics for medical applications: A picture for the next 20 years. *Journal of the European Ceramic Society*, 29:1245–1255.

Chin, P., Cheok, Q., Glowacz, A. and Caesarendra, W. 2020. A review of in-vivo and in-vitro real-time corrosion monitoring systems of biodegradable metal implants. *Applied Sciences*, 10:3141.

Choueka, J., Charvet, J. L., Koval, K. J., Alexander, H., James, K. S., Hooper, K. A. and Kohn, J. 1996. Canine bone response to tyrosine-derived polycarbonates and poly (L-lactic acid). *Journal of Biomedical Materials Research: An Official Journal of the Society for Biomaterials and the Japanese Society for Biomaterials*, 31:35–41.

Davis, J. R. 2003. Overview of biomaterials and their use in medical devices. In *Handbook of materials for medical devices*, 1–12. ASM International, USA.

De Aza, P. N., De Aza, A. H. and De Aza, S. 2005. Crystalline bioceramic materials. *Boletin De La Sociedad Española De Ceramica y Vidrio*, 44:135–145.

Dorozhkin, S. V. 2012. Calcium orthophosphates and human beings: A historical perspective from the 1770s until 1940. *Biomatter*, 2:53–70.

Eliaz, N. 2019. Corrosion of metallic biomaterials: A review. *Materials*, 12:407.

Fekry, A. M. and Ameer, M. A. 2011. Electrochemistry and impedance studies on titanium and magnesium alloys in Ringer's solution. *International Journal of Electrochemical Science*, 6, 1342–1354.

Geetha, M., Singh, A. K., Asokamani, R. and Gogia, A. K. 2009. Ti based biomaterials, the ultimate choice for Orthopaedic implants: A review. *Progress in Materials Science*, 54:397–425.

Gul, H., Khan, M. and Khan, A. S. 2020. Bioceramics: Types and clinical applications. In *Handbook of ionic substituted hydroxyapatites*, 53–83. Woodhead Publishing, Elsevier, Amsterdam, The Netherlands.

He, S., Yaszemski, M. J., Yasko, A. W., Engel, P. S. and Mikos, A. G. 2000. Injectable biodegradable polymer composites based on poly (propylene fumarate) crosslinked with poly (ethylene glycol)-dimethacrylate. *Biomaterials*, 21:2389–2394.

Hench, L. L. 1994. Bioactive ceramics: Theory and clinical applications. *Bioceramics*, Pergamon:3–14.

Hench, L. L. and Polak, J. M. 2002. Third-generation biomedical materials. *Science*, 295:1014–1017.

Hermawan, H. 2012. *Biodegradable metals: From concept to applications*. Springer Science & Business Media, Switzerland.

Hermawan, H., Ramdan, D. and Djuansjah, J. R. 2011. Metals for biomedical applications. *Biomedical Engineering-from Theory to Applications*,1:411–430.

Hollinger, J. O. and Battistone, G. C. 1985. *Biodegradable bone repair materials: Synthetic polymers and ceramics*. Army Institute of Dental Research, Washington, DC.

Jacobs, J. J., Gilbert, J. L. and Urban, R. M. 1998. Current concepts review-corrosion of metal Orthopaedic implants. *The Journal of Bone and Joint Surgery*, 80:268–282.

Juhasz, J. and Best, S. M. 2012. Bioactive ceramics: Processing, structures and properties. *Journal of Materials Science*, 47:610–624.

Kamachimudali, U., Sridhar, T. M. and Raj, B. 2003. Corrosion of bio implants. *Sadhana*, 28:601–637.

Kruger, J. 1979. Fundamental aspects of the corrosion of metallic implants. *Corrosion and degradation of implant materials*. ASTM, Philadelphia.

Kubasek, J. and Vojtěch, D. 2012. Zn-based alloys as an alternative biodegradable material. *Proceedings Metal*, 5:23–25.

Kumar, P., Saini, M., Dehiya, B. S., Sindhu, A., Kumar, V., Kumar, R., Lamberti, L., Pruncu, C. I. and Thakur, R. 2020. Comprehensive survey on nano-biomaterials for bone tissue engineering applications. *Nanomaterials*, 10.

Li, M., Liu, Q., Jia, Z., Xu, X., Cheng, Y., Zheng, Y., Xi, T. and Wei, S. 2014. Graphene oxide/hydroxyapatite composite coatings fabricated by electrophoretic nanotechnology for biological applications. *Carbon*, 67:185–197.

Liu, D., Yang, F., Xiong, F. and Gu, N. 2016. The smart drug delivery system and its clinical potential. *Theranostics*, 6:1306.

Liu, H., Cheng, J., Chen, F., Bai, D., Shao, C., Wang, J., Xi, P. and Zeng, Z. 2014. Gelatin functionalized graphene oxide for mineralization of hydroxyapatite: Biomimetic and in vitro evaluation. *Nanoscale*, 6:5315–5322.

Lodge, T. P. and McLeish, T. C. 2000. Self-concentrations and effective glass transition temperatures in polymer blends. *Macromolecules*, 33:5278–5284.

Mano, J. F., Silva, G. A., Azevedo, H. S., Malafaya, P. B., Sousa, R. A., Silva, S. S., Boesel, L. F., Oliveira, J. M., Santos, T. C., Marques, A. P. and Neves, N. M. 2007. Natural origin biodegradable systems in tissue engineering and regenerative medicine: Present status and some moving trends. *Journal of the Royal Society Interface*, 4:999–1030.

Mikrolegiranih, R. M. L. 2011. Investigation into the mechanical properties of micro-alloyed as-cast steel. *Materiali in Tehnologije*, 45:159–162.

Moghaddam, N. S., Andani, M. T., Amerinatanzi, A., Haberland, C., Huff, S., Miller, M., Elahinia, M. and Dean, D. 2016. Metals for bone implants: Safety, design, and efficacy. *Biomanufacturing Reviews*, 1:1.

Niinomi, M. and Nakai, M. 2011. Titanium-based biomaterials for preventing stress shielding between implant devices and bone. *International Journal of Biomaterials*, 2011:1–10.

Pal, T. K. 2015. Fundamentals and history of implant dentistry. *Journal of the International Clinical Dental Research Organization*, 7:6–12.

Parida, P., Behera, A. and Mishra, S. C. 2012. Classification of biomaterials used in medicine. *International Journal of Advances in Applied Sciences*, 1:31–35.

Park, G. E. and Webster, T. J. 2005. A review of nanotechnology for the development of better Orthopaedic implants. *Journal of Biomedical Nanotechnology*, 1:18–29.

Patel, N. R. and Gohil, P. P. 2012. A review on biomaterials: Scope, applications & human anatomy significance. *International Journal of Emerging Technology and Advanced Engineering*, 2:91–101.

Pattanayak, D. K., Srivastava, D., Gupta, H., Rao, B. T. and Mohan, T. R. 2005. Evaluation of epoxy/sodium Bioglass ceramic composites in simulated body fluid. *Trends BiomaterArtif Organs*, 18:225–229.

Pellier, J., Geringer, J. and Forest, B. 2011. Fretting-corrosion between 316L SS and PMMA: Influence of ionic strength, protein and electrochemical conditions on material wear: Application to orthopaedic implants. *Wear*, 271:1563–1571.

Pezzin, A. P. T., Van Ekenstein, G. A., Zavaglia, C. A. C., Ten Brinke, G. and Duek, E. A. R. 2003. Poly (para-dioxanone) and poly (L-lactic acid) blends: Thermal, mechanical, and morphological properties. *Journal of Applied Polymer Science*, 88:2744–2755.

Piconi, C. and Maccauro, G. 1999. Zirconia as a ceramic biomaterial. *Biomaterials*, 20:1–25.

Pina, S. and Ferreira, J. M. 2012. Bioresorbable plates and screws for clinical applications: A review. *Journal of Healthcare Engineering*, 3:243–260.

Poinern, G. E. J., Brundavanam, R. K. and Fawcett, D. 2013. Nanometer scale hydroxyapatite ceramics for bone tissue engineering. *American Journal of Biomedical Engineering*, 3:148–168.

Raghavendra, G. M., Varaprasad, K. and Jayaramudu, T. 2015. Biomaterials: Design, development and biomedical applications. In *Nanotechnology applications for tissue engineering*, 21–44. William Andrew Publishing, Oxford, UK.

Ratner, B. D. 2004. A history of biomaterials. *Biomaterials Science: An Introduction to Materials in Medicine*, 2.

Ratner, B. D. 2015. The biocompatibility of implant materials. In *Host response to biomaterials*, 37–51. Academic Press, Oxford, UK.

Rautray, T. R., Narayanan, R., Kwon, T. Y. and Kim, K. H. 2010. Surface modification of titanium and titanium alloys by ion implantation. *Journal of Biomedical Materials Research Part B: Applied Biomaterials*, 93:581–591.

Rikhari, B., Mani, S. P. and Rajendran, N. 2016. Investigation of corrosion behavior of polypyrrole-coated Ti using dynamic electrochemical impedance spectroscopy (DEIS). *RSC Advances*, 6:80275–80285.

Rikhari, B., Mani, S. P. and Rajendran, N. 2018. Electrochemical behavior of polypyrrole/chitosan composite coating on Ti metal for biomedical applications. *Carbohydrate Polymers*, 189:126–137.

Rikhari, B., Mani, S. P. and Rajendran, N. 2020. Polypyrrole/graphene oxide composite coating on Ti implants: A promising material for biomedical applications. *Journal of Materials Science*, 55:5211–5229.

Schinhammer, M., Hänzi, A. C., Löffler, J. F. and Uggowitzer, P. J. 2010. Design strategy for biodegradable Fe-based alloys for medical applications. *Actabiomaterialia*, 6:1705–1713.

Shrivastava, S. and Dash, D. 2009. Applying nanotechnology to human health: Revolution in biomedical sciences. *Journal of Nanotechnology*, 2009:1–14.

Sivakumar, M., Kamachimudali, U. and Rajeswari, S. 1994. Investigation of failures in stainless steel Orthopaedic implant devices: Fatigue failure due to improper fixation of a compression bone plate. *Journal of Materials Science Letters*, 13:142–145.

Sridhar, S. R., Rajagopal, R. V., Rajavel, R., Masilamani, S. and Narasimhan, S. 2003. Antifungal activity of some essential oils. *Journal of Agricultural and Food Chemistry*, 51:7596–7599.

Sung, H. J., Meredith, C., Johnson, C. and Galis, Z. S. 2004. The effect of scaffold degradation rate on three-dimensional cell growth and angiogenesis. *Biomaterials*, 25:5735–5742.

Syrett, B. C. and Wing, S. S. 1978. An electrochemical investigation of fretting corrosion of surgical implant materials. *Corrosion*, 34:378–386.

Tamilselvi, S., Murugaraj, R. and Rajendran, N. (2007). Electrochemical impedance spectroscopic studies of titanium and its alloys in saline medium. *Materials and Corrosion*, 58: 113–120.

Tavares, D. D. S., Castro, L. D. O., Soares, G. D. D. A., Alves, G. G. and Granjeiro, J. M. 2013. Synthesis and cytotoxicity evaluation of granular magnesium substituted β-tricalcium phosphate. *Journal of Applied Oral Science*, 21:37–42.

Vandrovcova, M. and Bacakova, L. 2011. Adhesion, growth and differentiation of osteoblasts on surface-modified materials developed for bone implants. *Physiological Research*, 60:403–417.

Viceconti, M., Muccini, R., Bernakiewicz, M., Baleani, M. and Cristofolini, L. 2000. Large-sliding contact elements accurately predict levels of bone: Implant micromotion relevant to osseo integration. *Journal of Biomechanics*, 33:1611–1618.

Vojtěch, D., Kubásek, J., Šerák, J. and Novák, P. 2011. Mechanical and corrosion properties of newly developed biodegradable Zn-based alloys for bone fixation. *Acta Biomaterialia*, 7:3515–3522.

Vroman, I. and Tighzert, L. 2009. Biodegradable polymers. *Materials*, 2:307–344.

3 Material Properties for Biomedical Applications

D. Ajith, K.G. Ashok and K. Aravind

CONTENTS

3.1 INTRODUCTION

The choice of biomaterial for a specific biomedical application is to be determined based on its physical and mechanical properties and biocompatibility. Traditionally biomaterials are designed to be inert and not to interact with our body. According to the American National Institute of Health, a biomaterial is defined as

> any substance or combination of substances, other than drugs, synthetic or natural in origin, which can be used for any period of time, which augments or replaces partially or totally any tissue, organ or function of the body, to maintain or improve the quality of life of the individual.
>
> [1]

Historically materials from nature such as wood were used to replace damaged body parts. For example, a prosthetic hip replacement was reported as early as 1890 [2]. With the advancement of biomaterials science in the twentieth century, engineered materials such as metal/alloys, polymer, ceramic, and composite materials have replaced the natural materials due to their higher performance, improved functionality and reproducibility. It is important to realize that although several materials that have excellent properties for specific scientific applications (for example aerospace, maritime), their performance inside the body is vastly different and is dictated by the local host environment (such as extracellular matrix, temperature, pH, host cells, etc.) and the immune response (both cellular and humoral). The basic requirement for any material that is intended for use as a biomaterial is that it should be biocompatible [3].

DOI: 10.1201/9781003173533-3

A biocompatible material is one which, when in contact with the body, does not produce any local and systemic toxicity. Again, depending upon their application, biomaterials may be classified based on the type of the material and the biologic response they elicit when implanted. For example, it can be either bioinert (e.g., alumina, zirconia, tantalum, niobium, and carbon) or bioactive (Ca-phosphate and hydroxyapatite). Biomaterials may be intended either for long-term applications (such as Ti implants) or short-term duration (stainless steel plates for fracture fixation). They may be either resorbable (such as chitosan and (PLLA) poly-L-lactic acid) or non-resorbable (such as acrylics and silicones).

The market potential of biomaterial is ever increasing in the last few decades. For example, materials used in eye lenses or those used as biosensors require suitable optical properties. Materials such as contrast mediums (iodine-based material, barium-sulfate, or gadolinium) are intended for short-duration imaging applications. Cardiac biomaterials have various applications, such as heart valves, endovascular stents, vascular grafts, catheters, heart assist devices, and hemodialyzers. These biomaterials are predominantly blood contacting and hence they must necessarily be blood compatible, not induce thrombosis, have appropriate physical and mechanical properties including tensile strength, friction and wear resistance, as well as flow resistance among others [4]. Materials used for drug delivery and tissue engineering applications usually require to be biodegradable, with the degraded products being non-toxic to the body and easily excreted out. While the property of biodegradation is desirable for scaffolds used for tissue engineering, their rate of degradation must match the rate of regeneration. Also, it must be noted that the physical and mechanical properties of these degradable materials deteriorate with progress in biodegradation. Hence, wherever required, appropriate modification of the scaffolds may be done either by adding functional groups (surface modification), creating blends (two or more polymers), cross-linking, and composites.

The present chapter focuses on the material properties used in the load-bearing implants such as orthopedic and dental applications. More specific orthopedic applications are fracture fixation, knee joints, hip replacements, and joint replacements (Figure 3.1). Metals, Ceramics, and Polymers can be considered as "primary" orthopedic and dental biomaterials. The knowledge of mechanical properties, uses and limitations of the primary biomaterials is indispensable in order to improve the performance of current implant materials [2].

3.1.1 Metals

In the orthopedic and dental industry, metals and alloys are used for fracture fixations (stainless steel and titanium plates and screws), dental amalgams, prosthesis (limb or jaw replacement/ reconstruction), and so forth [5], [6]. Its availability and advanced manufacturing methods provide metallic biomaterials a major share in the biomaterials market [7]. For example, SS316 L (3% Mb and 16% Ni; L stands for low carbon content 0.03%), used for screws in joint replacements for bone fixation, is stiff, hard, durable, ductile, and quite biocompatible. Stainless steel remains a popular choice for cable fixation components in total knee arthroplasty due to economic considerations and greater ductility compared to Ti and Co alloys. Ti-6Al-4V alloy forms an oxidative layer on the top. This process is called passivation and it protects the material from external damage. However, due to its low notch sensitivity, even a small amount of scratching in Ti or its alloys makes it lose all the wear characteristics and hence it is preferred for non-articular components (for example, tibia) and not used in femoral articulating components [8]. Co-Cr-Mo (ASTM F-75) alloys, due to their high corrosion resistance and remarkable wear characteristics compared with stainless steel, are preferred for total hip arthroplasties. Due to poor frictional properties of Co-Ni-Cr alloy and the biologic reactivity released from Ni from Co-Ni- Cr alloys, it is least preferred for total joint components. In summary, the desired material properties such as stiffness, ductility, fracture toughness, hardness, corrosion

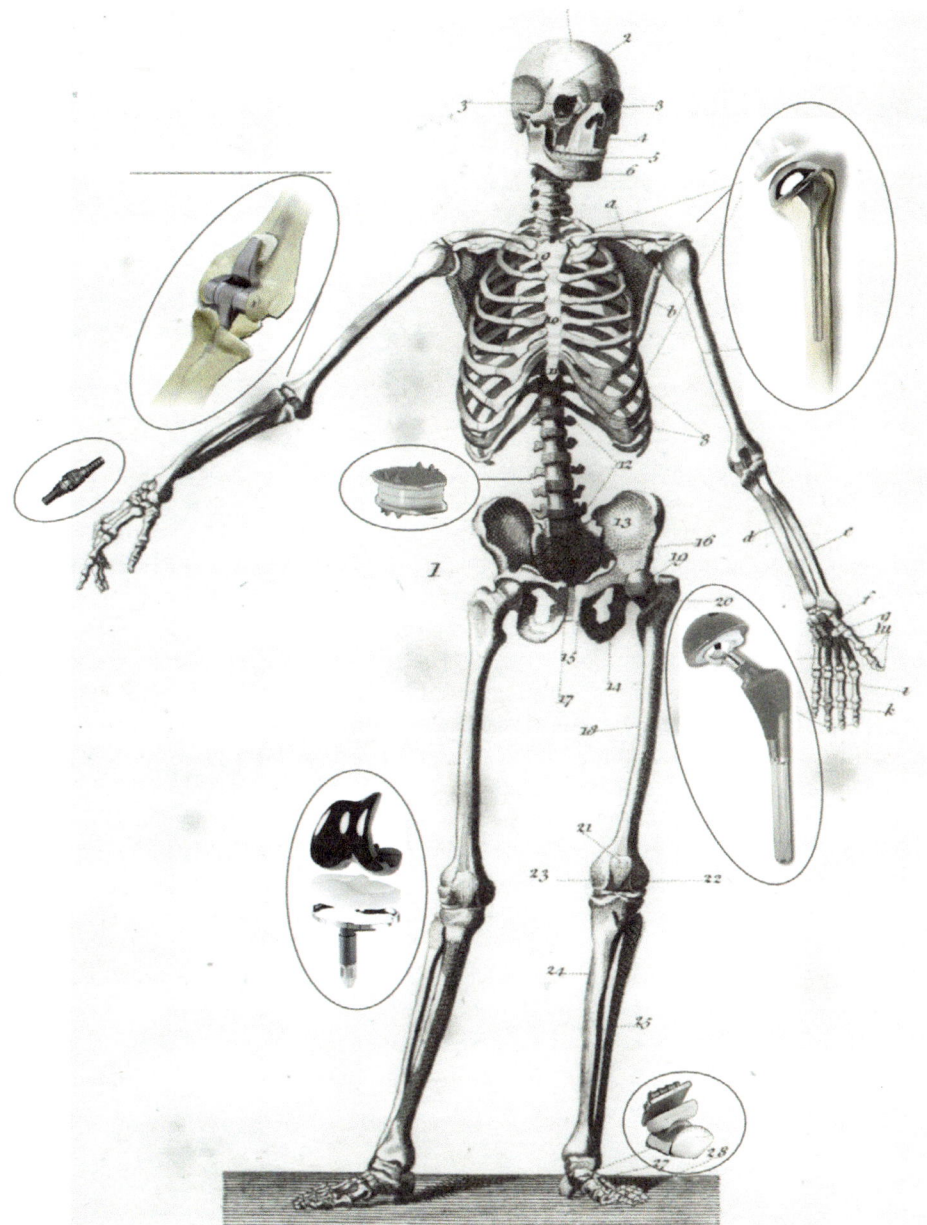

FIGURE 3.1   Joint replacement: knees, hip, shoulder, etc.

Source: Reproduced with permission from [2].

resistance, formability, and biocompatibility are provided by the metals and their alloys for the suitable medical device or implant designs [2].

In order to improve the short-term and long-term performance of implants by encouraging bone ingrowth and providing enhanced fixation, a variety of surface coating techniques are currently available. For example, roughened titanium, porous coatings made of cobalt chromium, titanium wire mesh, and bioactive nonmetallic materials such as hydroxyapatite, bioactive glass coating, or other calcium phosphate components [9].

Apart from the different metal implants described previously, the role of Mg, Fe, Zn (biodegradable metals) is also important. The main advantage of biodegradable metals is enhanced metabolic reactions and biological mechanisms of the body. Thus it reduces the healing duration compared to that of traditional Ti or SS wires [10]. It can be noted that several types of bone fixation devices in the form of screws, plates, nails, staples, and wires can be developed using biodegradable metals. With the recent introduction of biodegradable metallic biomaterials, the scope of applications of biodegradable metals has expanded even more [11]. Mg-based biomaterials have densities and elastic moduli that are fairly close to that of bone. However, the high corrosion rate of Mg-based materials and very low corrosion rate of Fe-based materials is a matter of concern in their applications for biomedical applications. Research is underway in several laboratories on improving the corrosion resistance of these materials [12].

3.1.2 Ceramics

The high rate of biodegradation of metals in the human body environment can be offset by coating with less reactive materials, for example, bioactive glasses or glass-ceramics. Bio ceramics are inorganic materials that are used for repair and regeneration of diseased and damaged parts of soft and hard tissues of the human body. Ceramics, because of their ionic bonds and chemical stability, are biocompatible and possess excellent wear resistance. The ceramic materials are manufactured under high temperature and pressure and hence minimum voids and are almost a perfect material of choice for implants. Further, the low coefficient of friction, hardness, and wettable properties makes it suitable for knee replacements. In recent years ceramics and glasses have played an important role in implants [13]. These ceramic materials are extensively used for orthopedic load-bearing applications, bone grafts, and dental implants. Like glasses, these are strong and brittle. They can be classified as (a) bioinert when they have no interaction with the surrounding tissue, (b) bioactive, when they bond directly with bone, and (c) bioresorbable, when they degrade gradually inside the body (*in vivo*). Some of the popular bio-ceramics are: alumina and zirconia which are bioinert; bioglasses and glass-ceramics which are bioactive; and calcium phosphates (CaPs) and calcium carbonates, which are bioresorbable. It is reported that the stability of all ceramic components is extremely dependent on manufacturing practices [2].

The particular interest to biomaterials scientists working in the area of orthopedics is the mineral hydroxyapatite, which is a natural mineral found in our bones. Being biocompatible as well as bioactive, it is extensively used as bone grafts, coating over metallic implants, and as a filler in polymer scaffolds for bone tissue engineering [14]. Some of the limitations of ceramic biomaterials are low resistance to fatigue and anisotropy.

3.1.3 Polymers

Polymer materials such as PTFE (polytetrafluorethylene), UHMWPE, and PEEK are widely studied and their advantages and limitations are documented in the literature [15]. For example, the crystallinity and superior packing of linear chains increased the mechanical property of the UHMWPE required for orthopedic applications.

Another material widely used in various medical applications is PMMA (comes in two phases, the solid and liquid form) due to its versatility and reliability. It is primarily used in dentistry for the fabrication of denture bases mainly used for (a) bone cements, (b) screw fixation in bones, (c) filler for bone cavities and skull defects, etc. Apart from load-bearing implants, polymer-based biomaterials are extensively used for several applications ranging from implants such as breast implants and ear prostheses (silicone) to scaffolds for drug delivery and tissue engineering, from heart valves (collagen) to orthopedic prosthesis (polyether ether ketone), and from dialysis membranes to meshes for hernias. The greatest advantage of polymers is that they can be easily prepared in various designs such as sheets, sponges, nanofibers, micelles, and so on through different fabrication processes such

TABLE 3.1
Summary of Primary Biomaterials Used in Orthopedic and Dental Applications

Materials	Examples	Applications
Metals	Stainless steel 316L	Fracture plates, screws, joint replacements
	Ti-6Al-4V (Titanium alloys)	Hip and knee prostheses, screws and pins for bone fixations
	Co-Cr-Mo alloy	Joint replacements, dental implants
Ceramics	High alumina ceramics (Al_2O_3)	Load-bearing surfaces, dental implants
	Zirconia (ZrO_2)	Bearing surface components
	Glass-ceramics	Dental implants
Polymers	Polymethylmethacrylate (PMMA)	Acrylic bone cement
	Ultra-high-molecular-weight-polyethylene (UHMWPE)	Low friction inserts for bearing surfaces

as solvent evaporation, sol-gel technique, 3-D printing, phase-inversion, injection molding, etc. Ease of fabrication, light weight, and biocompatibility are the main advantages. Low strength and degradation with time are the major drawbacks with polymeric materials. They can be used in heart valves, hip joint sockets, contact lenses, blood vessels, etc. The properties of polymers are governed by the molecular chain structure, molecular weight, and degree of branching or chain linearity. Table 3.1 summarizes the primary biomaterials used for implants discussed earlier.

3.2 MECHANICAL PROPERTIES

Engineering designs are largely associated with trying to prevent the part/product from failure. When the stress in the material exceeds the allowable limit (σ_{allow}), the design is subjected to failure. In order to design these allowable normal stresses, torsional stresses, we should have the understanding of how the material that we choose behaves and how it fails. These can be carried out by standard mechanical tests such as tensile, compressive, bending, and shear tests. In general, the material property can be quantified by the stress-strain curve (Figure 3.2). From this we can determine the stress at which the material breaks and the most common practice is the tension test. These tests are carried out in a universal testing machine (UTM). Interested readers can read further in the topic, and most of the characteristics of the metal implants are well documented in the literature. However, a brief description of the important mechanical properties of the important materials are compared in the following section.

3.2.1 TENSILE PROPERTIES

Figure 3.2 shows the characteristics of the stress-strain curve of a ductile material (e.g., metal) under tension. The initial straight line (linear elastic region) demonstrates the specimen can be loaded and unload and it will return to the original shape. After the yield point there is permanent deformation. After this point the specimen continuous to elongate in what is called its plastic region [16]. After the plastic region the material is able to take up more load, and stress increases until the peak value is reached. (highest point in the curve). This is called the ultimate tensile strength (σ_u). After this point stress decreases and finally the fracture occurs where the specimen physically breaks. In metals, mechanical properties such as yield strength, elongation to break, fatigue vary with both the alloy type, processing (microstructure), cryogenic treatments [17]. The yield strength (tensile) of titanium alloys varies between 500–1000 MPa. Compared to human bone the modulus value is higher. The elongation at break varies between approximately 10–20%. The mechanical properties of titanium alloys developed for implant materials can be found in [18].

The percentage of elongation of polymers for example, UHMWPE is higher than that of metals and offers high toughness. Ceramics usually exhibit poor tensile properties; however, they possess excellent compressive strength and are often used in coatings.

3.2.2 Young's Modulus

The slope in the elastic region is termed as elastic modulus or Young's modulus and represents the stiffness of a material. As discussed, the allowable stress can be calculated using $\sigma_{allow} = \sigma_{fail}/\text{factor}$ of safety; safety factor is typically prescribed by design codes. The amount of strain energy it takes to yield the specimen and the area under the curve is called the modulus of resilience. Materials that are very stiff (e.g., ceramics) have large values of E, whereas polymers (Figure 3.2) may have low values of E. Most of the metallic biomaterials have a higher modulus value than that of cortical bone. Co alloys are very stiff and modulus varies between 210–253 GPa. The modulus of titanium alloys is lower than that of other metallic biomaterials (~110 GPa) [18]. In orthopedic applications ideal implant materials are required to possess low Young's modulus and high strength [19].

It is interesting to note that Mg-based alloys have Young's modulus between 30–45 GPa, which is closer to that of natural bone which is between 0–27 GPa (measured in tension) in contrast with Ti-based alloys which show Young's modulus values around 110 GPa. In total hip arthroplasty and total knee arthroplasty, ceramics have been found to be more effective than metals due to their favorable mechanical properties in addition to the inertness of the wear debris. In terms of elastic modulus, zirconia has a value of around 200 GPa, while that of alumina is in between 350–370 GPa [20]. Among the different applications of biomaterials based on polymers and their composites, one that is very prominent is their use as mesh in hernia repair. This application requires that the material of choice be strong enough to bear the weight of the abdominal organs and be resistant to infection/colonization by microorganisms.

3.2.3 Fracture Toughness

The area under the entire stress-strain curve is called the toughness and is the amount of strain energy it takes to break or fracture the specimen. Materials that do not exhibit any yield behavior

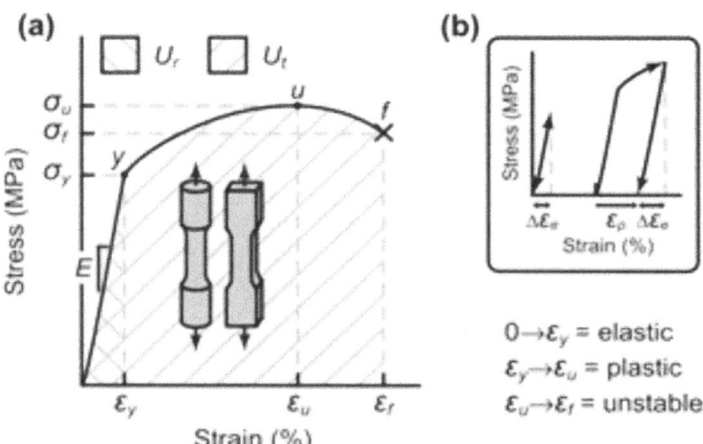

FIGURE 3.2 (a) Schematic stress-strain (σ–ε) curves for a ductile material (metal); (b) the important mechanical properties include: the elastic modulus (E), yield strength (σ_y), ultimate strength (σ_u), fracture strength (σ_f).

Source: Reproduced with permission from Elsevier [16].

before the fracture are termed as brittle materials (ceramics). Brittle materials fail well below their theoretical strength, especially if any surface cracks are present. This is due to increase in the stress concentration level. Fracture toughness of different dental crown materials like zirconia, alumina, lithium disilicate are reported in literature [21]. Polymer-infiltrated ceramic could be a better choice to prevent chipping of dental crowns, and adhesive cementations are recommended for durability.

The fracture toughness of human bone is around 3–6 MPa m$^{1/2}$. Metals show a higher value compared to the bone. However ceramic biomaterials like hydroxyapatite exhibits lower fracture toughness than the bone. Mode 1 fracture toughness of PMMA is reported in [22]. Their study concluded that addition of carbon nanotubes to PMMA reduces the fracture toughness. Polymers like UHMWPE (ASTM F 648–07 d) exhibit superior fracture toughness and are preferred for hip prostheses [15]. Cobalt-chromium has a very high Young's modulus, is durable and relatively hard. A comparison of the stress-strain curves of various biomaterials is shown in Figure 3.3.

3.2.4 Fatigue Strength

Most body parts are not subjected to simple tensile or compressive types of loading, but cyclic loading (i.e., alternative tensile and compressive stress). Fatigue strength can be explained with the help of the S-N curve. In this curve, the Y-axis represents the stress and the X-axis represents the number of cycles (alternative tensile and compressive) of loading. Fatigue limit or endurance limit is the stress level at which someone can withstand loading 10 million times before the material fails. If we could develop material that stays at the endurance limit, effectively the arthroplasty will last ten years. Hip replacements with round heads and round sockets and hence larger contact surfaces result in lower contact stress, and mostly these joints will operate below the endurance limit. But for the knee a round or flat design is used and hence higher contact stresses. Therefore knee joints will operate at or above the endurance limit. Hence the designer should aim for a material that has higher wear resistance, for hip joints and knee joints.

Most metallic implants have higher fatigue strength (in air) than the bone. Further, their fatigue strength can be improved by post processing treatments which introduce high compressive strength of the surfaces [23]. For example, a hot isostatic process improves the fatigue life of cobalt chromium alloys compared to the casting process. Long-term failure of polymers is associated with environmental stress cracking, creep and aging mechanisms, and finally results in brittle fracture. Various methods to improve the fatigue life of the dental implants are reported [24]. Table 3.2 shows the summary of mechanical properties of the materials discussed in the previous section.

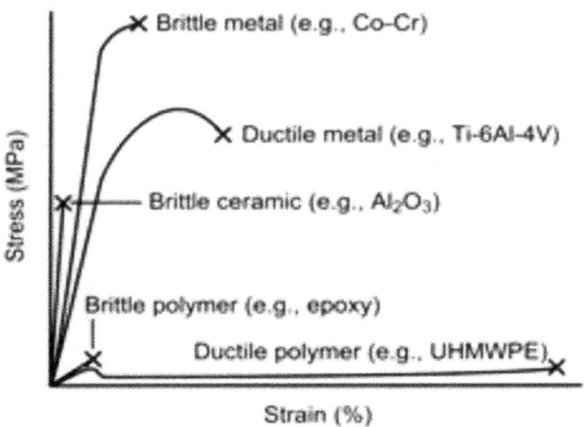

FIGURE 3.3   Comparison of stress-strain curves of different implant materials (not to scale).

Source: Reproduced with permission from Elsevier [16].

TABLE 3.2

Comparison of Mechanical Properties of Metallic, Ceramics, and Polymer-Based (Measured under Tension) Biomaterials Used in Implants.

Biomaterial	Young's Modulus (GPa)	Elastic Limit (MPa)	Ultimate Strength (MPa)	Fracture Toughness (MPa-m$^{1/2}$)	Endurance Limit (MPa)	% of Elongation
Stainless Steel	190	793	862	50–200	241–820	43–45
Pure Ti (ASTM F67)	110	485	785	97	300	14–18
Ti-6Al-4V (ASTM 136)	116	897–1034	965–1103	44–66	620–689	8
Co-Cr-Mo Alloy (ASTM F75)	210–253	448–841	655–1277	–	207–950	4–14
Magnesium Metal	44	162	250	15–40	–	18
Zinc Metal	96	30	37	–	–	–
Alumina	366	–	310	3–5	–	–
Zirconia	201	–	420	8	–	–
UHMWPE	0.5–1.3	20–30	30–40	–	13–20	130–500
PMMA	1.8–3.3	35–70	38–80	1.11	19–39	2.5–6
Cortical Bone	17	114	90	3–6	–	–

Source: [2, 25].

3.3 TRIBOCORROSION

3.3.1 MECHANICAL WEAR CHARACTERISTICS

Generally, implant materials should exhibit high wear resistance and low coefficient of friction. Friction is defined as the resistance to sliding when two surfaces are in contact. The effects of friction are energy loss, increased effort to move joints, and load on implant. Further, wear mechanism involves the loss of material in particulate form as a consequence of relative motion between two surfaces. Every articulating surface has a unique coefficient of friction and it depends on material combinations (Table 3.3).

It can be noted that when two materials are placed together, they result in an "apparent" and "true" contact area. At a microscopic level, the surface of a material can be seen as little projections (asperities). The mean height of asperities is known as the roughness (Ra) and each material has its own individual roughness value (Ra values of polymers are higher compared to metals). For example, articular cartilage has Ra value of 1–6 μm; polyethylene cup 0.25–2.5μm; metal femoral head 0.025μm; ceramic femoral head 0.02μm. The true contact area is proportional to the load and inversely proportional to the hardness (resistance to scratching). Increasing the load will increase the asperities and thus increase in friction and also result in the loss of material in the form of particles (wear debris). These wear debris may be lost from the system, transferred to the counter face, or remain between the sliding surfaces. Mechanical wear can be mainly classified into three types: (a) abrasion wear (e.g., metal on polymer)—when a hard surface slides over a soft surface and removal of surface material takes place; (b) adhesion wear—a softer material is spread onto a harder counter surface forming a transfer film through tearing/breaking; and (3) fatigue/subsurface delamination—detachment of surface particles due to cyclic loading. The generation of wear debris produced by different types of orthopedic materials, and the subsequent tissue reaction to such debris, is crucial to the durability of total joint replacements [2].

During the function, implant materials undergo wear when in contact with their articulating surfaces. Their articulating surface could be either the host's natural bone (polyether ether ketone (in case of knee replacement), or UHMWPE in the case of hip replacement). In these situations, the

TABLE 3.3
Coefficient of Friction of Articulations

Articulation	Coefficient of Friction
Native Knee	0.005–0.02
Native Hip	0.001–0.04
Metal—Polymer	0.02
Metal—Metal (dry)	0.8
Aluminum—Aluminum	2.0

amount of wear of natural bone due to metallic biomaterials is higher than when compared with UHMWPE. The joints also get loose if the articulating surfaces have low wear resistance and high coefficient of friction. The two-body wear results in debris that is cleared from the site by the phagocytes and lymphatic system. When there is a decreased clearance rate, the wear particles interfere in the proper articulation of the joints and further cause a three-body wear which could increase the wear rate of the biomaterials. These wear particles, though engulfed by the phagocytes, cannot be eliminated from the body through lymphatic drainage or the excretory system. In such situations, either these wear particles are walled off from the rest of the body through fibrous encapsulation or they end up accumulating in organs such as the liver and kidney. Hence, the investigation into the tribological properties and understanding the wear mechanism is important for improving the successful outcome of implant biomaterials.

Some of the common test methods used to analyze the tribological properties are block-on-disc, ball-on-disc, and pin-on-disc. The wear tests are conducted in an environment of simulated body fluids (Ringer's solution; Table 3.4). The volume loss in the wear track can be measured using advanced characterization techniques such as optical profilometer, atomic force microscope, etc. By using a depth-sensing indentation method, various mechanical properties discussed earlier can be measured. Researchers have investigated the wear mechanisms of various alloys and pure biomaterials made from different processing conditions. Development of composites material with the addition of zirconia particles yielded improvement in the wear characteristics of UHMWPE. However, this material needed to undergo *in vivo* studies before applying in clinical practices [26]. Many studies have revealed that heat treatment, alloying and surface modifications (nitriding, plasma spray coating, ion implantation) generally improves the wear resistance. The major failure mechanisms observed due to friction are abrasion, transfer layer formation, and cracking [27]. The ranking on articulating surface for joint prosthetics in terms of wear resistance from most preferred to least desired are ceramic on ceramic, ceramic on polymer, metal on polymer, and metal on metal respectively [7].

3.3.2 CHEMICAL WEAR CHARACTERISTICS

Chemical wear or corrosion is defined as the unwanted dissolution of a metal in a solution. It can be noted that the dissolution of metals happens at local (crevice, pitting) and at generalized levels (galvanic, fretting, stress, etc.) [28]. In crevice corrosion some sort of cavity in the implants is shielded from tissue fluids and has a local microenvironment where it is more acidic. Pitting results from the random pits on the surface or random imperfections causing micro environments where the concentration of hydrogen ions is greater and therefore the passive layer wears away, making it more susceptible to corrosion (oxidation layer on the surface) [29]. For example, TiO_2 (Figure 3.4) film in dental implants is susceptible to fluoride ions, present in the oral environment. Hence the corrosion resistance of these implants will decrease [30].

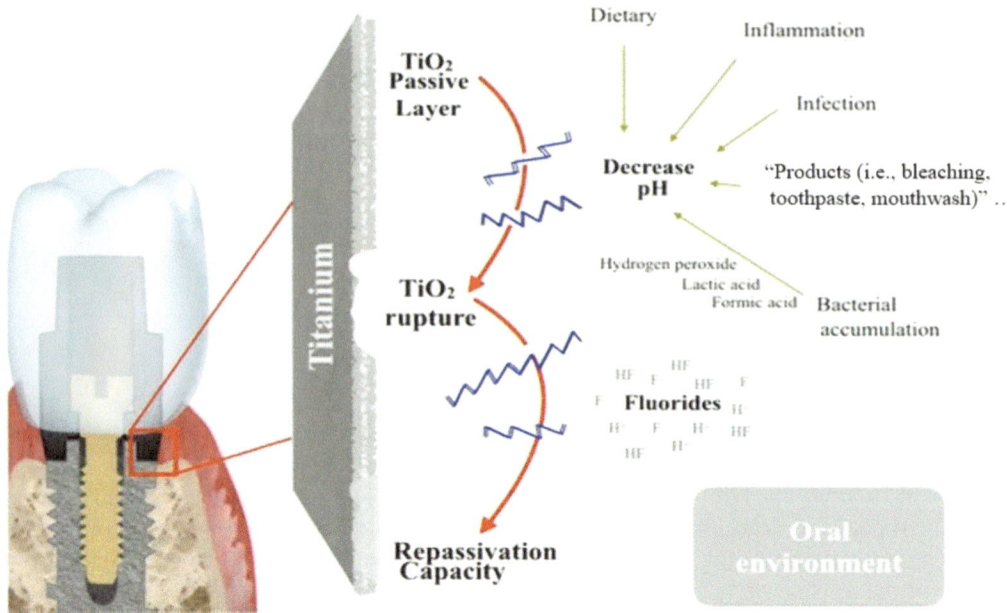

FIGURE 3.4   Factors affecting the corrosion behavior of Ti dental implants.

Source: Reproduced with permission from [30].

Therefore, corrosion in biomaterials is also attributed by chemical or electrochemical degradation of a material through interactions with its environment. Galvanic corrosion occurs when two dissimilar materials are present in a solution and one becomes cathode and the other becomes anode and the electrons will move from one to another causing the dissolution of the metal. Relative micromotion between two mating surfaces causes fretting corrosion. For example, SS 316L is also prone to wear and various corrosions such as crevice corrosion, intergranular corrosion, pitting corrosion, and fretting corrosion. Other widely used implants, Co-Cr alloys, are subjected to intergranular corrosion, etching, selective dissolution of cobalt, etc. The corrosion behavior of various biomaterials including metals, metallic glasses, and biodegradable metals are comprehensively discussed in [31]. Implant failure mechanisms, retrieval and failure analysis are highlighted in that study. Table 3.4 shows the concentration of ions in blood plasma extracellular fluid and is compared with Ringer's solution.

For efficient orthopedic implants, researchers are wieldy investigating the complex interaction of various factors such as mechanical loading, metallurgical behavior, and topology design and solution variables. Hence, knowledge about the chemical composition and its variation, its physical, chemical, and mechanical properties, and its effect on the local host environment is essential for developing novel metallic biomaterials [2, 32].

Noble metals are less susceptible to corrosion due to their high electrochemical potential. However, the metals that are at the bottom in the electrochemical series are likely to corrode. Highly reactive material like titanium, chromium, etc. immediately reacts with oxygen upon exposure with the atmosphere. It thus forms an impervious oxide layer on the surface of the implant and protects the underneath material. In the case of biodegradable metals, commonly used strategies adopted by the researchers to improve the corrosion resistance of Mg and its alloys are purification, surface coating, alloying, and rapid solidification. Commonly used coating materials are polymers, bioceramics, and their composites.

TABLE 3.4

Ionic Concentration of Human Blood Plasma, Extracellular Fluid, and Ringer's Solution.

Anion/Cation	Blood Plasma (mmol L^{-1})	Extracellular Fluid (mmol L^{-1})	Ringer's Solution (mmol L^{-1})
Cl$^-$	96–111	112–120	155.7
HCO$_3^-$	16–31	25.3–29.7	–
SO$_4^{2-}$	0.35–1	193–102	–
Na$^+$	131–155	141–15	147.2
K$^+$	3.5–5.6	3.5–4	4
Mg^{2+}	0.7–1.9	1.3	–

3.4 BIOCOMPATIBILITY

Traditionally in engineering design the selection of materials for the specific applications is governed by matching the requirements for the targeted applications. In the case of biomaterials, apart from the mechanical, physical, and/or chemical properties, biological properties need to be considered. The biological property of the material describes the host-biomaterial interaction or in other words, it is a study of how the biomaterial, when in contact with the tissues (both *in vitro* as well as *in vivo*), responds (Figure 3.5). Since the biomaterials are in contact with various biological tissues, knowledge on the host-biomaterial interaction is essential to not only select the appropriate material but also for enhancing the safety of the clinician and the patient as well. The materials could be either bioinert or bioactive. Bioinert materials such as alumina and zirconia do not elicit any reaction from the host cells and remain passive within the body. Whereas bioactive materials such as bioglasses and hydroxyapatite elicit a favorable response from the body.

Applications of bioinert materials include use as hip replacement (polyether ether ketone), and scaffolds for tissue engineering (collagen). Bioactive materials have predominant applications in orthopedics as implants (TiO2) or coatings for implants (bioglasses, hydroxyapatite, and silicates). Bioactivity is imparted by the presence of charges in the molecules that comprise the biomaterial. For example, chitosan has amine and amide groups which bear a negative charge. Similarly, minerals such as hydroxyapatite also have phosphate groups which also bear a negative charge. The presence of these charges induces cell signaling with the host cells and appropriate host response is initiated. The host response should be viewed from a greater perspective by taking into consideration the antigenicity of the biomaterial also. Some host proteins (that are present in body fluids) would get adsorbed onto the surface of the biomaterials which would further influence the host response.

For implant materials, the new material should neither adversely affect its biological environment nor return the affected by the surrounding host tissues and fluids. Thus, understanding of relationship between the structure and properties of the natural tissues that are being replaced is critical. Replicating the unique properties of bone (composed of collagen, anisotropic and inhomogeneous) with long-lasting engineered biomaterials still poses a great challenge to material scientists [2]. The biocompatibility, identifying the risk of any new biomaterial, must be substantiated by doing a series of tests regarding its genotoxicity, carcinogenicity, toxicity, etc. prior to doing any clinical trials in humans as per the standards. The biocompatibility of a biomaterial was investigated by considering the wound healing processes that happen after biomaterial implantation. The evaluation of biological responses to biomaterials consists of implanting biomaterials in the targeted tissues at certain time points and then studying the timely changes in the implanted tissue and its surroundings [1]. The biological evaluation plan should be gathering information from suppliers, literature, and we must ensure that the new material should not possess any risk to the patients. The success of

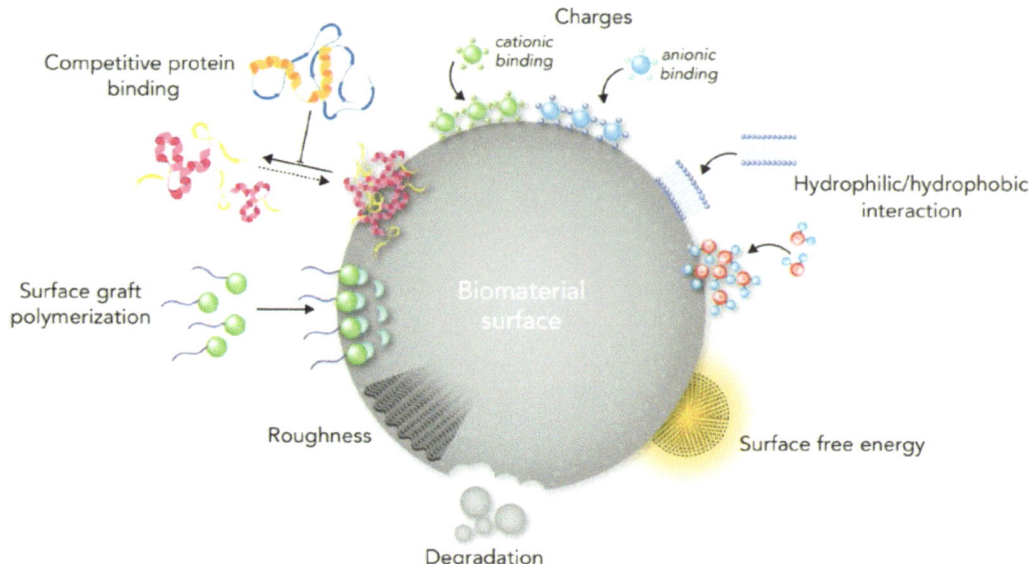

FIGURE 3.5 Schematic of important surface physicochemical properties in directing biological responses to biomaterials. Biomaterials can manipulate molecular and cellular signaling pathways through their surface physicochemical properties (e.g., topography, stiffness, functional groups, ions, charges, surface free energy).

Source: Reproduced with permission from [1].

a biomaterial is based on response to the foreign body of a biomaterial to stimulate minimal inflammatory responses. Hence, the focus in designing biomaterials is on reducing foreign body responses through directing macrophage responses [33].

3.4.1 OSSEOINTEGRATION

For the success of implants that are used as bone substitutes (such as femur replacements and dental implants), these implants should integrate with the bone at the site of implantation. Based on the ability of the implants to induce bone formation, they may be classified as osteoinductive and osteoconductive. Osteoinductive materials are those that induce new bone formation by providing stimulus for induction, stimulation, and differentiation of undifferentiated and pluripotent stromal cells into osteoblasts. Osteoconductive materials are different from osteoinductive materials as the former is associated with the process of bone cells proliferating on their surfaces. Materials that show osteoconduction are more successful as implant materials than those materials that exhibit only osteoinduction. Generally, the goal of materials scientists and clinicians is to achieve osseointegration of the implants with the host bone through osteoconduction, as in the case with dental implants [34].

Though pure Ti is bioinert, the formation of TiO_2 on the surface of the Ti-based dental implants imparts osteoconductive properties on the material. Recent studies have shown that titanium with porous structures has the ability to induce bone-like apatite formation and osteoinduction phenomenon [35]. Because of their highly biocompatible nature and their excellent osseointegration capacity, Ti-based dental implants, with the different surface modifications and implant design, have now become a globally accepted treatment modality with a more than 95% success rate for ten-year survivability. However, further research is in progress to investigate the effect of different surface modifications (both additive and subtractive) of Ti implants on osseointegration.

3.5 INSTRUMENTAL METHODS AND TECHNIQUES FOR BIOMATERIALS CHARACTERIZATION

Apart from the bulk properties (mechanical properties) of the material, biomaterial characterization requires the understanding of the knowledge of surface properties. It is recognized that interactions between most biomaterials are governed by their surface properties and manifest themselves at the interface formed between them. The chemical structure and physicochemical behavior of the material can be studied by a variety of instrumental techniques. Chemical characterization helps in identifying the configuration of the material, the material composition, and influence of manufacturing process, etc. There are different types of chemical characterization studies as compared to all other biocompatibility studies. It needs to be relevant for the medical device, how it is used, and what will be the analytical evaluation threshold. It can be noted that no single analytical technique can detect all potential hazardous compounds. The instrumentation techniques used for evaluating biomaterials is vast and cannot be exhaustively covered in this section. However, some of the relevant techniques shall be briefed for the benefit of readers.

The chemical structure and physiochemical properties of the materials can be studied by the following techniques [36]. (a) Thermal analysis: thermogravimetry (TGA), differential thermal analysis (DTA), differential scanning calorimetry (DSC); (b) particle size: particle size distribution (PSD), dynamic light scattering (DLS), single angle X-ray scattering; (c) structural and chemical analysis: contact angle analysis, energy dispersive X-ray (EDX), wavelength dispersive X-ray (WDX) spectroscopy, X-ray photoelectron spectroscopy (XPS), vibrational spectroscopy (Fourier-Transform Infrared spectroscopy (FTIR), Raman spectroscopy), nuclear magnetic resonance spectroscopy (NMR), electron spin resonance spectroscopy (ESR), atomic absorption spectroscopy and mass spectroscopy [37]; (d) morphology: scanning electron microscopy-energy disperse X-ray (SEM-EDX), transmission electron microscopy (TEM) and scanning probe microscopy (atomic force microscopy, AFM). The detailed description of all the methods is beyond the scope of the chapter. However, a few practical examples are presented in the following section.

X-ray photoelectron spectroscopy (XPS) and time of flight secondary ion mass spectroscopy (ToF-SIMS) are highly useful techniques for studying the biological responses to change in the surface chemistry of biomaterials. These techniques analyze the surface composition as well as the structure of biomaterials over a nanometer-sized scale and shed light on the biomaterial-biomolecule interactions [1].

Thermal characteristics studies of composites/blends are important. Using differential scanning calorimetry (DSC) and thermogravimetric analysis (TGA), the influence of hot pressing on the properties (processing method for ceramic implants), crystallization kinetics, and decomposition of HAp/PLLA composite biomaterial was analyzed. DSC curves can be used to determine the functionalization of polymers too. For example, sulphonate polyether ether ketone can be subjected to DSC analysis and the degree of sulphonation may be determined. TGA curves give useful information on the composition of the materials as well as display the thermal stability of materials [38].

Fourier Transform-Infrared spectroscopy (FTIR) can be used for both quantitative as well as qualitative analysis of the chemical nature of the biomaterial. Hence, analysis such as blend formation between two polymers, cross-linking, the interaction between the fillers and the matrix in a composite, and so on is possible through FTIR [39]. This technique can also be used for identifying foreign materials (particulates, fibers, residues), bulk material compounds, constituents in multilayered materials, etc.

The main advantage of scanning electron microscopy (SEM) is that sub-atomic resolutions (nano scale) can be obtained as compared to light microscopy. SEM is useful in determining not only the surface texture (coating evaluation, surface contaminations) of biomaterials, it can also be used for the cell structure or cell spreading on the surface of the biomaterial. To study the cell spreading on the material surface, the material is incubated in a cell culture plate with cell lines. At the end of the predetermined time, the material is retrieved from the cell culture plate, and it is treated in a serial dilution of alcohol so as to fix the cells adhering to its surface. Later, the sample surface is viewed under SEM at an appropriate magnification. The cell spreading gives us an idea of the affinity of the cells toward the scaffold (material).

3.6 SUMMARY AND FUTURE PERSPECTIVES

Advancement in the medical field has helped in improvement of people's quality of life. The ever-lasting quest for designing new biomaterials in the numerous medical fields is evident from the literature, especially in orthopedic and dental implants. Further the aging population around the globe has spurred the demand for implant materials. The issues faced by the current generation of implant materials in their long-term performance need further investigation, especially in evaluating their biological response. Efforts are being taken care to understand the physicochemical properties of biomaterials, but still there is large gap of knowledge that needs more attention. There is a wealth of theoretical knowledge about the required properties of the biomaterial that is available in the literature. However, studies pertaining to *in vivo* are rare.

For orthopedic implants the durability of the implant is governed by the complex interaction of geometric variables, metallurgical variables, mechanical variables, solution variables, and the mechanical loading environment. Hence, the focus in designing biomaterials is on reducing foreign body responses through directing macrophage responses is gaining attraction among researchers. Understanding implant degradation, material science, and also the resulting biological response is the key for developing efficient and new generation biomaterials. In conclusion, we must be confident that the new materials aimed at relieving pain and suffering of people possess no long-term risk in their health.

REFERENCES

1. Rahmati M, Silva EA, Reseland JE, et al. (2020) Biological responses to physicochemical properties of biomaterial surface. *Chem Soc Rev* 49:5178–5224. https://doi.org/10.1039/d0cs00103a
2. Nadim James Hallab JJJ (2020) Biomaterials science. In: Wagner WR, Sakiyama-Elbert SE, Zhang G, Yaszemski MJ (eds.) *Biomaterials Science*, Fourth Edi. Academic Press, pp. 1079–1118.
3. Navarro M, Michiardi A, Castaño O, Planell JA (2008) Biomaterials in orthopaedics. *J R Soc Interface* 5:1137–1158. https://doi.org/10.1098/rsif.2008.0151
4. Jaganathan SK, Supriyanto E, Murugesan S, et al. (2014) Biomaterials in cardiovascular research: Applications and clinical implications. *Biomed Res Int* 2014. https://doi.org/10.1155/2014/459465
5. Irena Gotman PD (1997) Characteristics of metals used in implants. *J Endourol* 11:383–389.
6. Hu CY, Yoon TR (2018) Recent updates for biomaterials used in total hip arthroplasty. *Biomater Res* 22:1–12. https://doi.org/10.1186/s40824-018-0144-8
7. Chen Q, Thouas GA (2015) Metallic implant biomaterials. *Mater Sci Eng R Reports* 87:1–57. https://doi.org/10.1016/j.mser.2014.10.001
8. Dick JC, Bourgeault CA (2001) Notch sensitivity of titanium alloy, commercially pure titanium, and stainless steel spinal implants. *Spine (Phila Pa 1976)* 26:1668–1672. https://doi.org/10.1097/00007632-200108010-00008
9. Baino F, Verne E (2017) Glass-based coatings on biomedical implants: A state-of-the-art review. *Biomed Glas* 3:1–17. https://doi.org/10.1515/bglass-2017-0001
10. Asgari M, Hang R, Wang C, et al. (2018) Biodegradable metallic wires in dental and orthopedic applications: A review. *Metals* 8:181–212. https://doi.org/10.3390/met8040212
11. Rahim MI, Ullah S, Mueller PP (2018) Advances and challenges of biodegradable implant materials with a focus on magnesium-alloys and bacterial infections. *Metals (Basel)* 8: https://doi.org/10.3390/met8070532
12. Bowen PK, Drelich J, Goldman J (2013) Zinc exhibits ideal physiological corrosion behavior for bioabsorbable stents. *Adv Mater* 25:2577–2582. https://doi.org/10.1002/adma.201300226
13. Heimann R (2002) Materials science of crystalline bioceramics: A review of basic properties and applications. *Chiang Mai Univ J* 1:23–46.

14. Alizadeh-Osgouei M, Li Y, Wen C (2019) A comprehensive review of biodegradable synthetic polymer-ceramic composites and their manufacture for biomedical applications. *Bioact Mater* 4:22–36. https://doi.org/10.1016/j.bioactmat.2018.11.003

15. Merola M, Affatato S (2019) Materials for hip prostheses: A review of wear and loading considerations. *Materials (Basel)* 12: https://doi.org/10.3390/ma12030495

16. Roeder RK (2013) *Mechanical Characterization of Biomaterials*. Elsevier.

17. Gu K, Zhang H, Zhao B, et al. (2013) Effect of cryogenic treatment and aging treatment on the tensile properties and microstructure of Ti-6Al-4V alloy. *Mater Sci Eng A* 584:170–176. https://doi.org/10.1016/j.msea.2013.07.021

18. Mitsuo N (1998) Mechanical properties of biomedical titanium alloys. *Mater Sci Eng A* 243:231–236.

19. Niinomi M, Nakai M (2011) Titanium-based biomaterials for preventing stress shielding between implant devices and bone. *Int J Biomater* 2011. https://doi.org/10.1155/2011/836587

20. Bahraminasab M, Sahari BB, Edwards KL, et al. (2013) Aseptic loosening of femoral components: Materials engineering and design considerations. *Mater Des* 44:155–163. https://doi.org/10.1016/j.matdes.2012.07.066

21. Rohr N, Märtin S, Fischer J (2018) Correlations between fracture load of zirconia implant supported single crowns and mechanical properties of restorative material and cement. *Dent Mater J* 37:222–228. https://doi.org/10.4012/dmj.2017-111

22. Simhi T, Banks-Sills L, Fourman V, Shlayer A (2015) Mode i fracture toughness of CNT-reinforced PMMA. *Strain* 51:474–482. https://doi.org/10.1111/str.12158

23. Khorasani AM, Goldberg M, Doeven EH, Littlefair G (2015) Titanium in biomedical applications: Properties and fabrication: A review. *J Biomater Tissue Eng* 5:593–619. https://doi.org/10.1166/jbt.2015.1361

24. Armentia M, Abasolo M, Coria I, Albizuri J (2020) Fatigue design of dental implant assemblies: A nominal stress approach. *Metals (Basel)* 10. https://doi.org/10.3390/met10060744

25. Mantripragada, VP, Lecka-Czernik B, Ebraheim NA and ACJ (2013) An overview of recent advances in designing orthopedic and craniofacial implants. *J Biomed Mater Res Part A* 101:3349–3364. https://doi.org/10.1002/jbm.a.34605.An

26. Plumlee K, Schwartz CJ (2009) Improved wear resistance of orthopaedic UHMWPE by reinforcement with zirconium particles. *Wear* 267:710–717. https://doi.org/10.1016/j.wear.2008.11.028

27. Hussein MA, Mohammed AS, Al-Aqeeli N (2015) Wear characteristics of metallic biomaterials: A review. *Materials (Basel)* 8:2749–2768. https://doi.org/10.3390/ma8052749

28. Hansen DC (2008) Metal corrosion in the human body: The ultimate bio-corrosion scenario. *Electrochem Soc Interface* 17:31–34. https://doi.org/10.1149/2.f04082if

29. Manivasagam G, Dhinasekaran D, Rajamanickam A (2010) Biomedical implants: Corrosion and its prevention: A review. *Recent Patents Corros Sci* 2:40–54. https://doi.org/10.2174/1877610801002010040

30. Souza JCM, Apaza-Bedoya K, Benfatti CAM, et al. (2020) A comprehensive review on the corrosion pathways of titanium dental implants and their biological adverse effects. *Metals (Basel)* 10:1–14. https://doi.org/10.3390/met10091272

31. Eliaz N (2019) Corrosion of metallic biomaterials: A review. *Materials (Basel)* 12. https://doi.org/10.3390/ma12030407

32. Diomidis N, Mischler S, More NS, Roy M (2012) Tribo-electrochemical characterization of metallic biomaterials for total joint replacement. *Acta Biomater* 8:852–859. https://doi.org/10.1016/j.actbio.2011.09.034

33. Gibon E, Córdova LA, Lu L, et al. (2017) The biological response to orthopedic implants for joint replacement. II: Polyethylene, ceramics, PMMA, and the foreign body reaction. *J Biomed Mater Res: Part B Appl Biomater* 105:1685–1691. https://doi.org/10.1002/jbm.b.33676

34. Guglielmotti MB, Olmedo DG, Cabrini RL (2019) Research on implants and osseointegration. *Periodontol 2000* 79:178–189. https://doi.org/10.1111/prd.12254

35. Yang K, Zhou C, Fan H, et al. (2018) Bio-functional design, application and trends in metallic biomaterials. *Int J Mol Sci* 19. https://doi.org/10.3390/ijms19010024

36. Mitić Ž, Stolić A, Stojanović S, et al. (2017) Instrumental methods and techniques for structural and physicochemical characterization of biomaterials and bone tissue: A review. *Mater Sci Eng C* 79:930–949. https://doi.org/10.1016/j.msec.2017.05.127

37. Lach S, Jurczak P, Karska N, et al. (2020) Spectroscopic methods used in implant material studies. *Molecules* 25. https://doi.org/10.3390/molecules25030579

38. Ignjatovic N, Suljovrujic E, Budinski-Simendic J, et al. (2004) Evaluation of hot-pressed hydroxyapa-tite/poly-L-lactide composite biomaterial characteristics. *J Biomed Mater Res: Part B Appl Biomater* 71:284–294. https://doi.org/10.1002/jbm.b.30093

39. Kaczmarek K, Leniart A, Lapinska B, et al. (2021) Selected spectroscopic techniques for surface analysis of dental materials: A narrative review. *Materials (Basel)* 14:2624. https://doi.org/10.3390/ma14102624

4 Biocompatibility Studies of Materials—An Overview

Ananda Babu Sairam and Nacchiappan Annamalai

CONTENTS

4.1 INTRODUCTION

Various metals are used for implants and are known as biocompatible materials like gold, lead, iridium, tantalum, stainless steel and cobalt alloy. Even though various polymers (polyvinylchloride, polyethene, polypropylene, polymethylmethacrylate, polystyrene, polytetrafluoroethylene, polyurethane, polyamide, polyethylenterephthalate, polyethersulfone, polycarbonates and polyetheretherketone) are used in implant manufacture, we can be confident that metals will dominantly be used as biomaterials in the near future (Manam et al. 2017). Due to the easy customization of polymers and fiber-reinforced polymer composites' incapability and mechanical properties compared to metals that can corrode and reach fatigue easily, they also do not release ions into the body, and, therefore, are preferred in the manufacture of biomaterials. There could be a change in biomedical applications' requirements rather than a general application. For this reason, there must be collaboration between the manufacturing community and the biomedical field researchers to obtain products that meet the specific needs of the human body.

Biocompatibility of metals/polymeric composites show the favorable response in a given biological environment. It depends on the corrosion resistance and cytotoxicity of the products. Types of corrosion and the resistance of implants to the surrounding environment are given in Figure 4.1.

Biocompatibility is the interaction of biomaterials with its environment without any undesirable reactions. Organic polymers are used in various materials like plasma extenders, tissue adhesives, bone cement and laboratory wares; and polymer implants are considered a carcinogenic risk (Bischoff 1972). Since one of the first implants in the 1960s, intricate designs and new materials are being introduced every day to reduce the drawbacks of previous versions, but the failure rate is very high, leaving us with a vast gap for improvement. The bulk and surface material properties of the implant and how it interacts with the tissues at the implant interface and the body are determined by the *in vitro* and *in vivo* characteristics (Bauer et al. 2013). According to the Food and Drug Administration (FDA), the definition of biocompatibility is vague because of the presence of a wide range of patents and grants. Biocompatible polymers are state of the art technology which fuels many upcoming start-ups (Ranganathan et al. 2018).

Materials like titanium and other functionally graded materials were fabricated for implant applications and there was a concentration gradient along the longitude of the cylinder. Mechanical

DOI: 10.1201/9781003173533-4

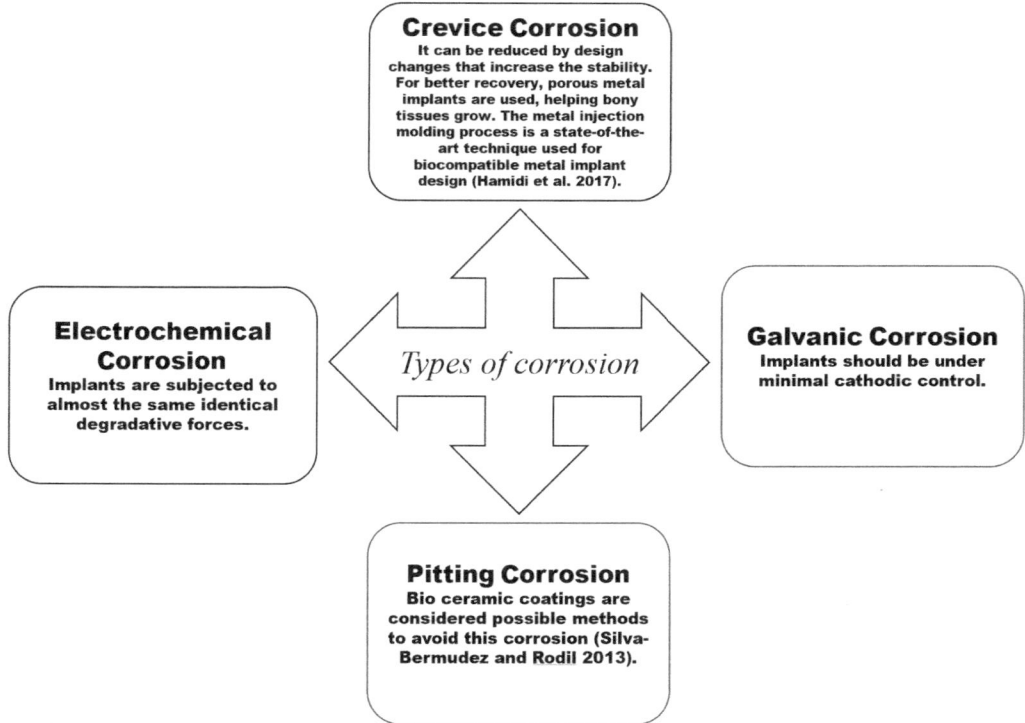

FIGURE 4.1   Types of corrosion in an implant.

properties and biocompatibility were optimized using powder metallurgy, which also was the cause of the change in concentration (Watari et al. 2004). The alloys of Mg and the metal per se have a novel biodegradable property. However, the material's fast degradation can cause osteolysis, implant tissue gap formation, mechanical loss, hydric bubble and this degradation property can be effectively controlled by surface modification according to the organisms (Wan et al. 2016). Mg can also be coated with a polymer, but the effect of this coating on controlling the metal degradation is lacking, various research is determining the polymer structure, thickness of the coating and the molecular weight to deepen our understanding. One such experiment was done with polycaprolactone, poly-ω-pentadecalactone and it was concluded that the coating increased the resistance for corrosion and that it had little to no effect on the cell growth, all the while improving the biocompatibility (Mahapatro et al. 2020).

Mg-based metallic glass/polycaprolactone composites were fabricated using coextrusion and alloying processes and were explored for their biomedical applications. These structures proved acceptable *in vivo* for mechanical and thermal properties (Sharma et al. 2020). There is a vast array of biomedical materials possibilities by combining natural biodegradable and synthetic polymers. One such composite was chitosan and polypyrrole through electrochemical polymerization in oxalic acid medium. There were many benefits like improved surface morphology, surface hydrophilicity and enhanced defensive performance towards corrosion.

Due to the better price for power utilization and wide availability of intricate dies, metal injection molding has a better outlook in the near future. A cranial implant is manufactured using incremental sheet formation (ISF), a biocompatible polymer, flexible technology. Cranioplasty can be better because of its mechanical, thermal properties and its being lightweight all the while being cost-effective (Bagudanch et al. 2018). There are various tests performed on the implant to determine the biocompatibility, for example, the factors responsible for the degradation process in Zn implants.

Tests like the analysis of degradation products, cytocompatibility testing on various products of degradation, determining the degradation rate and monitoring the effect of Zn ion and pH change on cytocompatibility (Su et al. 2019).

Medical implants that are used for structural support or device implants with electronics are used to treat clinical issues for mechanical and structural support in orthopedic and dental fields. These and other similar devices are continuously monitored for their proper functioning in the human body (Cha et al. 2019). There are various advancements done to improve the testing techniques themselves and since polyurethanes are actively studied for their potential in implants for tissue engineering, a good technique was used to test the effectiveness of animal research in implant integration. There is a possibility of bacterial infection due to implants, and 3-sulfopropyl methacrylate potassium salt proved to be biocompatible and effective in eliminating any infections related to implants (Li et al. 2020). There are rapid developments in the 3D and additive manufacturing fields and steps to commercialize the technology for practical use in implant application in the medical field. However, the 3D printed implants have various drawbacks because of their low fatigue strength and many other imperfections. Researchers are working on breakthroughs every day for inventing new biocompatible material and those that can be a solution for this sustainable and circular economy (Yadav et al. 2020). Even though there are a few drawbacks with 3D printing, which is continually improving, it has many advantages compared to conventional methods like the possibility of customizing intricate designs and doing it in less time (Tan et al. 2017). A patient need not wait for a donor, let alone a suitable donor; we can print whatever we want, and this is not a fantasy anymore—due to the advancements in technology, this future is now!

4.2 EXISTING TECHNOLOGY TO ASSESS THE BIOCOMPATIBILITY OF METALS/POLYMERIC COMPOSITE MATERIALS

This part of the chapter discusses some of the tests performed to determine a material's biocompatibility. The testing method and the material per se have been changing with time, and new methods of testing and improved materials come up every day and every time this happens, we are one step closer to being perfect.

Cytotoxicity tests: there are different types of cytotoxicity and cell viability assays that are classified according to types of endpoints (color changes, fluorescence, luminescence, etc.). An assay for *in vitro* viability and cytotoxicity evaluation should be a quick, efficient, reliable, safe, and time and cost-effective method. It should not interfere with the test compound. Acute systemic toxicity determines short-term effects that occur immediately after the material is ingested; this usually is tested using animals, but after serious concerns were raised on animals' well-being, tests like lethal doses were deleted (Botham 2004). There are tens of thousands of chemicals without toxicity data on them and it would cost a lot of money, time and animals to test all those.

This toxicity can be caused by three factors: cell-specific function toxicity, selective cytotoxicity, general cytotoxicity. Subchronic study and chronic study are done on animals to determine the dosage of test substance that can be ingested without any problems or at what dosage the test material causes subchronic problems to the human body. A test for polyvinyl alcohol was performed on rats for a subchronic toxicity study and even though many tests were performed previously, the test methods and standards keep changing according to the latest research results (Kelly et al. 2003). Subchronic toxicity tests are performed for various chemicals and mostly on rats. Ageing is a natural phenomenon that causes various problems in an organism; on the other hand, an immature or underdeveloped organism is at a susceptible stage. To evaluate the maximum dosage that can be safe for these age groups, rats that are immature and old are being used in the tests to determine acute and subchronic toxicity studies. The studies concluded that the substance causes more damage when the subject is old and can develop hepatotoxicity (Yang et al. 2020).

TABLE 4.1
List of Existing Tests for Biocompatibility.

Types of Tests	Description
Cytotoxicity Tests	There are different tests to determine cytotoxicity and there are three tests to determine this: direct contact, indirect contact or extracts (Denstedt and Atala 2009).
Device Implantation Tests	It involves the examination of tissues in order to study the demonstration of disease.
Genetic Toxicology	There are various tests performed to assess the genotoxicity of material and the comet assay or the single-cell electrophoresis is a highly sensitive testing method. Surface topography does not affect the genotoxicity.
Hemocompatibility	Biocompatibility with blood is called hemocompatibility. The devices are tested in *in vitro* models using fresh human blood under controlled conditions.
Irritation/Intracutaneous Reactivity	The tissue damage or inflammation, specifically the level and the recovery from any topical reaction near the implant, is monitored.
Phototoxicity Tests	It is a toxic response elicited by topically or systemically administered photoreactive chemicals after the body's exposure to environmental light.
Pyrogenicity Tests	The rabbit pyrogen test is considered the best for medical device pyrogenicity testing. The *in vitro* monocyte activation test is also used.
Sensitization Tests	This test is essential in evaluating a device's potential to cause an allergic response following (repeated) exposure.

Device implantation test (histopathology): it's evaluating the implants by microscopically examining tissue to study the demonstration of disease. The medical devices are planned to be implanted for either short-term or long-term periods. This is very essential to provide a reasonable assessment of safety: the implant study should closely approximate the intended clinical uses. When examined, the cochlear implant had a formation of fibrosis and subsequent ossification, but it did not affect the implant performance in most cases, but, when the implant was a hybrid, there was a chance of implant function being affected. It was also found that the ganglion cells formation was negligible even when the device was used for decades.

Genetic toxicology: titanium and its alloys are widely used as implant materials because they are corrosion-resistant and have other biocompatible properties; they can corrode under certain circumstances like low pH or in the presence of fluorine. However, it also depends on the type of alloy, and first-generation Ta alloy is highly toxic. Ta particles can be cytotoxic and can cause DNA damage (genotoxic). Ta ions and its alloy particles can cause genotoxicity and are hypothesized that the partially filled and low energy D-orbital is the reason for genotoxicity (Markowska-Szczupak et al. 2020), and patients with implant joints have a higher cancer risk due to genotoxic materials used as implants (Qin et al. 2020).

There are various tests performed to assess the genotoxicity of material and the comet assay. The single-cell electrophoresis is a highly sensitive testing method which is also a rapid testing method. For a long time studies showed that Ta-35Nb alloy did not have any cytotoxicity and recent studies using the tests mentioned previously show that it is not genotoxic, as damages to DNA were not observed. It was also proved that the surface topography did not affect the genotoxicity (Dennia Perez et al. 2020). Implantation tests evaluate the implants' pathological effects and this is done as short- and long-term tests on animals like rats, pigs and the other usual lab animals (Raghavendra et al. 2015). Various parameters are tested like inflammation, tissue degeneration, necrosis or the death of tissues or cells, tissue alterations, quantity and quality of the formed tissues (Ramakrishna et al. 2015; Silva-Bermudez and Rodil 2013).

Hemocompatibility: blood is a complex fluid and the biocompatibility with blood is called hemocompatibility, which is one of the significant factors that are and should be considered for biocompatibility. An international standard indicates that thrombosis, platelets, coagulation, immunology and hematology are considered biocompatible. The devices are tested in *in vitro* models using fresh human blood under controlled conditions. The test device is subjected to various preliminary procedures like thorough cleaning of the device to ensure no external factors affect the test results. *In vitro* models like the static flow model, agitated blood incubation model and shear flow model, the test results in changes of platelets, erythrocytes and leukocytes, deposition of proteins and cells on the material, generation of activation products in plasma are obtained, the test category is mentioned saline in Table 4.1 (Weber et al. 2018). Even though there are various methods to improve the hemocompatibility of synthetic materials, plasma treatment and endothelialization with the human umbilical vein remains one of the best methods.

Irritation/intracutaneous reactivity: this is for medical devices or materials in contact with living tissues or body fluid. The extract (example: saline solution) from the medical devices managed in the animal model's skin and the tissue damage or inflammation induction were evaluated. Specifically, the level and recovery from any topical reactions at the administration site were measured.

Phototoxicity testing: it is defined as a toxic response elicited by topically or systemically administered photoreactive chemicals after the exposure to environmental light. As the awareness of the safety increases, combined effects of sunlight, chemicals and their intrinsic toxicity draw attention from the general public, which supports the initiative to study the mechanism of phototoxicity and use the 3T3 Neutral Red Uptake (NRU) test to evaluate it. The 3T3 NRU phototoxicity test is a cytotoxicity-based test that utilizes mouse fibroblasts to measure the concentration-dependent reduction in neutral red uptake by the cells after exposure to test material.

Pyrogenicity testing: the rabbit pyrogen test RPT is considered the best for medical device pyrogenicity testing; however, the *in vitro* monocyte activation test (MAT) can be used instead, because the MAT can screen for many classes of pyrogenic substances since it measures cytokines released by human blood cells. This method can detect non-bacterial as well as bacterial endotoxin pyrogens.

Sensitization tests: sensitization is one of the three most common tests that ensure the medical devices' safety. This test is essential in evaluating a device's potential to cause an allergic response following (repeated) exposure. *In vivo* and *in vitro* assays are available based upon the specific route of exposure to the body and acceptance of the notified body.

Systemic acute, subacute, subchronic and chronic toxicity: it is essential to make sure any medical substance is biocompatible and to monitor and ensure this, there are various norms to be followed and tests to be performed on the material before it is out for commercial use. The material is subjected to some initial tests and then the material used with non-human living beings like rats or dogs and performed over different periods, with more extended periods helping with the understanding of healing and regenerative processes. When that material passes all the previous tests and the materials are to be used and tested on humans, various organizations like the International Organization for Standardization (ISO) and the FDA are the bodies that look over and standardize the process (Swetha et al. 2015).

Toxic substances can cause damage to the cell membrane, hinder the metabolic activities and even damage the genetic material. Biomaterials that often cause damages are surfactants, plasticizers and residual monomers. When a toxic substance damages the gene, it's called genotoxicity and various tests are used to determine this. When the toxic substances that cause genotoxicity are at an alarming level, they can only cause cytotoxicity. There are direct and indirect testing methods for evaluating the biomaterial's toxicity. During the direct contact method, a single layer of cell bed is made, and without causing any physical damage the test material is placed over this bed. A similar process is followed for the indirect contact method, but there is a permeable layer present between the cell bed and the test material. The assessment of cell death is done to determine the toxicity. There is also a huge advantage of choosing an *in vitro* method over *in vivo* because of cost effectiveness and more relatable results, as when the component is tested on animals the result may not be accurate.

We must have an understanding of the metabolic ability of the target cell in order to perform *in vitro* cytotoxicity on it. We must also consider the exposure time and the composition of the environment as the chemicals used and the amount of time of exposure can affect the results. Many first-generation metals on metal implants are considered carcinogens due to the release of metal ions, and there was a brief period where metal-on-polyethene implants were used. Due to certain factors causing implant failure, they returned to metal-on-metal implants, but this time it was improved. CoCr is a second-gen metal-on-metal implant that is widely used for hip replacements and among implants; the hip implant is one of the most used. Carcinogenicity of the Co/Cr ion was evaluated and it was concluded that the concentration of these ions in the blood or on tissues is deficient in posing a possible systemic cancer risk. It also was found that Cr ions from the implant did not cause any tumorigenic or genotoxic effects in the human body regardless of the dose (Christian et al. 2014). After all the tests were done, the material was proven safe; the implant was subjected to *in vitro* and *in vivo* preclinical safety testing. There are various metals and implants under preclinical safety tests like a minimally invasive implant device for apnea or a Zr metal as a replacement for Ta. While the studies for the implant device mentioned previously were conducted through various animals and various parameters according to international standards, the results were approved for the next phase of trials, which is on humans, meaning it did not have any adverse effects. Clinicians testing the Zr metal as a clinical replacement for Ta had many questions like its long-term stability, soft tissue response, implant design and many more, requiring further testing before it is proved safe in preclinical tests (Nishihara et al. 2019).

Safety is the first concern with usage of any material by living beings, be it any edibles that we consume or the products that we wear—anything like cosmetics that people wear on their bodies to the implants that are used in the body. These mentioned parameters and tests are only some of the ways to examine any material's biocompatibility. Several national and international organizations monitor and lay down guidelines and acceptable results for these tests to ensure humans' safety.

4.3 METHODOLOGIES ADAPTED TO DESIGN METAL/ POLYMERIC COMPOSITE PROP-UP MEDICAL DEVICES

Biomaterials have been made in a nano structural version through primary novel techniques, and their biological process is being studied for the improvement of biocompatibility of biomaterials. To optimize the biocompatibility of implants, the concentration of titanium/hydroxyapatite or other functionally graded material (FGM) in implants was changed along the cylindrical directions and they were fabricated using powder metallurgy and sedimentation in a solvent or by packing the powder into a mold to create a concentration gradient. These were compressed and subjected to sintering (Watari et al. 2004).

The deformation effect on peripheral blood mononuclear cells on a scaffold regarding macrophage polarization was evaluated; it showed that all groups had inflammatory macrophages. There is a vast array of possibilities with the latest 3D printing technology, especially in printing tissues; it offers high flexibility and the actual fabrication throughput. The animal testing of these is underway in a laboratory (Malinauskas et al. 2012). Polydopamine and acrylamide were used in membranes by incorporating them in the matrix to improve the bio-functionality and the mechanical property, respectively, under continuous flow conditions (Obiweluozor et al. 2019). The biocompatibility of polymers was evaluated by making them into a braided structure in porcine vasculature with the polymers mass more significant than the mass of the expected vascular stent.

To develop, manufacture and keep any product in the market for a period of time requires various steps and regulations to be followed, and for implants, the process is tedious and has to pass even more crucial steps to ensure safe usage. The product is first designed and then it's manufactured as a prototype and there are various standards to be met, along with the input of medical device engineers. After the usage of guidelines from previous studies, the product is cleared by quality control when it is satisfiable for clearance from boards and mass manufacture.

TABLE 4.2
Designing of a Biomaterial.

Preferably, Designing Begins with the Following Steps:

Step 1	Design and development planning, identifying the project's scope and how all the aspects of the standards will be met. A strong product definition is extracted by analyzing the market needs.
Step 2	Design inputs, product requirements, marketing requirements and regulatory requirements (regional or international applicable standards) for compliance. Medical device classification is based on the risk associated with its use. Also, we need to look for any pre-existing intellectual property on the proposed idea or similar; it might prevent us from using that mechanism or technology. A strong development team is crucial to carry out the complete discovery phase as it may be a disadvantage if the team is not well experienced. In any case, we can decide to proceed with an existing team, a consultant or a company to get help with medical device engineering.
Step 3	The next step is to convert the idea into the discovery phase; this phase consists of initial designing, prototyping and driven redesign. Once we complete the product conceptualization and discovery phase, we may proceed further for standards approval and commercialization.

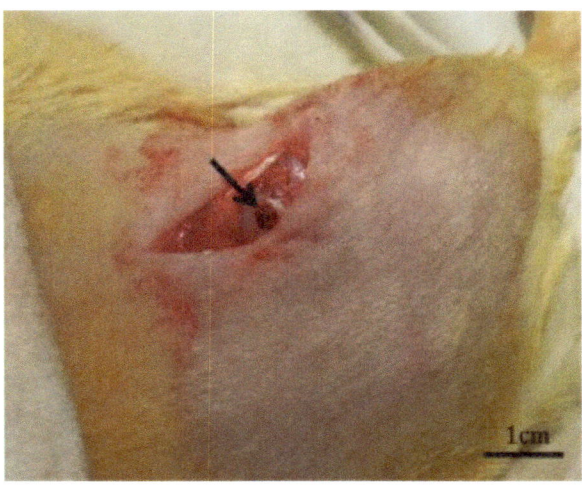

FIGURE 4.2   Scaffold implantation in the lateral epicondyle.

4.4   FINDINGS WITH A SUGGESTION FOR AN IMPLANT MANUFACTURER

Implants are biocompatible metals, polymers, composites including glasses, glass-ceramics and composites. They are used in this field in different forms: particles, fibers, thin films, coatings, dense and non-porous, porous, etc. Researchers usually accepted a durable, stable and free flow of fluid (blood) that resists retrograde reflux as one of the features of a biomaterial used for medical applications. New valve designs are being tested every day and there are certain things like a little reflux that can be advantageous or there are also designs where the valve does not open or close completely, preventing valve cusps. The sail valve showed no blockage after short-term follow-up and no stent fracture, suggesting a viable stent design was adequate with only minor changes (Boersma et al. 2017). Balloon angioplasty can be used to treat inferior vena cava aneurysms. A biocompatible device should not affect or damage the tissues when brought in contact with a device. Many developments have been made in biomaterials like the cell-friendly materials with calculated degeneration, particular designs for contact areas and many more advantages. These developments have opened a pathway to many possibilities and raised questions that need to be re-examined (Brown and Badylak 2013).

Developing implants and tissue engineering technologies are improving or restoring diseased organs, but there are still various challenges, inflammation being the major one, and immunomodulation seems to be a possible solution. A type of nanofibrous mesh manufactured using 3D technology with specific surface chemistry using polymers, protein repellents and hydrogel-coated protein repellents serves to induce healing.

It was found that the hydrophilic carbon nanofibers (CNFs) resulted in an inadequate inflammatory response from macrophages by provoking less pro-inflammatory cytokine secretion than when compared to hydrophobic CNFs (Chun et al. 2011). Selective surface flow carbon membrane optimization for hydrocarbon-nitrogen separation was done by taking into account the polymer precursor, carbonization temperature and coating conditions while using polyvinylidene chloride deposits of macroporous ceramic supports. There are various traditional requirements for a medical device; development of a database containing information of existing devices will be useful in implementing existing and new biomaterials in this field. There are various developments in biomaterials and much advancement in biodegradable implants has made the need for the removal of the implants unnecessary; usually, polymers are used. However, they cannot usually bear the mechanical stress due to body weight and for this reason, Mg is a suitable biodegradable material. Metallic biomaterials have been predominant in implants due to their ability to withstand high static or cyclic loads and be biodegradable (Aslam et al. 2016).

There are reports of degradation of polymer screws in patients who underwent arthroscopy anterior cruciate ligament reconstruction. There is a constant development of coatings for biomaterials to improve biocompatibility, improve the balance between cell adhesions, low cytotoxicity and improve mechanical properties. Due to the advancements in biomaterials, micro-scale scaffolds are used for bone replacement or even regenerations, though their design is very basic and the regenerated microstructures are different from that of the actual tissues. The interaction between the

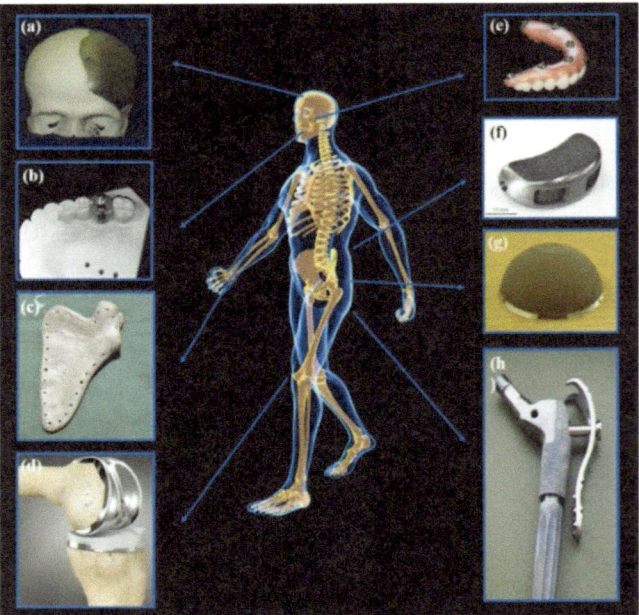

FIGURE 4.3   Examples of 3D printed implants.

Source: Ni et al. (2019).

implants and the nearby tissue is essential, and it depends on factors like shape, surface property, loading conditions and tissue properties. The manufacture of a unique polymer blend system can be advantageous in providing a robust coating and controlling the drug release, demonstrated in an *in vivo* experiment.

The suggestions in this section are apt, as manufacturers would agree that implant materials should not cause unfavorable tissue response, implant strength must provide lifetime support and axial compressive mechanical supports and bone implants should have a porous interface with the bone to help in tissue growth.

This part of the chapter discussed the various advancements in biocompatible materials every day, like different types of valves and 3D printed implants and how they are modifying the industry as we know it and how far we have come since the first generation of implants. We also discussed various new materials in the market and how they are helping manufacturers to improve the biocompatibility of materials to work in adverse environments, helping the growth of tissues at the material tissue interface and customizing the biomaterials' shape and size according to individual requirements. The many day-to-day developments should be taken into account by the manufacturers and with the help of scientists in this field the products for consumers' use shall be improved, which both the consumers and the manufactures can benefit from.

4.5   DISCUSSION, LIMITATIONS, FUTURE RESEARCH AND CONCLUSIONS

Biocompatibility is the determination of the extent to which a material can be used in a given environment and this chapter discusses the various test methods to determine factors like phototoxicity, cytotoxicity, hemocompatibility and a few others. We have also seen about the steps involved in the designing of biomaterials according to the market needs and also the standards set by various international organizations and how the manufacturers may consult with experts in this field and use the previously available clinical data to provide more compatible materials. A significant opening exists in both our research and the existing test methods that are useful in understanding the challenges in the biocompatibility of medical implants with our tissue. Various research studies are undertaken in every part of the world, providing more suitable materials, and there are vast opportunities in 3D technology for it to ease the manufacturing and testing methods, improve biocompatibility and facilitate the design of the customized implant. We are not far from a near perfect implant specific to each individual and one day, that will be the real deal.

REFERENCES

Aslam, Muhammad, Faiz Ahmad, Puteri Sri Melor Binti Megat Yusoff, Khurram Altaf, Mohd Afian Omar, and Randall M. German. 2016. "Powder Injection Molding of Biocompatible Stainless Steel Biodevices." *Powder Technology*. https://doi.org/10.1016/j.powtec.2016.03.039.

Asri, R. I. M., W. S. W. Harun, M. Samykano, N. A. C. Lah, S. A. C. Ghani, F. Tarlochan, and M. R. Raza. 2017. "Corrosion and Surface Modification on Biocompatible Metals: A Review." *Materials Science and Engineering C*. https://doi.org/10.1016/j.msec.2017.04.102.

Bagudanch, Isabel, María Luisa García-Romeu, Ines Ferrer, and Joaquim Ciurana. 2018. "Customised Cranial Implant Manufactured by Incremental Sheet Forming Using a Biocompatible Polymer." *Rapid Prototyping Journal*. https://doi.org/10.1108/RPJ-06-2016-0089.

Bauer, Sebastian, Patrik Schmuki, Klaus von der Mark, and Jung Park. 2013. "Engineering Biocompatible Implant Surfaces: Part I: Materials and Surfaces." *Progress in Materials Science*. https://doi.org/10.1016/j.pmatsci.2012.09.001.

Bischoff, F. 1972. "Organic Polymer Biocompatibility and Toxicology." *Clinical Chemistry*. https://doi.org/10.1093/clinchem/18.9.869.

Boersma, Doeke, Aryan Vink, Frans L. Moll, and Gert J. De Borst. 2017. "Proof-of-Concept Evaluation of the Sail Valve Self-Expanding Deep Venous Valve System in a Porcine Model." *Journal of Endovascular Therapy*. https://doi.org/10.1177/1526602817700120.

Botham, P. A. 2004. "Acute Systemic Toxicity: Prospects for Tiered Testing Strategies." *Toxicology in Vitro.* https://doi.org/10.1016/S0887-2333(03)00143-7.

Brown, Bryan N., and Stephen F. Badylak. 2013. "Expanded Applications, Shifting Paradigms and an Improved Understanding of Host-Biomaterial Interactions." *Acta Biomaterialia.* https://doi.org/10.1016/j.actbio.2012.10.025.

Cha, Gi Doo, Dayoung Kang, Jongha Lee, and Dae Hyeong Kim. 2019. "Bioresorbable Electronic Implants: History, Materials, Fabrication, Devices, and Clinical Applications." *Advanced Healthcare Materials.* https://doi.org/10.1002/adhm.201801660.

Christian, Whitney V., Lindsay D. Oliver, Dennis J. Paustenbach, Marisa L. Kreider, and Brent L. Finley. 2014. "Toxicology-Based Cancer Causation Analysis of CoCr-Containing Hip Implants: A Quantitative Assessment of Genotoxicity and Tumorigenicity Studies." *Journal of Applied Toxicology.* https://doi.org/10.1002/jat.3039.

Chun, Young Wook, Wenping Wang, Jungil Choi, Tae Hyun Nam, Yong Hee Lee, Kwon Koo Cho, Yeon Min Im, et al. 2011. "Control of Macrophage Responses on Hydrophobic and Hydrophilic Carbon Nanostructures." *Carbon.* https://doi.org/10.1016/j.carbon.2011.01.044.

Dennia Perez, de Andrade, Silva Carvalho Isabel Chaves, Godoi Bruno Henrique, da Silva Newton Soares, Alves Cairo Carlos Alberto, Soares Cristina Pacheco, and Carvalho Yasmin Rodarte. 2020. "In Vitro Genotoxic Study Reinforces the Use of Titanium-35niobium Alloy in Biomedical Implants." *International Journal of Oral and Craniofacial Science.* https://doi.org/10.17352/2455-4634.000044.

Denstedt, John, and Anthony Atala. 2009. *Biomaterials and Tissue Engineering in Urology.* https://doi.org/10.1533/9781845696375.

Kelly, C. M., C. C. DeMerlis, D. R. Schoneker, and J. F. Borzelleca. 2003. "Subchronic Toxicity Study in Rats and Genotoxicity Tests with Polyvinyl Alcohol." *Food and Chemical Toxicology.* https://doi.org/10.1016/S0278-6915(03)00003-6.

Li, Pengfei, Zhichao Ding, Yue Yin, Xiaojie Yu, Yucheng Yuan, Maria Brió Pérez, Sissi de Beer, G. Julius Vancso, Yunlong Yu, and Shiyong Zhang. 2020. "Cu2+-Doping of Polyanionic Brushes: A Facile Route to Prepare Implant Coatings with Both Antifouling and Antibacterial Properties." *European Polymer Journal.* https://doi.org/10.1016/j.eurpolymj.2020.109845.

Mahapatro, Anil, Kayla Jensen, and Shang You Yang. 2020. "Effect of Polymer Coating Characteristics on the Biodegradation and Biocompatibility Behavior of Magnesium Alloy." *Polymer-Plastics Technology and Materials.* https://doi.org/10.1080/25740881.2019.1634728.

Malinauskas, Mangirdas, Daiva Baltriukiene, Antanas Kraniauskas, Paulius Danilevicius, Rasa Jarasiene, Raimondas Sirmenis, Albertas Zukauskas, et al. 2012. "In Vitro and In Vivo Biocompatibility Study on Laser 3D Microstructurable Polymers." *Applied Physics A: Materials Science and Processing.* https://doi.org/10.1007/s00339-012-6965-8.

Manam, N. S., W. S. W. Harun, D. N. A. Shri, S. A. C. Ghani, T. Kurniawan, M. H. Ismail, and M. H. I. Ibrahim. 2017. "Study of Corrosion in Biocompatible Metals for Implants: A Review." *Journal of Alloys and Compounds.* https://doi.org/10.1016/j.jallcom.2017.01.196.

Markowska-Szczupak, Agata, Maya Endo-Kimura, Oliwia Paszkiewicz, and Ewa Kowalska. 2020. "Are Titania Photocatalysts and Titanium Implants Safe? Review on the Toxicity of Titanium Compounds." *Nanomaterials.* https://doi.org/10.3390/nano10102065.

Ni, J., H. Ling, S. Zhang, Z. Wang, Z. Peng, C. Benyshek, R. Zan, et al. 2019. "Three-Dimensional Printing of Metals for Biomedical Applications." *Materials Today Bio.* https://doi.org/10.1016/j.mtbio.2019.100024.

Nishihara, Hironobu, Mireia Haro Adanez, and Wael Att. 2019. "Current Status of Zirconia Implants in Dentistry: Preclinical Tests." *Journal of Prosthodontic Research.* https://doi.org/10.1016/j.jpor.2018.07.006.

Obiweluozor, Francis O., Arjun Prasad Tiwari, Jun Hee Lee, Tumurbaatar Batgerel, Ju Yeon Kim, Dohee Lee, Chan Hee Park, and Cheol Sang Kim. 2019. "Thromboresistant Semi-IPN Hydrogel Coating: Towards Improvement of the Hemocompatibility/Biocompatibility of Metallic Stent Implants." *Materials Science and Engineering C.* https://doi.org/10.1016/j.msec.2019.02.054.

Qin, Hong Min, Denise Herrera, Dian Feng Liu, Chao Qian Chen, Armen Nersesyan, Miroslav Mišík, and Siegfried Knasmueller. 2020. "Genotoxic Properties of Materials Used for Endoprostheses: Experimental and Human Data." *Food and Chemical Toxicology.* https://doi.org/10.1016/j.fct.2020.111707.

Raghavendra, Gownolla Malegowd, Kokkarachedu Varaprasad, and Tippabattini Jayaramudu. 2015. "Biomaterials: Design, Development and Biomedical Applications." *Nanotechnology Applications for Tissue Engineering*, no. January 2015: 21–44. https://doi.org/10.1016/B978-0-323-32889-0.00002-9.

Ramakrishna, Seeram, Lingling Tian, Charlene Wang, Susan Liao, and Wee Eong Teo. 2015. "Safety Testing of a New Medical Device." *Medical Devices*. https://doi.org/10.1016/b978-0-08-100289-6.00006-5.

Ranganathan, Balu, Charles Miller, and Anthony Sinskey. 2018. "Biocompatible Synthetic and Semi-Synthetic Polymers: A Patent Analysis." *Pharmaceutical Nanotechnology*. https://doi.org/10.2174/2211738505666 171023152549.

Sharma, Adit, Alexey Kopylov, Mikhail Zadorozhnyy, Andrei Stepashkin, Vera Kudelkina, Jun Qiang Wang, Sergey Ketov, et al. 2020. "Mg-Based Metallic Glass-Polymer Composites: Investigation of Structure, Thermal Properties, and Biocompatibility." *Metals*. https://doi.org/10.3390/met10070867.

Silva-Bermudez, P., and S. E. Rodil. 2013. "An Overview of Protein Adsorption on Metal Oxide Coatings for Biomedical Implants." *Surface and Coatings Technology*. https://doi.org/10.1016/j.surfcoat.2013.04.028.

Su, Yingchao, Hongtao Yang, Julia Gao, Yi Xian Qin, Yufeng Zheng, and Donghui Zhu. 2019. "Interfacial Zinc Phosphate Is the Key to Controlling Biocompatibility of Metallic Zinc Implants." *Advanced Science*. https://doi.org/10.1002/advs.201900112.

Swetha, B., Sylvia Mathew, B. V. Sreenivasa Murthy, N. Shruthi, and Shilpa H. Bhandi. 2015. "Determination of Biocompatibility: A Review." *International Dental & Medical Journal of Advanced Research— VOLUME 2015*. https://doi.org/10.15713/ins.idmjar.2.

Tan, X. P., Y. J. Tan, C. S. L. Chow, S. B. Tor, and W. Y. Yeong. 2017. "Metallic Powder-Bed Based 3D Printing of Cellular Scaffolds for Orthopaedic Implants: A State-of-the-Art Review on Manufacturing, Topological Design, Mechanical Properties and Biocompatibility." *Materials Science and Engineering C*. https://doi.org/10.1016/j.msec.2017.02.094.

Wan, Peng, Lili Tan, and Ke Yang. 2016. "Surface Modification on Biodegradable Magnesium Alloys as Orthopedic Implant Materials to Improve the Bio-Adaptability: A Review." *Journal of Materials Science and Technology*. https://doi.org/10.1016/j.jmst.2016.05.003.

Watari, Fumio, Atsuro Yokoyama, Mamoru Omori, Toshio Hirai, Hideomi Kondo, Motohiro Uo, and Takao Kawasaki. 2004. "Biocompatibility of Materials and Development to Functionally Graded Implant for Bio-Medical Application." *Composites Science and Technology*. https://doi.org/10.1016/j.compscitech.2003.09.005.

Weber, Marbod, Heidrun Steinle, Sonia Golombek, Ludmilla Hann, Christian Schlensak, Hans P. Wendel, and Meltem Avci-Adali. 2018. "Blood-Contacting Biomaterials: In Vitro Evaluation of the Hemocompatibility." *Frontiers in Bioengineering and Biotechnology*. https://doi.org/10.3389/fbioe.2018.00099.

Yadav, Dinesh, Ramesh Kumar Garg, Akash Ahlawat, and Deepak Chhabra. 2020. "3D Printable Biomaterials for Orthopedic Implants: Solution for Sustainable and Circular Economy." *Resources Policy*. https://doi.org/10.1016/j.resourpol.2020.101767.

Yang, Dong, Wan Yi Huang, Yan Qiao Li, Shi Yu Chen, Si Yu Su, Yue Gao, Xian Li Meng, and Ping Wang. 2020. "Acute and Subchronic Toxicity Studies of Rhein in Immature and D-Galactose-Induced Aged Mice and Its Potential Hepatotoxicity Mechanisms." *Drug and Chemical Toxicology*. https://doi.org/10.1080/01480545.2020.1809670.

5 Fabrication Methods for 2D Materials with Heterostructures

Manoharan Arun Kumar and Mahalingam Shanmugam

CONTENTS

5.1 INTRODUCTION

For a decade, there has been rapid growth in material science research. Meanwhile, with the effective preparation method of graphene in 2004 [1], two-dimensional materials have attracted more research for its properties and applications. 2D materials are classified as single-element 2D materials such as black phosphorus (BP), germanene, graphene, silylene, etc., and compound 2D materials such as hBN, TMDs, TMCs compound semiconductor, III–V group elements, etc. [2]. 2D materials have many more interesting peculiar properties than bulk materials. The specialty of 2D materials is how the number of layers are strongly supported. The band structure of the monolayer graphene at the Dirac point is moderately dissimilar from the parabolic band structure of the double-layered graphene [3]. For black phosphorus (BP), the band structure and bandgap are dependent on the number of layers. When the number of layers increases, then the bandgap shows the redshift as evident. In TMD semiconductors, the bandgap transition can be observed as a direct to indirect manner from single layer to multilayer. The peculiar properties of the thin-layered 2D materials have been deliberately promising materials for fabricating optical and electronic sensor devices. Especially, 2D materials are resistant to the short channel effect and have improved mechanical strength which makes the thin structure develop flexibly and can be used in wearable sensor devices [4]. However, 2D materials are applicable for certain applications owing to their limitations. Direct deposition of metal on 2D materials creates high interaction resistance because of the Schottky barrier effect. Additionally, in single 2D-based semiconductor materials, the intralayer excitons are hard to use because of a very short lifetime. Hence, it is difficult to use for exciton devices. Therefore, 2D materials are used in heterostructures because of its interesting properties [5]. Researchers developed two categories of heterostructures: vertical arrangement of stacked heterostructures and epitaxially developed planar heterostructures. In this chapter, we focus on the fabrication methods, the properties, and the applications of 2D heterostructures.

5.2 FABRICATION PROCESS OF 2D HETEROSTRUCTURES

5.2.1 Fabrication Using Deterministic Transfer Method

The layer formation of 2D materials is achieved by different methods like mechanical exfoliation and Chemical Vapor Deposition. The layers are finally transformed to different substrates at the anticipated location. The process is usually based on the optical inspection system with proper $XYZ\theta$ direction for accurate placement on the substrate. Polymer carriers are used for fabricating the devices and are classified as (a) hybrid stamp composed of PDMS/PPC or PMMA/hBN [6], (b) PMMA/sacrifice layer [7], (c) PDMS [8], (d) thermoplastic polymer [9]. The transfer process is shown in Figure 5.1.

Step 1: The PMMA/sacrifice layer is developed onto Si/SiO$_2$ by using a spin-coating system. It is also developed on a glass slide. Then the thin flakes are achieved by the mechanical exfoliation method.

Step 2: The glass slides are kept in the wet process, then the flakes on polymer-carriers are prepared on the substrate in the desired location. The flakes are settled down till the van der Waals force of attraction present in the developed heterostructure.

Step 3: The target flakes combined with other layered material form the multilayers. The heterostructure layers are lifted by taking the glass slides and washing them with acetone solution.

During this process, the cleanness is dependent on the exposure of heterostructure to the polymer. The processing speed and easiness are dependent on the steps followed for the spin coating,

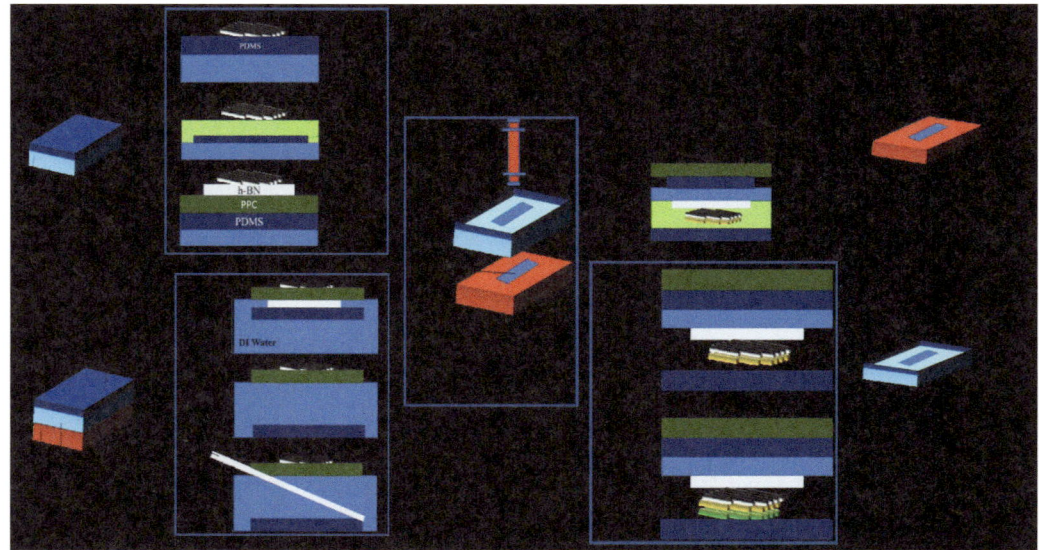

FIGURE 5.1 Schematic diagram shows the deterministic transfer method.

wet transferring, heating, and washing. The PDMS is more suitable for the quick process of coating, wet transfer, heating, and washing. Additionally, PDMS is the more commercialized and best way to prepare the multi-heterostructure. Therefore, the deterministic better transfer method shows more flexibility for fabricating the 2D heterostructures. The drawback of this method is that the 2D material substrates might be destroyed or it creates wrinkles in the samples during the transfer process.

5.2.2 Fabrication Using Chemical Vapor Deposit (CVD) Growth

The vertical 2D heterostructures are prepared by the bottom-up strategy named the CVD synthesis. The growth of multiple layers in the heterostructures can be observed. The main advantage of the CVD method is to achieve a sharp heterojunction structure and clean vertically stacked multilayer heterostructures. The fabrication done by the CVD method will be scalable and controllable in both planar and vertical structures.

5.2.2.1 The One-Step CVD Method

The schematic diagram of the one-step CVD method process is shown in Figure 5.2. It is the best suitable method to prepare the 2D heterostructures. Gong et al. [10] effectively prepared the MoS_2/WS_2 heterostructures by using the one-step CVD method. It confirmed the presence of epitaxial growth of WS_2 on the surface of MoS_2 at a lower temperature of about 650°C but it was not enough for the nucleation. To form the stable structure of WS_2 on the surface of MoS_2 as a stacked heterostructure a higher structure of about 850°C was used. These kinds of heterostructures have good van der Waals and are thermodynamically stable.

5.2.2.2 The Two-Step CVD Method

The two-step CVD method is specially used for extracting the second layer on the substrate. As with the one-step method, it produces the heterostructures vertically and the lateral stacked manner is affected by the high-rate flow of gas. Li et al. [11] fabricated the 2D GaSe/$MoSe_2$ heterostructure by using this two-step CVD method. It is also used to fabricate the stacked TMD/hBN heterostructure using Ni-Ga alloy without transitional operations.

5.2.2.3 The Multi-Step CVD Method

The multi-step CVD method was established by switching the gas flow components [12]. This type of approach is used to develop the sharp and sequential growth multi-junction heterostructures. Zheng et al. [12] developed the WS_2 to $WS_{2(1-x)}Se_{2x}$ alloy formation by controlling the Ar gas flow. Herein, the product x defines the ratio of the mixed WS_2/WSe_2 powders. Even the researchers tried WM_2 and MoX_2 mixed powders in the same boat to develop the multi-junction heterostructures. The multi-step CVD method has flexible features and is more controllable than the one-step and two-step CVD methods. Most of the optoelectronic devices approach the multi-step CVD method because it can separate the hole-electron pairs into different materials.

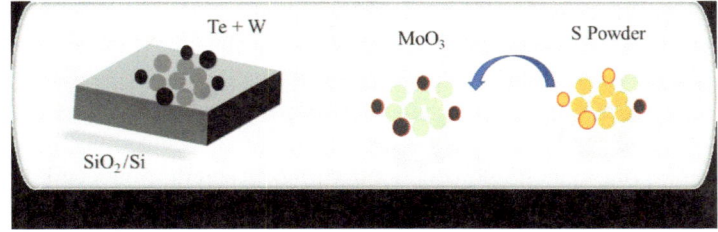

FIGURE 5.2 Schematic diagram shows the Chemical Vapor Deposit (CVD) growth.

5.3 PROPERTIES OF THE 2D MATERIALS

5.3.1 2D MATERIALS BAND ALIGNMENT

Band alignment is the basic property of two-dimensional heterostructures. In general, the behavior of the charge transportation and illumination are obtained from the band alignment. In the theory, the first principles are used to calculate the band energy and for measurement we use μ-XPS. The Perdew-Burke-Ernzerhof (PBE) with spin-orbit coupling (SOC) is used to calculate the band alignments of monolayer SnS_2 and monolayer semiconducting TMDs. The inference of the obtained result is maximum energy of the valance band and the conduction band is minimum for the monolayer TMDs in MX_2. Two dissimilar materials with sizable band gaps are grouped to form a new heterostructure at the interface without caring about the band bending. This introduces three different types of band alignments (I–III) known as the straddling gap, staggering gap, and broken gap. The straddling gap used in the luminescent devices allows quick recombination of holes and electrons. The staggering gap separates the holes and electrons and enhances the lifetime of interlayer excitations. It is highly recommended for optoelectronic devices. The broken gap heterostructures support band-to-band tunneling and empower the operation of the tunnel field-effect transistor. The electron and hole transportations are much faster compared to the type II heterostructure. This peculiar behavior of the type III-based heterostructures are more suitable for photovoltaic cells.

The more novel properties are obtained from the band structures at the interface while considering the band bending. The type I band alignment of two semiconductors form a PN junction with the valance band and the conduction band is bending together in the opposite direction.

The quality of the interface should be considered because many factors affect the band structure of a two-dimensional heterostructure. The impurities of the materials and defects at the interface cause impurities or sometimes defects in the energy level.

5.3.2 CHARGE TRANSPORTATION PROPERTY IN 2D HETEROSTRUCTURES

The various charge transport characteristics are obtained during the energy transfer process due to different band alignments. From one perspective, single-molecule transportation, including interlayer tunneling impact and charge catching wonders, has been generally announced and novel devices dependent on those transportation instruments have been created that display low energy-suspicion with the elite. Moreover, the many-body transport and the partition of electron-hole sets in two-dimensional heterostructures have likewise been hypothetically anticipated and exhibited, which empower the possibility of expansive applications in the optoelectronic field.

5.3.2.1 Transport of Single-Particle

The very small channel length and molecularly sharp interfaces can understand the band-to-band burrowing impact by electrostatic gating to keep away from the hardship of band-edge sharpness that came about because of synthetic doping [13]. The conduction band is minimum of WSe_2 and is quite larger than the valance band maximum of $SnSe_2$ when applying a small voltage Vg [14]. The electrons in $SnSe_2$ can't burrow into WSe_2, yet the dispersion of charges happens at the interface, comparing to channel current and high opposition state, individually, at the point when applied with a positive and negative bias voltage. The conduction band which is minimum of WSe_2 is set below the valance band maximum of $SnSe_2$ when increasing the voltage Vg. The layer with large bandgap results in spatial confinement of electron-hole pairs in the type I band alignment [15]. When contemplating the band bending, the interfacial charge catching can be identified in the PN intersection of the type II band arrangement. Band bending occurs between the conduction band minimum of WSe_2 and the lowest unoccupied molecular present in the orbital of p-MSB. The Fermi level of WSe_2 shifts downward. This will enhance the interfacial charge trapping process and interfacial energy barrier. Another approach to get charge catching is to put a floating gate layer, which is typically sandwiched with the burrowing and obstructing dielectric layers between the channel and control gate as a charge catching layer. The behavior of charge transportation occurs in the floating gate structure with graphene acting as the trapping layer and monolayer MoS_2 acting as the channel.

5.3.2.2 Generation of Interlayer Excitations in 2D Heterostructures

The conduction band minimum and valance band maximum exist in different layers of type II band alignments. The electrons and holes are considered spatially seared in the type II band alignment. The pump-probe technique is used to detect the superfast separation of electron-hole pairs [16]. The resonance excitation is induced in the MoS_2/WS_2 and isolated MoS_2 monolayer. The rising time of the signal obtained from the experiment is less than 50 fs. The generation of interlayer excitations occurs at 630 nm and 875 nm [17]. The hot excitons then patch up and spread the excess energy to shape the solidly bound interlayer excitons inside 800 fs. It addresses the presence of the last solidly bound interlayer excitons that could be particularly appeared in photoluminescence spectra. The photoluminescence signal was taken from number 1 and 2 areas contrasting with the A-exciton resonances in the monolayer MoS_2. However, the signs from the heterostructure territories showed exciton resonances in MoS_2 besides WSe_2, inciting an additional interlayer exciton top. The exciton photoluminescence signs of MoS_2, likewise WSe_2, were doused in the heterostructure due to the charge move measure.

5.3.3 THE PROPERTIES OF EXCITONS IN 2D HETEROSTRUCTURES

The interlayer excitons grip enormous restricting energy and the delayed lifetime [18] which is emphatically identified in vertical heterostructures with two layers' distance. The QEH (quantum electrostatic heterostructure) model dependent on the MoS_2/x-hBN/WSe_2 vertical heterostructures was built by Simone Latini et al. [19]. At room temperature, the interlayer excitons binding energy is determined up to 0.3 eV and diminishes with the increasing boundary x, demonstrating a steady presence of the interlayer excitation. In any case, the force backhanded nature and the oscillator with weak strength will be challenging to observe the interlayer excitons by resounding optical excitation.

The spectra of photoluminescence and electroluminescence for horizontal p–n intersections dependent on a vertical $MoSe_2$–WSe_2 hetero-bilayer. The electrode contacts playing a major role for the career injection and recombination at the hetero-bilayer edge at forward voltage (V_{sd}). The amplitude of the photocurrent at the resonant excitation of the intralayer exciton is around 200 times greater than the interlayer exciton.

Trions exist in addition to the interlayer and intralayer excitons in two-dimensional heterostructures. There is a presence of bound interlayer trions positioned below the neutral interlayer [20]. The binding energy for the positive and negative interlayer trions is found to be 18 and 28 meV while electrons and holes are situated on the similar layer.

5.3.4 MAGNETIC PROPERTIES IN 2D HETEROSTRUCTURES

The properties of low-dimensional magnetic materials like saturation magnetization, curie temperature, coercivity, and other parameters are subjected to grain size and the many layers are also available. Hence the two-dimensional materials have good magnetic properties compared to bulk form. The utilization of two-dimensional magnetic materials is an extraordinary significance for the investigation of electromagnetics, valleytronics, and spintronics in heterostructures [21]. The property of magnetism is helpful to the propagation of control and spin of the valley current at room temperature. A greater number of experimental and theoretical results show that the assembly of two-dimensional materials into a van der Waals heterostructure with ferromagnetic materials [22]. The magnetic trade field can intensify the impact of the outer magnetic field, which starts from the nearness impacts of the heterojunction [23].

5.4   APPLICATION IN DEVICES

5.4.1   ELECTRONIC DEVICES

Because of their automatically thin channels, 2D materials have been widely investigated as channel materials used for future electronic device applications. To improve the control in and resistance as well as the loss of band-edge sharpness, the 2D heterostructures band structures are doped with chemicals. Band engineering of the 2D heterostructure received a lot of attention for fabricating many electronic devices. TMDs with the loss of band-edge sharpness, as well as wide bandgap insulator thin layers of hBN, are commonly used as the dielectric layer in TFETs. Pinhole, interlayer defects, and oxygen doping can be avoided to replace conventional metallic oxides by 2D materials. The TFET device based on the $SnSe_2$/WSe_2 vertical heterostructure (Figure 5.3) was fabricated with a subthreshold swing and an ultrahigh ON/OFF ratio, and results show high performance, simply by tuning the back-gate voltage to shift the BTBT effect [14].

Likewise, memories dependent on the floating-gate structure are essential 2D heterostructure applications where the floating-gate layer must be chosen carefully. Gold nanoparticles (AuNPs) were considered for the trapping layer. The dark current in the system channel were suppressed by the trapping layer. Researchers recently developed a programmable memory system using a vertically stacked MoS_2/hBN/graphene heterostructure as the floating-gate material which has an extremely high on/off ratio with high stretchability (> 19%). The on/off ratio has been increased to over 109 by using an hBN dielectric layer of acceptable thickness. The memory cycles as programming (1), reading (2), erasing (3), and reading (4) operations, which are realized by repeated voltage pulses. The hysteresis activity in the I_{ds}-V_d plot can also be used to clarify certain functions and are the product of the hBN tunneling effect and the asymmetric potential decrease caused by MoS_2's high resistivity. Furthermore, Si et al. [24] built a nonvolatile memory system which reflects in transfer characteristics with a stable ferroelectric hysteresis loop by integrating a 2D ferroelectric insulator of $CuInP_2S_6$ on top of MoS_2.

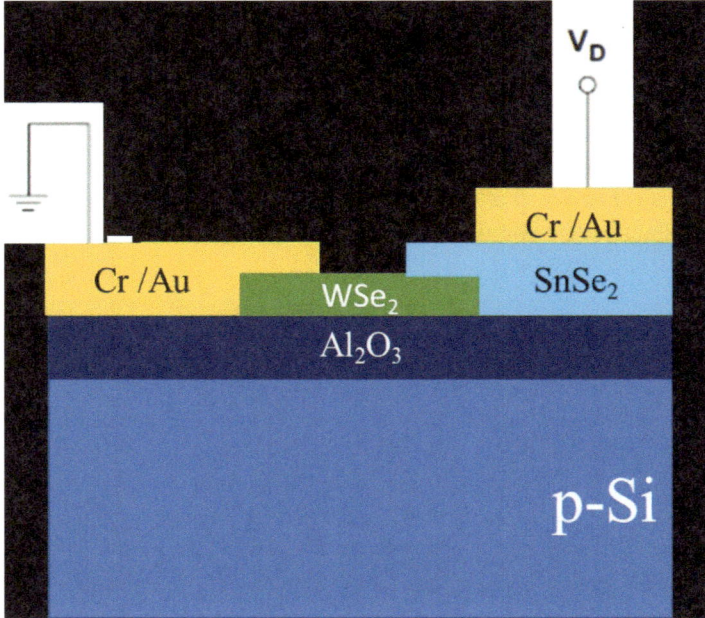

FIGURE 5.3   Electronic devices based on 2D heterostructure.

5.4.2 Optoelectronic Devices

The unique properties of 2D materials include its having good light-matter interaction, excellent flexibility, and broad response spectrum range. 2D materials are interesting applicants for optoelectronic devices [25]. 2D heterostructures have also been widely used in light source devices, photovoltaics, including optoelectronics and photodetectors due to the production of flexible band engineering and interlayer excitons. Many studies have shown that 2D heterostructures are used for photodetectors in a spectrum ranging from ultraviolet (UV) to near-infrared (NIR). Due to the production of a robust built-in electrical field in such photodetectors, atomically thin structures based on 2D heterostructures perform better. Researchers are likewise proceeding to explore a few different ways to regulate gadgets and multi-designs to upgrade framework execution. Since graphene, as a 2D material with ultrahigh transporter versatility, can diminish horizontal dispersion in the semiconductor, it has been exhibited that utilizing graphene as the contact terminals is a successful method for essentially speeding up a molecularly meager photodetector [26]. Tan et al. [27] as of late created WS_2/MoS_2 hetero-bilayer gadgets utilizing layered graphene anodes. The SBH at the graphene interface and the TMD layer can be changed by tuning the work capacity of graphene. The photogate impact brings down the boundary, making it simpler to re-infuse electrons into the TMD tube. With a normal explicit detectivity of 41011 Jones, these gadgets had the most noteworthy photoresponsivity of up to 2340 A W1 and a solid inner photoconductive increase of over 3.7104. Photodetectors made of MoS_2–WS_2 planar heterostructures have additionally been expressed to have a detection of 4.361013 Jones [28]. Since the shaped electron-opening sets can isolate underworked in potential at zero inclination, p–n heterostructures are normally self-fueled. Because of the challenges in dealing with the stacking direction, the manufacture of in-plane intersections is considerably more controllable and adaptable than that of vertical heterostructures. The specialists had the option to plan a bent picture sensor exhibit dependent on a MoS_2–graphene heterostructure [29] that can be utilized as a natural eye-roused delicate implantable optoelectronic framework for recognizing optical signs utilizing customized electrical incitement of optic nerves because of the remarkable adaptability of heterostructures. Besides, 2D heterostructure-based optoelectronic recollections that can collect and deliver photograph-created transporters under an electric field and light illumination have been created. Moreover, the utilization of 2D materials considers the advancement of little, adaptable, and low-energy optoelectronic capacity [29]. The gadget can be modified to peruse and delete by changing the door voltage and the light pulse.

Also turning on the lamp and removing the negative gate, positive charges can be stored in hBN. The computer has over 128 (7 bit) storage states and has a retention time of over 4.5104s. Interestingly, at different wavelengths, the storage states are highly distinct, suggesting that the WSe_2/hBN optoelectronic memory has excellent wavelength discriminating capability. Besides, the atomically thin crystals reduced Coulomb screening which resulted in a drastic increase in exciton binding energy which stabilized the excitons at room temperature. In the heterostructure, the interlayer exciton was excited, and recombination was positively observed at the heterostructure's edge. The computer was turned on and off the Vg is set at 1 to a ratio of 100 on/off. Besides, by modulating the bias voltage in which the excitons will drift to a low potential named as forward bias voltage or be confined in the potential hydrazine considered as reverse bias voltage, this allows the exciton diffusion distance to be changed to 5 m under a controlled forward bias voltage.

5.4.3 Spintronic and Valleytronic Devices

2D materials with great spin valley properties, for example, graphene, have incredible thermal, electrical, and mechanical properties. It likewise has a long spin dispersion term up to room temperature, considering turn infusion, control, and location in a solitary coordinated framework, bringing about versatile and ultrafast nonvolatile rationale circuits with ultralow energy scattering. The spin valve impact has as of late been seen in FM/2D-material/FM sandwich-like attractive intersections.

The grid confusion among graphene and FM is especially little in certain FM/G/FM intersections (FM = Ni, Co, and so forth), and a spin sifting impact is hypothetically conceivable. Because of the distinctive spin polarization at various interfaces, the presentation of 2D heterostructures in attractive burrowing intersections will effectively regulate the aftereffect of the spin channel. Under room temperature, Iqbal et al. [30] tracked down a negative passage magnetoresistance (TMR) of 0.85% in a NiFe/G–hBN/Co attractive intersection. The consistency of the interfaces and the grid direction between various layers, then again, are critical. Moreover, monolayer TMDs with broken reversal evenness have two inequivalent valleys that are associated with time-inversion balance. The particular physical science in TMDs, particularly the coupled twist and valley of opportunity, are because of this property and the solid twist circle coupling. TMDs with a direct bandgap empower to energize transporters specifically inside a valley utilizing circularly spellbound light and a particular valley pseudospin. Valley Lobby impacts have additionally been found in a doped TMD study. Nonetheless, outside control of valleytronic gadgets stays a test because the conditions for lifting the valley decadence in a solitary 2D material by Zeeman parting are amazingly difficult, requiring a solid attractive field and a low temperature. Albeit underlying designing procedures, for example, doping and absconds [31] can be utilized to tweak graphene attractive properties; they at last add to the intricacy of tasks. As of late, it was found that utilizing the interfacial attractive trade field (MEF) from a ferromagnetic substrate significantly improved valley parting in monolayer TMDs. The significant distance turn transport capacity of graphene has been exhibited at room temperature for graphene and TMDs in spintronics [32], yet the absence of the SOC made creating unadulterated spin current troublesome.

In monolayer TMDs like WS_2 and MoS_2, on the other hand, an external excitation field is used in spin/valley polarization. At room temperature, lateral spin or valley transport has recently been achieved using MoS_2/few-layer graphene hybrid spin valves. An external excitation field, on the other hand, can efficiently generate spin/valley polarization in monolayer TMDs like WS_2 and MoS_2. At room temperature lateral spin or valley transport has recently been achieved with the help of MoS_2/few-layer graphene hybrid spin valves.

5.5 APPLICATION OF 2D MATERIAL IN BIOFIELD

5.5.1 NANOMATERIAL FOR THE DETECTION OF DNA HYBRIDIZATION

In 2D material, graphene plays a vital role in sensor applications. Especially, monolayer graphene materials are more attractive for biosensor applications. The transfer characteristics of those devices illustrate the detection of DNA hybridization [33]. In transfer characteristics, the carrier concentration or V_{Dirac} of the device was determined with different treatments such as single-strand DNA (ssDNA), double-strand DNA (dsDNA), and DNA hybridization. Remarkably, the Dirac point shifting in transfer characteristics for ssDNA and complementary DNA towards the left side specifies an n-doping effect. Similarly, the shifting was analyzed with non-complementary DNA and it was observed that no shifting occurs during this process. Hence, this practical variation observed in the transfer characteristics between the probe DNA with the target DNA, and the probe DNA with the non-complementary DNA discloses that the graphene material was more sensitive in the detection of DNA nucleotides. Further development of these 2D-based sensor devices will explore future biosensor applications, particularly, in the detection of genetic diseases.

5.5.2 NANOMATERIAL FOR THE DETECTION OF pH

The 2D material graphene/MoS_2 is also suitable for measuring the stability of the device for the pH from acidic to basic with the addition of ions [34]. Herein, the pH detection was observed based on the shifting of Dirac point for each difference in carrier concentration of the device. Whenever the doping level of the pH ions was increased that would be reflected in the shifting of Dirac point for various pH values from acidic to basic. The electric response of the 2D materials is more attractive and gives a good platform for the enzyme-sensing-based devices for direct observation of the enzyme on the surface of the 2D material (graphene). The electrical response gained from the overall performance of the sensor devices exhibited a good sensitivity. The 2D material-based platform offers a hopeful route to fabricate high-performance sensor devices for biosensing applications.

5.5.3 NANOMATERIAL FOR PHOTOTHERMAL CANCER THERAPY

In an advanced biosensing platform, nanostructures play a major role especially in cancer therapy. Molybdenum disulphide nanostructures have been found more useful and interesting in the detection of cancer by photothermal therapy. Preparation of one spot synthesis of MoS_2 flakes by hydrothermal method influenced degradation properties for many critical biomedical applications (Figure 5.4). Decoration of MoS_2 nanostructures with polyethylene glycol has superior stability with exceptional photothermal properties. Hence, it is more suitable for photothermal cancer therapy. The degradation rates were observed for MoS_2–PPEG in different conditions [11]. Interestingly, MoS_2–PPEG degradation was slower in acidic conditions and hence it is suitable for tumor environment. Byproduct of MoS_2–PPEG is favorable for *in vitro* bio-compatibility and it was established with hemolysis and cytotoxicity studies. It was also suggested for *in vivo* destruction of tumor growth [35]. Table 5.1 shows the various diagnostic methods based on 2D nanomaterials for biosensor applications.

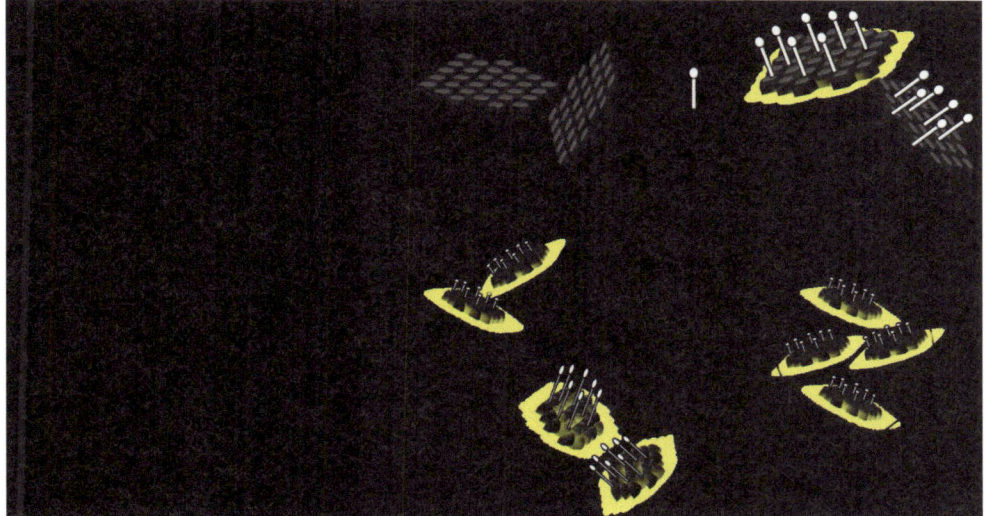

FIGURE 5.4   Schematic diagram illustrates the preparation of MoS_2-PPEG for photothermal cancer therapy.

TABLE 5.1
Biosensing Applications of 2D Materials.

Diagnostic	2D Nanomaterial	Functionalization	Target	References
Photothermal Therapy	MoS_2 nanosheets	-	HeLa cells	[36]
Photothermal Therapy	MoS_2 nanosheets	PEG	4TI cells and tumors	[37]
Drug Delivery and Photothermal Therapy	MoS_2 nanosheets	chitosan	KB-Panc-1 cells	[38]
Drug Delivery	MoS_2/GO	-	4TI cells and tumors	[39]
Sensing Biomolecules	MoS_2/Graphene	FET device	DNA, pH	[33], [40]
	MoS_2 nanosheets	Microfluidic biosensor	DNA	[41]

5.6 CONCLUSION

In the last decade, research in 2D materials has entered the mature era. However, the heterostructures of 2D materials has been triggering a new wave of research in bio applications. In this chapter, we have presented a complete overview of the 2D materials, such as TMDs, graphene, and MoS_2; heterostructures of 2D materials; properties of 2D materials; and energy excitations. We have included the process of 2D material using different fabrication techniques. The recent advances and challenges of these materials for various applications like DNA hybridization, detection of pH and in cancer therapy have also been highlighted. Moreover, how the biomaterials decorated on a graphene/MoS_2/WS_2 sheet and sensed by the transfer characteristics to explore the ambipolar behavior of the sensor devices have been deliberated in this chapter. The major properties of the 2D materials have also been highlighted. The electrical property and magnetic property of the 2D materials were attracted for implementation in electronic devices and advanced biosensor applications. Fabrication of compact 2D material-based devices enhances effective sensing and user-friendliness in the platform of biosensing applications. To develop the existing 2D materials in the practical bio-applications like cancer therapy, cancer diagnosis, and drug delivery, advanced feasible methods and technology are essential to be further investigated.

REFERENCES

[1]   K.S. Novoselov, A.K. Geim, S.V. Morozov, D. Jiang, Y. Zhang, S.V. Dubonos, I.V. Grigorieva, A.A. Firsov, Electric field effect in atomically thin carbon films. *Science* 306(5696), 666–669 (2004). https://doi.org/10.1126/science.1102896

[2]   M. Osada, T. Sasaki, Two-dimensional dielectric nanosheets: Novel nanoelectronics from nanocrystal building blocks. *Adv. Mater.* 24(2), 210–228 (2012). https://doi.org/10.1002/adma.201103241

[3]   L.M. Malard, J. Nilsson, D.C. Elias, J.C. Brant, F. Plentz et al., Probing the electronic structure of bilayer graphene by Raman scattering. *Phys. Rev. B* 76(20), 201401 (2007). https://doi.org/10.1103/PhysRevB.76.201401

[4]   L. Xie, M.Z. Liao, S.P. Wang, H. Yu, L.J. Du et al., Graphene-contacted ultrashort channel monolayer MoS_2 transistors. *Adv. Mater.* 29(37), 1702522 (2017). https://doi.org/10.1002/adma.201702522

[5]   W.J. Jie, Z.B. Yang, G.X. Bai, J.H. Hao, Luminescence in 2D materials and van der Waals heterostructures. *Adv. Opt. Mater.* 6(10), 1701296 (2018). https://doi.org/10.1002/adom.201701296

[6]   J.I.J. Wang, Y.F. Yang, Y.A. Chen, K. Watanabe, T. Tanigu-chi, H.O.H. Churchill, P. Jarillo-Herrero, Electronic transport of encapsulated graphene and WSe_2 devices fabricated by pick-up of prepatterned hBN. *Nano Lett.* 15(3), 1898–1903 (2015). https://doi.org/10.1021/nl504750f

[7]   C.R. Dean, A.F. Young, I. Meric, C. Lee, L. Wang et al., Boron nitride substrates for high-quality graphene electronics. *Nat. Nanotechnol.* 5(10), 722–726 (2010). https://doi.org/10.1038/nnano.2010.172

[8]   A. Castellanos-Gomez, M. Buscema, R. Molenaar, V. Singh, L. Janssen, H.S.J. van der Zant, G.A. Steele, Deterministic transfer of two-dimensional materials by all-dry viscoelastic stamping. *2D Mater.* 1(1), 011002 (2014).

[9]   P.J. Zomer, S.P. Dash, N. Tombros, B.J. van Wees, A transfer technique for high mobility graphene devices on commercially available hexagonal boron nitride. *Appl. Phys. Lett.* 99(23), 232104 (2011). https://doi.org/10.1063/1.3665405

[10]   Y.J. Gong, J.H. Lin, X.L. Wang, G. Shi, S.D. Lei et al., Vertical and in-plane heterostructures from WS_2/MoS_2 monolayers. *Nat. Mater.* 13(12), 1135–1142 (2014). https://doi.org/10.1038/Nmat4091

[11]   X.F. Li, M.W. Lin, J.H. Lin, B. Huang, A.A. Puretzky et al., Two-dimensional GaSe/$MoSe_2$ misfit bilayer heterojunctions by van der Waals epitaxy. *Sci. Adv.* 2(4), 1501882 (2016). https://doi.org/10.1126/sciadv.1501882

[12]   B.Y. Zheng, C. Ma, D. Li, J.Y. Lan, Z. Zhang et al., Band alignment engineering in two-dimensional lateral heterostructures. *J. Am. Chem. Soc.* 140(36), 11193–11197 (2018). https://doi.org/10.1021/jacs.8b07401

[13]   M. Li, D. Esseni, G. Snider, D. Jena, H.G. Xing, Single particle transport in two-dimensional heterojunction inter-layer tunneling field effect transistor. *J. Appl. Phys.* 115(7), 074508 (2014). https://doi.org/10.1063/1.4866076

[14] X. Yan, C.S. Liu, C. Li, W.Z. Bao, S.J. Ding, D.W. Zhang, P. Zhou, Tunable $SnSe_2/WSe_2$ heterostructure tunneling field effect transistor. *Small* 13(34), 1701478 (2017). https://doi.org/10.1002/smll.201701478

[15] T. Yamaoka, H.E. Lim, S. Koirala, X.F. Wang, K. Shinokita et al., Efficient photocarrier transfer and effective photo-luminescence enhancement in type i monolayer $MoTe_2/WSe_2$ heterostructure. *Adv. Funct. Mater.* 28(35), 1801021 (2018). https://doi.org/10.1002/adfm.201801021

[16] X.P. Hong, J. Kim, S.F. Shi, Y. Zhang, C.H. Jin et al., Ultra-fast charge transfer in atomically thin MoS_2/WS_2 hetero-structures. *Nat. Nanotechnol.* 9(9), 682–686 (2014). https://doi.org/10.1038/Nnano.2014.167

[17] Y.J. Gong, J.H. Lin, X.L. Wang, G. Shi, S.D. Lei et al., Vertical and in-plane heterostructures from WS_2/MoS_2 monolayers. *Nat. Mater.* 13(12), 1135–1142 (2014). https://doi.org/10.1038/Nmat4091

[18] P. Rivera, J.R. Schaibley, A.M. Jones, J.S. Ross, S.F. Wu et al., Observation of long-lived interlayer excitons in monolayer $MoSe_2$–WSe_2 heterostructures. *Nat. Commun.* 6, 6242 (2015). https://doi.org/10.1038/ncomms7242

[19] S. Latini, K.T. Winther, T. Olsen, K.S. Thygesen, Interlayer excitons and band alignment in $MoS_2/hBN/WSe_2$ van der Waals heterostructures. *Nano Lett.* 17(2), 938–945 (2017). https://doi.org/10.1021/acs.nanolett.6b04275

[20] T. Deilmann, K.S. Thygesen, Interlayer trions in the MoS_2/WS_2 van der Waals heterostructure. *Nano Lett.* 18(2), 1460–1465 (2018). https://doi.org/10.1021/acs.nanolett.7b05224

[21] Y.P. Liu, W.S. Lew, L. Sun, Enhanced weak localization effect in few-layer graphene. *Phys. Chem. Chem. Phys.* 13(45), 20208–20214 (2011). https://doi.org/10.1039/c1cp22250c

[22] L.Y. Ping, G. Sarjoosing, M. Chandrasekhar, L.W. Siang, W.S.J.A.N. Kai, Effect of magnetic field on the electronic transport in trilayer graphene. *ACS Nano* 4(12), 7087–7092 (2010). https://doi.org/10.1021/nn101296x

[23] H.X. Yang, A. Hallal, D. Terrade, X. Waintal, S. Roche, M. Chshiev, Proximity effects induced in graphene by magnetic insulators: First-principles calculations on spin filtering and exchange-splitting gaps. *Phys. Rev. Lett.* 110(4), 046603 (2013). https://doi.org/10.1103/PhysRevLett.110.046603

[24] M.W. Si, P.Y. Liao, G. Qiu, Y.Q. Duan, P.D.D. Ye, Ferro-electric field-effect transistors based on MoS_2 and $CuInP_2S_6$ two-dimensional van der Waals heterostructure. *ACS Nano* 12(7), 6700–6705 (2018). https://doi.org/10.1021/acsnano.8b01810

[25] Y.P. Liu, K. Tom, X.W. Zhang, S. Lou, Y. Liu, J. Yao, Alloying effect on bright: Dark exciton states in ternary monolayer $Mo_xW_{1-x}Se_2$. *New J. Phys.* 19, 073018 (2017). https://doi.org/10.1088/13672630/aa6d39

[26] W.G. Luo, Y.F. Cao, P.G. Hu, K.M. Cai, Q. Feng et al., Gate tuning of high-performance InSe-based photodetectors using graphene electrodes. *Adv. Opt. Mater.* 3(10), 1418–1423(2015). https://doi.org/10.1002/adom.201500190

[27] H.J. Tan, W.S. Xu, Y.W. Sheng, C.S. Lau, Y. Fan et al., Lateral graphene-contacted vertically stacked WS_2/MoS_2 hybrid photodetectors with large gain. *Adv. Mater.* 29(46), 1702917(2017). https://doi.org/10.1002/adma.201702917

[28] W.H. Wu, Q. Zhang, X. Zhou, L. Li, J.W. Su, F.K. Wang, T.Y. Zhai, Self-powered photovoltaic photodetector established on lateral monolayer MoS_2–WS_2 heterostructures. *Nano Energy* 51, 45–53 (2018). https://doi.org/10.1016/j.nanoen.2018.06.049

[29] C. Choi, M.K. Choi, S.Y. Liu, M.S. Kim, O.K. Park et al., Human eye-inspired soft optoelectronic device using high-density MoS_2: Graphene curved image sensor array. *Nat. Commun.* 8, 1664 (2017). https://doi.org/10.1038/s41467-017-01824-6

[30] M.Z. Iqbal, S. Siddique, G. Hussain, M.W. Iqbal, Room temperature spin valve effect in the NiFe/Gr–hBN/Co magnetic tunnel junction. *J. Mater. Chem. C* 4(37), 8711–8715 (2016). https://doi.org/10.1039/C6TC03425J

[31] L. Cai, J.F. He, Q.H. Liu, T. Yao, L. Chen et al., Vacancy induced ferromagnetism of MoS_2 Nanosheets. *J. Am. Chem. Soc.* 137(7), 2622–2627 (2015). https://doi.org/10.1021/ja5120908

[32] Y.P. Liu, H. Idzuchi, Y. Fukuma, O. Rousseau, Y. Otani, W.S. Lew, Spin injection properties in trilayer graphene lateral spin valves. *Appl. Phys. Lett.* 102(3), 033105 (2013). https://doi.org/10.1063/1.4776699.

[33] A.K. Manoharan, S. Chinnathambi, R. Jayavel, N. Hanagata, Simplified detection of the hybridized DNA using a graphene field effect transistor. *Sci. Technol. Adv. Mater.* 18, 43–50 (2017).

[34] M.H. Lee, B.J. Kim, K.H. Lee, I.S. Shin, W. Huh, J.H. Cho, M.S. Kang, Apparent pH sensitivity of solution-gated graphene transistors. *Nanoscale* 7, 7540–7544 (2015).

[35] L. Chen, Y. Feng, X. Zhou, Q. Zhang, W. Nie, W. Wang, Y. Zhang, C. He, One-pot synthesis of MoS_2 nanoflakes with desirable degradability for photothermal cancer therapy. *ACS Appl. Mater. Interfaces* 9, 17347–17358 (2017).

[36] S.S. Chou, B. Kaehr, J. Kim, B.M. Foley, M. De, P.E. Hopkins, J. Huang, C.J. Brinker, V.P. Dravid, Chemically exfoliated MoS2 as near-infrared photothermal agents. *Angew. Chem.* 125, 4254 (2013).

[37] S. Wang, K. Li, Y. Chen, H. Chen, M. Ma, J. Feng, Q. Zhao, J. Shi, Biocompatible PEGylated MoS2 nanosheets: Controllable bottom-up synthesis and highly efficient photothermal regression of tumor. *Biomaterials* 39, 206–217 (2015).

[38] W. Yin, L. Yan, J. Yu, G. Tian, L. Zhou, X. Zheng, X. Zhang, Y. Yong, J. Li, Z. Gu, High-throughput synthesis of single-layer MoS2 nanosheets as a near-infrared photothermal-triggered drug delivery for effective cancer therapy. *ACS Nano* 8, 6922–6933 (2014).

[39] Y. Liu, J. Peng, S. Wang, M. Xu, M. Gao, T. Xia, J. Weng, A. Xu, S. Liu, Molybdenum disulfide/graphene oxide nanocomposites show favorable lung targeting and enhanced drug loading/tumor-killing efficacy with improved biocompatibility. *NPG Asia Mater.* 10, e458 (2018).

[40] D. Sarkar, W. Liu, X. Xie, A.C. Anselmo, S. Mitragotri, K. Banerjee, MoS_2 field-effect transistor for next-generation label-free biosensors. *ACS Nano* 8, 3992–4003 (2014).

[41] Y. Huang, Y. Shi, H.Y. Yang, Y. Ai, A novel single-layered MoS_2 nanosheet based microfluidic biosensor for ultrasensitive detection of DNA. *Nanoscale* 7, 2245–2249 (2015).

6 The Manufacturing of Magnesium Degradable Biomedical Implants

Lifei Wang, Pengbin Lu, Qiang Zhang, Liangliang Xue, Xiaohuan Pan, Hua Chai, Srinivasan Arthanari and Maurizio Vedani

CONTENTS

6.1 INTRODUCTION

Metallic materials have attracted much attention as bio-devices such as osteons, intravascular stents, bone screws, etc. due to their good comprehensive performance, biocompatibility, and good corrosion resistance. In general, the 316L stainless steel, NiTi alloys, CoCr alloys, tantalum alloys, and other implants will be kept in the body for a long time. [1–3] Moreover, a potential second surgery is needed to remove the implants; as a result more cost and pain will be brought on patients. Degradable implants have advantages to overcome these problems. For example, bioabsorbable polymers are put into use in various surgeries widely. However, their mechanical properties are poor, and an acidic pH environment promotes infection, as well as the hydrolysis behavior generating formation of osteoclasts. [4] Magnesium-based alloys are also considered as degradable biomaterials because of their excellent mechanical property and biodegradability. The density and Young's modulus are close to natural bone, which removes the elastic mismatches. [5] The strength of traditional metal implants is usually much higher. For example, Young's modulus of the cobalt-chromium alloy is 100~200 GPa, while the Young's modulus of cancellous bone is only 60 GPa, which will cause serious stress shielding. Hard metal implants deprive the normal stress of bone growth, and bones may be severely damaged during the process of healing and growth. [6] Among the current metal implants, Young's modulus of Mg alloy is the closest to cortical bone (magnesium is 40–45 GPa and 10–27 GPa for cortical bone), while Young's modulus of 316L stainless steel and titanium-based alloys are as high as 193 and 110 GPa, respectively. [3, 7] Besides, Mg ions are necessary for the human body and participate in metabolism. [8] The first-time use of Mg alloy as a biomedical device can date from 1878. Huse used Mg wire ligatures to prevent bleeding vessels successfully. [9] Then due to the poor mechanical properties and higher degradation rate at that time, the application of Mg-based implants was stopped. However, it has been attracting attention again since

DOI: 10.1201/9781003173533-6

the year 2000 owing to the development of science and technology, [10] and new Mg-based alloys with high mechanical and degradation performance were proposed. Further, methods to modify the degradation rate and mechanical property such as micro-alloying, severer plastic deformation (SPD), texture control, and surface engineering, etc. have been developed.

Several research works conducted on magnesium alloys revealed that purity is an important factor to enhance the degradation rate. The high purity Mg generally exhibits a better corrosion resistance. Wang et al. [11] reported that the purity of 99.98% or 3N8 (mass fraction) magnesium was achieved by vacuum distillation where 96% Mg was as an initial material. Lam et al. [12] indicated that the ultra-high purity materials could be produced by the vacuum distillation method to increase purity by approximately 500 times in a single step. The Mg purity was increased from 99.95% to greater than 99.9999%. Besides, new Mg alloys are also developed by adding alloying elements. Compared with high purity Mg, any of the alloying elements in pure or intermetallic form are nobler. [13] As alloying elements, Al could promote both solid solution and precipitation strengthening. [14] However, it is poisonous and harmful to the nerve during degradation so its being added is usually avoided. Mn is mainly used to modify ductility. [15] In smaller amounts, Zn is used to improve the strength due to solid solution strengthening. [16] Calcium can be also contributed to a similar effect. [17] Moreover, the double peak non-basal texture will be generated after Ca addition which is beneficial for the improvement of ductility. [18] For rare earth elements (RE), some RE with a large solid solubility like Y, Gd, Er may be kept as a solid solution, so that the solid solution effect will play an important role. While RE with limited solubility in Mg such as Nd, La, Ce, Sm, the intermetallic phases will be generated early at the grain boundaries, thus, the precipitation strengthening has a significant effect to improve the mechanical property. [19–21] For example, Gui et al. [22] developed the extruded Mg-3.0Gd-2.7Zn-0.4Zr-0.1Mn (wt. %) alloys whose yield strength (YS) and fraction elongation (FE) could reach 315.2 MPa and 21.3%, respectively. The corrosion rate in Hank's solution at 37 ± 0.2 °C was less than 0.5 mm/year. Zhang et al. [23] indicated that the YS and FE of extruded Mg-3.08Nd-0.27Zn-0.46Zr (JDBM) could be 360 MPa and 14.6%. Also, the corrosion rate is much lower. Besides, various elements have a significant effect on health if the ions come into human body, which should be considered. Like Pr element adding, serious damage may be induced to human lives. [24] Sr can increase the activity of osteoblasts and promote bone formation significantly. [25] Ca can promote bone growth. [17] Zr can also improve the degradation resistance of Mg alloys. [26] It has shown that Mg in the human body is mainly distributed in bones. The presence of Mg can enhance the adhesion of osteoblasts and promote optimal osteogenesis. [26] Besides, Mg supplementation increases bone density and strength significantly as well. [27] After a Mg alloy implant enters the human body, hydrogen bubbles are found in one week, and the hydrogen is self-absorbed, then disappears. [28] Previous studies have shown that the hydrogen generated during the degradation of Mg alloy implants is acceptable to the human body. However, complications during bone healing will occur if the hydrogen is too much. Therefore, it is necessary to control the degradation rate of the Mg alloy to reduce the risk of gas accumulation. [29] Except as described earlier, a large number of similar works are reported in previous literature. Therefore, this chapter introduces the manufacturing processes of various degradable Mg biomedical implants in detail. Besides, the affecting factors and future trends are also analyzed.

6.2 MANUFACTURING OF MG ALLOYS USED AS DEGRADABLE IMPLANTS

6.2.1 FORMING OF CARDIOVASCULAR STENTS

Implanting vascular stents is one of the most effective ways to cure coronary heart disease. [30, 31] To support the narrowed blood vessel, the stent is generally expressed as a cylindrical mesh structure with a diameter of < 2 mm. [32, 33] In the past ten years, metal biodegradable stents have been studied widely. [34] Especially Mg alloys, as Mg ions participate in the metabolism of the human body and promote the cure of blood vessels. [26, 35, 36] However, due to its close-packed hexagonal crystal structure (HCP), Mg alloys express poor plasticity and formability. Thus, the successful fabrication of the stents from Mg alloys is one of the most significant aspects. This part summarizes the processing technologies for Mg vascular stents, mainly focusing on extrusion, drawing, laser cutting, etc. Besides, the factors affecting the fabrication methods are discussed.

6.2.1.1   Raw Thin-Wall Tube Extrusion

Extrusion is one of the most important processing techniques to produce Mg vascular stents. However, the deformation is difficult to be performed at lower temperatures owing to the poor extrudability, so the extrusion is usually carried out at higher temperatures. Besides, dynamic recrystallization (DRX) happens at an elevated temperature so that fine and uniform equiaxed grains are generated which helps to obtain excellent properties on extruded Mg thin-wall tube. [36]

Ge et al. [37, 38] successfully fabricated the thin-wall tube on ZM21 Mg alloy through hot direct extrusion in one step. Before extrusion, severe plasticity deformation (SPD) was conducted. As well known, grain refinement is an efficient method to enhance the mechanical property, not only strength but also ductility. Especially, the grain boundary sliding (GBS) can be activated so that the ductility is improved and super-plasticity may be achieved when the grains are refined to a critical value. [39–41] In the research of Ge et al., [37, 38] the relatively low-temperature ECAP process (200 °C for 8 passes then 150 °C for 4 passes) was conducted first. Compared with the initial ZM21 alloy, the grains were refined (0.52 μm) and the YS was also improved to 340 MPa. Then the ultra-fine-grained (UFG) Mg billets were used to produce mini-tubes with an outer diameter of 4 mm through a hot extrusion process (temperature range from 150 °C to 300 °C). The thin-wall tube was successfully obtained even at 150 °C due to GBS, and also no grain coarsening happened. Successively, the outer diameter of the tube was reduced to 2.4 mm and the thickness was about 0.4 mm.

New Mg alloys were also used to fabricate the raw thin-wall tube through hot extrusion. Wang et al. [33] achieved the thin-walled hollow tubes from Mg-2Zn-0.46Y-0.5Nd (wt.%) alloys through the hot direct extrusion. Initially, the as-cast Mg-2Zn-0.46Y-0.5Nd (wt.%) alloy billets were extruded into bars at 350 °C with an extrusion ratio of 7. Then solution treatment was conducted for the extruded bars at 430 °C for 72 hours. Finally, the solution-treated Mg bars were machined into small hollow billets and extruded at 390 °C into thin-walled hollow tubes with an outer diameter and wall thickness of 3.2 mm and 0.4 mm, respectively. The extrusion ratio is 49:1 and molybdenum disulfide (MoS_2) was used as a lubricant. The researchers from Shanghai Jiao Tong University [42–44] processed three types of Mg alloy rods (JDBM, AZ31, and WE43) into raw thin-wall tubes with outer diameter and thickness of 8 mm and 0.5 mm, respectively, through hot extrusion at 350 °C with an extrusion ratio of 20:1.

The thin-wall tube could be produced not only by direct extrusion but also indirect extrusion process. Wang et al. [45] designed a laboratory-scale horizontal press device to produce Mg alloy thin-wall tube by an indirect extrusion process, as shown in Figure 6.2(b). On this device, the indirect extrusion on ZM21 Mg alloys was conducted at 480 °C. The extrusion speed was about 0.13 mm/s and the reduction ratio was 50. Finally, the raw thin-wall tube with an outer diameter of 3.16 mm and thickness of 0.25 mm was obtained.

Therefore, the raw thin-wall tube could be processed by the extrusion process in one step. However, most of the extrusion deformation is conducted at higher temperatures (more than 400 °C) owing to the large strain and limited plasticity. Because of the high temperature, the grains may grow up fast and coarsening structures are obtained, which results in the reduction of strength as well as ductility according to the Hall-Petch relationship. In order to enhance the comprehensive properties of the Mg alloy tube, the server plastic deformation technology may be applied to refine the grain size. Faraji et al. [46, 47] developed the tubular channel angular extrusion (TCAP, as shown in Figure 6.2(c)) to refine the microstructure of the Mg alloy tube. After TCAP at 300 °C on AZ91 Mg alloys, the grains could be refined to an average size of about 1 μm. However, this method is conducted on the thick wall tubes, which can be considered to apply on a thin-wall tube as well in

(a)

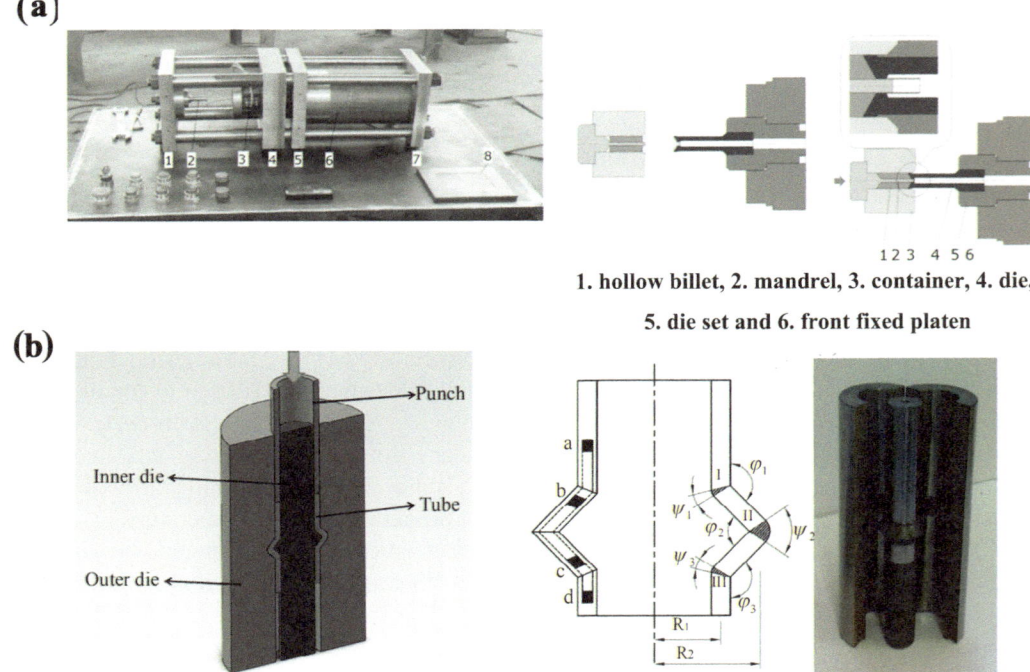

(b)

1. hollow billet, 2. mandrel, 3. container, 4. die,
5. die set and 6. front fixed platen

FIGURE 6.1 The thin-wall tube extrusion processes: (a) direct extrusion and the mini-tube sectors, [37] (b) indirect extrusion and the schematic diagram, [45] (c) tubular channel angular extrusion. [46, 47]

the future. Due to the requirement of coaxially and the minimal outer diameter (≦2 mm), most of the raw extruded tube cannot fit the demand. Thus, the subsequent multi-pass drawing is applied to reduce the thickness and diameter to fit the requirement of a vascular stent. This will be described in the following section.

6.2.1.2 Thin-Wall Tube Multi-Pass Drawing

Tube drawing is also one of the typical methods to produce thin-wall tubes. In theory, due to the relatively high deformation force, and low plasticity, cold deformation is usually not applied on Mg alloys. However, large amounts of experiments have proved that the multi-pass cold drawing is an efficient method to manufacture Mg alloy mini-tubes. [36] Hanada et al. [48] conducted the multi-pass cold drawing on the extruded AZ61 and Mg-0.8% Ca alloy raw tubes (outer diameters of 1.9–2.9 mm and approximately 0.2 mm thickness extruded at 450 °C, extrusion ratio: 68) with a settled mandrel to reduce the thickness and diameter. Finally, the mini-tube was obtained with an outer diameter of 1.5–1.8 mm and thickness of 150 μm. During each pass drawing, the intermediate annealing at 300 °C for 30 min was conducted to remove the work-hardening. The dimensional error of the obtained tube was only between 0.02–2.5%. A schematic diagram of cold drawing with fixed mandrel and the Mg alloy mini-tube product are shown in Figure 6.2(a).

Fang et al. [49] developed a multi-pass cold drawing process with moving mandrel and the attempts on ZM21 Mg alloy were conducted. The raw ZM21 tubes were cast and extruded (indirect extrusion) to solid billets. Then the thin-wall tube with outside diameter and thickness of 3.13 mm and 0.315 mm, respectively, was obtained. After cold drawing for 5 passes and annealing between each pass on extruded ZM21 tube, a mini-tube with an outer diameter of 2.9 mm and thickness of 0.217 mm was successfully achieved. The cross-section area reduction for each pass was less than 15%.

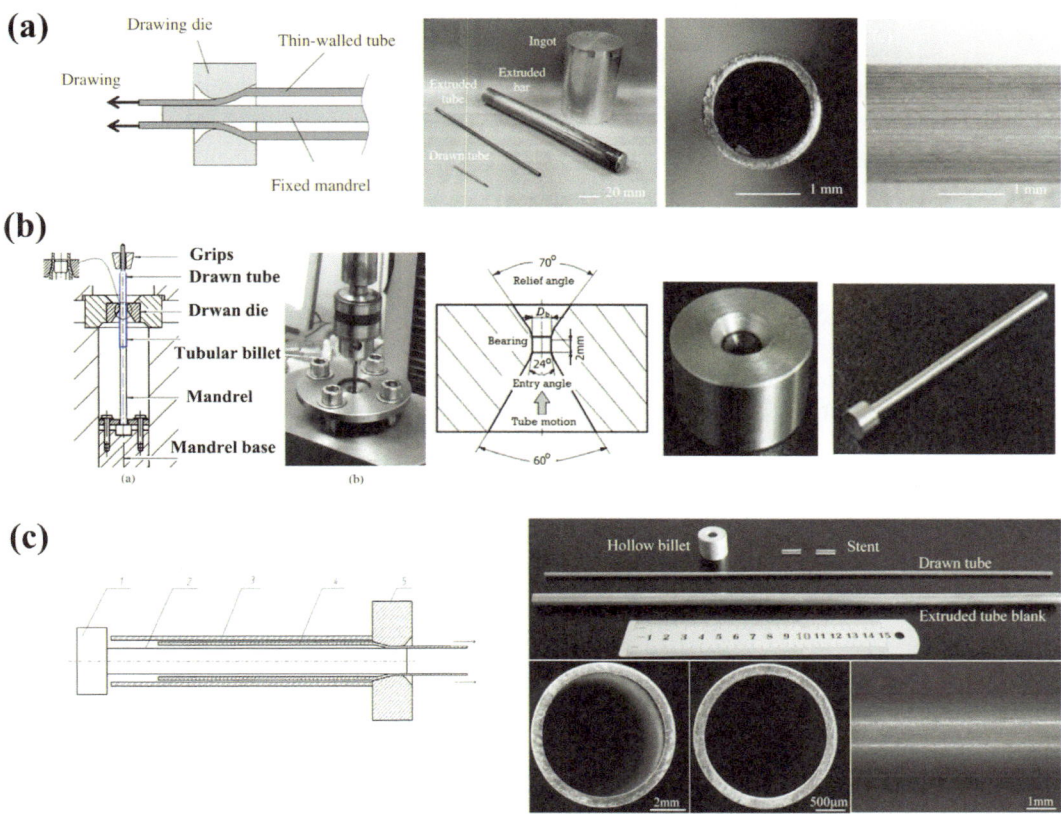

FIGURE 6.2 The thin-wall tube drawing processes: (a) cold drawing with settled mandrel and the Mg-0.8Ca alloy mini-tubes, [48] (b) multi-pass cold drawing as well as the relative dies, [49] (c) microtubes after cold drawing on JDBM Mg alloys. [42]

The drawing speed used was 6 mm/min and molybdenum disulfide (MoS$_2$) was applied as a lubricant. Figure 6.2(b) shows the schematic diagram of the tube drawing dies in operation and more details about the die as well as the mandrel.

Wang et al. [33] also reduced the outer diameter and thickness of the extruded ZM21 Mg alloy tube which was presented in the last part through multi-pass cold drawing. The sample cross-section was reduced by less than 10% in each pass. A total of 17 passes drawing were performed. Annealing was conducted at 3–4 drawing intervals; finally, a mini-tube with the outer diameter and wall thickness of 2.0 mm and 0.15 mm, respectively, was prepared.

Liu et al. [42] fabricated the raw thin-wall tubes with the diameter and thickness of 8.0 mm and 0.5 mm, respectively, on JDBM, AZ31, and WE43 Mg alloys by hot extrusion. Then the extruded initial tubes were rolled by three-roller tube rolling between 10% and 15% every pass. Finally, the subsequent fixed mandrel drawing was conducted with an area reduction of less than 13% for each pass when the rolled tube size reached the outer diameter of 3.5–4.0 mm and thickness of 225–250 μm. During every pass of cold rolling or drawing, annealing was performed at 350 °C for 30 min. In the rolling and drawing processes, machine oil and MoS$_2$ were set as the lubricant. Finally, the outer diameter and thickness of the Mg alloy mini-tubes were 3.00 mm and 180 μm, respectively. The schematic map of the multiple drawing processes is shown in Figure 6.2(c). In which, 1—fixed table, 2—the mandrel, 3—the outer shock-absorbing tube, 4—the magnesium alloy tube, and 5—the drawing die.

6.2.1.3 Laser-Cutting Process

Most vascular stents are processed by a laser-cutting process on Mg alloy mini-tubes. With the help of the computer, the size of the obtained stents is usually precise. [34] However, the quality and shape structure of the stent will have a significant effect on the biocompatibility of the stent. Therefore, fabrication of a high-quality stent with an optimized shape by laser cutting is important.

In the University of Politecnico di Milano, Ge et al. [37] successfully achieved the stent by a laser cutting on the extruded ZM21 Mg alloy thin-wall tube with an outer diameter of 2.4 mm. Under the action of active fiber laser source, the average power and spot of the nanosecond pulse were 7 W and 19 lm, respectively. Through laser cutting, a scaffold network was formed on the tube. Then, chemical etching was carried out in HNO_3/ethanol solution under ultrasonic conditions, so that a semi-finished stent prototype was obtained. In the same group, Demi et al. [50, 51] explored the detailed laser-cutting procedures and the conditions on AZ31 Mg alloy thin-wall tube. To obtain a good quality Mg alloy stent, the process should include laser micro-cutting with a Q-switched fiber laser and subsequent chemical etching and surface finish. The laser devices and the cutting procedures are shown in Figure 6.3. Similarly, Hua et al. [52] used a solid-state fiber laser processing machine to

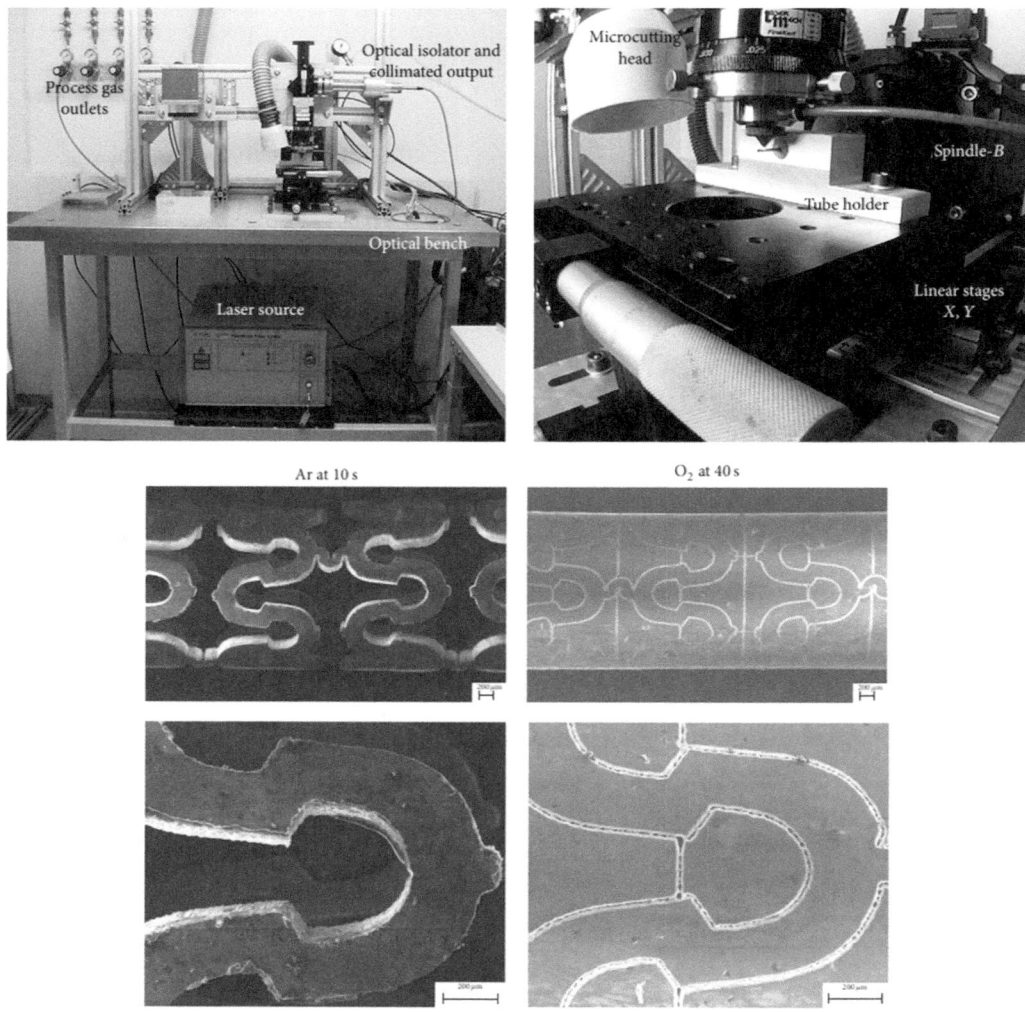

FIGURE 6.3 Laser-cutting machine and the detail procedures. [50, 51]

fabricate the engrave Mg alloy microtubes. The shape of the support beam of the stent was in the S form, which was proved to exhibit excellent properties.

6.2.1.4 Affecting Factors

To manufacture the stent with a good property successfully, there are several important affecting factors, like temperature, alloy composition, microstructure, and processing parameters, etc. The appropriate conditions have a significant effect not only on the processing but also in the subsequent service properties.

6.2.1.4.1 Temperature

Temperature plays a significant role in the forming of vascular stents by affecting the formability and microstructure of Mg alloy thin-walled tube. Ge et al. [37] reported that when the billet was processed directly at 150 °C using the ECAP process, the billet cracked in a wide range. However, the formability of the alloy improved significantly and a defect-free billet was obtained when the experiment was conducted at 200 °C. It was because of non-basal slips activating at the increased temperature. Besides, the dynamic recrystallization (DRX) will happen and the softening effect is enhanced so that the plasticity and formability are modified as well. However, it should avoid grain coarsening at higher temperatures.

6.2.1.4.2 Alloy Composition

The mechanical and corrosion properties of Mg alloy vascular stents can be controlled by micro-alloying. [53] Zn, Mn, Ca, Zr, and some RE elements have been proven to have the effect of solid solution and precipitation strengthening as well as the texture softening to improve the mechanical properties of Mg alloys, both strength and ductility. [36] Polmear et al. [54] reported that a more random texture would be generated on Mg alloys containing Y and Nd during the extrusion process so that the ductility was enhanced. Luo et al. [55] reported that the $Mg_{12}Ce$ particles could be formed and distributed uniformly and the grain structure was refined greatly after 0.2% Ce addition. Compared with pure Mg, the mechanical properties were improved greatly owing to the finer grain size and dispersion strengthening provided by $Mg_{12}Ce$ particles. The yield strength could be enhanced by about 35%, and the total elongation increased more than 4 times. However, the toxicity of certain rare earth elements to the human body is still under further research, and this factor should also be considered when designing and manufacturing vascular stents. [56] Besides, the elements that participate in the metabolism could be investigated more, like Zn, Mn, Ca elements, etc.

6.2.1.4.3 Microstructure

The microstructure with different features such as texture, grain size, and twins has a great influence on the performance of formed thin-walled tubes. Due to the crystal structure of Mg alloys, the basal texture is formed during extrusion. Therefore, there are limited slips to be activated and the ductility is poor. The research on how to promote the basal texture weakening or to obtain a favored grain orientation for slips is much more important. Micro-alloying, induced shear deformation as well twins do help to control the texture. Wang et al. [57] indicated that the ductility was improved almost about 100% at a low temperature when the grains tilted away from the basal pole induced by the shear deformation after 4 pass ECAP. Twins can rotate the grains by a given angle which has a function to weaken the basal texture as well, especially tensile twins who rotate the grains by 86° so that the formability improves significantly. Wang et al. [45] reported that a large number of twins were generated in the 3rd and 4th passes of cold drawing on Mg alloy mini-tubes. The corresponding forming force reduced, and the wall thickness was uniform. Grain refinement strengthening is another important way to improve the strength as well as the plasticity, especially the activation of GBS. Various severe plastic deformation technologies are developed to produce the ultra-fine grain (UFG) microstructure. The traditional, accumulative roll-bonding process can be used to achieve the bulk UFG Mg sheet. The high-pressure torsion process is much more effective to refine the grain size to nano-scale. [58] Multi-directional forging (MDF) can be also be used to produce the fine-grained bulk Mg samples. [59] ECAP is widely investigated by researchers to refine the grain size, and not only the texture control. The strain path, processing parameters have a great effect on the microstructure evolutions. [60, 61] Through increased ECAP process, the grains can be reduced to nano-scale as well. However, the size of samples produced by traditional SPD deformation is relatively small which limits the application. Thus, continued SPD technologies are attractive and should be developed urgently, like continued shear-extrusion processes [62, 63] and so on. Due to the grain refinement, GBS can be promoted to improve the plasticity and the strength will be enhanced as well. Ge et al. [37] indicated that the formability of the UFG grained ZM21 Mg alloys was enhanced greatly due to the activation of GBS. Sikand et al. [64] reported that the mechanical property of tube extruded by porthole die was much better than those extruded by conical die under similar conditions owing to the grain refinement.

Besides, the second phases also play an important effect on the properties of Mg implants. In cast and extruded alloys, second phases are often formed. Usually, the strength will be enhanced; however, the corrosion resistance of cast and extruded alloys become worse. Azzeddine et al. [65] indicated that the appearance of second phase would lead to a poor corrosion resistance by providing cathodic phases and causing more dissolution of the Mg matrix. Besides, the potential difference between the second phase and the Mg matrix has a great effect on the corrosion rate. With a smaller difference, the micro galvanic corrosion will be reduced. Lu et al. [66] showed that the grain size and second phase were two important factors that affect the degradation rate of Mg alloy implants. The grain refinement was beneficial to the improvement of corrosion resistance while the second phase played the opposite role. The sample with the smallest grain size but the largest amount of the second phase showed the fastest degradation rate. This was mainly because the galvanic corrosion effect caused by the second phase exceeded the beneficial effects of grain refinement; besides, the degradation rate of the sample is also faster when the second phase number is the lowest but the grain size is the largest, which was due to the larger crystal grains promoting the degradation rate of the magnesium alloy.

6.2.1.4.4 *Processing Parameters*

At present, processing of Mg alloy microtubes is mainly based on traditional plastic deformation, such as extrusion, rolling, drawing, and so on. Forming the extruded or drawn tube is affected significantly by the processing parameters, including extrusion ratio, strain speed, temperature, lubricant, heat treatment, etc. Generally, dynamic recrystallization is the key deformation mechanism affected by the processing parameters, especially at high temperatures. Du et al. [67] reported that the dynamic recrystallization (DRX) behavior of Mg alloys was promoted by the increase of hot extrusion ratio so that the grain refinement and growth were affected. The number of extrusions passes also has a significant effect on the DRX process, the DRXed crystal grains are refined more as the passes increase. Thus, mechanical and degradable performances are optimized effectively. Besides, the DRX at the evaluated temperature is also affected by the extrusion speed. The DRX behavior can be completed with a lower extrusion speed and the stress concentration will be released so that the formability is modified. Otherwise, a strong stress concentration is generated and fracture happens at an early stage. This also suits the cold deformation, like cold drawing. The annealing process is usually conducted as the intermediate heat treatment to eliminate the work-hardening during cold deformation so that the subsequent plasticity is enhanced. Appropriate lubrication can promote manufacturing successfully at a lower temperature.

6.2.1.5 Studies on Implant Samples

Various in-vivo tests of Mg alloy stents were carried out on animals. Kandala et al. [68] implanted the diamond-shaped AZ31 Mg alloy stent into the peripheral arteries of two domestic pigs for in-vivo testing. Within 28 days of implanting tests, the stent was almost intact. The in-vivo corrosion rate was about 0.75 mm/year measured by micro-CT. Besides, there is no life-threatening during the increase of degradation time. Waksman et al. [69] implanted Mg alloy vascular stents in 63 patients with coronary heart disease. It was shown that the Mg alloys vascular stent could be completely degraded within 4 months. Followed up for 12 to 28 months on 8 patients, they did not need to rebuild the vascular stent after 4 months. Besides, Mg alloy vascular stents did not show any unusual findings except that the degradation rate was still too fast.

All in all, the degradable Mg alloy vascular stent can be manufactured by hot extrusion, subsequent multi-pass drawing then laser cutting. However, most of the products are on commercial AZ31 or pure Mg to explore the forming technology. The properties may not fit the requirement. Thus, the new Mg alloy with higher strength should be considered to apply for manufacturing the stent. Further, how to reduce the degradation rate is the most important factor to facilitate the application of Mg stents.

6.2.2 Forming of Bone Screws and Osteons

6.2.2.1 Forming Procedures

The bone screw is one of the potential implants, and several researchers have introduced various manufacturing processes. Usually, bone screws can be formed by the conventional casting then extrusion method, powder metallurgy, and metal injection molding, etc. The metal injection molding has great potential in near-net-shape production. The researchers have already realized that Mg alloy implants molded in this way have a better load-bearing capacity. Wolff et al. [70] used metal injection molding to produce Mg alloy bone screws successfully. First, the prepared metal powder and binder were mixed, followed by the de-binding and sintering operations. It was worth noting that degreasing and sintering were carried out in a combined hot fireplace, and both were completed in a protective atmosphere, as shown in Figure 6.4(a).

The bone screws can also be formed by extrusion processes as well. Huehnerschulte et al. [71] indicated that the cylindrical pins from ZEK100 and AX30 Mg alloys could be fabricated by gravity die casting and further processed by direct extrusion at 380 °C. The diameter and length of the final implants were 2.5 mm and 25 mm, respectively. Erdmann et al. [72] reported that screws could be completed by extruding bar stock in several steps through turning operations using Mg-0.8% Ca (wt.%) alloys. The cylinder with diameter of 80 mm was centered and clamped in the CNC turning center. The shaft (major diameter 4.0 mm, length 6.0 mm) and head of the screw are formed by several turnings. Then the thread outline (length of 5.0 mm, and core diameter of 3.0 mm) on the billets was tapped. The in-vivo tests were conducted on 40 female New Zealand white rabbits. The bone screws and the position during in-vivo tests are shown in Figure 6.4(b). On the other hand, Han et al. [73] suggested that the bone screws could also be formed by extrusion and three-roll milling on high pure Mg. During which, the as-cast pure Mg was applied to hot extrusion with an extrusion ratio of 148 at 200 °C, then the diameter of the extruded rod was reduced to 7.5 mm with a three-roll mill at room temperature. Annealing at 160 °C for 20 min was necessary to eliminate the internal stress; finally, the required bone screws were obtained, as shown in Figure 6.4(b)-ii. Above all, conventional extrusion and casting processes require turning in a CNC to produce threads.

Osteon is also one of the important implants which can be processed by degradable Mg alloys. Naujokat et al. [74, 75] reported that the osteon of WZ43 Mg alloy implant (prototypes of standard-size, four-hole plates (1 mm in thickness, and 22 mm in length) and cortical bone screws (2 mm in diameter, and 5 mm in length)) could be manufactured by powder molding, followed by hot extrusion and computer-aided turning process. The in-vivo experiment was carried out on miniature pigs, as shown in Figure 6.4(c).

Powder metallurgy (PM) technology is also widely used in the preparation of Mg alloys. Xie et al. [76] used ball milling and spark plasma sintering (SPS) processes to develop magnesium-iron alloys with excellent mechanical properties, ultra-high hardness and strength,

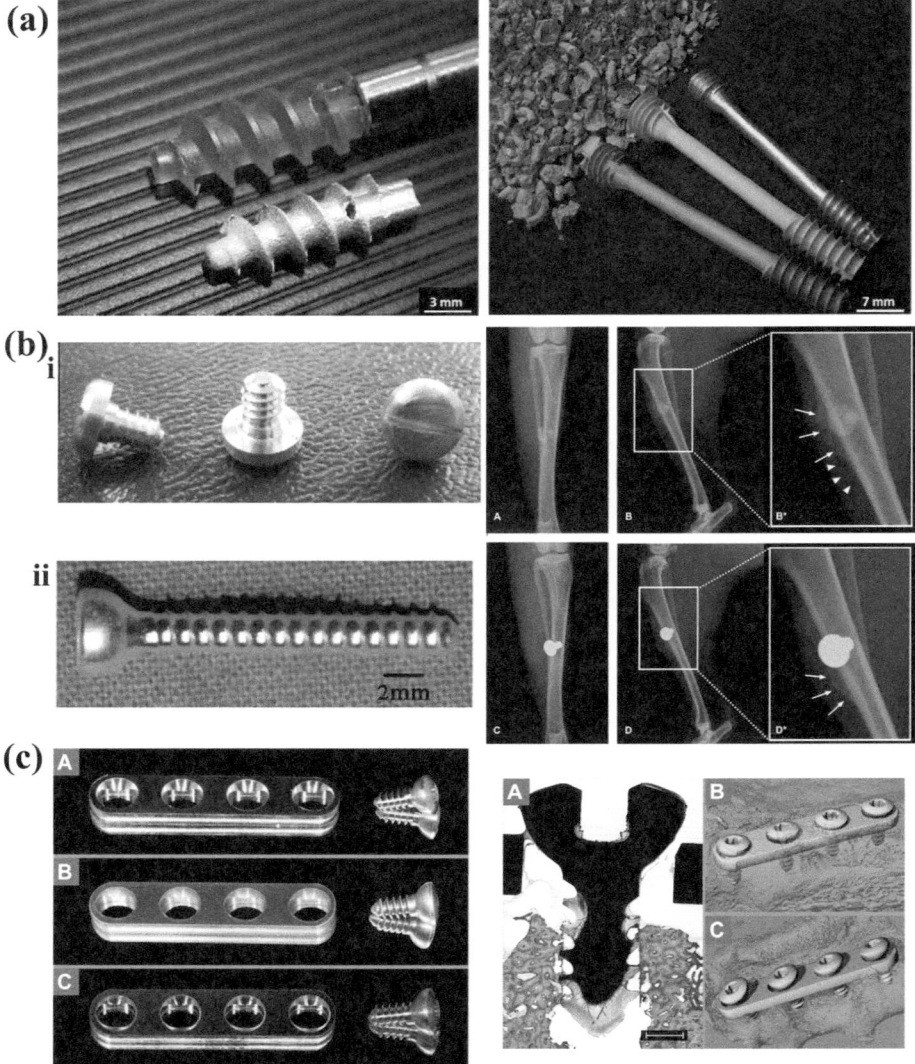

FIGURE 6.4 The Mg implants produced by various methods: (a) screws metal injection molding, [70] (b) bone screws by hot extrusion, [72, 73] (c) osteon by powder molding, followed by hot extrusion. [74, 75]

and improved biodegradability. Wolff et al. [70] successfully prepared Mg-0.9Ga biodegradable implants with ultimate tensile strength (UTS) of 141 MPa, tensile yield strength (YS) of 73 MPa, and Young's modulus of 38 GPa using metal injection molding (MIM) technology. Alizadeh et al. [77] studied the super-plastic behavior and microstructure of Mg-5Gd-4Y-0.4Zr alloy using the SPD process of extrusion (equal channel angular pressing (ECAP) and high-pressure steering (HPS)).

6.2.2.2 Major Influencing Factors

To manufacture the bone screws and osteon successfully, there are many factors that should be considered. In the case of powder metallurgy, the applied pressure and heating temperature are the two main factors. Seyedraoufi et al. [78] indicated that the highest elastic modulus and compressive strength were obtained at 550 °C on porous Mg-Zn alloys. Tahmasebifar et al. [79]

found that compaction pressure could increase relative density, and bending strength, but sintering time has little effect on the properties of AZ91D alloys. For the SPS process, the sintering temperature and time need to be carefully designed. [80] In addition, for the metal injection process, the selection of a suitable polymer binder component is the main factor in the molding of Mg alloy implants. [70] To ensure the minimum mechanical load of the billet during tapping, the maximum cutting depth of the Mg alloy implant is usually 0.1 mm. [72] For extrusion, the affecting factors are similar to vascular stents. However, only limited literatures are available on the fabrication of Mg bone screws and osteons. And orthopedic biodegradable implants must have a degradation rate that matches the process of vascular healing or regeneration. [76] If the degradation rate of the implant is too fast or too slow, it will affect the growth and healing of bones. Tomac et al. [81] found that the processing technology will have a greater impact on the surface and subsurface of the processed product, which in turn affects the corrosion behavior of the alloy. Based on this, Lucas et al. found that the degradation kinetics can be preset for degradable Mg alloys. Besides, these Mg alloy implants can be customized according to the location of the implant and the health of the patient. [6] Most surface modification techniques can also alleviate the corrosion rate.

6.2.2.3 In-Vivo Studies of Mg Implants

The in-vivo experiments of Mg bone implants have been also carried out. Han et al. [82] put the high pure Mg bone implants into New Zealand white rabbits. It was reported that high-purity Mg bone implants in rabbits had broad prospects for load-bearing fracture fixation. Henderson et al. [83] studied the in-vivo degradation rate of bone screw made by pure Mg, AZ31, and other comparative materials on 15 New Zealand White rabbits for approximately 1 year. It pointed out that alloying could change the degradation curve of Mg. This means that the composition of the initial materials is much more important during degradation.

Naujokat et al. [74] implanted the WE43 Mg alloy osteosynthesis plates and screws in pig recipients. The results have shown that the surgical procedure and the osteosynthesis material were tolerated well by animals; besides, the bone healing of the osteoplasty was undisturbed. Chaya et al. [29] used the high pure Mg plates (20 × 4.5 mm with a thickness of 1–1.5 mm) and screws (the length, shaft outer, and inner diameter was 7, 1.75, and 1 mm, respectively) as implants in 12 New Zealand White rabbits. It was found that the degradation of Mg did not prevent fracture healing, and even promoted bone formation near the implant.

6.2.3 Forming of Degradable Wire

6.2.3.1 Process Procedure of Medical Wire

In 1878, Huse stopped bleeding vessels by using an Mg wire. From then on, Mg wires were gradually applied to surgical sutures. [84] The degradable Mg wire is usually produced by hot extrusion combined drawing which is similar to other materials. However, due to its poor plastic performance and the large strain induced by suture operation, how to produce Mg wires with favorable properties is explored.

The procedures to manufacture Mg thin wires usually include casting, hot extrusion, and multi-pass cold drawing combined with intermediate annealing. Bai et al. [85] has successfully fabricated fine wire by extruding the original thick wires with a diameter of 2.7 mm at 480 °C, then the multi-pass cold drawing was conducted gradually at room temperature on Mg-4% RE (Gd/Y/Nd)-0.4%Zn alloys. Finally, the wire with a diameter of < 0.4 mm was obtained, as shown in Figure 6.5(a). The grain size of various wires was ~3.8, 3.8, and 3.6 μm, respectively. Yan et al. [86] prepared the Mg-6Zn alloy thin wire with a diameter of 0.31 mm via extrusion, ECAP, and hot drawing together. The peak strength and ductility could reach 300 MPa and 11%, respectively, which was enhanced greatly compared with only hot-drawn samples with the same diameter. Besides, there was no fracture that happens during bending and three knots tests which expressed an excellent ductility,

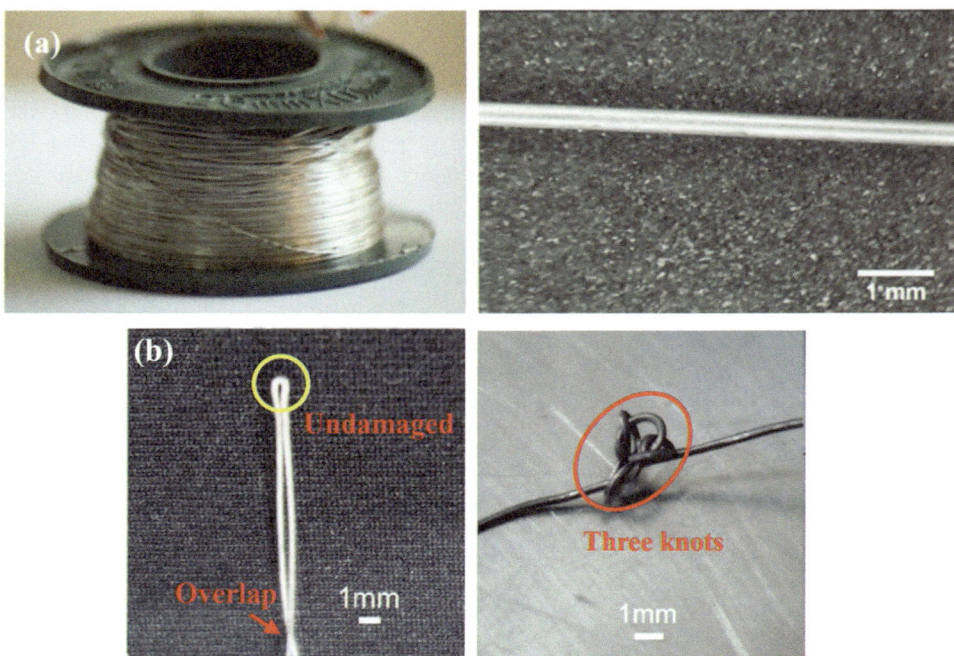

FIGURE 6.5 The magnesium biodegradable thin wire: (a) cold drawn Mg-4%RE-0.4% Zn alloy wire with diameter of 0.3 mm, [85] (b) bending and knotting test on Mg-6Zn wire with diameter of 0.31 mm after ECAP and drawing. [86]

as shown in Figure 6.5(b). This means that grain refinement is also very important in Mg wire fabrication.

Sharifzadeh et al. [87] obtained Mg fine wires via a new process procedure called friction stir extrusion (FSE). In this process, the extrusion axis was parallel to the plunge die to provide the force on the original cylindrical Mg alloy chips. The plunge was revolving with the rotational speeds of 180, 250, and 355 r/min when it moved towards the chips under a constant speed of 120 mm/min. Finally, the chips with 6–10 mm in length, 1–4 mm in width, and 0.2 mm in thickness were obtained. Besides, the grain of the fine wire was much smaller, which could be investigated further.

6.2.3.2 Factors Influencing Mg Wire Manufacturing

The manufacture of Mg wire is difficult under room temperature because of its low ductility; hence, choosing suitable parameters is quite important. It is similar to other Mg implants; the processing parameters like temperature, speed, etc. play an important effect. Milenin et al. [88] studied a series of Mg-Ca alloy wires fabricated under a series of hot drawing temperatures with an initial diameter of 1.8 mm to a final diameter of 1.5 mm. The results demonstrated that the Mg-Ca alloys performed higher ductility at high drawing temperatures. The more stress concentration will be produced and fracture may occur when the drawing speed is larger. During multi-drawing, intermediate annealing is usually carried out to remove the strain hardening. Besides, the precision of mold is also an important factor. Dodyim et al. [89] utilized the finite element method analysis to investigate the optimum drawing conditions and the Cockroft-Latham fracture criterion was used to clarify it. The results indicated that 6° of die half-angle and 13% of pass reduction were the suitable conditions to prevent an internal crack on pure Mg.

6.3 FORMING OF DEGRADABLE IMPLANTS BY ADDITIVE MANUFACTURING

As indicated, most of the degradable biomedical implants are manufactured by plastic deformation. However, there are many complex procedures to be carried out and it requires more manufacturing time and cost. Recently, the additive manufacturing method (AM, and 3D printing) is proved to be a new approach to produce the degradable sectors. Through the AM process, the implants with complex geometry that are difficult to fabricate can be effectively produced. The raw materials for AM are usually in the form of powder, liquid resin, or wire. For Mg alloys, the powder bed fusion (PBF) is proved to be effective. [90] However, it is difficult to be printed owing to the highly reactive nature. Special devices and an inert atmosphere are needed to prevent the oxidation of raw powder. The laser is the most commonly used for PBF on Mg alloys. The rapid heating and quenching lead to rapid solidification; thus, grains are refined. Laser power, scanning speed plays a significant effect on the melting pool and the vaporization so that the deposition on PBF happens. Previous reports suggested that optimal AM parameters for Mg alloys were low energy densities, where the density of the resultant part was high and vaporization of the alloying elements was low. [91] For example, AZ91D alloy was printed at about 200 W and 0.09 m/min scanning speed. The energy density was in the range from 83 J/mm^3 to 167 J/mm^3. [92] The quality of depositions is dependent on the powder particle size and the alloying elements. Printing smaller powder particles could raise the melting pool temperature and result in aggressive oxidation. [93]

Many practical attempts were conducted to fabricate Mg implants. For example, Yang et al. [94] developed a selective laser melting (SLM) system to fabricate the degradable Mg sectors. An MFSC-100 fiber laser with a max power of 100 W was selected as the energy source and Mg powder (99.99%), which had a spherical shape with a size of 1–10 μm, was applied as the original materials. The Mg powder was paved on a substrate with a thickness of 0.1 mm. The flow rate of Argon gas was 1.2 L/min. The laser beam spot size was 50 μm, the laser power was 20–100 W and scanning speed was set to 100–900 mm/min. During scanning by the laser beam, the powder melted and deposited on the substrate. Fresh powder was filled when a previous layer was finished. Subsequently, the powder bed was lowered again, a new layer was deposited, and the third layer of the part started to scan. The process was repeated until all slices of the 3D model were indicated into the powder bed and the manufacturing of the 3D part was finished. Dong et al. [95] applied the solvent-cast 3D printing (SC-3DP) process to build biodegradable scaffolds on pure Mg. During printing, an ink containing a binder system and various amounts of Mg powder loading were prepared and transformed into plastic syringes with a 410 μm tapered nozzle. The pure Mg powder particles had a purity of 99.8 wt.% with a size from 25 to 80 μm. Then, the prepared ink was extruded under a pressure from 1 and 600 KPa, and the printing speed between 1 and 18 mm/s. Finally, the scaffold samples were becoming pure Mg scaffolds through thermal treatment.

Thus, the AM method is a potentially effective method to fabricate the complex implants and new designs of Mg alloys. However, the properties of biomedical sectors are usually not excellent comparing with those by plastic deformation which needs to be investigated more in the future.

6.4 FUTURE TRENDS ON MG-BASED IMPLANTS MANUFACTURING

Due to technology development, the corrosion and mechanical properties of Mg alloys have been enhanced greatly, which fits the requirement of biodegradable implants. Various Mg alloy-based biomedical implants are tried to be manufactured. The products (stents, bone screws, wires, etc.) are mainly fabricated through alloy casting, hot extrusion, multi-passes drawing, post polishing, and other treatments. In most cases, the attempts like hot extrusion to achieve the raw billet are conducted at a higher temperature, usually above 400 °C. Due to the thermal activation effect, the non-basal slips are activated so that the formability is enhanced greatly and the fabrications are completed successfully. Nevertheless, the microstructure coarsening happens very fast which reduces the subsequent service performance at the same time. Thus, how to avoid grain growth during fabrication is significant. Manufacturing at low temperature (avoid the grain growth) is a potential approach to reach this goal and the low-temperature super-plasticity is important. Texture softening and grain refinement to induce grain boundary sliding is proving to be feasible. When the grains are refined to a critical value, the grain boundary sliding starts, and the plasticity is improved greatly. Ge et al. [37] extruded the thin-wall tube even at 150 °C when the grains are refined to 0.5 μm by the ECAP process. The grain coarsening does not happen. On the other hand, texture controlling is also an effective way. The plasticity can be improved almost to 100% with a 45° inclined orientation away from the basal pole at 150 °C. [57] However, more studies about manufacturing the implants at lower temperatures are rarely found, but needs to be investigated further, including the affecting factors.

On the other hand, in order to obtain good properties, new Mg alloy systems are developed to fit the requirement. Most of them contain elements that are not necessary for the human body. How they react during degradation is still unknown. More in-vivo tests should be conducted to ensure safety. Besides, the elements that participate in metabolism are suggested to be used. New Mg-Zn-Mn-Ca alloys with excellent mechanical and corrosion properties can be designed as well and the parameters to manufacture the implants should be explored.

6.5 SUMMARY

The present chapter mainly discusses the various forming methods for Mg alloys. Due to the good comprehensive properties, Mg and its alloys can be fabricated as degradable implants, such as vascular stents, bone screws, osteons, and thin wires. For vascular stents, the raw thin-wall tube will be produced first by hot extrusion on as-cast Mg alloys. Then the subsequent multi-pass cold drawing is conducted to decrease the diameter as well as the thickness. After laser cutting on the mini-tube with a diameter of less than 2 mm, the vascular stent with an optimized shape will be obtained. However, due to the crystal structure, tube extrusion is usually performed and the grain coarsening will happen. Texture modifying or grain refinement to improve ductility is a potential method to manufacture the raw tube at low temperatures to avoid grain growth. For bone screws and osteons, various technologies are possible to be used, such as hot extrusion, powder metallurgy, and metal injection molding. However, the turning to produce the thread is necessary. Multi-pass hot drawing is usually used to fabricate the Mg thin wire with a diameter less than 0.5 mm and SPD technology can be applied, like the FSE method. For all the Mg implants during manufacturing, it can be affected by the microstructure, forming parameters, alloy composition, and so on. Especially the composition and grain size which are the key factors to achieve the implants successfully with an excellent service property.

REFERENCES

[1]   Mythili, P., Janis, L., Kristine, S.A., et al., 2017. Biodegradable materials and metallic implants: A review. *Journal of Functional Biomaterials* 8(4): 44.

[2]   Wu, S., Liu, X., Yeung, K.W.K., et al., 2013. Surface nano-architectures and their effects on the mechanical properties and corrosion behavior of Ti-based orthopedic implants. *Surface & Coatings Technology* 233: 13–26.

[3]   Biesiekierski, A., Wang, J., Gepreel, A.H., et al., 2012. A new look at biomedical Ti-based shape memory alloys. *Acta Biomaterialia* 8(5): 1661–1669.

[4]   Rattier, B.D., Hoffman, A.S., Schoen, F.J., Lemons, J.E., 2004. *Biomaterials science: An introduction to materials in medicine.* Elsevier Academic Press.

[5]   Chen, Y., Xu, Z., Smith, C., et al., 2014. Recent advances on the development of magnesium alloys for biodegradable implants. *Acta Biomaterialia* 10(11): 4561–4573.

[6]   Denkena, B., et al., 2007. Biocompatible magnesium alloys as absorbable implant materials: Adjusted surface and subsurface properties by machining processes. *CIRP Annals: Manufacturing Technology* 56: 113–116.

[7]   Li, J., Wan, P., et al., 2015. Study on microstructure and properties of extruded Mg-2Nd-0.2Zn alloy as potential biodegradable implant material. *Materials Science & Engineering C Materials for Biological Applications* 49: 422–429.

[8]   Saris, N.-E.L., Mervaala, E., Karppanen, H., et al., 2000. Magnesium: An update on physiological, clinical and analytical aspects. *Clinica Chimica Acta* 294(1–2): 1–26.

[9]   Witte, F., 2015. Reprint of: The history of biodegradable magnesium implants: A review. *Acta Biomaterialia* 23: S28–S40.

[10]   Jiang, P., Blawert, C., Zheludkevich, M.L., 2020. The corrosion performance and mechanical properties of Mg-Zn based alloys: A review. *Corrosion and Materials Degradation* 1(1): 7.

[11]   Wang, Y.C., Tian, Y., Qu, T., et al., 2014. Purification of magnesium by vacuum distillation and its analysis. *Materials Science Forum* 788: 52–57.

[12]   Lam, R.K.F., Marx, D.R. 1996. Ultra high purity magnesium vacuum distillation purification method. US5582630 A.

[13]   Witte, F., Hort, N., Vogt, C., et al., 2008. Degradable biomaterials based on magnesium corrosion. *Current Opinion in Solid State and Materials Science* 12(5): 63–72.

[14]   Xiao, D.H., Geng, Z.W., Chen, L., et al. 2015. Effects of alloying elements on microstructure and properties of magnesium alloys for tripling ball. *Metallurgical & Materials Transactions A* 46(10): 4793–4803.

[15]   Kaviania, M., Ebrahimi, G.R., Ezatpour, H.R. 2019. Improving the mechanical properties and biocorrosion resistance of extruded Mg-Zn-Ca-Mn alloy through hot deformation. *Materials Chemistry and Physics* 234: 245–258.

[16]   Salleh, E.M., Zuhailawati, H., Ramakrishnan, S., et al., 2015. A statistical prediction of density and hardness of biodegradable mechanically alloyed Mg-Zn alloy using fractional factorial design. *Journal of Alloys & Compounds An Interdisciplinary Journal of Materials Science & Solid State Chemistry & Physics.*

[17]   Li, Z., Gu, X., Lou, S., et al., 2008. The development of binary Mg-Ca alloys for use as biodegradable materials within bone. *Biomaterials* 29(10): 1329–1344.

[18]   Zhang, B., Wang, Y., Geng, L., et al., 2012. Effects of calcium on texture and mechanical properties of hot-extruded Mg-Zn-Ca alloys. *Materials Science & Engineering A* 539(none): 56–60.

[19]   Liu, D., Yang, D., Li, X., et al., 2018. Mechanical properties, corrosion resistance and biocompatibilities of degradable Mg-RE alloys: A review. *Journal of Materials Research and Technology* 8(1): 1538–1549.

[20]   Imandoust, A., Barrett, C.D., Al-Samman, T., et al., 2017. A review on the effect of rare-earth elements on texture evolution during processing of magnesium alloys. *Journal of Materials Science* 52(1): 1–29.

[21]   You, S., Huang, Y., Kainer, K.U., et al., 2017. Recent research and developments on wrought magnesium alloys. *Journal of Magnesium & Alloys* 5(3): 239–253.

[22]   Gui, Z., Kang, Z., Li, Y., 2016. Mechanical and corrosion properties of Mg-Gd-Zn-Zr-Mn biodegradable alloy by hot extrusion. *Journal of Alloys & Compounds* 222–230.

[23]   Zhang, X.B., Yuan, G.Y., Wang, Z.Z., 2013. Effects of extrusion ratio on microstructure, mechanical and corrosion properties of biodegradable Mg-Nd-Zn-Zr alloy. *Materials Science & Technology* 29(1): 111–116.

[24]   Nakamura, Y., Tsumura, Y., Tonogai, Y., Shibata, T., Ito, Y., 1997. Differences in behavior among the chlorides of seven rare earth elements administered intravenously to rats. *Fundamental and Applied Toxicology: Official Journal of the Society of Toxicology* 37(2): 106.

[25]   Li, Y., Wen, C., Mushahary, D., et al., 2012. Mg-Zr-Sr alloys as biodegradable implant materials. *Acta Biomaterialia* 8(8): 3177–3188.

[26]   Ramsden, J.J., Allen, D.M., Stephenson, D.J., et al., 2007. The design and manufacture of biomedical surfaces. *CIRP Annals: Manufacturing Technology* 56(2): 687–711.

[27]   Zreiqat, H., Howlett, C.R., Zannettino, A., et al., 2010. Mechanisms of magnesium-stimulated adhesion of osteoblastic cells to commonly used orthopaedic implants. *Journal of Biomedical Materials Research Part A* 62(2).

[28]   Witte, F., Kaese, V., Haferkamp, H., et al., 2005. In vivo corrosion of four magnesium alloys and the associated bone response. *Biomaterials* 26(17): 3557–3563.

[29]   Chaya, A., Yoshizawa, S., Verdelis, K., et al., 2015. In vivo study of magnesium plate and screw degradation and bone fracture healing. *Acta Biomaterialia* 18: 262–269.

[30]   Serruys, P.W., 2006. Fourth annual American College of Cardiology international lecture: A journey in the interventional field. *Journal of the American College of Cardiology* 47(9): 1754–1768.

[31] Ni, L., Chen, H., Luo, Z., et al., 2020. Bioresorbable vascular stents and drug-eluting stents in treatment of coronary heart disease: a meta-analysis. *Journal of Cardiothoracic Surgery* 15.

[32] Lally, C., Kelly, D.J., Prendergast, P.J. 2006. *Stents.* Wiley Encyclopedia of Biomedical Engineering.

[33] Wang, J., Zhou, Y., Yang, Z., et al., 2018. Processing and properties of magnesium alloy micro-tubes for biodegradable vascular stents. *Materials Science & Engineering C* S0928493117337153.

[34] Moravej, M., Mantovani, D., 2011. Biodegradable metals for cardiovascular stent application: Interests and new opportunities. *International Journal of Molecular Sciences* 12(7): 4250–4270.

[35] Hu, T.Z., Yang, C., et al., 2018. Biodegradable stents for coronary artery disease treatment: Recent advances and future perspectives. *Materials Science & Engineering: C, Materials for Biological Applications* 91: 163–178.

[36] Liu, Y., Lu, B., Cai, Z., 2019. Recent progress on Mg- and Zn-based alloys for biodegradable vascular stent applications. *Journal of Nanomaterials* 2019(6): 1–16.

[37] Ge, Q., Dellasega, D., Demir, G.A., et al., 2013. The processing of ultrafine-grained Mg tubes for biodegradable stents. *Acta Biomaterialia* 9(10): 8604–8610.

[38] Ge, Q., Vedani, M., Vimercati, G., 2012. Extrusion of magnesium tubes for biodegradable stent precursors. *Materials and Manufacturing Processes* 27(2): 140–146.

[39] Koike, J., Ohyama, R., Kobayashi, T., et al., 2005. Grain-boundary sliding in AZ31 magnesium alloys at room temperature to 523 K. *Materials Transactions* 44(4): 445–451.

[40] Watanabe, H., Mukai, T., Higashi, K., 2007. Low temperature superplasticity in a ZK60 magnesium alloy. *Materials Transactions Jim* 40(4): 315–317.

[41] Somekawa, H., Singh, A., 2018. Superior room temperature ductility of magnesium dilute binary alloy via grain boundary sliding. *Scripta Materialia* 150: 26–30.

[42] Liu, F., Chen, C., Niu, J., et al., 2015. The processing of Mg alloy micro-tubes for biodegradable vascular stents. *Materials Science and Engineering: C* 48: 400–407.

[43] Zhang, X.B. 2011. *The properties of biodegradable biological magnesium alloy and preparation technology of cardiovascular scaffold.* Shanghai Jiaotong University.

[44] Lu, W., Yue, R., Miao, H., et al., 2019. Enhanced plasticity of magnesium alloy micro-tubes for vascular stents by double extrusion with large plastic deformation. *Materials Letters* 245: 155–157.

[45] Wang, L.X., Fang, G., Qian, L.Y., Leeflang, S., Duszczyk, J., Zhou, J., 2014. Forming of magnesium alloy microtubes in the fabrication of biodegradable stents. *Progress in Natural Science: Materials International* (5): 500–506.

[46] Faraji, G., Mashhadi, M.M., Abrinia, K., et al., 2012. Deformation behavior in the tubular channel angular pressing (TCAP) as a noble SPD method for cylindrical tubes. *Applied Physics A* 107(4): 819–827.

[47] Faraji, G., Mashhadi, M., Dizadji, A., et al., 2012. A numerical and experimental study on tubular channel angular pressing (TCAP) process. *Journal of Mechanical Science and Technology* 26(11): 3463–3468.

[48] Hanada, K., Matsuzaki, K., Huang, X., et al., 2013. Fabrication of Mg alloy tubes for biodegradable stent application. *Materials Science & Engineering C Materials for Biological Applications* 33(8): 4746–4750.

[49] Fang, G., Ai, W.J., Leeflang, S., et al., 2013. Multipass cold drawing of magnesium alloy minitubes for biodegradable vascular stents. *Materials Science & Engineering C Materials for Biological Applications* 33(6): 3481–3488.

[50] Demir, A.G., Previtali, B., Ge, Q., et al., 2014. Biodegradable magnesium coronary stents: Material, design and fabrication. *International Journal of Computer Integrated Manufacturing* 27(10): 936–945.

[51] Demir, A.G., Previtali, B., Biffi, C.A., 2013. Fibre laser cutting and chemical etching of AZ31 for manufacturing biodegradable stents. *Advances in Materials Science and Engineering* 2013(ID692635): 1–11.

[52] Hua, Y.L., Li, W., et al., 2015. Semi-solid extrusion thixoforming die and method for degradable magnesium alloy microtube. China's Invention Patent, ZL 201310154125.5.

[53] Yee, D.T.W., Koon, J.N.C., Huang, Y., et al., 2020. Bioresorbable metals in cardiovascular stents: Material insights and progress. *Materialia* 12: 100727.

[54] Polmear, I.J., 2006. *Light alloys: From traditional alloys to nanocrystals.* Oxford and Burlington, MA: Elsevier/Butterworth-Heinemann.

[55] Luo, A.A., Wu, W., Mishra, R.K., et al., 2010. Microstructure and mechanical properties of extruded magnesium-aluminum-cerium alloy tubes. *Rare Metal Materials & Engineering*.

[56] Hermawan, H., Dubé, D., Mantovani, D. 2009. Developments in metallic biodegradable stents. *Acta Biomaterialia* 6(5): 1693–1697.

[57] Wang, L., Mostaed, E., Cao, X., et al., 2016. Effects of texture and grain size on mechanical properties of AZ80 magnesium alloys at lower temperatures. *Materials & Design* 89(Jan.): 1–8.

[58] Matsunoshita, H., Edalati, K., Furui, M., et al., 2015. Ultrafine-grained magnesium: Lithium alloy processed by high-pressure torsion: Low-temperature superplasticity and potential for hydroforming. *Materials Science & Engineering A* 640: 443–448.

[59] Yang, X.Y., Sun, Z.Y., Xing, J., et al., 2008. Grain size and texture changes of magnesium alloy AZ31 during multi-directional forging. *Transactions of Nonferrous Metals Society of China* 18(supp-S1): s200–s204.

[60] Gzyl, M., Rosochowski, A., Yakushina, E., et al., 2013. Route effects in I-ECAP of AZ31B magnesium alloy. Route effects in I-ECAP of AZ31B magnesium alloy. Trans Tech Publications.

[61] Gautam, P.C., Biswas, S., 2021. On the possibility to reduce ECAP deformation temperature in magnesium: Deformation behaviour, dynamic recrystallization and mechanical properties. *Materials Science and Engineering A* 141103.

[62] Hu, H.J., Wang, H., Zhai, Z.Y., et al., 2014. The influences of shear deformation on the evolutions of the extrusion shear for magnesium alloy. *International Journal of Advanced Manufacturing Technology* 74(1–4): 423–432.

[63] Hu, H.J., Sun, Z., Ou, Z.W., et al., 2016. Wear behaviors and wear mechanisms of wrought magnesium alloy AZ31 fabricated by extrusion-shear. *Engineering Failure Analysis* 72: 25–33.

[64] Sikand, R., Kumar, A.M., Sachdev, A.K., et al., 2009. AM30 porthole die extrusions: A comparison with circular seamless extruded tubes. *Journal of Materials Processing Technology* 209(18–19): 6010–6020.

[65] Azzeddine, H., Hanna, A., Dakhouche, A., et al., 2020. Impact of rare-earth elements on the corrosion performance of binary magnesium alloys. *Journal of Alloys and Compounds* 829: 154569.

[66] Lu, Y., Bradshaw, A.R., Chiu, Y.L., et al., 2015. Effects of secondary phase and grain size on the corrosion of biodegradable Mg-Zn-Ca alloys. *Mater Sci Eng C Mater Biol Appl* 48: 480–486.

[67] Du, B., Hu, Z., Wang, J., et al., 2020. Effect of extrusion process on the mechanical and in vitro degradation performance of a biomedical Mg-Zn-Y-Nd alloy. *Bioactive Materials* 5(2): 219–227.

[68] Kandala, B.S.P.K., Zhang, G., Lcorriveau, C., et al., 2020. *Modelling, fabrication by photo-chemical etching and in vivo study of magnesium AZ31 stents.* Social Science Electronic Publishing.

[69] Waksman, R., Erbel, R., Mario, C.D., et al., 2009. Early- and long-term intravascular ultrasound and angiographic findings after bioabsorbable magnesium stent implantation in human coronary arteries. *JACC: Cardiovascular Interventions* 2(4): 312–320.

[70] Wolff, M., Schaper, J., et al., 2016. Magnesium powder injection molding (MIM) of orthopedic implants for biomedical applications. *JOM* 68: 1191–1197.

[71] Huehnerschulte, T.A., Reifenrath, J., Rechenberg, B.V., et al., 2012. In vivo assessment of the host reactions to the biodegradation of the two novel magnesium alloys ZEK100 and AX30 in an animal model. *Biomedical Engineering Online* 11(1): 14.

[72] Erdmann, N., Angrisani, N., Reifenrath, J., et al., 2011. Biomechanical testing and degradation analysis of MgCa0.8 alloy screws: A comparative in vivo study in rabbits. *Acta Biomaterialia* 7(3): 1421–1428.

[73] Han, P., Cheng, P., Zhang, S., et al., 2015. In vitro and in vivo studies on the degradation of high-purity Mg (99.99wt.%) screw with femoral intracondylar fractured rabbit model. *Biomaterials* 64: 57–69.

[74] Naujokat, H., Seitz, J.M., Ail, Y., et al., 2017. Osteosynthesis of a cranio-osteoplasty with a biodegradable magnesium plate system in miniature pigs. *Acta Biomaterialia* 62: 434–445.

[75] Naujokat, H., Ruff, C.B., Klüter, T., et al., 2019. Influence of surface modifications on the degradation of standard-sized magnesium plates and healing of mandibular osteotomies in miniature pigs. *International Journal of Oral and Maxillofacial Surgery* 49(2): 272–283.

[76] Xie, G., Takada, H., Kanetaka, H., 2016. Development of high performance MgFe alloy as potential biodegradable materials. *Materials Science & Engineering* A 671: 48–53.

[77] Alizadeh, R., Mahmudi, R., et al., 2017. Microstructural evolution and superplasticity in an Mg-Gd-Y-Zr alloy after processing by different SPD techniques. *Materials Science and Engineering: A* 682: 577–585.

[78] Seyedraoufi, Z.S., Mirdamadi, S., 2013. Synthesis, microstructure and mechanical properties of porous Mg-Zn scaffolds. *Journal of the Mechanical Behavior of Biomedical Materials* 21: 1–8.

[79] Tahmasebifar, A., Kayhan, S.M., Evis, Z., et al., 2016. Mechanical, electrochemical and biocompatibility evaluation of AZ91D magnesium alloy as a biomaterial. *Journal of Alloys & Compounds* 687: 906–919.

[80] Sezer, N., Evis, Z., Kayhan, S.M., et al., 2018. Review of magnesium-based biomaterials and their applications. *Journal of Magnesium & Alloys* 6: 23–43.

[81] Tomac, N., et al., 1991. Formation of flank build-up in cutting magnesium alloys. *CIRP Annals: Manufacturing Technology* 40(1): 79–82.

[82] Han, P., Cheng, P.F., Zhao, C.L., et al., 2017. Comparative study about degradation of high-purity magnesium screw in intact femoral intracondyle and in fixation of femoral intracondylar fracture. *Journal of Materials Science & Technology* 33(3): 305–310.

[83] Henderson, S.E., Verdelis, K., Maiti, S., et al., 2014. Magnesium alloys as a biomaterial for degradable craniofacial screws. *Acta Biomaterialia* 10(5): 2323–2332.

[84] Witte, F., 2010. The history of biodegradable magnesium implants: A review. *Acta Biomaterialia* 6(5): 1680–1692.

[85] Bai, J., Yin, L.L., Lu, Y., et al., 2014. Preparation, microstructure and degradation performance of biomedical magnesium alloy fine wires. *Progress in Natural Science: Materials International* 24(5): 523–530.

[86] Yan, K., Sun, J., Bai, J., et al., 2019. Preparation of a high strength and high ductility Mg-6Zn alloy wire by combination of ECAP and hot drawing. *Materials Science and Engineering: A* 739: 513–518.

[87] Sharifzadeh, M., ali Ansari, M., Narvan, M., et al., 2015. Evaluation of wear and corrosion resistance of pure Mg wire produced by friction stir extrusion. *Transactions of Nonferrous Metals Society of China* 25(6): 1847–1855.

[88] Milenin, A., Kustra, P., Wojcik, D.B., et al., 2020. The influence of the parameters of hot drawing of MgCa alloys wires on the mechanical properties that determine the applicability of the material as a high strength biodegradable surgical thread. *Procedia Manufacturing* 50: 804–808.

[89] Dodyim, N., Yoshida, K., Murata, T., et al., 2020. Drawing of magnesium fine wire and medical application of drawn wire. *Procedia Manufacturing* 50: 271–275.

[90] Niu, X., Shen, H., Fu, J., 2018. Microstructure and mechanical properties of selective laser melted Mg-9wt.%Al powder mixture. *Materials Letters* 221: 4–7.

[91] Karunakaran, R., Ortgies, S., Tamayol, A., et al., 2020. Additive manufacturing of magnesium alloys. *Bioactive Materials* 5(1): 44–54.

[92] Wang, Z.M., Zeng, X.y., et al., 2014. Effect of energy input on formability, microstructure and mechanical properties of selective laser melted AZ91D magnesium alloy. *Materials Science & Engineering, A. Structural Materials: Properties, Microstructure and Processing* 611: 212–222.

[93] Hu, D., Wang, Y., Zhang, D., et al., 2015. Experimental investigation on selective laser melting of bulk net-shape pure magnesium. *Materials and Manufacturing Processes* 30(11): 1298–1304.

[94] Yang, Y., Wu, P., et al., 2016. System development, formability quality and microstructure evolution of selective laser-melted magnesium: Virtual and Physical Prototyping. *Virtual & Physical Prototyping* 11(3): 173–181.

[95] Dong, J., Li, Y., Lin, P., et al., 2020. Solvent-cast 3D printing of magnesium scaffolds. *Acta Biomaterialia* 114: 497–514.

7 Polymeric Materials and Their Components for Biomedical Applications

Lakshmanan Saravanan

CONTENTS

7.1 INTRODUCTION

7.1.1 OVERVIEW OF POLYMERIC MATERIALS

Polymeric components will be the significant material for the invention of green, sustainable, energy-efficient, high quality, economical parts in the new millennium. Polymers exist naturally in the form of DNA, RNA, proteins and polysaccharides in plants and animals. The words 'poly' and 'mers' means 'many' and 'units' respectively, which are taken from Greek. Single molecules linking together 3-dimensionally (3D) through the polymerization process are called polymers. Monomers bonded via different molecular interactions produce polymers with different properties. Artificially produced (man-made) polymers are called synthetic polymers.

Biomaterials are the materials used in device form that are projected to make contact with humans or any biologic matter [1]. Polymeric biomaterials, both synthetic and natural, having larger molecular weight are intended for interfacing with biological systems to regenerate and repair tissue

DOI: 10.1201/9781003173533-7

or human organs. Compared to metals or ceramics, polymeric biomaterials have the advantage of flexibility, biocompatibility and easy fabrication of various shapes. Synthetic polymers gained interest for medicinal applications in drug delivery, vascular stents, sutures, clot removal, orthodontic therapy and so on [2]. Biopolymers are used as a packaged material on implanted devices, to shield them from the moisture and the ions present in the body. Packaging the implants and devices with polymeric materials is also necessary for proper functioning and to avoid thermal influence from the body, and the electrical interference, and to protect internal organs. In WWII the warplane pilots injured by the aircraft canopy made from polymethyl methacrylate (PMMA), did not face any reaction due to those fragments. Later, this PMMA was considered broadly for corneal replacement, as a membrane for blood dialyzer and for damaged skull bones. Earlier in 1947, Ingraham and his co-workers used polyethylene (PE) as an implant material [3].

This chapter discusses some widely used natural and man-made polymeric materials, polymer types, followed by the major applications of various polymeric components in the medical field. This chapter further discusses the modern development of polymeric components, and their classifications, including conducting polymers, supramolecular polymers and polymer matrix composites. It also covers current and potential aspects of polymeric tools in the biomedical market. Additionally, the chapter discusses selective examples of biodegradable, advanced polymeric composites and organic-inorganic hybrid nanocomposites presented in the literature and the enduring research for biomedical applications.

7.1.2 CLASSIFICATION OF POLYMERS

Biopolymers or polymeric biomaterials with complex molecular 3D structures are merely obtained from bio-sources such as proteins, amino acids, fats, plants or trees (resins) [4]. Generally, the polymeric biomaterials are biologically inert, bioresorbable (slowly degraded and substituted for the natural tissue), and bioactive, based on the tissues responding to the implant material placed inside the body. The physical properties of biopolymers (e.g. polysaccharides, polyesters, polyamides) produced from microorganisms depend on the molecular weight and its composition. Polymers are classified based on the source, structure, molecular forces and the polymerization method, and those classifications are illustrated in Figure 7.1.

Natural polymers, cellulose, starch, proteins, resins, etc. are found in living things. Natural polymers which are chemically modified are called semi-synthetic polymers; examples include cellulose derivatives. Linear polymers are formed when monomers couple together as straight chains end to end (polythene and polyvinyl chloride). Branched-chain polymers are linear polymers that contain

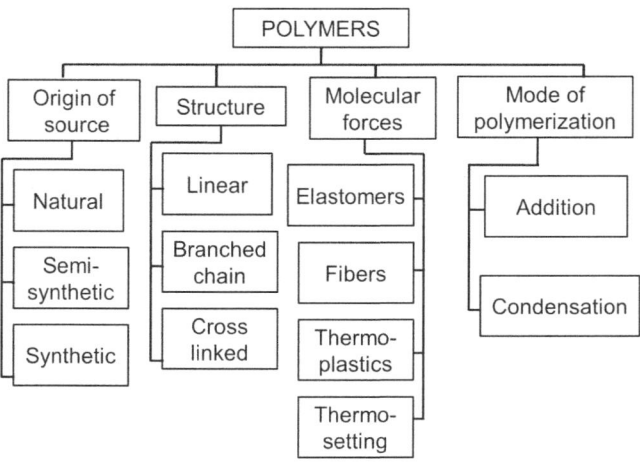

FIGURE 7.1 Classification of polymers.

branches which are replaced by a substituent such as hydrogen atom monomers. The combination of bi- and tri-functional monomers or crosslinking between different polymers chains by the covalent bond is called crosslinked polymers. Additional polymers are prepared by repeatedly adding the monomers. Condensation polymers are formed by the condensation (elimination of water, alcohol) mechanism of different monomeric units. Elastomers are polymers having weak intermolecular forces which possess both viscosity and elasticity. Thermosetting polymers are soft solids that irreversibly transform into insoluble polymers and are not reusable. The conventional and advanced techniques used to fabricate the biomedical devices with different polymeric composites are shown in Figure 7.2.

Poly(α-esters) have an aliphatic ester bond in their backside and some are degradable with short aliphatic chains. Examples are polyglycolide or poly(glycolic acid) (PGA), polylactide (PLA), poly(lactide-co-glycolide) copolymer (PLA-PGA), poly(3-hydroxybutyrate) (PHB), polycaprolactone (PCL), poly(propylene fumarate) (PPF), which is most predominantly used for tissue engineering, drug delivery and as fillers for bone defects. *Polyanhydrides* are polymers comprising two carbonyl groups bounded by an ether bond and are notably used in drug delivery (chemotherapeutics, antibiotics, vaccines and proteins) applications. Degradable *polyacetals* are polymers having one carbon molecule connected with two ether bonds, to focus on delivering drugs directly for heart and cancer diseases. *Poly(ortho esters)* are hydrophobic, having three ether bonds connected with a single carbon molecule and exclusively extended for drug (analgesics, anti-proliferative drugs) delivery. *Polycarbonates* are linear polymers that have one carbon molecule connected with two ether bonds and one carbonyl bond, such as poly(trimethylene carbonate) (PTMC). *Polyurethanes* (PUs) are strong, biocompatible and biostable and contain ester bonds with two amide bonds connected with a carbon molecule. *Polyphosphazenes* are the unique degradable polymers with phosphorous and nitrogen as the backbone, linearly bonded single and double alternatively. *Polyphosphoesters* and their composites are encompassing phosphorous-integrated monomers which are the attractive class of biomaterials for chemotherapy and DNA delivery devices and in bone tissue engineering.

Polyhydroxyalkanoates (PHAs) are used in controlled drug release, surgical sutures, bone plates and wound care. Devices made by the combination of polylactide and polyglycolic acid or polyglycolide (PLA-PGA) copolymers have been applicable for controlled release of antibiotics, as agents for anticancer and anti-malarial drugs, and also they are essential for generating hormones, narcotic antagonists and insulin. The composite contains poly(lactic acid)-poly(ethylene glycol)-chitin (PLA/PEG/Chitin), which is used in bone and dental implants [5]. The polymeric materials employed in different biomedical applications and their advantages are listed in Table 7.1 [6, 7].

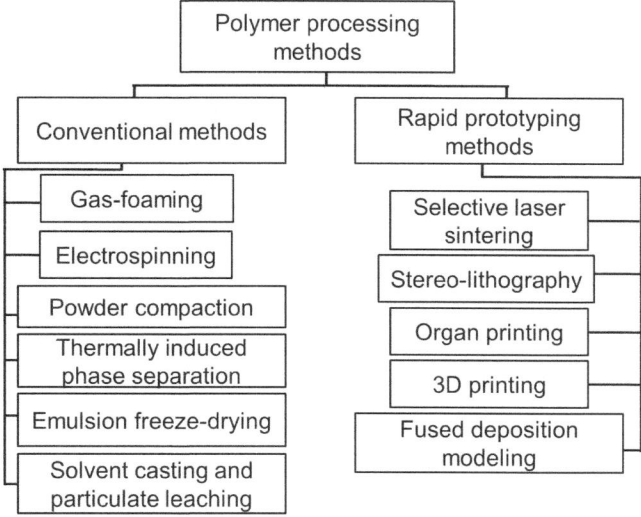

FIGURE 7.2 The processing techniques of polymer composites.

TABLE 7.1
Biomedical Applications of Different Polymers and Their Advantages

Polymers	Biomedical Applications	Advantages
Polyphosphazenes	Tissue engineering	Synthetic flexibility,
	Vaccine adjuvant	Controllable mechanical properties
Polyanhydrides	Drug delivery	Significant monomer flexibility,
	Tissue engineering	Controllable degradation rates
Polyacetals	Drug delivery	Mild pH degradation products,
		pH sensitive degradation
Poly(ortho esters)	Drug delivery	Controllable degradation rates,
		pH sensitive degradation
Polyphosphoesters	Drug delivery	Biomolecule compatibility,
	Tissue engineering	Highly biocompatible degradation products,
Polycaprolactone	Tissue engineering	Highly processable
Polyurethanes	Prostheses	Mechanically strong
	Tissue engineering	
Polylactide	Tissue engineering	Highly processable
	Drug delivery	
Polycarbonates	Drug delivery	Chemically-dependent mechanical properties
	Tissue engineering fixators	Surface eroding
Polyamides	Drug delivery	Highly biocompatible degradation products
	Epidural catheters	Has long-lasting tensile strength and high elasticity
	Orbital implant	
Polyethylene	Retinal prosthesis	Good chemical resistance
	Artificial intraocular lens	Low melting temperature
	Orthopedic implants	Quick drying characteristics
	Cognitive prostheses	
Polypropylene	Synthetic blood vessels	Non-toxic
	Artificial urinary sphincter implant	Good dielectric properties
	Pacemaker	High melting point
Polydimethylsiloxane	Nasal implants for nose	Non-flammable
	Reconstruction, breast implants	Good biocompatibility
	Cognitive prostheses	Good electrical insulation
	Artificial blood vessels	
Polymethylmetacrylate	Glaucoma valve, orbital implant	Poor thermal and electrical conductivity
	Catheters	Acceptable biocompatibility
	Dental implants	Radiolucency
Polytetrafluoroethylene	Pacemaker, epidural catheters	Chemically inert
	Lip implant	Electrically inert
	Central venous access device	Hydrophobic
Polyhydroxyalkanoates	Orthopedic implants	Biocompatibility
	Neuroprosthetics	Biodegradability
	Cognitive prostheses	Excellent barrier properties
	Foley catheter	Flexible strengths
Polyvinylchloride	Blood and solution bag,	Economic
	surgical packaging, IV sets, dialysis	Light weight, flexible durability
	devices, catheter bottles, connectors	

Source: [6, 7].

Synthetic polymers like polyvinylidene fluoride (PVDF), polydimethylsiloxane (PDMS), polytetrafluoroethylene and polymethylmethacrylate (PMMA), have also been employed in biomedical device fabrication.

7.2 PROPERTIES AND TYPES OF POLYMERS

Polymers having their own physical, mechanical properties, and their non-toxic nature are the key benefits for the fabrication of medical device parts and instruments [8]. Based on their electrical property and technical characteristics, the specific application in medical devices is obtained. Analogous to the properties and the bonding mechanism, the polymer materials exist in different types. Examples are polymeric materials with conjugated chain structures called conducting polymers (CPs). The polymer structure resulting from the hydrogen bonding motifs are called supramolecular polymers (SPs) and form an incredible class of synthetic units to illustrate immense application in materials science. SPs are the monomeric arrays having non-covalent hydrogen bonds held together by the highly directional secondary interactions, which are reversible too. The combined system of supramolecules and CPs also gained much attention for the expansion of unique immobilized biomolecular matrices.

7.2.1 CONDUCTING POLYMERS

Conducting polymers have substantial optical and electrical properties comparable with metals and semiconductors and include their own mechanical properties. The advantages of CPs are that they are flexible, anti-corrosive and lightweight. Because of their biocompatibility, CPs have swiftly developed in recent years to augment a range of practical medical purposes [9–11]. The first CPs/oligomers of transition metal thiolates was reported by John Reynolds in 1987 [12]. CPs are materials having electrons in their unsaturated backbones, and these delocalized π-electrons move without restraint within the backbone to develop an electrical trajectory for charge carriers [13]. Polyaniline (PANI) is a well-known conducting polymer. Polypyrrole-doped dodecylbenzenesulphonate (PPy-DBS) is used to produce soft actuators, artificial muscle fibers with respect to the volume change during the redox process. Poly(3,4-ethylenedioxythiophene) (PEDOT) has high chemical stability and superior conductivity for biosensing applications. George et al. attached drug molecules on the CP surface rather than incorporating them into its structure [14]. CPs, owing to their advantage of imitating the natural environment, are considered to be fascinating materials for the immobilization of biomolecules in biosensors. Soylemez et al. prepared a biosensor using a CP, poly(2-(2-octyldodecyl)-4,7-di(selenoph-2-yl)-2H-benzo-[d][1,2,3]triazole)) (PSBTz)-bearing β-cyclodextrin with enhanced performance, and with improved immobilization platform [15]. Recently, the composite containing PPy, PANI, polydopamine, PEDOT for photothermal therapeutic (PTT) applications has been reviewed by different groups [16, 17]. An amperometric glucose biosensor derived with graphite rod as working electrode modified by poly(pyrrole-2-carboxylic acid) (PCPy) particles has been reported by Minkstimiene [18]. CPs incorporated with hydrogels to form porous conducting hydrogels and the CP composites contain CNT, graphene and oxide nanoparticles exhibit significant features for biosensors.

7.2.2 SUPRAMOLECULAR POLYMERS

Assembling the biomaterials through non-covalent interactions by supramolecular chemistry is emerging to be a promising method of mimicking the unique dynamics of nature. Monomeric building blocks self-assembled into polymeric nanostructures by the reversible non-covalent interactions and highly directional, are termed as supramolecular polymers, having a vital self-healing property. Supramolecular block copolymers have their advantages of exceptional biodegradability, low cytotoxicity and smart environmental responsiveness. SPs can be used as thermoplastic elastomers for short-term biomaterials in regenerative medicine and in tissue engineering. Covalent polymers with crosslinked non-covalent interactions are represented as supramolecular polymer networks (SPNs) [19]. Some examples of SPN formed between the covalent polymers are polyethylene glycol (PEG), polypropylene glycol (PPG) and cyclodextrin, a water-soluble host, and SPN constructed from covalent polymers that exploit calix[n]arene (phenol-formaldehyde cyclic oligomers) based host [20–22].

7.2.3 POLYMER COMPOSITES

Composites containing one continuous and intermittent part are termed as reinforcement, while a continuous phase is called a matrix. It is a vital part of numerous products/devices due to its low weight, corrosion resistive, high wearable strength and rapid assembly. Human organs are the natural composites in which the micro-structure and properties conclude its behavior. When different types of polymers merge to form a thin layer of composite, the properties of each layer differ compared to their parent materials. The key advantage of polymers composites compared to metals and ceramics is that they are non-magnetic and radio transparent for X-ray radiography and magnetic resonance imaging (MRI). Polymer composites will deliver high strength and are osteo-conductive, which is more suitable for load bearing and high potential for (bone) tissue engineering applications.

7.3 POLYMER MATRIX COMPOSITES

7.3.1 PREAMBLE

Polymeric composites are most preferable in biomedical device production, owing to the ease of manufacture, flexibility and biocompatibility. By precisely tuning the properties such as mechanical, chemical, biological and combined with functional materials to form polymeric composites, will open an innovative platform for the fabrication of modern integrated biomedical devices. Polymeric composites, which include reinforced plastics, and advanced composites, which have low density, light weight and high strength, are used extensively in the invention of medical devices such as prostheses and implants, surgical instruments and orthopedic products.

CNTs incorporation into polymer matrices containing collagen, polylactide (PLA), PMMA or poly(N-isopropyl acrylamide-co-methacrylic acid) (PNIPAAM) increases their mechanical strength and permits new functional properties [23]. Combining the inorganic materials properties such as high hardness, refractive index, thermal stability, chemical stability, etc., with the features of organic polymers, like easy processing ability, flexible nature, low weight, can extend these hybrids in a broad spectrum of biomedical applications. Aromatic polyamides (aramids) fiber reinforced low dense composites exhibit low weight, better strength and modulus, high elongation property to demonstrate cost-effective behavior.

7.3.2 REINFORCED PLASTICS

Reinforced plastics or reinforced polymer composites consist of blended polymer matrix with reinforcing materials. Some are fibers made of basalt, carbon, glass and aramid [24]. Plastics are being exploited to make medical products ranging from small tools to implants devices. Carbon and glass fibers are added with resins to make reinforced plastics with improved mechanical properties. Polyester, epoxy, vinylester and phenolic based resins are employed to obtain these reinforced plastics. Carbon-fiber reinforced plastic (CFRP) material is used in prosthetics and trauma devices— implants that embrace fractured bones. Carbon-fiber reinforced polyetheretherketone (CFR-PEEK) materials are used as implants for orthopedics and as perfect material for articulating implants, knee replacement tools, and are useful to produce leads for cardiology and neurology [25, 26]. Biological activity of PEEK can be improved by incorporating bioactive hydroxyapatite particles onto the surface of implants to accelerate the growth of bone and develop the attachment between implant and bone tissue. A high-modulus thermoplastic containing 30% carbon fiber reinforced polycarbonate and polyetherimide (PEI) resin, exhibits high stiffness to make disposable surgical instruments, fixation devices, medical device housings and drug delivery components [27].

7.3.3 Advanced Polymeric Nanocomposites

Advanced polymeric nanocomposites have an advantage of combining individual properties of both polymeric material and nanoparticles, and this led to employment in several therapies and diagnostics. Carbon materials, for example carbon black, graphene, carbon nanotubes are the materials used to prepare advanced functional polymer nanocomposites. Both single-walled carbon nanotubes (SWCNTs) and multi-walled (MWCNTs) and carbon nanofibers (CNFs) are considered to be promising reinforcing materials for acrylic polymers like PMMA. To improve the biological property together with mechanical and electrical properties and to minimize the toxicity, the CNT-polymer composite has been developed for neural tissue-, cardiac tissue- and bone tissue-engineering applications. Graphene-polymer hybrids exhibit considerable improvements in the physical properties [28, 29]. Nanoclay-based polymer composites are another grade of biodegradable plastics made by the polymer solution- and melt-embedding, and in-situ polymerization.

Polymeric composites are made from the combination of metal oxides (e.g. TiO_2, ZnO, CuO, Fe_2O_3 and Fe_3O_4) and the polymers are reported to be safe and prominent materials for various biomedical applications. Integration of iron oxide nanoscale particles and fluorine-containing compounds into the polymer matrix contains PEG-dimethacrylate (PEG-DMA), reported to have the unique capabilities to remote 3D magnetic function and MRI localization for in vitro tissue engineering [30], and also stated that they can stay long for years before being hydrolyzed [31]. A medical grade, layered S-nitroso-acetylpenicillamine (SNAP) doped with silicon-based CarboSil polymer blended with selenium (Se) polymer composite was recently developed, to prevent the constraint of infection and thrombosis associated with device implants, by the integration of nitric oxide release and generation [32]. Poly (l-lactic acid) (PLLA), PGA, poly lactic-co-glycolic acid (PLGA), poly(d,l-lactic Acid) (PDLLA), poly(ethylethylene-b-ethylene oxide) (PEE-PEO) and polybutylene terephthalate (PBT) are the artificial polymers used to produce porous scaffolding for medical applications. Scientists have developed adhesive surgical glue having methacryloyl-substituted tropoelastin named MeTro, a hybrid non-toxic elastic protein with less degradation rate that could alter emergency treatments by sealing up damaged skin or the organs, without requiring any staples or sutures [33]. Polycaprolactone (PCL) is biodegradable polyester, prepared by polymerization of ε-caprolactone which has a low melting point, superior thermal property, and low viscous nature. It also is combined with other polymers to raise adhesive properties and stress resistance.

7.3.4 Organic-Inorganic Polymer Hybrids

Functional hybrids with combined properties such as luminescence (imaging), molecular recognition (targeting) and non-linear optical (NLO) (therapy), together with size/shape control and biocompatibility (cells diffusion, solubility, biochemical stability), are developed for biomedical applications. A wide range of organic, natural or synthetic polymers and ceramic species have been combined to produce inorganic-organic hybrids which are focused to fabricate medicinal devices. The biocompatibility and biodegradability was achieved by the polymer (organic) part, and the intended mechanical properties were achieved through the inorganic content, by tuning the type, concentration and distribution. Polysiloxanes having Si-O backbone attached with two organic mono-valence radicals of each silicon atom such as PDMS is important in bridging between inorganic and organic polymers [34]. An-isometric particles, carbon black, graphene, CNTs are added to make polysiloxane electrically conductive due to their low percolation threshold values by create conducting pathways [35]. With their intrinsic electrical properties, the ultrasensitive glucose biosensor was made by the composite containing MWCNTs in poly(acrylonitrile-co-acrylic acid) (PANCAA) for diabetics [36]. The organic-inorganic hybrid nanocomposite consists of poly(2-hydroxyethyl acrylate) and silica nanoparticles (biocompatible and bioactive) and displayed significant enhancement in the mechanical properties [37]. Yang et al. synthesized hybrids by combining polystyrene, poly(methyl methacrylate), (PMMA) and their copolymers with 3-(trimethoxysilyl) propyl methacrylate for dental applications [38]. A group recently studied Gd-containing micelles for neutron capture therapy, in which the micelles contain inorganic-organic hybrid composites with calcium phosphate as a core, and PEG as shell, and incorporated with Gd-diethylenetriaminepentaacetic acid [39]. Cao et al. fabricated nanoscale hybrid structures with SnO_2 as a core material coated with porous silica, and polyethyleneimine (PEI) for drug delivery applications [40]. Shirosaki reported about the synthesis techniques of different types of biocompatible and biodegradable polymeric components containing silicon-based, organic-inorganic hybrids by incorporating chitosan and γ-glycidoxypropyltrimethoxysilane (GPTMS) for biomedical applications [41].

7.3.5 Synthesis and Process Parameters of Polymeric Composites

Feuser et al. synthesized PMMA nanoparticles by the low temperature mini-emulsion process, followed by polymerization at 70 °C with a biocompatible and biodegradable surfactant, lecithin, to achieve high stability and without cytotoxicity, which can potentially be used as an encapsulation agent to treat leukemia and solid tumors [42]. Novel PEEK copolymers were obtained using diphenyl sulfone as a solvent by solution polymerization process and later produced PEEK/CF and PEEK/GO composites by suspension blending method. The synthesized PEEK composites deliver superior mechanical strength the same as natural bone, showing non-toxic behavior with better thermal property for spinal cage applications in orthopedics [43]. Reduced graphene oxide incorporated gelatin methacryloyl (rGO-GelMA) hybrid conductive hydrogels was prepared by photo-crosslinking process for the construction of tissue engineering scaffolds for cardiac treatment. Here, the UV exposure time was extended to attain sufficient crosslinking, with higher rGO content and the hydrogel thickness was controlled by varying the quantity of rGO nanoparticles which block the transmission of UV inside the scaffold [44]. The β-G/BC nanocomposite was synthesized by free-radical polymerization process using β-glucan (extracted from barley flour) and bacterial cellulose using N',N-methylene-bis-acrylamide as a crosslinker. HAp nanoparticles and GO were added as reinforcement into this polymer network. Further, the biocompatible porous scaffold fabricated via freeze-drying (−40 °C) technique was used to be a potential candidate for the treatment of damaged bones in orthopedic (bone) tissue engineering [45]. GelMA/PEDOT:PSS composite hydrogels was synthesized at 4 °C by combining poly(3,4-ethylenedioxythiophene):polystyrene sulfonate and gelatin methacryloyl by dielectrophoresis (DEP) method. Here gelatin was taken from cold water fish skin. GelMA containing higher methacryloyl content was first synthesized by photo-crosslinking process, using methacrylic anhydride. The synthesized electroconductive composite was employed as a drug delivery vehicle and implantable neural electrodes [46]. Yabuta et al. synthesized organically modified silicates (ORMOSILs) using sucrose template, based on PDMS and tetraethoxysilane (TEOS) by sol-gel process, for the application for bone and soft-tissue generation. The porous bioactive sample was tested with body fluid, equivalent to human blood plasma for the formation of a biologically active apatite layer. Ca(II) was incorporated into the porous silicates which is essential for the efficient bioactivity and apatite growth [47]. It was reported that the production of RGO coated PDMS 3D scaffold by the salt porogen at 4 °C and by dip-coating process, followed by curing at 60 °C and leaching method. High conductivity was attained by the optimal RGO coating. Highly porous 3D scaffold achieved with high durability and cytocompatibility for better growth of adipose stem cells (ADSCs) in humans, to be used in orthopedic surgery [48].

7.4 MAJOR BIOMEDICAL APPLICATIONS OF POLYMERIC COMPONENTS

Several bio- and artificial polymers widely used in medical devices correspond to tissue engineering, drug delivery, biosensors, bio-imaging, implants, etc. Some of the practical applications of polymeric components in the medicinal field are displayed in Figure 7.3. Practically, the biomaterials are not directly used as a final product or device, but they are integrated into other medical products, to make the accurate biological response. Fabrication of such medical devices utilizing polymer biomaterials ranges from a small tongue depressor to the complex artificial heart or tissue regeneration. An example is drug vials made from polyolefins, high-density polyethylene (HDPE) and polypropylene (PP).

Developing synthetic functional hybrid biomaterial for bone regeneration is to mimic the bone composition and to have the ability to bond the natural bone and to further stimulate the cells for bone regeneration [49]. Synthesis types of some polymeric materials and their applications in biomedical research are listed in Table 7.2. Some of the important medical applications consuming polymers and its matrix composites are reviewed in this section.

7.4.1 Tissue Engineering

Tissue engineering is crucial in medicinal applications which involve cells manipulation to promote the regeneration and healing of defective tissues in the body. It also means developing the biological substitutes which can restore, or improve functions of the human organ. Tissue engineering describes the combination of living and functional cells with biocompatible materials, called scaffolds. The vital path in tissue engineering applications is to develop suitable bio-interfaces, that is, integrating and controlling appropriate tissues and their properties. The polymer matrix composites

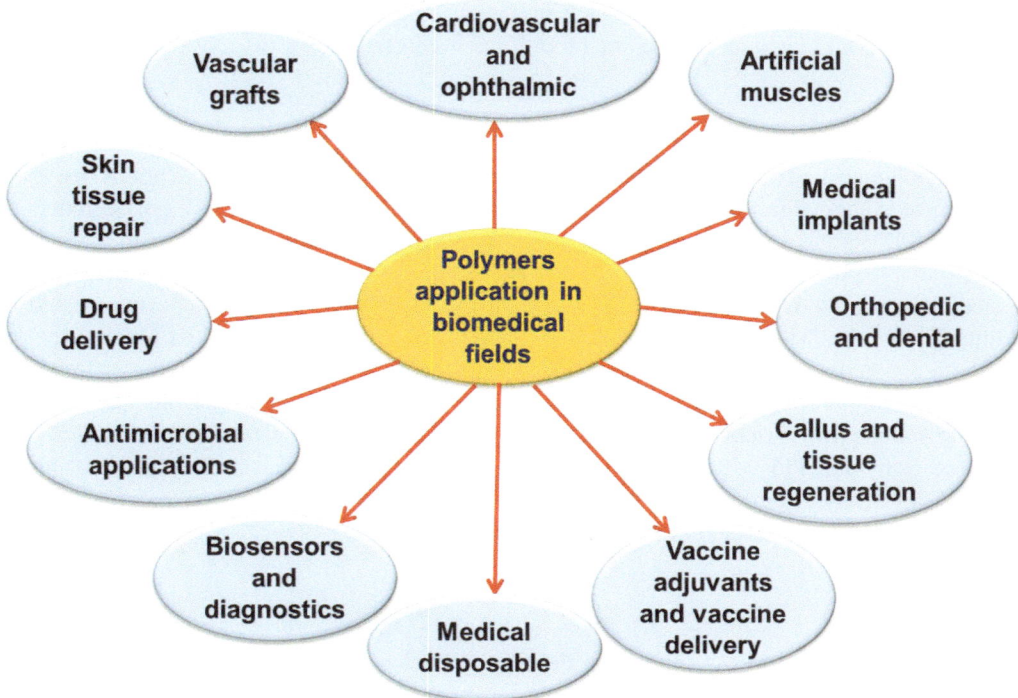

FIGURE 7.3   The possible applications of polymeric materials in the biomedical research areas.

TABLE 7.2
Synthesis Types of Some Polymeric Materials and Their Biomedical Applications.

Synthesis Type	Polymeric Material	Biomedical Applications	Reference
Electrospraying	PCL, PMMA microparticles	Drug delivery carriers, tissue engineering	[65]
Electrospinning	Polyhydroxyalkanoates (PHA-based blends) micro/nanofibers—(eg. P3HB). PCL/PPy nanofibrous scaffolds	Scaffolds for tissue engineering Bone tissue engineering	[66, 67]
Phase separation freeze-drying method	Fe_3O_4 blended gelatin/hydroxyapatite 3D magnetic nanocomposite	Scaffolds for bone replacement and drug delivery device	[68]
3D printing	MoS_2-PLGA composite coated borosilicate bioactive glass	3D Scaffolds for anti-tumor/bone tissue repair	[69]
Polymerization reaction	Crosslinked Tragacanth gum (TG)/PVP-PAAc hydrogels Crosslinked bacterial cellulose and gelatin composite hydrogels	Drug delivery system	[70, 71]
Gel-spinning process	Dialdehyde carboxymethyl cellulose (DCMC) crosslinked gelatin-PEG composite hydrogel fibers	Wound dressings	[72]

composed of synthetic polymers such as polyglycolic acid (PGA), polylactic acid (PLA), polycapro-lactone (PCL), poly (N-isopropylacrylamide) (PNIPAM), etc., or native polymers such as chitosan, alginate, collagen, hyaluronic acid, gellan gum, gelatin and fibrin are mostly used. For example, the combination of collagen-chitosan and graphene scaffold is developed as a conductive composite film, considering chitosan for its blood clotting factor and graphene with angiogenic properties as an antibacterial, making it an ideal biomaterial for skin regeneration [50].

7.4.2 ORTHOPEDIC

Recently, polymers and their components have been employed in orthopedic applications due to their resemblance to the natural bone tissues. PMMA, polyaryletherketones (PAEK), polysulfone, poly-propylene, PDMS and ultrahigh-molecular weight polyethylene (PE) are the most widely employed non-degradable polymers for orthopedic application. PMMA has been reported to have been used in medical implants as bone cement in total joint replacements [51]. Early reports revealed that adding CNTs into PMMA drastically improves the mechanical properties and reduces the exo-thermic reaction during polymerization of the bone cement [52]. Another researcher examined the sulfur-containing non-degradable thermoplastic polymers such as polyethersulfone and sulfonated polyether sulfone (SPES) nanofibers for bone regeneration [53]. Poly(α-esters), polyfumarates (poly-propylene fumarate (PPF)) [54], PUs, polyanhydrides, poly(amino acids) such as poly-γ-(glutamic acid) (γ-PGA), poly(aspartic acid) (PAA), poly(arginine), poly(lysine) [55] and polyphosphazenes are the few well-known biodegradable polymers employed in orthopedics.

7.4.3 DENTAL

Polymers also have significant interest in dental application owing to their excellent biocompat-ibility, mechanical strength and easy processing. Tissue engineering based methodologies have also been demonstrated to attract diverse strategies for dental pulp repair or regeneration. Polymers made from hydroxy-apatite (or hydroxylapatite) are a kind of calcium phosphate with the formula

$Ca_{10}(PO_4)_6$ $(OH)_2$ and are used to produce fillings for teeth. To treat different types of oral diseases and to improve tissue regeneration, the biodegradable polymer components made of PLA or PLGA can be employed, which is used to load drugs or bioactive macromolecules. The organic-inorganic complex scaffolds gain more interest in dental applications than single scaffold polymers. Na-hyaluronate/chitosan polyelectrolyte complex scaffold and microbeads made of collagen/fibrin containing Ag-doped bioactive glass are a few examples of such complex polymer scaffolds [56].

7.4.4 DRUG DELIVERY

Formerly, different nanoscale structures such as nanoparticles, nanogels and micelles have been focused to develop drug delivery systems due to their longer retention time, less toxicity, excellent biodegradable behavior and better drug accumulation. Most of the drug delivery carriers include magnetic nanoparticles to improve the safe drug release, and to reduce its toxicity. Recently, Mona et al. prepared iron oxide nanoparticles coated with inorganic layered double (Mg/Al) hydroxides (LDHs) along with polyvinyl alcohol (PVA) and PEG for efficient and controlled anticancer drug delivery [57]. Tajau et al. synthesized poly(APO-b-polyol ester) nanoparticles using acrylated palm olein (APO) and polyol ester as a drug delivery system for breast cancer therapy [58]. Li et al. developed a dual pH- and reduction-responsive system for anticancer drug delivery, based on conjugating methoxy-poly(ethylene glycol) and polysuccinimide using a disulfide linked block copolymer, which showed excellent cytocompatibility and enhanced cytotoxicity against cancer cells [59].

7.4.5 CARDIOVASCULAR

Worldwide cardiovascular disease is reported to be the major cause of death, and it is becoming a major global threat in the 21st century [60]. Cardiovascular devices are therapeutic supports for heart and blood vessels. For cardiac tissue engineering, two classes of materials, scaffolds and hydrogels, are commonly used [61]. Some of the synthetic polymeric materials mostly used to develop scaffolds for cardiovascular prostheses are polycaprolactone (PCL), polyhydroxyalkanoate, polytetrafluoroethylene, PU, polyglycerol sebacate, PEG, poly(N-isopropylacrylamide) and the natural polymeric materials used in cardiac tissue engineering applications are collagen, fibrin, chitosan, gelatin, alginate, hyaluronic acid and a combination of these polymers [62].

7.5 PRESENT AND FUTURE TRENDS OF POLYMERS IN BIOMEDICAL DEVICES

Recently, the healthcare industry is under pressure in developing modern devices that deal with an increased rate of chronic diseases and injuries while simultaneously maintaining special standards and properties with a reasonable cost. A leading research analyst C.B. Jaganathan, from Technavio©, a top technological research and consulting company, projected that "the composite materials in medical market in North America was valued at US$1.77 billion in 2015, and is projected to reach US$2.42 billion by 2020". The factors that are contributing to the growth in the global market are the production of lightweight and thermal-resistant composites, fiber-reinforced composites and advanced composite materials for biomedical devices. There is another eminent analysis report, which predicts that the biomedical polymers market globally will grow with 9.17% of CAGR from 2017 to 2025. This study covers the analysis of the chief geographical regions such as Europe, North America, Asia-Pacific and the rest of the world on the medical polymers market in the years 2017 to 2025 [63]. Biodegradable plastics production is progressively increasing, with an estimation of 2.6 million tons of production projected in 2023. In the future it is expected that polyethylene furanoate (PEF), a new bioplastic, will be in the market by 2023, for beverage bottles production (plasticstoday.com) [64]. Continuing research on biodegradable plastics is to explore the alternatives to develop novel nanocomposites with better mechanical strength which are biodegradable and possible to withstand higher temperatures. This includes a range of artificial polymers, like ultrahigh molecular weight polyethylene (UHMWPE), often used in hip and knee repair; PMMA for bone cements; and polyethylene terephthalate (PET), commonly used in vascular prostheses and sutures. In the future, it is also possible to employ a range of synthetic polymers to play a vital role in pharmaceutical industries too. Further, introducing the rapid prototyping techniques like 3D printing/additive manufacturing, the demands to fabricate implants by the use of functional polymeric composites with distinctive properties can be easily achieved.

7.6 SUMMARY

In summary, this chapter presented the properties, importance and applications of several polymers in medicine. This chapter discussed the utilization of several polymeric components and polymer matrix composites, biopolymers and organic-inorganic polymer hybrids which are applicable to develop various biomedical devices. The basic properties of polymers are super hydrophobicity, adhesion and self-healing and found to be useful in biomedical applications. The polymer composites should possess non-toxic, biocompatible and biodegradable properties, because they have contact with the tissues or organs of human body directly. Significant improvement in the antibacterial activity without cytotoxicity was observed in several graphene-based hybrids, suggesting a potential biomaterial for medicinal applications. Some examples are biosensing, bioimaging, gene or drug delivery, cancer therapy and tissue engineering. CNTs are used as a filler to improve mechanical and electrical properties and allow the fabrication of polymer-based scaffolds for neural, cardiac and bone tissue engineering. Importantly, CNT in polymer matrix promotes the conductivity, elasticity and biological responses of cardiac related scaffolds. The ongoing and imminent research directions in the biomedical sciences are projected to develop several multifunctional polymer-based biomaterials which are tuned to match the required properties of desired tissues or organs of the human body and are appropriate for the fabrication of medicinal devices. Natural and synthetic polymeric materials, both degradable and non-degradable, could actively take part in the development of devices for medical implants. With the development of rapid prototyping techniques, like 3D-printing, numerous polymeric components required for standard implants can be easily made as well.

REFERENCES

1. D.F. Williams, Definitions in biomaterials, Proceedings of a Consensus Conference of the European Society for Biomaterials, Chester, England, 3–5 March 1986, 4, New York, Elsevier (1987).
2. A. Lendlein, M. Behl, B. Hiebl, C. Wischke, Shape-memory polymers as a technology platform for biomedical applications, *Expert Rev Med Device*, 7 (2010) 357–379.
3. F.D. Ingraham, E. Alexander Jr, D.D. Matson, Polyethylene, a new synthetic plastic for use in surgery: Experimental applications in neurosurgery, *J Am Med Assoc*, 135 (2) (1947) 82–87.
4. N. Hernandez, R.C. Williams, E.W. Cochran, The battle for the "green" polymer: Different approaches for biopolymer synthesis: Bioadvantaged vs. bioreplacement, *Org Biomol Chem*, 12 (2014) 2834–2849.
5. M.B. Coltelli, P. Cinelli, V. Gigante, L. Aliotta, P. Morganti, L. Panariello, A. Lazzeri, Chitin nanofibrils in Poly(Lactic Acid) (PLA) nanocomposites: Dispersion and thermo-mechanical properties, *Int J Mol Sci* 20 (2019) 504.
6. B.D. Ulery, L.S. Nair, C.T. Laurencin, Biomedical applications of biodegradable polymers, *J Polym Sci B Polym Phys*, 49 (12) (2011) 832–864.
7. A. Teo, A. Mishra, I. Park, Y.J. Kim, W.T. Park, Y.J. Yoon, Polymeric biomaterials for medical implants and devices, *ACS Biomater Sci. Eng*, 2 (4) (2016) 454–472.
8. R.A. Perez, J.E. Won, J.C. Knowles, H.W. Kim, Naturally and synthetic smart composite biomaterials for tissue regeneration, *Adv Drug Deliv Rev* 65 (4) (2013) 471–496.
9. W. Wu, G.C. Bazan, B. Liu, Conjugated-polymer-amplified sensing, imaging, and therapy, *Chem* 2 (6), (2017) 760–790.

10. G. Kaur, R. Adhikari, P. Cass, M. Bown, P. Gunatillake, Electrically conductive polymers and composites for biomedical applications, *RSC Adv* 5 (47) (2015) 37553–37567.

11. R. Balint, N.J. Cassidy, S.H. Cartmell, Conductive polymers: Towards a smart biomaterial for tissue engineering, *Acta Biomater* 10 (6) (2014) 2341–2353.

12. J.R. Reynolds, J.C.W. Chien, C.P. Lillya, Intrinsically electrically conducting poly(metal tetrathiooxalates), *Macromolecules* 20 (6) (1987) 1184–1191.

13. T. Nezakati, A. Seifalian, A. Tan, A.M. Seifalian, Conductive polymers: Opportunities and challenges in biomedical applications, *Chem Rev* 118 (14) (2018) 6766–6843.

14. P.M. George, D.A. LaVan, J.A. Burdick, C.Y. Chen, E. Liang, R. Langer, Electrically controlled drug delivery from biotin-doped conductive polypyrrole, *Adv Mater* 18 (5) (2006) 577–581.

15. S. Soylemez, S.O. Hacioglu, M. Kesik, H. Unay, A. Cirpan, L. Toppare, A novel and effective surface design: Conducting polymer/β-cyclodextrin host-guest system for cholesterol biosensor, *ACS Appl Mater Interfaces*, 6 (2014) 18290–18300.

16. Y. Wang, H.M. Meng, G. Song, Z. Li, X.B. Zhang, Conjugated polymer based nanomaterials for photothermal therapy, *ACS Appl. Polym Mater*, 2 (10) (2020) 4258–4272.

17. F. Lin, Q.Y. Duan, F.G. Wu, Conjugated polymer-based photothermal therapy for killing microorganisms, *ACS Appl. Polym Mater*, 2 (10) (2020) 4331–4344.

18. A.K. Minkstimiene, L. Glumbokaite, A. Ramanaviciene, E. Dauksaite, A. Ramanavicius, An amperometric glucose biosensor based on poly (pyrrole-2-carboxylic acid)/glucose oxidase biocomposite, *Electroanal* 30 (8) (2018) 1642.

19. D. Xia, P. Wang, X. Ji, N.M. Khashab, J.L. Sessler, F. Huang, Functional supramolecular polymeric networks: The marriage of covalent polymers and macrocycle-based host-guest interactions, *Chem Rev*, 120 (13) (2020) 6070–6123.

20. R.M. Ariza, L.G. Suarez, W. Verboom, J. Huskens, Cyclodextrin-based supramolecular nanoparticles for biomedical applications, *J. Mater. Chem. B* 5 (2017) 36–52.

21. X. Ma, Y. Zhao, Biomedical applications of supramolecular systems based on host-guest interactions, *Chem Rev* 115 (15) (2015) 7794–7839.

22. G. Liu, Q. Yuan, G. Hollett, W. Zhao, Y. Kang, J. Wu, Cyclodextrin-based host-guest supramolecular hydrogel and its application in biomedical fields, *Polym. Chem.* 9 (2018) 3436–3449.

23. M.C. Serrano, M.C. Gutie´rrez, F.D. Monte, Role of polymers in the design of 3D carbon nanotube-based scaffolds for biomedical applications, *Prog. Polym. Sci.* 39 (7) (2014) 1448–1471.

24. M.P. Groover, *Fundamentals of modern manufacturing materials, processes, and systems*, 4th Ed. New York, USA: John Wiley & Sons; 2010.

25. C.S. Li, C. Vannabouathong, S. Sprague, M. Bhandari, The use of carbon-fiber-reinforced (CFR) PEEK material in orthopedic implants: A systematic review, *Clin Med Insights Arthritis Musculoskelet Disord.* 8 (2015) 33–45.

26. N. Bonnheim, F. Ansari, M. Regis, P. Bracco, L. Pruitt, Effect of carbon fiber type on monotonic and fatigue properties of orthopedic grade PEEK, *J Mech Behav Biomed Mater* 90 (2019) 484–492.

27. © 2020 Plastics.gl.

28. D.K. Patel, Y.R. Seo, K.T. Lim, Stimuli-responsive graphene nanohybrids for biomedical applications, *Stem Cells International* (2019) 9831853.

29. T.B. Rouf, J.L. Kokini, Biodegradable biopolymer-graphene nanocomposites, *J Mater Sci* 51 (2016) 9915–9945.

30. J.R. Pinney, G. Melkus, A. Cerchiari, J. Hawkins, T.A. Desai, Novel functionalization of discrete polymeric biomaterial microstructures for applications in imaging and three-dimensional manipulation, *ACS Appl Mater Interfaces* 6 (16) (2014) 14477–14485.

31. S.L. Gibson, S. Bencherif, J.A. Cooper, S.J. Wetzel, J.M. Antonucci, B.M. Vogel, F. Horkay, N.R. Washburn, Synthesis and characterization of PEG dimethacrylates and their hydrogels, *Biomacromolecules* 5 (4) (2004) 1280–1287.

32. A. Mondal, M. Douglass, S.P. Hopkins, P. Singha, M. Tran, H. Handa, E.J. Brisbois, Multifunctional S-Nitroso-N-acetylpenicillamine-incorporated medical-grade polymer with selenium interface for biomedical applications, *ACS Appl Mater Interfaces* 11 (38) (2019) 34652–34662.

33. N. Annabi, Y.N. Zhang, A. Assmann, E.S. Sani, G. Cheng, A.D. Lassaletta, A. Vegh, B. Dehghani, G.U.R. Esparza, X. Wang, S. Gangadharan, A.S. Weiss, A. Khademhosseini, Engineering a highly elastic human protein: Based sealant for surgical applications, *Sci Transl Med*, 9 (2017) eaai7466.

34. J.A.G. Calderón, D.C. López, E. Pérez, J.V. Montesinos, Polysiloxanes as polymer matrices in biomedical engineering: Their interesting properties as the reason for the use in medical sciences, *Polymer Bulletin*, 77 (2020) 2749–2817.

35. M. Norkhairunnisa, A. Azizan, M. Mariatti, H. Ismail, L.C. Sim, Thermal stability and electrical behavior of polydimethylsiloxane nanocomposites with carbon nanotubes and carbon black fillers, *J Compos Mater* 46 (8) (2012) 903–910.

36. Z.G. Wang, Y. Wang, H. Xu, G. Li, Z.K. Xu, Carbon nanotube-filled nanofibrous membranes electrospun from poly(acrylonitrile-co-acrylic acid) for glucose biosensor, *J Phys Chem C* 113 (7) (2009) 2955–2960.

37. J.C.R. Hernández, A.S. Aroca, J.L.G. Ribelles, M.M. Pradas, Three-dimensional nanocomposite scaffolds with ordered cylindrical orthogonal pores, *J Biomed Mater Res Part B Appl Biomater* 84 (2) (2008) 541–549.

38. J.M. Yang, H.S. Chen, Y.G. Hsu, F.H. Lin, Y.H. Chang, Organic-inorganic hybrid sol-gel materials, 2-application for dental composites, *Angew Makromol Chem* 251 (1997) 61–72.

39. P. Mi, N. Dewi, H. Yanagie, D. Kokuryo, M. Suzuki, Y. Sakurai, Y. Li, I. Aoki, H. Ono, H. Takahashi, H. Cabral, N. Nishiyama, K. Kataoka, Hybrid calcium phosphate-polymeric micelles incorporating gadolinium chelates for imaging-guided gadolinium neutron capture tumor therapy, *ACS Nano* 9 (6) (2015) 5913–5921.

40. N. Cao, M. Li, Y. Zhao, L. Qiu, X. Zou, Y. Zhang, L. Sun, Fabrication of SnO_2/porous silica/polyethyleneimine nanoparticles for pH-responsive drug delivery. *Mater Sci Eng C* 59 (2016) 319–323.

41. Y. Shirosaki, Preparation of organic: Inorganic hybrids with silicate network for the medical applications, *J Ceram Soc Jpn*, 120 (1408) (2012) 555–559.

42. P.E. Feuser, P.C. Gaspar, E.R. Júnior, M.C.S. da Silva, M. Nele, C. Sayer, P.H.H. de Araújo, Synthesis and characterization of poly(methyl methacrylate) PMMA and evaluation of cytotoxicity for biomedical application, *Macromol Symp* 343 (1) (2014) 65–69.

43. J.W. Chon, X. Yang, S.M. Lee, Y.J. Kim, I.S. Jeon, J.Y. Jho, D.J. Chung, Novel PEEK copolymer synthesis and Biosafety. I: Cytotoxicity evaluation for clinical application, *Polymers* 11(11) (2019) 1803.

44. S.R. Shin, C. Zihlmann, M. Akbari, P. Assawes, L. Cheung, K. Zhang, et al., Reduced graphene oxide-GelMA hybrid hydrogels as scaffolds for cardiac tissue engineering, *Small* 12 (27) (2016) 3677–3689.

45. M.U.A. Khan, S. Haider, A. Haider, S.I.A. Razak, M.R.A. Kadir, S.A. Shah et al., Development of porous, antibacterial and biocompatible GO/n-HAp/bacterial cellulose/β-glucan biocomposite scaffold for bone tissue engineering, *Arab J Chem*, 14 (2) (2021) 102924.

46. A.R. Spencer, A. Primbetova, A.N. Koppes, R.A. Koppes, H. Fenniri, N. Annabi, Electroconductive gelatin methacryloyl-PEDOT:PSS composite hydrogels: Design, synthesis, and properties, *ACS Biomater Sci Eng*, 4 (5) (2018) 1558–1567.

47. T. Yabuta, E.P. Bescher, J.D. Mackenzie, K. Tsuru, S. Hayakawa, A. Osaka, Synthesis of PDMS-based porous materials for biomedical applications, *J Sol-Gel Sci Techn* 26 (1–3) (2003) 1219–1222.

48. J. Li, X. Liu, J.M. Crook, G.G. Wallace, Development of a porous 3D graphene-PDMS scaffold for improved osseointegration, *Colloids Surf B Biointerfaces* 159 (2017) 386–393.

49. E. O'Neill, G. Awale, L. Daneshmandi, O. Umerah, K.W.H. Lo, The roles of ions on bone regeneration, *Drug Discovery Today* 23 (4) (2018) 879–890.

50. T. Liu, W. Dan, N. Dan, X. Liu, X. Liu, X. Peng, A novel grapheme oxide-modified collagen-chitosan bio-film for controlled growth factor release in wound healing applications, *Mater Sci Eng*, C 77 (2017) 202–211.

51. A. Rivkin, A prospective study of non-surgical primary rhinoplasty using a polymethylmethacrylate injectable implant, *Dermatol Surg* 40 (3) (2014) 305–313.

52. R.W. Ormsby, M. Modreanu, C.A. Mitchell, N.J. Dunne, Carboxyl functionalised MWCNT/polymethyl methacrylate bone cement for orthopaedic applications, *J Biomater Appl* 29 (2) (2014) 209–221.

53. I. Shabani, V.H. Asl, M. Soleimani, E. Seyedjafari, S.M. Hashemi, Ion-exchange polymer nanofibers for enhanced osteogenic differentiation of stem cells and ectopic bone formation, *ACS Appl Mater Interfaces* 6 (1) (2014) 72–82.

54. K.U. Lewandrowski, S.P. Bondre, D.L. Wise, D.J. Trantolo, Enhanced bioactivity of a poly(propylene fumarate) bone graft substitute by augmentation with nano-hydroxyapatite, *Biomed Mater Eng* 13 (2003) 115–124.

55. H. Li, S. Tao, Y. Yan, G. Lv, Y. Gu, X. Luo, L. Yang, J. Wei, Degradability and cytocompatibility of tricalcium phosphate/poly(amino acid) composite as bone tissue implants in orthopaedic surgery, *J Biomater Sci Polym Ed* 25 (11) (2014) 1194–1210.

56. P. Coimbra, P. Alves, T.A. Valente, R. Santos, I.J. Correia, P. Ferreira, *Int J Biol Macromol* 49 (2011) 573–579.

57. M. Ebadi, S. Bullo, K. Buskara, M.Z. Hussein, S. Fakurazi, G. Pastorin, Release of a liver anticancer drug, sorafenib from its PVA/LDH- and PEG/LDH-coated iron oxide nanoparticles for drug delivery applications, *Sci Rep* 10 (1) (2020) 21521.

58. R. Tajau, R. Rohani, S.S. Abdul Hamid, Z. Adam, S.N.M. Janib, M.Z. Salleh, Surface functionalisation of poly-APO-b-polyol ester cross-linked copolymers as core: Shell nanoparticles for targeted breast cancer therapy, *Sci Rep* 10 (1) (2020) 21704.

59. B. Li, M. Shan, X. Di, C. Gong, L. Zhang, Y. Wang, G. Wu, A dual pH- and reduction-responsive anti-cancer drug delivery system based on PEG-SS-poly(amino acid) block copolymer, *RSC Adv* 7 (2017) 30242–30249.

60. W.H. Zimmermann, T. Eschenhagen, Cardiac tissue engineering for replacement therapy, *Heart Failure Rev*, 8 (3) (2003) 259–269.

61. L. Saludas, S.P. Gil, F. Prósper, E. Garbayo, M.B. Prieto, Hydrogel based approaches for cardiac tissue engineering, *Int J Pharm* 523 (2) (2017) 454–475.

62. S.M. Nasr, N. Rabiee, S. Hajebi, S. Ahmadi, Y. Fatahi, M. Hosseini, M. Bagherzadeh, A.M. Ghadiri, M. Rabiee, V. Jajarmi, T.J. Webster, Biodegradable nanopolymers in cardiac tissue engineering: From concept towards nanomedicine, *Int J Nanomedicine*, 15 (2020) 4205–4224.

63. Medical polymers market: Global industry analysis, trends, market size, and forecasts up to 2025, Infinium Global Research, Report, (2019) 4840275, 100 p.

64. R. Shah, R. Chen, H. Wong, Present and future trends in biodegradable polymers, *Plastics Today*, October 20, 2020.

65. Y. Wu, R.L. Clark, Electrohydrodynamic atomization: A versatile process for preparing materials for biomedical applications, *J. Biomater. Sci. Polymer Edn* 19 (5) (2008) 573–601.

66. C. Sanhueza, F. Acevedo, S. Rocha, P. Villegas, M. Seeger, R. Navia, Polyhydroxyalkanoates as bioma-terial for electrospun scaffolds, *Int J Biol Macromol* 124 (2019) 102–110.

67. B. Maharjan, V.K. Kaliannagounder, S.R. Jang, G.P. Awasthi, D.P. Bhattarai, G. Choukrani, et al., In-situ polymerized polypyrrole nanoparticles immobilized poly(ε-caprolactone) electrospun conduc-tive scaffolds for bone tissue engineering, *Mat Sci Eng C-Mater* 114 (2020) 111056.

68. A. Hajinasab, S.S. Samandari, S. Ahmadi, K. Alamara, Preparation and characterization of a biocom-patible magnetic scaffold for biomedical engineering, *Mater Chem Phys* 204 (2018) 378–387.

69. H. Wang, X. Zeng, L. Pang, H. Wang, B. Lin, Z. Deng, E.L.X. Qi, N. Miao, D. Wang, P. Huang, H. Hu, J. Li, Integrative treatment of anti-tumor/bone repair by combination of MoS2 nanosheets with 3D printed bioactive borosilicate glass scaffolds, *Chem. Eng. Technol* 396 (2020) 125081.

70. B. Singh, V. Sharma, Crosslinking of poly(vinylpyrrolidone)/acrylic acid with tragacanth gum for hydrogels formation for use in drug delivery applications, *Carbohydr Polym* 157 (2017) 185–195.

71. W. Treesuppharat, P. Rojanapanthu, C. Siangsanoh, H. Manuspiya, S. Ummartyotin, Synthesis and characterization of bacterial cellulose and gelatin-based hydrogel composites for drug-delivery systems, *Biotechnol Rep (Amst)*, 15 (2017) 84–91.

72. D. Li, Y. Ye, D. Li, X. Li, C. Mu, Biological properties of dialdehyde carboxymethyl cellulose crosslinked gelatin-PEG composite hydrogel fibers for wound dressings, *Carbohydr Polym*, 137 (2016) 508–514.

8 Perspectives of 3D Printing Technology on Polymer Composites for Biomedical Applications

Lakshmanan Saravanan

CONTENTS

8.1 INTRODUCTION

8.1.1 OVERVIEW OF POLYMER COMPOSITES FOR BIOMEDICAL DEVICES

Polymers are considered to be the major class of biomaterials which are comprehensively employed in a large number of biomedical applications for accurately designing and fabricating complex structures. In general, polymer-based composites are employed in numerous fields including automobile, sports, and aerospace and recently have been used to fabricate surgical components and devices for tissue engineering, drug delivery [1], and in repairing cardiovascular stents, soft, and hard tissues. Composites made of elastomers, hydrogels, polymer blends, thermoplastic and thermosetting polymers are widely used polymers to develop medical devices or parts, to replace metals and glass [2]. Depending on the required medical application, those polymer composites could be designed with precise geometrical structure and with specific mechanical and chemical properties. The main spotlight in medicinal applications is to develop the biodegradable and biocompatible polymeric biomaterials with better flexibility, and high strength, and to prevent toxicity, especially for tissue engineering and scaffolds fabrication. Presently, the 3D printing (3DP) technology signifies an immense opening to help the pharmaceutical and health industries to produce more precise drugs, and to enable quick production of devices for medical implants.

DOI: 10.1201/9781003173533-8

8.1.2   3D PRINTING TECHNOLOGIES FOR POLYMERIC SYSTEMS

Additive manufacturing (AM) is a quite innovative technology comprising a range of layer-by-layer processes to fabricate several minor or major components. AM is thriving in the biomedical sector owing to its quick and efficient design and fabrication of complex medical parts. Developing a 3D printing system is a key progress in biomedical engineering, as the materials used here include soft substances, for example living cells, polymers, and silicones. Due to their ease in the processability and inexpensiveness, polymeric materials are the most largely consumed materials for 3D printing. 3DP became popular following the stereolithography method launched by Hull in 1986; using the method he created several three-dimensional objects [3]. AM is already a technique used in several healthcare products, for example hearing aids, models to aid tumor surgery, artificial limbs, etc. Developing the AM tools to support tissue manufacturing, designing and fabricating composites having a complex structure, and manufacturing direct 3D polymer parts, will be the upcoming research directions of AM technology.

Polymers used to fabricate tissue and organs must have the following functions: (a) allowing cells to attach and migrate, (b) transferring growth parts and residues, (c) retaining shape during cell growth and (d) sustaining mechanical property. The man-made polymers like poly (ethylene glycol) diacrylate or natural, gelatin methacrylate are applied in 3DP [4]. Other synthetic polymers such as poly(D, L-lactic-co-glycolic acid (PLGA), and poly(ε-caprolactone (PCL) also are advantageous in 3D printing technology. Adding metal particles into polymers composites used to eliminate from deformation of parts, due to thermal expansion and increase the thermal stability. For good quality printing, polypropylene (PP) is used to accurately fix the initial layer to the substrate. This can also be enhanced by the addition of glass fiber in the polypropylene [5]. Fillers such as CNTs and graphene, and additives are also included for a better mechanical and electrical property. Poly(lactides) (PLA), and its composites are commonly used in the AM techniques for the production of scaffolds for bone support, cartilage and regeneration of adipose tissue. Two-photon polymerization is an AM procedure able to print the parts in the length scale of 100 nm.

8.1.3   CLASSIFICATION OF 3D PRINTING METHODOLOGIES

AM or 3DP techniques are known for their intelligent manufacturing technology which aims to take advantage of all the advanced information to make diverse products. Different 3D printing methods developed so far to construct complex polymers components have been recently evaluated by Ligon et al. [6], for example light-based 3D printing, extrusion-based printing, inkjet 3D printing, powder bed fusion-based methods, and laminated method. The most commonly applied additive manufacturing technologies which are classified based on the input materials, like solid, liquid, and powder, are shown in Figure 8.1.

Recent investigations on 3DP technology in the medical applications are based on: (a) manufacturing pathological organ models to assist pre-operative planning and the analysis on surgical treatment; (b) customized fabrication of permanent implants; (c) scaffolds fabrication and (d) printing tissues and organs directly, with complete life functions. AM continuously grows and blooms into a surplus of processes, which include stereolithography (SLA), laminated object manufacturing (LOM), fused deposition modeling (FDM), selective laser sintering (SLS), inkjet printing, selective laser melting (SLM), laser metal deposition (LMD), and so on. This chapter deals with the importance of additive manufacturing and progression in the 3DP technology and lists the major AM techniques available for developing the multifunctional polymeric components, which is typically employed in the easy fabrication of biomedical devices. Some of the major AM processes related to developing the parts made from polymer-based composites in the application of medicinal research and biomedical device fabrication are discussed in the following section. It also highlights the major advantages of several additive manufacturing techniques, the current and the future perspectives of 3D printing techniques in the field of biomedical research and applications.

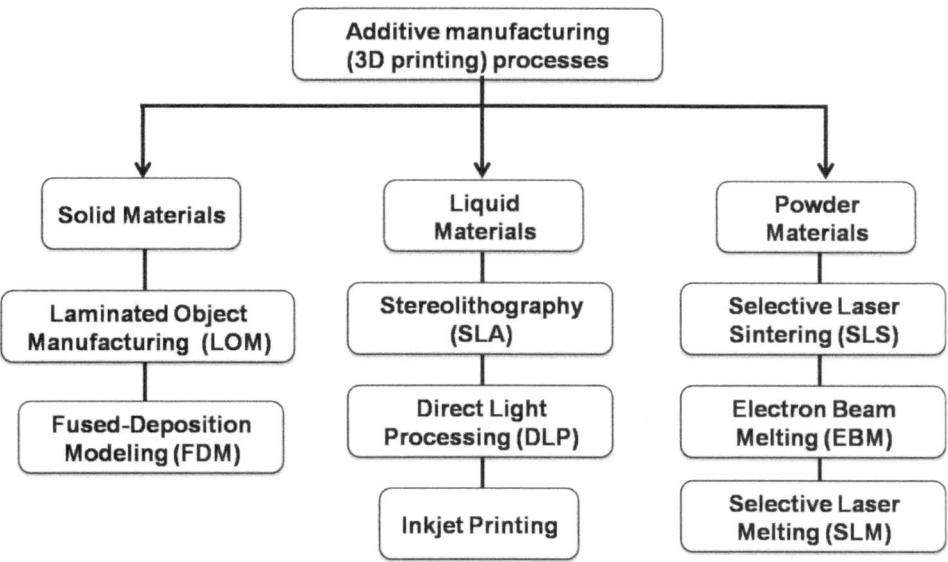

FIGURE 8.1 Classification of additive manufacturing processes.

8.2 DIFFERENT TYPES OF 3D PRINTING TECHNIQUES

8.2.1 FUSED DEPOSITION MODELING

Fused deposition modeling (FDM) belongs to the class of AM which follows the material extrusion principle and is suitable for the production of medical instruments and devices, and rapid prototyping exoskeleton. The schematic illustration of FDM is shown in Figure 8.2(a) [7].

In this method, the vicinity is heated to 80°C and the polymeric material is inserted into a nozzle and heated above the melting point and becomes hardened after it comes out from the nozzle to the printing bed. This method has an advantage of producing components with high strength, good precision, and small raw materials usage, without the need of cleaning and finishing products [8, 9]. Polystyrene (PS), PLA, polycarbonate (PC), and acrylonitrile butadiene styrene (ABS) are the most commonly used thermoplastic polymer materials in FDM process [10] to fabricate polymer scaffolds. The biocompatible and biodegradable polymers such as PLA, polybutylene terephthalate (PBT), polyglycolic acid (PGA) and PCL are widely used for biomedical applications [11]. Incorporating electrically conductive graphene, carbon nanotubes [12], and carbon black [13] into the polymer matrix are used to produce conductive polymer composites by the FDM technique. Wei et al. [14] introduced a solution intercalation approach in the FDM method, using n-methylpyrolidone as a solvent to disperse graphene oxide and ABS (rGO/ABS), to prepare filament to develop device parts. Using the same methodology, Chen and his coworkers in 2017 utilize TPU and PLA added with different GO ratios to fabricate nanocomposite filament and studied their biocompatibility to produce tissue engineering scaffolds [15]. Zhong et al. have studied about processing of different glass contents on fiber-reinforced ABS matrix in FDM, to increase the surface rigidness and mechanical strength [16]. Another group [17] prepared PCL bonded with Nd-Fe-B composites filament for the FDM method and reported a new method using a magnetic platform to achieve an anti-gravitational printing process. FDM is beneficial for using multi-nozzles to load different materials to produce polymer matrix composites.

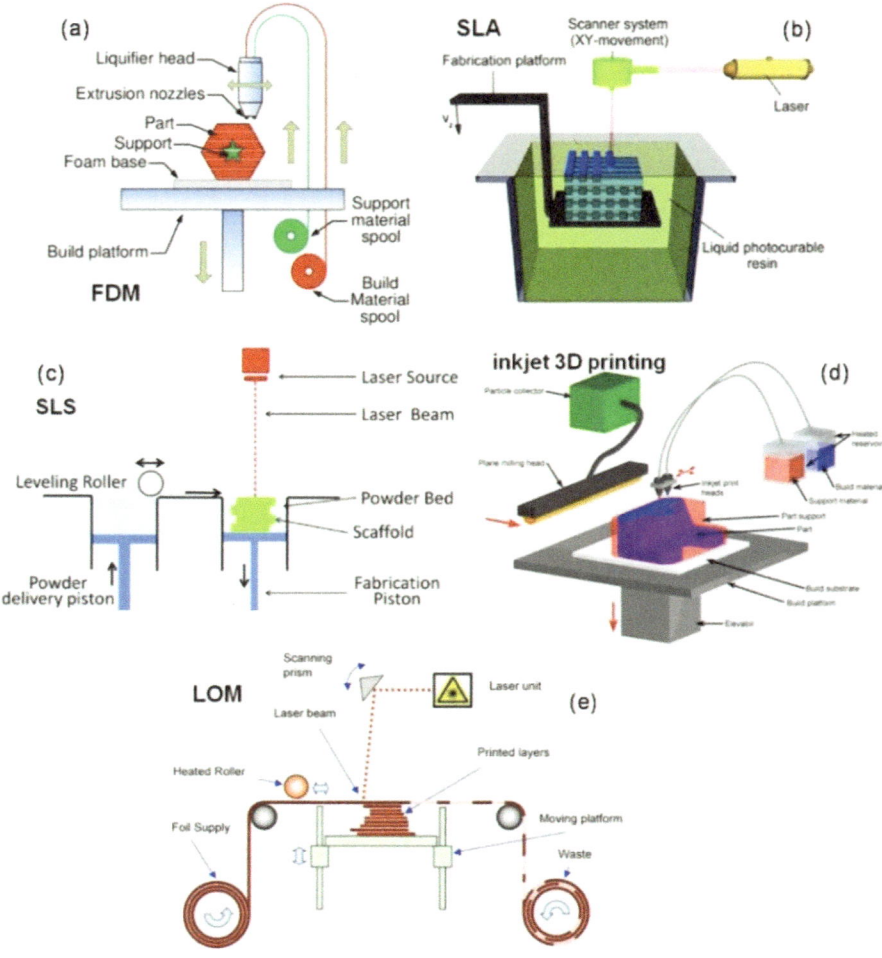

FIGURE 8.2 The schematic diagram of widely used AM techniques.

8.2.2 STEREOLITHOGRAPHY

Stereolithography (SLA) is a solution-based process of solidifying the photosensitive polymer resin using an ultraviolet laser. SLA is the first system which was commercialized earlier than other 3D techniques. The process uses both single and multiple UV laser beams to treat photopolymer thin resin layers [18]. The schematic diagram of SLA is given in Figure 8.2(b) [19]. Mostly available resins for the SLA method have low molecular weight, and monomers have multi-functional properties, for example polyacrylate and epoxy [20]. The quality depends on the duration of laser exposure and power intensity, and scan speed. Fabrication of products made from polymer-ceramic composites, alumina or hydroxyapatite (HA) particles are homogeneously mixed in the resin and photo-polymerized using SLA [21, 22]. At present, most hearing aids are made by SLA, by an automated, patient-reliant, and fast process [23]. Researchers [24] developed a PMMA/TiO_2 nanocomposite to fabricate complete denture sets by SLA with improved antibacterial activity. It was also reported that the addition of organically modified fillers (nano SiO_2) into the aliphatic urethane acrylate oligomer resin, boosted the tensile strength of the SLA printed composite parts [25]. Sandoval et al. has developed MWCNT reinforced epoxy resin to fabricate complex 3D parts with increased tensile and fracture strength using vector scanning based stereolithography [26]. The technique is vastly suitable to make bone, dental models, dental implant guides, and hearing aids.

To construct a hard (bone) tissue engineering scaffold, strong and rigid biodegradable materials are requisite. Opting the exact material (particle and binder), is a critical part in increasing the mechanical and biological properties of the printed scaffolds. Biodegradable composite scaffolds (Figure 8.3(a)) with osteopromotive properties were fabricated by the UV stereolithography using photo-crosslinkable polymer, poly(trimethylene carbonate) (PTMC) added with HA for bone tissue engineering without additives. In this method described by Guillaume et al., the three-armed PTMC synthesized initially, then the photo-resins were prepared using the optimal content of nano-HA and finally the 600 micro-scale sized porous scaffolds were designed by the means of UV-light irradiation (180 mW/dm^2), which are suitable to allow the formation of bone (in-growth) [27]. Methacrylated poly(D,L-lactide) (PDLLA) based resin was prepared with ethyl lactate used as a non-reactive diluent. A porous PDLLA scaffold (Figure 8.3(b)) with excellent mechanical properties, with gyroid architecture, is free of reactive diluents. The scaffold was designed and fabricated by the UV stereolithography process, in the power of 20 mW/cm^2 and the peak wavelength was 440 nm [28]. The developed polymer scaffolds in this method were successfully applicable for use in the bone tissue engineering. In both these methods, the primary polymer oligomers were prepared by the ring opening polymerization method using stannous octoate as a catalyst.

8.2.3 SELECTIVE LASER SINTERING

Selective laser sintering (SLS) is the powder bed fusion process invented by Carl Deckard in 1992. The SLS process requires a high intense laser to cure or fuse polymeric powder materials. The working procedure of this technique was described by Mazzoli [36] and the schematic was shown in Figure 8.2(c) [29]. In the biomedical field of research this SLS technique is remarkably important, to fabricate surgical tools, and customize implantable parts. In this powder-based AM process, a scanning laser (e.g., CO$_2$ laser) beam is used to sinter the materials. Poly(3-hydroxybutyrate) (PHB) was printed into porous scaffolds by SLS and the osteogenic growth peptide used to functionalize the scaffolds, to indicate the osteogenesis [30]. In this method, the polymers that can be used include poly(aryl ether ketone) (PAEK), polystyrenes, poly(ethyl ether ketone) (PEEK), and its derivatives, polyamides, elastomers, and semi-crystalline thermoplastics. Tortorici et al. investigated altering

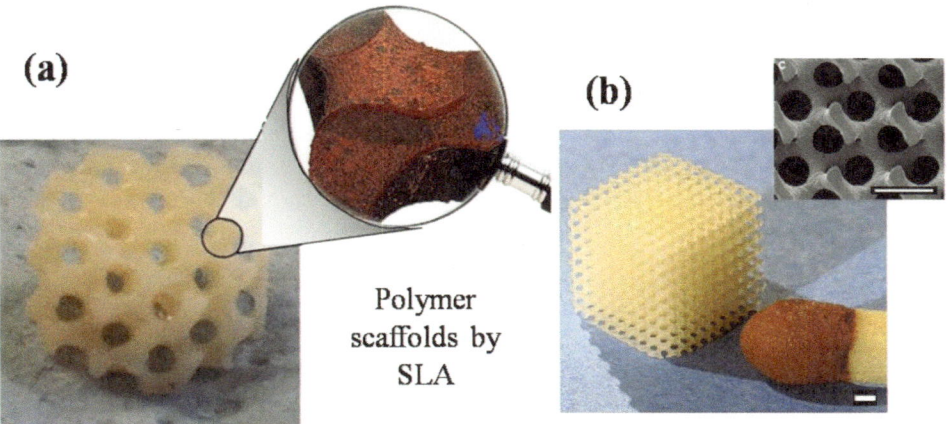

FIGURE 8.3 Images of (a) scaffold of PTMC/HA and (b) porous PDLLA scaffolds with a gyroid architecture, printed by stereolithography.

Source: © 2009, 2017 Elsevier Ltd.

the power of the laser beam, beam diameter, and beam compensation on the precision of dimension and mechanical rigidity of the PCL scaffolds for bone TE using SLS [31]. Yuan et al. prepared CNT-reinforced PU and PA composites to make complex 3D parts via SLS technique with enhanced electrical and thermal conductivities [32]. Natural polymers cannot be employed in SLS, generating high temperature by the high power laser.

8.2.4   INKJET 3D PRINTING

This method is similar to the inkjet printers, which operate either in continuous or drop on demand (DoD) approaches, [33] providing better accuracy and fine surface finishes. This technique is based on powder processing and a jet was used for material deposition on the substrate. The material is prepared as a solution and dispersed as droplets by a controlled valve. During the process, fine polymers are deposited on a surface by the droplets to build up the structures gradually. A schematic of 3D inkjet printing process is presented in Figure 8.2(d) [34]. The valve is used to print (drop) directly with cells, which can further be added with polymers to form an extracellular matrix, to prepare a scaffold. This method also used to print tissue engineering scaffolds [35]. This 3DP technique keeps the cells alive, to create an artificial organ alike to an organoid, tissue-engineered, bio-artificial organ. Koch et al. applied inkjet printing to develop artificial 3D printed skin using the laser beam, in which the printing ink comprises fibroblast/keratinocytes/collagen and blood plasma/alginate solution [36]. Earlier, the polyvinyl pyrrolidine (PVP) stabilized copper nanoparticles with 2-(2-butoxyethoxy) ethanol was used as ink in the inkjet printing [37]. High accuracy performance of inkjet printing depends on ink material, substrate properties, printing platform, and generation of droplets. It was demonstrated a 3D printed material with an ink made of graphene/polylactide-co-glycolide composite would exhibit high flexibility and improved strength, and with high electrical conductivity > 800 S/m [38].

8.2.5   DIRECT INK WRITING

Direct-ink-writing (DIW) is an extrusion-based AM used to fabricate 3D structures layer by layer, without the necessity of expensive accessories [39]. In DIW the controlled ink filament is made of colloidal particles, nanoparticles, and other organic materials. Polymer nanocomposites including polymer matrix and nanomaterials such as CNTs, graphene flakes are also used. The DIW technique uses visco-elastic polymer liquids, hydrogels, as the printing ink to fabricate various complex 3D designs. It has advantages of having a low fabrication temperature and integrating multiple polymeric materials [40, 41]. The distinctive feature of inks has their ability to quickly self-heal after the extrusion. Barry and his co-workers prepared a polymeric ink for DIW, which consists of physically tangled and cross-linked poly(acrylamide) chains in an acrylamide solution for photo polymerization [42]. Lewicki et al. prepared a mesostructured carbon fiber reinforced polymer (bisphenol-F epoxy resin oligomer) composite resin as an ink with excellent orthotropic mechanical and electrical properties using DIW technology [43].

8.2.6   LAMINATED OBJECT MANUFACTURING

Laminated object manufacturing (LOM) method is used to make 3D objects by depositing the layers of laminated materials [44]. Adhesion fixing, holding, and ultrasonic welding are the three major steps in the LOM printing method [45] without requiring any structural support. The schematic representation of the LOM method is given in Figure 8.2(e) [46]. The LOM technique is chosen when rapid and large-size printing is required. In this process, heat and pressure are used to fuse or laminate the plastic layers, which are then sliced into the desired pattern by a computer-controlled laser. Each layer is coated with the adhesive which used in determining the strength and thickness of the developed part and the excess material is cut away. Starting sheet materials are cellulose, paper, metals, and plastic or reinforced fiber materials. The LOM technique is mostly suitable to fabricate orthopedic modeling of bone surfaces. It is an easy and inexpensive method to produce parts ranging from millimeter size to meter size, in less time. The LOM process can be employed to generate low-cost polymeric (films) products with high precision and the device does not utilize any toxic materials.

8.3   ADVANTAGES OF ADDITIVE MANUFACTURING (3DP) METHODS

Additive manufacturing or a 3D printing system is the advanced technology which is considered to have great potential and prospects in various industries. 3DP has been applied in a wide range of biomedical applications, cardiology, ophthalmology, vascular surgery, orthopedics, plastic surgery, and neurosurgery, etc. Some of the advantages of additive manufacturing in fabricating medical devices are listed in Table 8.1.

Various bio-printing techniques have been employed to regenerate human muscles, blood vessels, bone, cartilage, nerves, skin, or corneas, by fabricating 3D scaffolds with or without cells [47]. Initially, AM methods allow scanning and building a physical model of damaged bones and provide doctors an outline about the procedure, to save time and money and to help yield a better outcome [48]. In comparison to traditional fabrication methods 3D printing has the advantage of printing layer by layer (bottom up) of a predesigned object, is prompt in the development of complex objects, is light-weight, and can substantially print with new and assorted geometric designs. The inkjet printing method is faster than screen printing. The 3D printing technique has the advantages of replacing, restoring, retaining, and recovering the tissue functions. The replaced tissue has an interconnected pore network, is biocompatible, and has good mechanical properties [49]. The advantages of the polymer-based powder 3DP technique includes the high rate of materials consumption, free structure support, vast freedom in designing, and the large availability of materials. Efforts were taken to develop various polymer composites using the SLS method with higher laser wavelengths, without dropping the material properties. Chemical functionalization of thermoplastic polymers with supramolecular interactions is an additional method to enhance the ability of 3D printing.

TABLE 8.1
Advantages of Additive Manufacturing Processes in Medical Device Fabrication.

Sl. No	Production Criteria	AM Advantages	References
1.	Production of surgical aid tools, bio-models, and implants	AM plays a vital role in designing, creating, and upgrading the surgical tools, bio-models, and implants.	[73, 74]
2.	Production of scaffolds for tissue engineering	AM is used to manufacture different scaffolds for the restoration of tissues and replaces the conventional methods. AM helps to print organs, produced cells, cell-loaded biomaterials.	[75, 76]
3.	Medical devices and surgical training models	AM is used to develop medical models, surgical training models for medical education.	[77–80]
4.	Complex geometries	AM has potential to fabricate complex geometries implants.	[81]
5.	Cost reduction	Compared to other manufacturing processes AM technology helps to reduce production cost of medical devices.	[82]
6.	Rapid production	AM has the ability to produce medical model/prototype in a short time.	[83]
7.	Light-weight	Weight reduction made by the AM technology by the choice of material.	[84, 85]

Source: [72].

Designing and successful fabrication of heterogeneous hydrogels consisting of both weak and/ or hard materials at specific locations are added advantages [50]. It is beneficial that the LOM technique has a larger working area, the ability to deposit multiple materials, a high volumetric build rate, and can integrate hybrid manufacturing systems. Material extrusion has the advantage of low-temperature process, small equipment size, wider material selection, and rapid printing parts. Stereolithography, digital light processing techniques have the advantages of high-level accuracy and high-quality finishing, relatively quick process, and large build areas. The advantages of SLS are its range of polymers that can be used, its manufacturing speed, and its capability of producing complex geometries [51]. Its polymers are processed traditionally prior to printing, and for PAEKs the porous scaffolds are without the defects [52].

8.4 PRODUCTION OF POLYMERIC MATERIALS BY ADDITIVE MANUFACTURING FOR MEDICAL APPLICATIONS

In the recent decades, the materials selection for the AM process has been rapidly developing printable polymer nanocomposites, which creates more possibility to produce multifunctional parts with intricate structures more precisely. Polymeric materials and their composites are the class of materials with a high prospective for exploitation in several 3DP techniques for medical device applications (Table 8.2).

TABLE 8.2
Few Examples of the Polymeric Materials Utilized in Different 3DP Technologies, and Their Possible Medical Applications.

Sl. No.	AM Technology	Polymeric Materials	Medical Applications	Improved Properties	Ref.
1.	SLA	(Photopolymer resin), e.g., poly(d,l-lactide), fumaric acid monoethyl ester-functionalized poly(d,l-lactide)/N-vinyl-2-pyrrolidone resins, graphene-polymer composites	Bone, tissue engineering, scaffolds	Mechanical properties (tensile strength, modulus)	[87, 88]
2.	FDM	(Plastics, ABS, nylon, PC), PEEK, e.g., PCL scaffold integrated with cell-laden chitosan hydrogels	Medical instruments and devices, prototyping exoskeleton, bone TE	Mechanical strength	[89–91]
3.	SLS	PP, polyethylene (PE), polystyrene, PMMA, polycarbonate, acrylonitrile butadiene styrene polyamide, PE, polyether ether ketone, (PCL), (TPU), PEEK/HA	Medical models, dental casts, dental implant guides, Scaffolds	Bioactivity, biocompatibility, mechanical property	[92, 93]
4.	3D inkjet printing	Collagen-calcium phosphate composites.	Scaffolds for bone regeneration	Biocompatibility and osteo-conductivity	[94]
5.	DIW	Poly(isopropyl glycidyl ether)-block-poly(ethylene oxide)-block-poly(isopropyl glycidyl ether) ABA triblock copolymers, inks made of gelatin methacryloyl (GelMA), poly(ethylene glycol) diacrylate, or norbornene-functionalized HA, carbon black-TPU/PDMS/Ag-PA/PLA, or ABS	Tissue engineering, scaffolds, soft robots, cardiac device	Mechanical properties, rapid self-healing	[39, 42, 95, 96]
6.	LOM	Paper, plastic (PVC)	Orthopedic modeling of bone surfaces	Flexural strength, surface roughness	[97, 98]

Source: [86].

Polymers such as urethane dimethacrylate, diisopropyl acrylamide are biocompatible and the polymers, polypropylene fumarate, linear poly(D,l-lactide)-methacrylate are biodegradable, which have been recently developed for biomedical applications. Combining polymer nanocomposites and fiber reinforcement could increase the inter-laminar strength to yield strong AM parts. Several materials, for example zeolites [53], metal oxides [54, 55] and ceramics [56], are employed as fillers, for photopolymerization. CNTs are added with ABS polymer as an additive for printing in FDM [57]. It was reported that the researchers have used 3DP technology with HeLa cells and gelatin/alginate/fibrinogen hydrogels to construct *in vitro* cervical tumor models [58]. The chief goal in selection of biomaterials and 3D printing techniques is to develop the materials catalog initially to accurately transform a choice of additive manufacturing processes. Kim et al. prepared a hybrid PCL/collagen-based uniaxially aligned 3D printed scaffold for tissue regeneration [59]. Campos et al. produced extracellular matrix via 3DP, by developing cells loaded in a hydrophobic high-density fluid, perfluorotributylamine (C12F27N) used as hydrogels [60].

It has been reported that incorporating graphene oxide as nanofillers increased the mechanical strength and ductility of polymer composite in the mask projection based stereolithography system [61]. The cytotoxic acrylates are replaced with the methacrylates, thiol-ene systems, and the photo-reactive monomers in the lithographic methods (SLA, polyjet). Conducting polymers such as polyaniline (PANI) and poly(3,4-ethylenedioxy-thiophene):poly(styrene sulfonate) (PEDOT:PSS) have attracted increasing attention, due to their high electrical conductivity [62, 63]. Darabi et al. synthesized novel conductive self-healing (CSH) hydrogels, mechanically mimicking human skin and polymers with self-healing property based on cross-linked networks [64]. Sun et al. reported that the advancement in the LOM process enabled the fabrication of microchips based on PC/PMMA for lab-on-a-chip immunoassays [65].

8.5   CURRENT AND FUTURE PROSPECTS OF 3D PRINTING TECHNIQUES

Modern advancements in the biomedical research exclusively on tissue engineering, drug delivery, fabrication of medical parts and devices guarantee that 3D printing techniques will perform an essential role in the future of healthcare. 3D printing is employed as "an essential ingredient" in Industry 4.0 and is being indicated as part of the "Fourth Industrial Revolution" [66]. Medical devices that can be developed via 3D printing devices include custom-made knee and hip implants, hearing aids, and prostheses. Some of the biomedical parts made using different 3DP techniques were shown in Figure 8.4. A recent report describes the drivers, marketable trends, and the major confrontations for the AM and forecasts that the AM market will increase from 2014 with an estimation of $11,145.1 million by 2020, at a CAGR of 20.9% [67]. The market of 3D printing is now valued around $9.3 billion and a report by Smithers Pira predicted that the worth of the AM industry will be increased to $55.8 billion by 2027 [68]. Deloitte Global also predicts the 3D printing sales by large companies will exceed in 2019 US$2.7 billion and will reach US$3 billion in 2020 [69], as represented in Figure 8.5.

It has also been projected that 3D printing in the field of medicine will be valued at $3.5 billion by 2025, much more than the worth of $713.3 million in 2016 [70]. The industry's compound annual growth rate is supposed to reach 17.7% between 2017 and 2025. The major progress in the production of 3D polymeric components is focused effectively in the (a) powder bed fusion processes, (b) extrusion-based technologies, and (c) photopolymer-based printing methods. However, the supreme prospect for 3D bio-printing lies in its possibility on fabricating a fully functioning organ that could be transplanted into a human body. The choice of a suitable fabrication technique is also a challenging position. Polymers have high geometric complexity, but the requirements of better mechanical strength are often limiting the applications in AM. For low cost and processing flexibility, the polymers in a liquid state having a low melting point and low molecular weight are highly preferable to use in AM. At present, the key raw material for 3D printing is polymers, compounds that are largely synthetic and which use inks to create three-dimensional objects in accordance with the models computers use to execute the three-dimensional printing. In future research, various functional materials will be combined to achieve preferred mechanical strength and biocompatibility, which is a promising way to solve the present challenges in AM with polymer products.

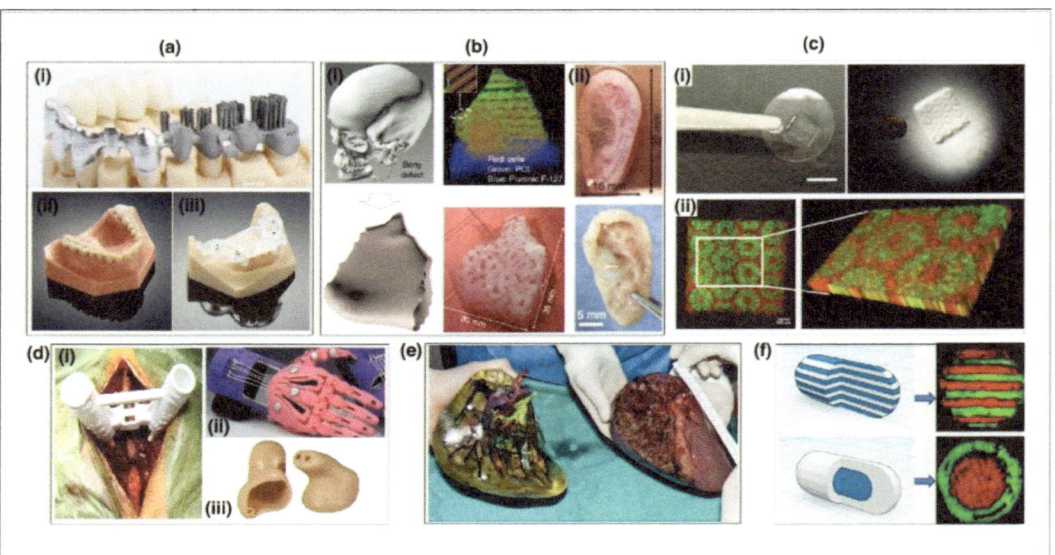

FIGURE 8.4   Products made by 3D printing process (a) dental models, (b) bio-printed tissues, (c) tissue constructs, (d) medical utensils, (e) surgical mocks, and (f) tablets holding drugs [71].

Source: © 2017 IOP Publishing Ltd.

FIGURE 8.5   Representation of global growth rate of 3D printing industry [69].

Source: © 2021 Deloitte.

8.6 SUMMARY

AM processes are currently increasing in demand in the medical industry due to its unique features of producing value-added components more easily than conventional techniques. Presently, several industries are moving toward implementing the 3D printing technologies, owing to its easy designing of numerous complex structures and rapid prototypes production. Medical models are customized, modified, and printed based on the data source from the diseased person, which vary for different patients. Surprisingly, AM is contributing more in the designing and production of implants, designing and development of the medical tools, diagnosis, surgical simulation of novel orthopedic products, and in dental applications. Several AM technologies developing recently with their own manufacturing pros and cons depend on material behavior and structure by layers. The continuing and future research on AM (3DP) technologies will give us the ability to produce typical functional organs with reasonable cost, with a better capability to enhance product customization which can be further augmented with electronic or mechanical elements. Employing 3DP technology in biomedical applications will be successful to solve the shortage crisis of donors for organ transplantations. Accordingly AM is considered to be a rapidly developing interdisciplinary area that integrates material science, biology, and clinical science.

REFERENCES

1. M.M. Zagho, E.A. Hussein, A. Elzatahry, Recent overviews in functional polymer composites for biomedical applications, *Polymers* 10 (2018) 739.
2. L. Li, Q. Lin, M. Tang, A.J.E. Duncan, C. Ke, Advanced polymer designs for direct-ink-write 3D printing, *Chem. Asian J* 25 (46) (2019) 10768.
3. C.W. Hull, Apparatus for production of three-dimensional objects by stereolithography, U.S. Patent: US4575330A, 11 March 1986.
4. T. Billiet, E. Gevaert, T. De Schryver, M. Cornelissen, P. Dubruel, The 3D printing of gelatin methacrylamide cell-laden tissue-engineered constructs with high cell viability, *Biomaterials* 35 (1) (2014) 49–62.
5. G. Sodeifian, S. Ghaseminejad, A.A. Yousefi, Preparation of polypropylene/short glass fiber composite as fused deposition modeling (FDM) filament, *Results Phys* 12 (2019) 205–222.
6. S.C. Ligon, R. Liska, J. Stampfl, M. Gurr, R. Mulhaupt, Polymers for 3D printing and customized additive manufacturing, *Chem Rev* 117 (15) (2017) 10212.
7. F. Ning, W. Cong, J. Wei, S. Wang, M. Zhang, Additive manufacturing of CFRP composites using fused deposition modeling: Effects of carbon fiber content and length, *MSEC2015–9436* V001T02A067 (2015) 7.
8. T. Srivatsan, T. Sudarshan, *Additive manufacturing: Innovations, advances, and applications*, CRC Press, Taylor & Francis group, Boca Raton, 2016.
9. F. Ning, W. Cong, J. Qui, J. Wei, S. Wang, Additive manufacturing of carbon fiber reinforced thermoplastic composites using fused deposition modeling, *Composites Part B* 80 (2015) 369–378.
10. L.N. Marcincinova, I. Kuric, Basic and advanced materials for fused deposition modeling rapid prototyping technology, *Manuf and Ind Eng* 11 (1) (2012) 24–27.
11. H.N. Chia, B.M. Wu, Recent advances in 3D printing of biomaterials, *J Biol Eng* 9 (1) (2015) 4.
12. G. Postiglione, G. Natale, G. Griffini, M. Levi, S. Turri, Conductive 3D microstructures by direct 3D printing of polymer/carbon nanotube nanocomposites via liquid deposition modeling, *Composites Part A* 76 (2015) 110–114.
13. S.J. Leigh, R.J. Bradley, C.P. Purssell, D.R. Billson, D.A. Hutchins, A simple, low-cost conductive composite material for 3D printing of electronic sensors, *PLoS One* 7 (11) (2012) e49365.
14. X. Wei, D. Li, W. Jiang, Z. Gu, X. Wang, Z. Zhang, Z. Sun, 3D printable graphene composite, *Sci Rep* 5 (2015) 11181.
15. Q. Chen, J.D. Mangadlao, J. Wallat, A. De Leon, J.K. Pokorski, R.C. Advincula, 3D printing biocompatible polyurethane/poly(lactic acid)/graphene oxide nanocomposites: Anisotropic properties, *ACS Appl. Mater. Interfaces* 9 (4) (2017) 4015–4023.
16. W. Zhong, F. Li, Z. Zhang, L. Song, Z. Li, Short fiber reinforced composites for fused deposition modeling, materials science and engineering, *Mater. Sci. Eng. A* 301 (2) (2001) 125–130.

17. J. Wang, H. Xie, L. Wang, T. Senthil, R. Wang, Y. Zheng, Anti-gravitational 3D printing of polycaprolactone-bonded Nd-Fe-B based on fused deposition modeling, *J Alloys Compd* 715 (2017) 146–153.

18. X. Wang, M. Jiang, Z. Zhou, J. Gou, D. Hui, 3D printing of polymer matrix composites: A review and prospective, *Comp Part B: Eng* 110 (2017) 442–458.

19. X. Tian, J. Jin, S. Yuan, C. K. Chua, S. B. Tor, Kun Zhou, Emerging 3D-printed electrochemical energy storage devices: A critical review, *Adv. Energy Mater* 7 (17) (2017) 1700127.

20. F.P. Melchels, J. Feijen, D.W. Grijpma, A review on stereolithography and its applications in biomedical engineering, *Biomaterials* 31 (24) (2010) 6121–6130.

21. V.K. Popov, A.V. Evseev, A.L. Ivanov, V.V. Roginski, A.I. Volozhin, S.M. Howdle, Laser stereolithography and supercritical fluid processing or custom-designed implant fabrication, *J Mater Sci Mater Med* 15 (2) (2004) 123–128.

22. J.W. Lee, G. Ahn, D.S. Kim, D.W. Cho, Development of nano- and microscale composite 3D scaffolds using PPF/DEF-HA and micro-stereolithography, *Microelectron Eng* 86 (4–6) (2009) 1465–1467.

23. C. Sandstrom, Adopting 3D printing for manufacturing: The case of the hearing aid industry, Chalmers University of Technology, The Ratio Institute, Stockholm, Sweden, 2015.

24. E.E. Totu, A.C. Nechifor, G. Nechifor, H.Y. Aboul-Enein, C.M. Cristache, Poly (methyl methacrylate) with TiO_2 nanoparticles inclusion for stereolitographic complete denture manufacturing- the future in dental care for elderly edentulous patients? *J. Dent.* 59 (2017) 68–77.

25. Z. Weng, Y. Zhou, W. Lin, T. Senthil, L. Wu, Structure-property relationship of nano enhanced stereolithography resin for desktop SLA 3D printer, *Composit. Part A Appl. Sci. Manufact* 88 (2016) 234–242.

26. J.H. Sandoval, K.F. Soto, L.E. Murr, R.B. Wicker, Nanotailoring photocrosslinkable epoxy resins with multiwalled carbon nanotubes for stereolithography layered manufacturing, *J Mater Sci* 42 (2007) 156–165.

27. O. Guillaume, M.A. Geven, C.M. Sprecher, V.A. Stadelmann, D.W. Grijpma, T.T. Tang, L. Qin, Y. Lai, M. Alini, J.D. de Bruijn, H. Yuan, R.G. Richards, D. Eglin, Surface-enrichment with hydroxyapatite nanoparticles in stereolithography-fabricated composite polymer scaffolds promotes bone repair, *Acta Biomater* 54 (2017) 386–398.

28. F.P.W. Melchels, J. Feijen, D.W. Grijpma, A poly(D,L-lactide) resin for the preparation of tissue engineering scaffolds by stereolithography, *Biomaterials* 30 (2009) 3801–3809.

29. A. Mazzoli, Selective laser sintering in biomedical engineering. *Med Biol Eng Comput.* 51 (3) (2013) 245–256.

30. P. Rider, Ž.P. Kačarević, S. Alkildani, S. Retnasingh, R. Schnettler, M. Barbeck, Additive manufacturing for guided bone regeneration: A perspective for alveolar ridge augmentation, *Int. J. Mol. Sci* 19 (11) (2018) 3308.

31. S. Saska, L.C. Pires, M.A. Cominotte, L.S. Mendes, M.F. de Oliveira, I.A. Maia, J.V.L. da Silva, S.J.L. Ribeiro, J.A. Cirelli, Three-dimensional printing and in vitro evaluation of poly(3-hydroxybutyrate) scaffolds functionalized with osteogenic growth peptide for tissue engineering, *Mater. Sci. Eng. C Mater. Biol. Appl* 89 (2018) 265–273.

32. M. Tortorici, C. Gayer, A. Torchio, S. Cho, J.H. Schleifenbaum, A. Petersen, Inner strut morphology is the key parameter in producing highly porous and mechanically stable poly(ε-caprolactone) scaffolds via selective laser sintering, *Mater. Sci. Eng. C* 123 (2021), 111986.

33. S. Yuan, Y. Zheng, C. Kai Chua, Q. Yan, K. Zhou, Electrical and thermal conductivities of MWCNT/ Polymer composites fabricated by selective laser sintering, *Compos. Part A* 105 (2018) 203–213.

34. M. Singh, H.M. Haverinen, P. Dhagat, G.E. Jabbour, Inkjet printing-process and its applications, *Adv. Mater* 22 (6) (2010) 673–685.

35. B. Derby, Printing and prototyping of tissues and scaffolds, *Science* 338 (2012) 921–926.

36. L. Koch, A. Deiwick, S. Schlie, S. Michale, M. Gruene, V. Coger, et al., Skin tissue generation by laser cell printing, *Biotechnology and Bioengineering* 109 (2012) 1855–1863.

37. Y. Lee, J. Choi, K.J. Lee, N.E. Stott, D. Kim, Large-scale synthesis of cooper nanoparticles by chemically controlled reduction for applications of inkjet-printed electronics, *Nanotechnology* 19 (2008) 415604–415611.

38. S. Hong, D. Sycks, H.F. Chan, S. Lin, G.P. Lopez, F. Guilak, K.W. Leong, X. Zhao, 3D printing of highly stretchable and tough hydrogels into complex, cellularized structures, *Adv Mater* 27 (27) (2015) 4035–4040.

39. Y.S. Zhang, A. Khademhosseini, Advances in engineering hydrogels, *Science* 356 (6337) (2017) eaaf3627.

40. M. Cianchetti, C. Laschi, A. Menciassi, P. Dario, Biomedical applications of soft robotics, *Nat Rev Mater* 3 (2018) 143–153.

41. A.E. Jakus, E.B. Secor, A.L. Rutz, S.W. Jordan, M.C. Hersam, R.N. Shah, Three dimensional printing of high-content graphene scaffolds for electronic and biomedical applications, *ACS Nano* 9 (2015) 4636–4648.

42. R.A. Barry, R.F. Shepherd, J.N. Hanson, R.G. Nuzzo, P. Wiltzius, J.A. Lewis, Direct-write assembly of 3D hydrogel scaffolds for guided cell growth, *Adv. Mater.* 21 (23) (2009) 2407–2410.

43. J.P. Lewicki, J.N. Rodriguez, C. Zhu, M.A. Worsley, A.S. Wu, Y. Kanarska, et al., 3D-Printing of mesostructurally ordered carbon fiber/polymer composites with unprecedented orthotropic physical properties, *Sci Rep* 7 (2017) 43401.

44. J. Park, M.J. Tari, H.T. Hahn, Characterization of the laminated object manufacturing (LOM) process, *Rapid Prototyp J.* 6 (2000) 36–50.

45. A. Pilipovic, P. Raos, M. Sercer, Experimental testing of quality of polymer parts produced by laminated object manufacturing-lom, *Tehnicki Vjesnik* 18 (2) (2011) 253–260.

46. J.R.C. Dizon, A.H. Espera Jr., Q. Chen, R.C. Advincula, Mechanical characterization of 3D-printed polymers, *Addit Manuf*, 20 (2018) 44–67.

47. H. Lee, Y. Koo, M. Yeo, S. Kim, G.H. Kim, Recent cell printing systems for tissue engineering, *Int. J. Bioprint* 3 (1) (2017) 27–41.

48. C. Chaput, J.B. Lafon, 3-D printing methods, *Ceram Ind* 161 (9) (2011) 15–16.

49. Y. Qian, D. Hanhua, S. Jin, H. Jianhua, S. Bo, W. Qingsong, S. Yusheng, A review of 3D printing technology for medical applications, *Engineering*, 4 (5) (2018) 729–742.

50. Z. Jiang, B. Diggle, M.L. Tan, J. Viktorova, C.W. Bennett, L.A. Connal, Extrusion 3D printing of polymeric materials with advanced properties, *Adv Sci*, 7 (2020) 2001379.

51. B. Chen, Y. Wang, S. Berretta, O. Ghita, Poly Aryl Ether Ketones (PAEKs) and carbon-reinforced PAEK powders for laser sintering, *J Mater Sci* 52, (10) (2017) 6004–6019.

52. S. Berretta, K. Evans, O. Ghita, Additive manufacture of PEEK cranial implants: Manufacturing considerations versus accuracy and mechanical performance, *Mater Des* 139 (2018) 141–152.

53. Y. Zhang, L. Josien, J.P. Salomon, A.S. Masseron, J. Lalevée, Photopolymerization of zeolite/polymer-based composites: Toward 3D and 4D printing applications, *ACS Appl Polym Mater* 3 (1) (2021) 400–409.

54. A. Malas, D. Isakov, K. Couling, G.J. Gibbons, Fabrication of high permittivity resin composite for Vat photopolymerization 3D printing: Morphology, thermal, dynamic mechanical and dielectric properties, *Materials* 12 (2019) 3818–3830.

55. A. Kobyliukh, K. Olszowska, U. Szeluga, S. Pusz, Iron oxides/graphene hybrid structures: Preparation, modification, and application as fillers of polymer composites, *Adv Colloid Interface Sci*, 285 (2020) 102285.

56. J. Janga, P.V.W. Sasikumar, F. Navaee, L. Hagelüken, G. Blugan, J. Brugger, Electrochemical performance of polymer-derived SiOC and SiTiOC ceramic electrodes for artificial cardiac pacemaker applications, *Ceram Int* 47 (6) (2021) 7593–7601.

57. M.L. Shofner, K. Lozano, F.J.R. Macías, E.V. Barrera, Nanofiber-reinforced polymers prepared by fused deposition modeling, *J Appl Polym Sci* 89 (11) (2003) 3081–3090.

58. Y. Zhao, R. Yao, L. Ouyang, H. Ding, T. Zhang, K. Zhang, S. Cheng, W. Sun, Three-dimensional printing of Hela cells for cervical tumor model in vitro, *Biofabrication* 6 (3) (2014) 035001.

59. W. Kim, M. Kim, G.H. Kim, 3D-printed biomimetic scaffold simulating microfibril muscle structure, *Adv Funct Mater* 28 (26) (2018) 1800405.

60. D.F.D. Campos, A. Blaeser, M. Weber, J. Jakel, S. Neuss, W. Jahnen-Dechent, H. Fischer, Three-dimensional printing of stem cell-laden hydrogels submerged in a hydrophobic high density fluid, *Biofabrication* 5 (1) (2013) 015003.

61. D. Lin, S. Jin, F. Zhang, C. Wang, Y. Wang, C. Zhou, G.J. Cheng, 3D stereolithography printing of graphene oxide reinforced complex architectures, *Nanotechnology* 26 (2015) 434003.

62. C.O. Baker, X. Huang, W. Nelson, R.B. Kaner, Polyaniline nanofibers: Broadening applications for conducting polymers, *Chem. Soc. Rev.* 46 (5) (2017) 1510–1525.

63. X.Y. Wang, G.Y. Feng, M.J. Li, M.Q. Ge, Effect of PEDOT:PSS content on structure and properties of PEDOT:PSS/poly(vinyl alcohol) composite fiber, *Polym Bull* 76 (4) (2018) 2097–2111.

64. M.A. Darabi, A. Khosrozadeh, R. Mbeleck, Y. Liu, Q. Chang, J. Jiang, J. Cai, Q. Wang, G. Luo, M. Xing, Skin-inspired multifunctional autonomic-intrinsic conductive self-healing hydrogels with pressure sensitivity, stretchability, and 3D printability, *Adv Mater* 29 (31) (2017) 1700533.

65. S. Sun, M. Yang, Y. Kostov, A. Rasooly, Elisa-loc: Lab-on-a-chip for enzyme-linked immuno detection, *Lab Chip* 10 (2010) 2093–2100.

66. U.M. Dilberoglu, B. Gharehpapagh, U. Yamana, M. Dolen, The role of additive manufacturing in the era of Industry 4.0, *Procedia Manuf* 11 (2017) 545–554.

67. MarketsandMarkets Analysis, ©2021 MarketsandMarkets Research Private Ltd.

68. www.jabil.com/, www.smithers.com/.

69. HP and Deloitte announce alliance to accelerate digital transformation of US$12 trillion global manufacturing industry, *Press Release*, www2.deloitte.com, August 24 (2017).

70. www.medicaldevice-network.com/.

71. C.Y. Liaw, M. Guvendiren, Current and emerging applications of 3D printing in medicine, *Biofabrication* 9 (2) 2017, 024102.

72. M. Javaid, A. Haleem, Additive manufacturing applications in medical cases: A literature based review, *Alexandria J. Medicine* 54 (4) (2018) 411–422.

73. S. Singare, L. Dichen, L. Bingheng, G. Zhenyu, L. Yaxiong, Customized design and manufacturing of chin implant based on rapid prototyping, *Rapid Prototyp J* 11 (2005) 113–118.

74. C. Song, Y. Yang, Y. Wang, J. Yu, D. Wang, Personalized femoral component design and its direct manufacturing by selective laser melting, *Rapid Prototyp J* 22 (2016) 330–337.

75. R. de Azevedo Gonçalves Mota, E.O. da Silva, F.F. de Lima, L. de Menezes, A. Thiele, 3D printed scaffolds as a new perspective for bone tissue regeneration: Literature review, *Materials Sciences and Applications* 7 (8) (2016) 430–452.

76. I.T. Ozbolat, Y. Yu, Bioprinting toward organ fabrication: challenges and future trends, *IEEE Trans Biomed Eng* 60 (2013) 691–699.

77. C. Wang, W. Huang, Y. Zhou, L. He, Z. He, Z. Chen, X. He, S. Tian, J. Liao, B. Lu, Y. Wei, M. Wang, 3D printing of bone tissue engineering scaffolds, *Bioact Mater* 5 (1) (2020) 82–91.

78. S. K. Malyala, A. Manmadhachary, Y. Ravi Kumar, A. Alwala, Manufacturing of patient specific AM medical models for complex surgeries, *Materialstoday: Proceedings, Part A* 4 (2) (2017) 1134–1139.

79. M.P. Bartellas, Three-dimensional printing and medical education: A narrative review of the literature, *University of Ottawa J Medicine* 6 (1) (2016) 38.

80. A. Ganguli, G.J. Pagan-Diaz, L. Grant, C. Cvetkovic, M. Bramlet, J. Vozenilek, T. Kesavadas, R. Bashir, 3D printing for preoperative planning and surgical training: A review, *Biomed Microdevices* 20 (3) (2018) 65.

81. N. Shahrubudin, T.C. Lee, R. Ramlan, An overview on 3D printing technology: Technological, materials, and applications, *Procedia Manuf* 35 (2019) 1286–1296.

82. S.D. Nath, S. Nilufar, An overview of additive manufacturing of polymers and associated composites, *Polymers*, 12 (11) (2020) 2719.

83. P. Ahangar, M.E. Cooke, M.H. Weber, D.H. Rosenzweig, Current biomedical applications of 3D printing and additive manufacturing, *Appl Sci* 9 (8) (2019) 1713.

84. M.B. Burn, G.R. Gogola, Three-dimensional printing of prosthetic hands for children, *J Hand Surg* 41 (5) (2016) 103–109.

85. M.S. Mannoor, Z. Jiang, T. James, Y.L. Kong, K.A. Malatesta, W.O. Soboyejo, N. Verma, D.H. Gracias, M.C. McAlpine, 3D printed bionic ears, *Nano Lett* 13 (6) (2013) 2634–2639.

86. A. Aimar, A. Palermo, B. Innocenti, The role of 3D printing in medical applications: A state of the art, *J Healthc Eng* 2019 (2019) 5340616.

87. J. Jansen, F.P.W. Melchels, D.W. Grijpma, J. Feijen, Fumaric acid monoethyl ester-functionalized poly(D,L-lactide)/N-vinyl-2-pyrrolidone resins for the preparation of tissue engineering scaffolds by stereolithography, *Biomacromolecules* 10 (2009) 214–220.

88. Z. Feng, Y. Li, L. Hao, Y. Yang, T. Tang, D. Tang, W. Xiong, Graphene-reinforced biodegradable resin composites for stereolithographic 3D printing of bone structure scaffolds, *J Nanomater* 2019 (2019) 9710264.

89. L. Dong, S.J. Wang, X.R. Zhao, Y.F. Zhu, J.K. Yu, 3D-printed poly (ϵ-caprolactone) scaffold integrated with cell-laden chitosan hydrogels for bone tissue engineering, *Sci Rep* 7 (2017) 13412.

90. P. Honigmann, N. Sharma, R. Schumacher, J. Rueegg, M. Haefeli, F. Thieringer, In-hospital 3D printed scaphoid prosthesis using medical-grade polyetheretherketone (PEEK) biomaterial, *BioMed Research International* 2021 (2021) 1301028.

91. S. Wickramasinghe, T. Do, P. Tran, FDM-based 3D printing of polymer and associated composite: A review on mechanical properties, defects and treatments, *Polymers* 12 (7) (2020) 1529.

92. K.H. Tan, C.K. Chua, K.F. Leong, M.W. Naing, C.M. Cheah, Fabrication and characterization of three-dimensional poly(ether-ether-ketone)/hydroxyapatite biocomposite scaffolds using laser sintering, *Proc Inst Mech Eng, Part H: J Eng Med* 219 (3) (2005) 183–194.

93. J.M. Williams, A. Adewunmi, R.M. Schek, C.L. Flanagan, P.H. Krebsbach, S.E. Feinberg, S.J. Hollister, S. Das, Bone tissue engineering using polycaprolactone scaffolds fabricated via selective laser sintering, *Biomaterials* 26 (23) (2005) 4817–4827.

94. J.A. Inzana, D. Olvera, S.M. Fuller, J.P. Kelly, O.A. Graeve, E.M. Schwarz, S.L. Kates, H.A. Awad, 3D printing of composite calcium phosphate and collagen scaffolds for bone regeneration, *Biomaterials* 35 (13) (2014) 4026–4034.

95. M. Zhang, A. Vora, W. Han, R.J. Wojtecki, H. Maune, A.B.A. Le, L.E. Thompson, G. M. McClelland, F. Ribet, A.C. Engler, A. Nelson, Dual-responsive hydrogels for direct-write 3D printing, *Macromolecules*, 48 (18) (2015) 6482–6488.

96. L. Li, Q. Lin, M. Tang, A.J.E. Duncan, C. Ke, Advanced polymer designs for direct-ink-write 3D Printing, *Chemistry a European Journal*, 25 (46) (2019) 10768–10781.

97. D. Olivier, J.A.T. Rodriguez, S. Borros, G. Reyes, R.J. Mesa, Influence of building orientation on the flexural strength of laminated object manufacturing specimens, *J Mech Sci Technol* 31 (2017) 133–139.

98. J. Kechagias, An experimental investigation of the surface roughness of parts produced by LOM process, *Rapid Prototyp J* 13 (1) (2007) 17–22.

9 Laser Powder Bed Fusion of Ti6Al4V Alloy for Biomedical Applications
Prospects and Challenges—Process Details, Powder Characteristics and Laser Irradiation

T.S.N. Sankara Narayanan and Hyung Wook Park

CONTENTS

9.1 INTRODUCTION

Conventional manufacturing (CM) technologies are time intensive and energy consuming. CM involves multi-stage processing with huge amounts of material loss. Unlike CM, which is characterized by several stages of subtractive manufacturing, additive manufacturing (AM) involves building up of parts by a layer-by-layer approach. AM processes assume significance due to their ability to fabricate near-net-shaped complex parts such as cellular lattices, struts and porous structures, high material utility and requirement of minimal machining. The process variables of LPBF process are highly interrelated, which makes it difficult to get a better understanding of their influence in building Ti6Al4V alloy parts with the desired characteristics (Gu et al., 2012; Vanmeensel et al., 2018; Jang et al., 2020). The utility of AM for the manufacture of medical devices has received considerable attention. The possibility of producing personalized patient-specific implants (e.g., fixation plates, jaw implants, cranial implants) that meet individual requirements is the best part of the AM process. The role of AM in cranio-maxillofacial surgery has been highlighted in terms of achieving high precision and considerable savings in surgical time (Martelli et al., 2016; Louvrier et al., 2017). The ability of the AM process to develop porous structures has considerable advantages in preparing biomedical implants for joint replacement surgery and bone grafting. Fabrication of porous lattices instead of bulk solid materials could help reduce both the modulus and weight besides improving bone cell in-growth.

9.2 CLASSIFICATION OF AM PROCESSES

AM processes are categorized in seven different types, namely, binder jetting, material jetting, powder bed fusion, direct energy deposition, sheet lamination, vat photopolymerization and material extrusion. Among them, powder bed fusion and direct energy deposition are considered suitable for production of metallic parts. Based on the source of energy, powder bed fusion processes can be distinguished as laser powder bed fusion (LPBF) and electron beam powder bed fusion (EBPBF). Both LPBF as well as EBPBF processes offer exceptional freedom of design and both of them are widely used for the manufacture of biomedical devices.

9.3 LASER POWDER BED FUSION (LPBF) PROCESS

Development of a 3D model of the part to be produced using computer aided design (CAD) is the key for success of the LPBF process. Subsequently, the entire part design is sub-divided into various layers. In accordance with the sliced layers, usually with thickness ranging from 20 μm to 60 μm, the metal/alloy powder is spread on the build plate. To achieve uniform distribution, a levelling system or a re-coater blade is used. A laser beam with power ranging from 200 W to 400 W is directed at a scanning speed between 0.5 m/s and 7 m/s across the powder bed. Since the laser beam carries no electric charge, the galvanometer mirror should be suitably adjusted to regulate the laser beam path. The laser diameter can be adjusted between 50 μm and 180 μm using a lens or galvanometer mirror under defocus mode. LPBF process is performed inside a closed chamber under inert gas atmosphere. Ar or N is passed inside the chamber to prevent oxidation of Ti6Al4V alloy powder as well the molten metal. Maintaining a proper flow of Ar could help removing process by-products such as fumes and spatters and avoids contamination (Herzog et al., 2016). The incidence of laser irradiation enables melting of the powders and re-solidification of the molten metal, thus generating a first layer over the build plate. Subsequently, the build plate is decreased so as to apply another layer of powders, which is spread uniformly by the re-coater arm. The process of spreading of powders, laser irradiation, melting and solidification continues until the whole part is built as per the 3D model generated by CAD. Support structures are essential for heat dissipation, for fixing the parts that are horizontally oriented or for those involving overhanging surfaces and to prevent deformation of the parts. Pre-heating of the powders between 200 °C and 500 °C is suggested as a remedy to prevent distortion of Ti6Al4V alloy parts (Herzog et al., 2016). The degree of freedom in part design, ability to fabricate near-net-shaped lattice structures with complex geometries, better dimensional accuracy and high material recycling rate are the prime benefits of the LPBF process (Ni et al., 2019). Lack of control of melt pool stability, higher surface roughness, higher residual stress and difficulty in scaling up are the most important drawbacks of the LPBF process (Attar et al., 2014).

The laser power (P), laser scanning speed (v), powder layer thickness (t) and hatch spacing (h) are the most common process parameters in the LPBF process (Oliveira et al., 2020). During the process, it is imperative to achieve a higher density of the Ti6Al4V alloy parts as close to its theoretical density. The relation between laser energy density (LED, laser power (P), laser scanning speed (v), powder layer thickness (t) and hatch spacing (h) are given in the following equation:

$$\text{LED (J/mm}^3) = \text{P (W)}/\text{v (mm/s)} \times \text{t (μm)} \times \text{h (μm)} \ldots (1)$$

Low laser power leads to improper melting of the powders. Use of a sufficiently high laser power enables easy melting of the Ti6Al4V alloy powder. Proper melting enables a reduction in surface tension and viscosity of the molten metal and facilitates its infiltration inside the grooves and pits, thus increasing the density of the fabricated parts. However, very high laser power causes over-burn and vaporization, resulting in a large decrease in relative density (Zhang and Attar, 2015). The optimum range of laser power suggested for the production of Ti6Al4V alloy parts by the LPBF process is 150–200 W.

When the laser scanning speed is very low, the interaction between the laser irradiation and the powder bed becomes much longer. This condition would promote absorption of very high energy by the powders, leading to vaporization and formation of pores, resulting in a lower density. At sufficiently large scanning speed, the right amount of energy will be absorbed by the powders leading to adequate melting of the powders, which would reduce the porosity and increase the density of the parts (Majumdar et al., 2019). The optimum scanning speed suggested for the production of Ti6Al4V alloy parts by LPBF process is 200–600 mm/s.

When the hatch spacing is small, sufficient overlap between the melt channels increases the relative density of the Ti6Al4V alloy parts. Higher hatch spacing decreases the extent of overlap, leading to unevenness in the melt channels, resulting in a decrease in density of the parts. Too much of overlap obviously leads to splashing of the powders, leading to a decrease in density of the parts (Majumdar et al., 2019). The optimum range of hatch spacing suggested for the production of Ti6Al4V alloy parts by LPBF process is 60–100 μm.

The choice of a low powder layer thickness enables melting of all the powders and remelts the top layer of the previously solidified layer, thus providing a higher density for the parts. With an increase in powder layer thickness, the laser irradiation becomes inadequate to cause sufficient melting, resulting in a decrease in the density of the parts (Majumdar et al., 2019). Many research reports recommend 30 μm as the optimum powder layer thickness for the production of Ti6Al4V alloy parts by the LPBF process. Hence, it is evident that the process parameters should be carefully optimized to fabricate Ti6Al4V alloy parts with densities > 99.9%.

9.3.1   Characteristics of the Melt Pool

The choice of a higher laser power increases the temperature, length, width and depth of the melt pool. However, a higher scanning speed increases the length of the melt pool but decreases the temperature and depth of the melt pool. Accordingly, the lifetime of the melt pool becomes higher with an increase in laser power and a decrease in laser scanning speed. Since a higher laser power increases the heat input, it is also reflected in the temperature gradient. The conditions of the melt pool should facilitate uniform spreading but at the same time it should prevent balling. The melt pool stability exerts a significant influence on the characteristics of the parts produced by the LPBF process (Sun et al., 2017). During the LPBF process, laser scanning of several neighboring layers leads to the creation of a heat affected zone (HAZ). The rapid cooling of the molten mass leads to a narrow HAZ around the melt pool. It is essential to minimize the HAZ by a careful choice of process parameters which would otherwise deleteriously influence the homogeneity of the parts.

9.3.2   Defect Generation

It is difficult to avoid the formation of defects during fabrication of metallic parts by the LPBF process. In case of Ti6Al4V alloy, the lower thermal conductivity of the alloy powder bed when compared to its solidified counterpart has compounded the effect on defect generation. When the laser irradiation interacts with the powder bed, depending on the choice of laser power and scanning speed, improper fusion, improper melting, Marangoni convection, melt pool instability, material spattering, evaporation and gas entrapment could occur (Pal et al., 2019, 2020). Both high and low laser power could lead to the generation of defects. Very high laser power promotes keyhole formation while low laser power causes improper melting. Defect generation becomes higher at higher laser scanning speed. Gong et al. (2014) have proposed a process window for fabrication of Ti6Al4V alloy parts by LPBF process (Figure 9.1). The process window is segmented into four regions. Due to insufficient energy and high scanning speed, incomplete melting prevails in zone III. On the contrary, the higher energy and insufficient speed promotes over melting in zone II. Zone OH is unsuitable for fabrication due to excessive heating and heavy deformation. Zone I is optimal in terms of number of obtaining fully dense parts.

9.3.2.1 Pores

Generation of pores is quite common during fabrication of metallic parts by the LPBF process. Unless and otherwise the process conditions are carefully optimized, the LPBF process leads to uncontrolled porosity. For Ti6Al4V alloy parts, the porosity is ~0.35% and larger pores are usually detected in the layer ranging from 0.4 mm from the top surface. Among the various process parameters, laser power and laser scanning velocity are considered to be critical in the development of pores. The pores could have emerged due to gas bubble entrapment, improper melting and keyhole formation (Liu and Shin, 2019; Pal et al., 2020). According to Gong et al. (2014), the evolution of porosity of Ti6Al4V alloy parts produced by the LPBF process can be correlated using the process window shown in Figure 9.1. Parts prepared using conditions defined in zone II and zone III consists of large number of pores due to excess energy in zone II and insufficient energy for melting in zone III. There are fewer pores when the parts are produced under conditions of zone I.

Gas pores are generated following the entanglement of gas in the molten pool. The gas pores are mostly small, spherical in shape and randomly distributed across the part. A higher energy and insufficient scan speed could promote the formation of gas pores. Entrapment of gas is a function of the consistency of the molten pool; the lower the scanning speed, the lower the viscosity of the molten pool and vice versa. The possibility of gas entrapment becomes high at higher scanning speed. The presence of gas pores up to 1 vol. % is generally acceptable whereas when it exceeds 5 vol. %, the mechanical properties would be drastically affected. The presence of unmelted particles within the pores would decrease the strength of the parts. Lack-of-fusion (LOF) pores exhibit different characteristics from the gas pores. They are usually large, irregular-wedge shaped with sharp tips. LOF pores are generated under conditions of insufficient melting and they are mainly identified at the boundaries between two adjacent layers. The choice of dense feedstock powders instead of sponge type ones could help decrease the formation of gas pores. An increase in energy

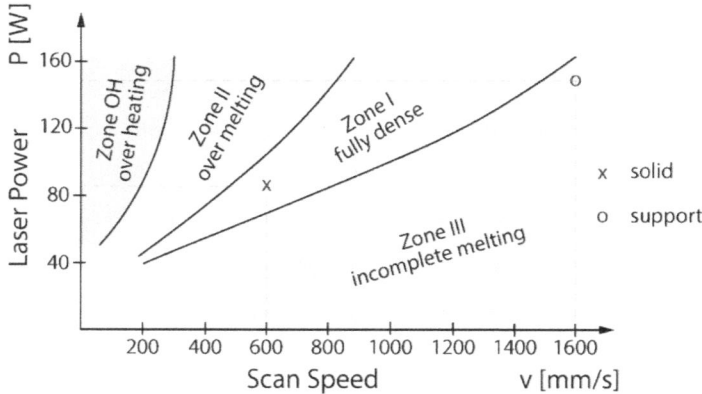

FIGURE 9.1 Process parameter relationships for L-PBF manufactured Ti-6Al-4V (originally proposed by Gong et al. (2014)).

Source: Weber et al. (2020). Under Creative Commons CC-BY license.

density could help alleviate the LOF pores (Liu and Shin, 2019). When the volume fraction of pores is higher, the tensile strength, ductility, fatigue strength and corrosion resistance of the parts would be drastically affected and among them LOF pores cause a major effect. The pores cannot be closed by simple heat treatment (HT) whereas both the volume and size of the pores could be considerably decreased by hot isostatic pressing (HIP).

9.3.2.2   Balling

During LPBF, the molten metal is expected to wet the substrate. In the absence of proper wetting, the molten metal pool shrinks and breaks up in the form of spheres (balling) of varying size, ranging from 10 mm to 500 mm and shape, varying from elliptical to spherical (Figure 9.2). The occurrence of balling could be due to instability, high velocity, splashing and oxidation of the molten metal mass.

Very low viscosity of the melt pool or excessive generation of molten mass promotes balling. The difference between balling and unmelted powder particles is identified based on their grain structures. Balling results in the formation of a very rough surface with higher porosity, which makes uniform spreading of powders in subsequent layers by the re-coater arm difficult. Balling would drastically reduce the fatigue life of the parts. It is possible to minimize the incidence of balling by decreasing the length-to-width ratio and maintaining a good consistency of the melt pool by an appropriate choice of laser exposure time. Surface remelting by a subsequent laser scan is found to be an effective method in eliminating the balling (Sun et al., 2017).

FIGURE 9.2   Scanning electron micrographs showing the balling effect.

Source: Reprinted from Pal et al. (2020) with permission from Elsevier.

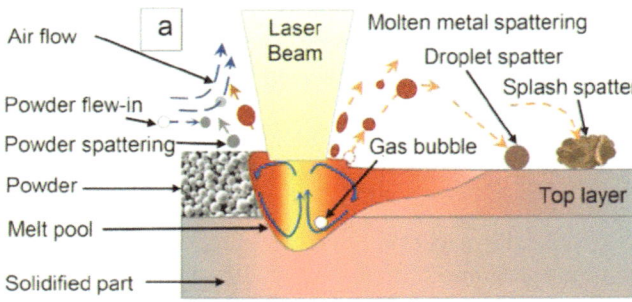

FIGURE 9.3   Schematic representation of the spattering phenomena.

Source: Reprinted from Pal et al. (2020) with permission from Elsevier.

9.3.2.3   Spattering

At higher LED, the melt pool is overheated and when its recoil pressure becomes much higher, the unmelted powder particles are expelled, resulting in spattering (Figure 9.3). The spatters are much larger in size than the powder feed. The deposition of oxides of volatile elements on spatters is commonly referred to as satellites (Sun et al., 2017). Spattering occurs since powder particles could absorb more amounts of laser irradiation than the solidified molten mass. Distributing the laser irradiation over a large surface and rescanning of the solidified layer are suggested to avoid spattering (Chen et al., 2018a; Pal et al., 2020).

Among the various types of defects, LOF defect, balling and surface roughness are considered to be more detrimental than gas pores and hot tears. Among the various process parameters, laser power and laser scanning speed leads to the generation of most of the defects, which warrants careful selection of these two process parameters to build parts with minimal defects.

9.4   INFLUENCE OF POWDER CHARACTERISTICS, POWDER SPREADING AND PRE-HEATING OF THE POWDER BED

9.4.1   POWDER CHARACTERISTICS

The ultimate quality of the as-built part is determined by the features of the feedstock powders such as size, shape, distribution, flowability and packing density. Good packing of the powders is recommended to reduce porosity, surface roughness and distortion of the as-built parts. The packing density is determined by the morphology of the particles and particle size distribution (PSD). Spherically shaped particles exhibit a better flowability, which would enable a high packing density. A wider PSD would enable excellent packing since smaller particles could effectively fill the gaps between the larger particles. However, the powder particles should not be too small as they would increase the surface area, promote extent of absorption of laser irradiation and increase the temperature of the molten pool. Particle agglomeration that could decrease the flowability is a concern in using smaller powder particles. In addition, use of smaller particles would lead to generation of large gaps, resulting higher porosity of the as-built part (Sun et al., 2017). A variety of methods are proposed for the preparation of Ti alloy feedstocks for LPBF process. Among them, water atomization (WA) is the most simplest and cost effective method. However, irregular shape, low flowability, poor density and high amount of O_2 makes WA unsuitable for production of Ti alloy feedstocks for the LPBF process (Eskandari Sabzi, 2019). Gas atomization (GA) uses Ar as atomizing medium, enables preparation of Ti alloy powder particles with a wide PSD with high production efficiency. However, formation of a large fraction of satellite powders and hollow powders are the major limitations in preparing Ti alloy feedstocks by GA (Tang et al., 2019; Eskandari

Sabzi, 2019). Plasma atomization (PA) uses wires as feedstocks and plasma as the melting source and it is suitable for the production of highly reactive Ti alloy powder feedstocks for LPBF process. The Ti alloy powder particles prepared by PA are highly spherical with reasonable fluidity and high filling density. Since PA uses metallic wire as the source, the lower productivity and higher cost are the major limitations in preparing of Ti alloy feedstocks for the LPBF process (Tang et al., 2019; Eskandari Sabzi, 2019). The plasma rotating electrode process (PREP) uses rotating bars as feedstocks and performs under vacuum. The particles prepared by PREP are pure, highly spherical with excellent flowability. They have fewer satellite particles and negligible hollow particles. However, the particles are coarser (50–350 μm), which limits their use as feedstocks for the LPBF process (Tang et al., 2019; Eskandari Sabzi, 2019). Chen et al. (2018b) have compared the characteristics of Ti6Al4V powders prepared by GA, PREP and PA. The Ar content and porosity within the powders becomes high when the size of the particles is increased. The sphericity of the pores in powders prepared by GA and PA is relatively higher than those prepared by PREP due to variation in gas pressures inside the powders.

9.4.2 Powder Spreading

During the fabrication of Ti6Al4V alloy parts by the LPBF process, powder spreading is a critical factor. The packing density and homogeneity of the powders determines the extent of absorption of the laser irradiation, heat transfer and melt pool behavior, which ultimately decides the merit of the as-built part (Wang et al., 2020; Yao et al., 2021). Hence, it is imperative to use a dense and uniform powder bed. The packing density and flowability of the powders are determined by PSD and particle adhesion (Wang et al., 2020). Bimodal powder mixture with proper size ratio increases the packing density and flowability of the powders. Adhesion of small sized particles with larger ones could reduce the packing density and uniform distribution of the powders (Wang et al., 2020). Powder spreading is effective using a roller instead of a rigid blade. A roller enables dense and smooth packing of powders with good particle rearrangement during spreading rather than a blade. The moving speed of the roller/blade also influences the powder spreading. If the moving speed is kept low then the quality of the powder bed becomes good, but it lowers the process efficiency. During particle spreading, interaction between the particles and the re-coater causes the segregation of finer and coarser particles (Lee et al., 2020). It would be possible to control segregation of particles by suitably altering the moving speed of the roller/blade.

9.4.3 Pre-Heating the Powder Bed

Pre-heating of the powders is suggested as a viable option to reduce the residual stress of the parts produced by the LPBF process. When Ti6Al4V alloy powders are pre-heated to 550 °C, distortion at the top layer of the part is decreased by 46.31% while the density of the part is increased by 2.88%. However, it leads to an increase in O_2 and hydrogen content beyond the limits specified by ASTM B348. Pre-heating of the Ti6Al4V alloy powders from 100 °C to 770 °C has led to a decrease in volume fraction of the pores (Pedrazzini et al., 2020). An improvement in ductility from ~6% to ~10% was observed when the powders are heated from 100 °C to 570 °C, whereas the ductility is considerably reduced at 770 °C (~ 0%). Since the volume fraction of the pores is decreased with an increase in temperature from 100 °C to 570 °C, it is obvious that the reduction in ductility is due to the change in microstructure. When the powders are heated to 100 °C, α and martensitic α′ are identified as the predominant phases with a very small fraction of retained β phase. The presence of high density of microtwins and dislocations are considered responsible for the lower ductility of parts produced at a pre-heating temperature of 100 °C. When the powders are pre-heated to 570 °C, the volume fraction of the martensitic α′ phase is decreased concurrent with a relatively more amount of retained β phase. At 770 °C, the fraction of the martensitic α′ phase is decreased further with a higher amount of the retained β phase. Pre-heating the Ti6Al4V alloy powders from 100 °C to 770 °C has increased the O_2 content of the as-built part, which suppresses the formation of the α' martensite phase. The higher O_2 content of the Ti6Al4V alloy part produced at 770 °C has deleteriously influenced the ductility of the part.

9.5 INFLUENCE OF LASER IRRADIATION

9.5.1 EFFECT OF LASER SCANNING MODE

The choice of laser scanning mode, continuous laser based (CLB) or pulsed laser based (PLB), strongly influences the melt pool characteristics, microstructure, porosity and fatigue behavior of struts made of Ti6Al4V alloy (Karami et al., 2020). The higher energy and larger exposure time has increased the width and depth of the melt pool in the pulsed laser mode. A comet-like melt pool is generated under CLB mode whereas an ellipse-like melt pool is emerged under PLB mode. The microstructure of struts processed under CLB mode reveals hierarchical hexagonal α' laths within the elongated columnar prior β grains whereas a uniform basket-weave morphology consisting of fine equiaxed dendritic ὰ grains with no preferential orientation is observed for struts processed under PLB mode. The fraction of retained prior β grains also varies with the scanning mode; 5% under CLB mode and 2.5% under PLB mode. The porosity of Ti6Al4V alloy struts is higher when processed under PLB mode than those produced using the CLB mode. The higher cooling rate in PLB mode has enabled a refinement in grain size, resulting in a moderate increase in hardness when compared to those fabricated using the CLB mode. Ability to control the heat input facilitates the formation of struts with uniform thickness and diameter for samples processed in PLB mode. In contrast, in CLB mode, the strut thickness and diameter lack uniformity. Under a constant laser power, the CLB mode provides a more stable melt pool that limits the quantity of partially molten particles. In contrast, melt pool instability and intermittent melting under the pulsed laser irradiation has increased the volume fraction of partially molten particles. The higher porosity and higher volume fraction of partially molten particles promote fatigue crack initiation, resulting in lower cycles of failure for struts processed in PLB mode when compared to those produced in CLB mode. The fatigue crack initiation site is found to be different for struts processed by CLB and PLB modes—at the strut junctions in CLB mode while it is between subsequent beads in PLB mode. Hence, it is imperative to adopt a suitable post-treatment to improve the fatigue behavior of these struts. Karami et al. (2020) have suggested a series of post-treatments, namely HIP, sand blasting (SB) and chemical etching (CE) to increase the fatigue life of the struts. According to them, HIP could decrease the porosity, transform the acicular α' martensite phase (brittle) to lamellar α/β phase with a higher fraction of the β phase (ductile). SB and CE could effectively get rid of most of the partially molten particles, which act as stress concentration points. The findings of the study suggest that CLB mode can be recommended for processing bigger parts, whereas PLB mode can be suggested for fabrication of lattice structures.

9.5.2 USE OF MULTIPLE LASERS

The LPBF process is time intensive and it might take several hours to complete building of a single part. Use of multiple lasers has been suggested as a viable option to increase the rate of production of Ti6Al4V alloy parts (Masoomi et al., 2017). Use of four lasers instead of one could offer considerable savings in processing time by about 75%. However, the change in locality and speed of fusion might significantly increase the residual stress in the as-built part, warranting a suitable post-treatment. The degree of change in temperature gradient becomes much higher with the use of multiple lasers. Hence, it is sensible to adopt a suitable scanning strategy when multiple lasers are used to reduce the processing time.

9.5.3 In Situ Laser Remelting

Karimi et al. (2020) have compared the effect of single, double and triple laser melting of Ti6Al4V alloy. In single melting, the powders are melted once. In double melting, the solidified first layer is melted again after rotating the sample by 72°. In triple melting, the first solidified layer and the second re-solidified layer is melted again after rotating the sample by another 72° to treat the solidified and re-solidified layer. When the number of melting steps is increased, the elemental constitution of the alloy is not altered. The acicular α' martensite phase is retained due to the higher cooling rate, which causes an increase in its width and a decrease in its length. The LOF pores are eliminated and the theoretical density is increased. Metallurgical pores, however, are still evident. Nevertheless, the pore size is considerably reduced. The elimination of LOF pores and decrease in size of the metallurgical pores enable an increase in effective energy absorbed. Consequently, the temperature of the melt pool is increased and during fast cooling the grain size is refined. During single melting, the concave-shaped melt pool is unstable, resulting in a wavy surface with a high surface roughness. In contrast, during triple melting, the stable and smooth melt pool track enables the formation of a smooth surface finish, which facilitates uniform spreading of powders in the next layer. The crystallite size and lattice parameters ('a' and 'c') are decreased, the intensity of the (002), (102) and (110) planes are increased with a strong texture towards the (102) plane. The dislocation density and hardness are increased. In addition, the hardness is found to be uniform throughout. The residual stress is reduced. The ultimate tensile strength is increased, but the ductility is decreased which signifies a more brittle nature. The decrease in grain size helps to prevent crack nucleation, which is likely to increase the fatigue life.

The scanning velocity employed for laser remelting is also an important factor in determining the merit of the parts fabricated by LPBF process. At relatively low scanning velocities, laser remelting decreased the roughness of the top layer. However, at high scanning velocities, the top surface roughness becomes much higher than those without remelting. At optimized conditions, laser remelting could improve the top surface roughness by 60%. Nevertheless, the side surface warrants significant improvement. Although laser remelting enables removal of the adhered particles, use of a higher laser power could lead to attachment of more number of newly adherent particles at the side surface, thus affecting the surface finish. Ability to achieve better homogeneity, improved surface finish, elimination of LOF pores, decrease in size of the metallurgical pores, grain refinement, decrease in residual stress, increase in hardness, increase in ultimate tensile strength (UTS) are the beneficial effects of in situ laser remelting during the LPBF process. In situ laser remelting provides an extended window of opportunity for in situ alloying using elemental metal powders for the development of new alloys in which melting of certain elements such as Mo possesses difficulty. The increase in cost, production time, decrease in ductility and formation of a thick oxide layer are the major limitations in employing in situ laser remelting.

9.6 INFLUENCE OF SHIELDING GAS

To prevent oxidation of Ti6Al4V alloy, it is essential to maintain an inert atmosphere during the LPBF process. The high affinity of Ti, Al and V towards O_2 and N_2 and the high sensitivity of the Ti6Al4V alloy to humidity facilitate the formation of corresponding oxide and nitride inclusions. The presence of such inclusions at the interface between the layers built one over the other deleteriously influences the inter-layer bonding. Hence, it is imperative to control the concentration of O_2, N_2 and moisture content of the powder feedstock. An increase in O_2 concentration (2 ppm–97 ppm) in the build environment could increase the O_2 content of the part by about 65% and decrease the fatigue resistance. In spite of an increase in yield strength (YS) and UTS, higher concentration of O_2 could also lead to a decrease in the elongation to failure. The higher strength and hardness as well as decrease in elongation is due to the development of acicular α' martensitic phase following rapid cooling during the LPBF process. It is imperative to decrease the interstitial elements to realize a higher ductility, fracture toughness and improved crack growth resistance. An increase in concentration of O_2 in the build environment could influence the Marangoni flow and make the melt pool unstable.

Presence of moisture in the pre-alloyed Ti6Al4V powder feedstock could also increase the O_2 content of the as-built alloy sample. In order to decrease the moisture content, the powders should be vacuum dried prior to use. Pauzon et al. (2019) have used an Ar-He gas mixture environment during the manufacture of Ti6Al4V alloy by LPBF process. The use of a 50–50 mixture of Ar and He has enabled an increase in build rate by 40%, a higher process stability and a relatively higher density than the one fabricated in Ar atmosphere alone. The major limitation in using the Ar-He gas mixture is the higher residual stress in the as-built part. Besides the environment, maintaining an adequate gas flow is required to remove process by-products such as spatter, ejected powders and condensates. Interaction between the laser ray and the process by-products could attenuate the laser irradiation leading to a change in the focal position, resulting in a 'splashy process'. Re-deposition of process by-products could increase the thickness of the powder bed and beyond a threshold value of thickness, it is likely to induce porosity and LOF defects in the as-built parts. Use of 50–50 mixture of Ar and He with an adequate gas flow rate has decreased the accumulation of by-products. Hence, it is clear that the nature of the build environment is very critical and unless and otherwise the gas composition and its flow rate are precisely controlled, it would be difficult to build parts with desired characteristics.

9.7   IN SITU ALLOYING

In situ alloying by LPBF is a promising approach for preparing new materials. It assumes significance since many pre-alloyed powders are either not available or not cost-effective. However, achieving homogeneity is the biggest challenge in the manufacture of alloys using elemental alloy powders through in situ alloying by LPBF. An essential requirement of in situ alloying is total melting of all the elemental powders, which can be accomplished by increasing the laser power. In addition, the molten metal mass must remain in solution for a sufficiently longer period of time to ensure complete mixing of the alloy parts (Yadroitsev et al., 2017). Variations in nature of the molten metal mass, incomplete melting, behavior of particles in the melt pool, convection, Marangoni flows, density and viscosity gradients are the controlling factors that determine the homogeneity of the alloy part.

Commercially available Ti6Al4V alloy powder feedstock that can be used for LPBF is expensive. Dong et al. (2020a) have made an alternative approach for preparing Ti6Al4V alloy parts by LPBF using elemental powders. Instead of pure Ti, they have used dehydrogenated Ti (HDH-Ti) powders as the source of Ti. However, HDH-Ti powders possess a highly irregular morphology and exhibit low flowability, which limits their use as feedstocks for LPBF process. In addition, the higher impurity level of O_2 (0.32 wt. %) in the fabricated part would drastically decrease the mechanical properties. Dong et al. (2020a) have used ball milling to redesign the HDH-Ti powder, which increased the sphericity and flowability of the redesigned HDH-Ti powder. Subsequently, the redesigned HDH-Ti powder is mixed with suitable proportions of elemental Al and V powders to prepare the Ti6Al4V alloy through in situ alloying by the LPBF process. The cost of the Ti6Al4V alloy powder prepared using this approach is merely one-seventh of that of the pre-alloyed Ti6Al4V alloy powder. The chemical composition analysis of the as-built part indicates a decrease in Al and V content of the alloy, possibly due to vaporization. A decrease in the content of the alloying elements could reduce the strength of the as-printed Ti6Al4V alloy part. Hence, it is imperative to add additional quantities of alloying elements for in situ alloying through LPBF process. The deleterious influence of higher amounts of O_2 impurity in the as-built part is circumvented with the addition of Y, an O_2 scavenger, during HT. The utility of ball mill modified HDH-Ti powders with better sphericity and improved flowability to produce pure Ti parts by the LPBF process has also been established by Dong et al. (2020b). Solution hardening of the as-printed Ti part at appropriate temperatures improved the mechanical properties that match with that of Ti6Al4V alloy parts. The excellent corrosion resistance, biocompatibility, cell adhesion and proliferation point out that these Ti parts can be effectively used as an alternative for Ti6Al4V alloy parts for orthopedic and dental implant applications.

In situ alloying of Ti6Al4V ELI powder with 3 at. % of Cu powder has been explored by LPBF (Vilardell et al., 2020a). Addition of Cu has promoted the development of α' martensite phase with fine needle-like morphology and decreased the β-transus temperature following its β-stabilizing effect. In spite of promoting a larger fraction of retained β-phase, addition of Cu has promoted the formation of an intermetallic CuTi$_2$ phase. Since liquid Cu has a higher density, it segregates near the fusion boundaries, thus affecting its uniform distribution (Yadroitsev et al., 2017; Vilardell et al., 2020b). Inhomogeneity has also been reported as a major problem in preparing Ti-15Mo alloys through in situ alloying by the LPBF process. Since the Mo powders could reflect the laser radiation to a larger extent than the Ti powders, the latter easily absorbs the laser radiation and melts completely whereas the Mo powder undergoes only partial melting (Yadroitsev et al., 2017).

The use of Ti-12Mo-6Zr-2Fe (TMZF), a β-phase Ti alloy assumes significance for implant applications because of its very good corrosion resistance and biocompatibility. Since the pre-alloyed β-phase TMZF alloy powders are very expensive, Duan et al. (2020) have explored fabrication of this alloy by in situ LPBF using the corresponding elemental powders. Adopting a chess scanning strategy and suitable post-heat treatment has enabled fabrication of TMZF alloy with high strength, low elastic modulus and good ductility. However, the low interdiffusion between Ti and Mo poses difficulty in achieving homogeneity. Complete homogenization requires sufficient time, but melting and cooling stages are very fast during the LPBF process. Hence, ensuring a better chemical homogeneity remains unsolved for fabrication of TMZF alloy by in situ alloying using corresponding elemental powders.

9.8 SUMMARY

- The process variables of LPBF process are highly interrelated, which makes it difficult to get a better understanding of their influence in building Ti6Al4V alloy parts with the desired characteristics. Although numerous efforts are constantly being made to solve many process related problems, much remains to be explored, thus making the field of research on LPBF process wide open.
- Generation of defects is inevitable during the LPBF process and eliminating them is highly challenging.
- Spherical powders with excellent flowability are suitable for the LPBF process. Powders prepared by PA and PREP assume significance.
- Re-use of powders is suggested as an option to reduce the cost. However, not all the remnant powders from the process can be used again. The powder particles have to be sieved to remove unwanted particles and only a particular sieve fraction can be used again for the LPBF process. With the re-use of powders, O$_2$ pick-up and its impact on mechanical properties of the Ti6Al4V alloy parts is a matter of concern.
- Segregation of finer and coarser particles during spreading of the powder by the re-coater arm is a concern. Although segregation of powders can be controlled by reducing the moving speed of the re-coater arm, it reduces the process efficiency. Ensuring good packing density and homogeneity with the use of bimodal powder mixture with proper size ratio holds the key to achieve a good quality powder bed.
- Pre-heating of the powders helps to decrease distortion as well as porosity and increase the density of the parts. However, the higher amount of α' martensite phase formed would decrease the ductility in spite of an increase in mechanical properties. In addition, an increase in O$_2$ content beyond the limits specified by ASTM B348 would deleteriously influence the ductility of the Ti6Al4V alloy parts. Since ductility is an essential requirement of implants, suitable post-treatments should be adopted to increase the ductility of the Ti6Al4V alloy parts if pre-heating of powders is adopted.

- The choice of laser scanning mode is very critical. Continuous laser scanning mode provides a more stable melt pool and limits the extent of incompletely melted particles. However, the thickness and diameter of struts lack uniformity. PLB mode enables the formation of struts with uniform thickness and diameter. The higher cooling rate helps to refine the grains and provides a moderate increase in hardness but it also increases the porosity. The lack of stability of the molten metal mass and intermittent melting has increased the fraction of incompletely melted particles. The fraction of retained prior β grains is also reduced. The higher porosity and presence of a larger fraction of incompletely melted particles promote fatigue crack initiation of Ti6Al4V alloy parts prepared under pulsed laser scanning mode. CLB mode can be recommended for processing bigger parts, whereas PLB mode can be suggested for fabrication of lattice structures.
- In situ remelting has been suggested to eliminate LOF pores, to reduce the residual stress and to decrease the surface roughness of Ti6Al4V alloy parts. The decrease in grain size would help to prevent crack nucleation and increase the fatigue life. However, the higher cost, higher production time and lower ductility are the major concerns in employing in situ laser remelting.
- The presence of an inert gas in the build environment is very critical and unless and otherwise the gas composition and its flow rate are precisely controlled, it would be difficult to build Ti6Al4V alloy parts with desired characteristics. It is essential to use Ar or N, to reduce the O_2 concentration of the build environment to $< 0.1\%$, which would improve the ductility and minimize crack growth. Maintaining a proper flow rate of the inert gas is warranted to prevent re-deposition of the process by-products.
- The lack of availability of many pre-alloyed powders opens up the avenue of in situ alloying by LPBF using elemental powders. In situ alloying by LPBF is indeed a viable option for the development of new alloys. However, achieving homogeneity is the biggest challenge in the manufacture of alloys using elemental alloy powders through in situ alloying by LPBF.
- An essential requirement of in situ alloying is complete melting of all the elemental powders, which can be achieved by an increase in laser power. In addition, the molten pool should remain in solution for a sufficiently longer period of time to ensure complete mixing of the alloy parts.
- An increase in laser power to achieve complete melting could also lead to vaporization of alloying elements, warranting the use of additional quantities of alloying elements for in situ alloying through LPBF process. Complete homogenization requires sufficient time; but melting and cooling stages are very fast during LPBF process. Ensuring a better chemical homogeneity remains unsolved for the fabrication of alloy parts through in situ alloying by LPBF using corresponding elemental powders.

REFERENCES

Attar, H., M. Calin, L. Zhang, S. Scudino and J. Eckert. 2014. Manufacture by selective laser melting and mechanical behavior of commercially pure titanium. *Materials Science and Engineering A*, 593:170–177.

Chen, G., S. Y. Zhao, P. Tan, J. Wang, C. S. Xiang and H. P. Tang. 2018b. A comparative study of Ti-6Al-4V powders for additive manufacturing by gas atomization, plasma rotating electrode process and plasma atomization. *Powder Technology*, 333:38–46.

Chen, Z., X. Wu, D. Tomus and C. H. J. Davies. 2018a. Surface roughness of selective laser melted Ti-6Al-4V alloy components. *Additive Manufacturing*, 21:91–103.

Dong, Y. P., Y. L. Li, S. Y. Zhou, Y. H. Zhou, M. S. Dargusch, H. X. Peng and M. Yan. 2020a. Cost-affordable Ti-6Al-4V for additive manufacturing: Powder modification, compositional modulation and laser in-situ alloying. *Additive Manufacturing*, 101699.

Dong, Y. P., J. C. Tang, D. W. Wang, N. Wang, Z. D. He, J. Li, D. P. Zhao and M. Yan. 2020b. Additive manufacturing of pure Ti with superior mechanical performance, low cost, and biocompatibility for potential replacement of Ti-6Al-4V. *Materials & Design*, 196:109142.

Duan, R., S. Li, B. Cai, W. Zhu, F. Ren and M. M. Attallah. 2020. A high strength and low modulus metastable β Ti-12Mo-6Zr-2Fe alloy fabricated by laser powder bed fusion in-situ alloying. *Additive Manufacturing*, 101708.

Eskandari Sabzi, H. 2019. Powder bed fusion additive layer manufacturing of titanium alloys. *Materials Science and Technology*, 35(8):875–890.

Gong, H., K. Rafi, H. Gu, T. Starr and B. Stucker. 2014. Analysis of defect generation in Ti-6Al-4V parts made using powder bed fusion additive manufacturing processes. *Additive Manufacturing*, 1–4:87–98.

Gu, D. D., W. Meiners, K. Wissenbach and R. Poprawe. 2012. Laser additive manufacturing of metallic components: Materials, processes and mechanisms. *International Materials Reviews*, 57(3):133–164.

Herzog, D., V. Seyda, E. Wycisk and C. Emmelmann. 2016. Additive manufacturing of metals. *Acta Materialia*, 117:371–392.

Jang, T.-S., D. E. Kim, G. Han, C.-B. Yoon and H.-D. Jung. 2020. Powder based additive manufacturing for biomedical application of titanium and its alloys: a review. *Biomedical Engineering Letters*, 10:505–516.

Karami, K., A. Blok, L. Weber, S. M. Ahmadi, R. Petrov, Ksenija Nikolic, E. V. Borisov, S. Leeflang, C. Ayas, A. A. Zadpoor, M. Mehdipour, E. Reinton, V. A. Popovich. 2020. Continuous and pulsed selective laser melting of Ti6Al4V lattice structures: Effect of post-processing on microstructural anisotropy and fatigue behaviour. *Additive Manufacturing*, 36:101433.

Karimi, J., C. Suryanarayana, I. Okulov and K. G. Prashanth. 2020. Selective laser melting of Ti6Al4V: Effect of laser re-melting. *Materials Science and Engineering: A*, 140558.

Lee, Y., A. K. Gurnon, D. Bodner and S. Simunovic. 2020. Effect of particle spreading dynamics on powder bed quality in metal additive manufacturing. *Integrating Materials and Manufacturing Innovation*, 9:410–422.

Liu, S. and Y. C. Shin. 2019. Additive manufacturing of Ti6Al4V alloy: A review. *Materials & Design*, 164:107552.

Louvrier, A., P. Marty, A. Barrabé, E. Euvrard, B. Chatelain, E. Weber and C. Meyer. 2017. How useful is 3D printing in maxillofacial surgery? *Journal of Stomatology, Oral and Maxillofacial Surgery*, 118(4):206–212.

Majumdar, T., T. Bazin, E. Massahud Carvalho Ribeiro, J. E. Frith and N. Birbilis. 2019. Understanding the effects of PBF process parameter interplay on Ti-6Al-4V surface properties. *PLoS One*, 14(8):e0221198.

Martelli, N., C. Serrano, H. van den Brink, J. Pineau, P. Prognon, I. Borget and S. El Batti. 2016. Advantages and disadvantages of 3-dimensional printing in surgery: A systematic review. *Surgery*, 159(6):1485–1500.

Masoomi, M., S. M. Thompson and N. Shamsaei. 2017. Quality part production via multi-laser additive manufacturing. *Manufacturing Letters*, 13:15–20.

Ni, J., H. Ling, S. Zhang, Z. Wang, Z. Peng, C. Benyshek, R. Zan, A. K. Miri, Z. Li and X. Zhang. 2019. Three-dimensional printing of metals for biomedical applications. *Materials Today Bio*, 3:100024.

Oliveira, J. P., A. LaLonde and J. Ma. 2020. Processing parameters in laser powder bed fusion metal additive manufacturing. *Materials & Design*, 193:108762.

Pal, S., G. Lojen, N. Gubeljak, V. Kokol and I. Drstvensek. 2020. Melting, fusion and solidification behaviors of Ti-6Al-4V alloy in selective laser melting at different scanning speeds. *Rapid Prototyping Journal*, 26(7):1209–1215.

Pal, S., G. Lojen, V. Kokol and I. Drstvenšek. 2019. Reducing porosity at the starting layers above supporting bars of the parts made by selective laser melting. *Powder Technology*, 355:268–277.

Pauzon, C., P. Forêt, E. Hryha, T. Arunprasad and L. Nyborg. 2019. Argon-helium mixtures as laser-powder bed fusion atmospheres: Towards increased build rate of Ti-6Al-4V. *Journal of Materials Processing Technology*, 116555.

Pedrazzini, S., M. E. Pek, A. K. Ackerman, Q. Cheng, H. Ali, H. Ghadbeigi, K. Mumtaz, T. Dessolier, T. B. Britton, P. Bajaj, E. Jägle, B. Gault, A. J. London and E. Galindo-Nava. 2020. Effect of bed temperature on solute segregation and mechanical properties in Ti-6Al-4V produced by selective laser melting arXiv.org > cond-mat > arXiv:2006.08288

Sun, S., M. Brandt and M. Easton. 2017. Powder bed fusion processes: An overview. In *Laser Additive Manufacturing Materials, Design, Technologies, and Applications*, ed. Milan Brandt, Chapter 2, pp. 55–77. Woodhead Publishing Series in Electronic and Optical Materials: Number 88, Elsevier.

Tang, J., Y. Nie, Q. Lei and Y. Li. 2019. Characteristics and atomization behavior of Ti-6Al-4V powder produced by plasma rotating electrode process. *Advanced Powder Technology*, 30(10) 2330–2337.

Vanmeensel, K., K. Lietaert, B. Vrancken, S. Dadbakhsh, X. Li, J. -P Kruth, P. Krakhmalev, I. Yadroitsev and J. Van Humbeeck. 2018. Additively manufactured metals for medical applications. In *Additive Manufacturing: Materials, Processes, Qualifications and Applications*, ed. J. Zhang and Y.-G. Jung, pp. 261–309. Butterworth-Heinemann.

Vilardell, A. M., G. Fredriksson, F. Cabanettes, A. Sova and P. Krakhmalev. 2020b. Surface integrity factors influencing fatigue crack nucleation of laser powder bed fusion Ti6Al4V alloy. *Procedia CIRP*, 94:222–226.

Vilardell, A.M., I. Yadroitsev, I. Yadroitsava, M. Albu, N. Takata, M. Kobashi, P. Krakhmalev, D. Kouprianoff, G. Kothleitner and A. du Plessis. 2020a. Manufacturing and characterization of in-situ alloyed Ti6Al4V (ELI)-3 at.% Cu by Laser Powder Bed Fusion. *Additive Manufacturing*, 36:101436.

Wang, L., E. L. Li, H. Shen, R. P. Zou, A. B. Yu and Z. Y. Zhou. 2020. Adhesion effects on spreading of metal powders in selective laser melting. *Powder Technology*, 363:602–610.

Weber, S., J. Montero, C. Petroll, T. Schäfer, M. Bleckmann and K. Paetzold. 2020. The fracture behavior and mechanical properties of a support structure for additive manufacturing of Ti-6Al-4V. *Crystals*, 10:343.

Yadroitsev, I., P. Krakhmalev and I. Yadroitsava. 2017. Titanium alloys manufactured by in situ alloying during laser powder bed fusion. *JOM*, 69:2725–2730.

Yao, D., X. An, H. Fu, H. Zhang, X. Yang, Q. Zou and K. Dong. 2021. Dynamic investigation on the powder spreading during selective laser melting additive manufacturing. *Additive Manufacturing*, 37:101707.

Zhang, L.-C. and H. Attar. 2015. Selective laser melting of titanium alloys and titanium matrix composites for biomedical applications: A review. *Advanced Engineering Materials*, 18(4):463–475.

10 Laser Powder Bed Fusion of Ti6Al4V Alloy for Biomedical Applications

Prospects and Challenges— Characteristic Properties of As-Built Parts and Post-Treatments

T.S.N. Sankara Narayanan and Hyung Wook Park

CONTENTS

10.1 CHARACTERISTIC PROPERTIES OF THE LPBF PROCESSED TI6AL4V ALLOY PARTS

10.1.1 DIMENSIONAL ERRORS

Dimensional errors are much higher on down-skin surfaces due to excessive melting and dross formation. Laser energy density exerts a strong influence in determining the dimensional deviations followed by powder bed thickness, scanning speed and interaction between laser radiation and powders while scanning pattern causes only a marginal effect (Charles et al., 2020).

10.1.2 SURFACE ROUGHNESS

Higher surface roughness is a serious limitation of as-built parts produced by the LPBF process and Ti6Al4V alloy parts are no exception to this. The staircase effect developed with increasing number of build layers, attachment of insufficiently melted particles and presence of pores is the major reason for the development of higher roughness of the as-built Ti6Al4V alloy parts (Chen et al., 2018a; Liu and Shin, 2019). The evolution of surface roughness has been correlated to the laser energy density (LED). At low LED with a low heat input, the time for solidification of the melt is short. This condition would lead to higher number of unmelted particles, higher porosity and balling, all of which is likely to increase the surface roughness. At high LED with a high heat input, melt pool instability, strong Marangoni convection and metal evaporation cause a large increase in surface roughness. It would be possible to restrain the development of a rough surface by increasing the lifetime of the molten metal so that it totally wets the surface and spreads uniformly, by reducing the extent of Marangoni convection and by reducing the extent of particle attachment. The surface roughness of as-built Ti6Al4V alloy parts can be reduced by decreasing the powder feed rate, increasing the scan speed and using a large overlap among the scans. Orientation of the part with reference to the build plate also influences the surface roughness. Inclined surfaces are more prone to melt pool extension and consequently become rougher. Although both the upward- and downward-facing surfaces exhibit a staircase effect, the later exhibits a higher surface roughness (Chen et al., 2018a).

Parts built on different locations under the same experimental conditions exhibit variation in surface roughness. This is primarily due to the variation in particle size distribution (PSD) caused by the re-coater arm and the effect of gas flow on the process by-products (Chen et al., 2018a). During spreading of the powders by the recoater, large particles are usually carried across the surface. Insufficient gas flow rate leads to re-deposition of by-products, leading to an increase in layer thickness, resulting in balling effect. In addition, scattering and attenuation of the laser rays either by the plasma or by the powder particles present in the molten metal mass and change in incidence angle of the laser rays with respect to surface orientation could cause considerable variation in the shape of the melt pool at various locations on the build plate. A laminar gas flow is recommended as an effective way to remove the process by-products. A higher surface roughness could drastically affect the mechanical properties of the as-built Ti6Al4V alloy parts manufactured by LPBF, particularly the fatigue behavior. Residual stress is also responsible for the poor fatigue behavior of Ti6Al4V alloy parts. Stress relief treatments could eliminate the residual stress and increase the fatigue life. However, surface roughness could promote cracking (Vilardell et al., 2019). Günther et al. (2018) have pointed out that parts with a smooth surface finish could only offer higher fatigue limits.

10.1.3 Microstructure

During LPBF of Ti6Al4V alloy parts, the quick rate of cooling has led to the development of an acicular α' martensitic structure with long columnar prior β grains in the direction of growth (Figure 10.1). The α' martensite is a metastable phase consisting of high densities of dislocations and twins. The columnar shape of the prior β grains points out the occurrence of an epitaxial growth of the successive layers and directional cooling in the direction of growth. The microstructure of as-built Ti6Al4V alloy parts is inhomogeneous with high residual stress, which is highly undesirable. The needle shaped α' martensite is highly anisotropic, brittle, difficult to deform and accommodate strains evenly (Wu et al., 2016; Vilardell et al., 2019). An increase in LED increases the width of the α' martensite needles and decreases the distance between them. The formation of an acicular α' martensitic structure would enable an increase in strength, which is beneficial for aerospace applications. However, for biomedical applications, fatigue strength and ductility are required.

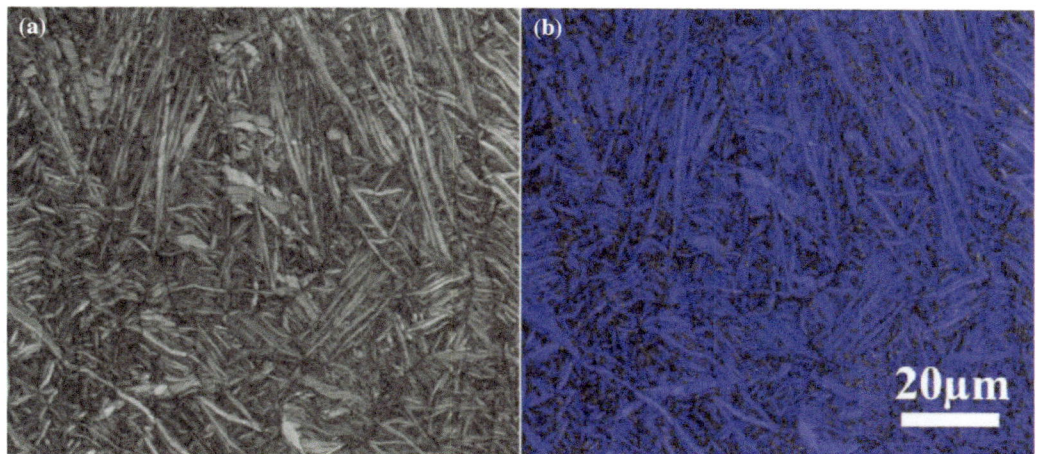

FIGURE 10.1 (a) EBSD band contrast and (b) phase contrast micrographs of as-built Ti-6Al-4V alloy sample fabricated by the LPBF process (α/α′-phase: blue color; β-phase: red color).

Source: Reprinted from Vilardell et al. (2019) with permission from Elsevier.

10.1.4 HARDNESS AND MECHANICAL PROPERTIES

LPBF processed Ti6Al4V alloy parts exhibit a hardness of ~409 Hv, which is higher than that of the parts manufactured by conventional routes. Their yield strength (YS) (~1110 MPa) and ultimate tensile strength (UTS) (~1267 MPa) exceeds that of the wrought alloy parts (Vanmeensel et al., 2018). The higher hardness, YS and UTS can be attributed to the development of the α′ martensite phase following rapid cooling. The formation of the α′ martensite phase is also accounted for the higher YS (~1440 MPa) and UTS (~1220 MPa) of LPBF processed Ti6Al4V alloy parts when compared to those produced by electron beam powder bed fusion (EBPBF) process (YS: ~870 MPa; UTS: ~928 MPa). However, the ductility of the LPBF processed Ti6Al4V alloy parts is ~5%, which is 50% lesser than those produced by the EBPBF process (Gokuldoss et al., 2017). The lower extent of work hardening and smaller elongation (< 3%) could lead to cracking of the Ti6Al4V alloy parts. ASTM F136–13 (for wrought Ti6Al4V (ELI)) and ASTM F1108–14 (for cast Ti6Al4V alloys) standards for surgical implants specify that the material must possess at least 8% of elongation. Hence, it is imperative to adopt suitable post-heat treatment (HT) and hot isostatic pressing (HIP) to increase the elongation in spite of a moderate decrease in strength. Due to layer-by-layer building of the parts, the mechanical properties show considerable variation along the direction of growth, questioning the repeatability of the LPBF process. Besides microstructural features, the presence of interstitial elements could also influence the mechanical properties of Ti6Al4V alloy parts. When the O_2 content is increased beyond 0.2 wt. %, the ductility is reduced in spite of an increase in strength. It is essential to decrease the O_2 content to < 0.1% so as to improve the ductility and to prevent crack growth (Vanmeensel et al., 2018).

10.1.5 RESIDUAL STRESS

Large thermal gradients, rapid cooling and non-uniform shrinkage promote residual stress in LPBF processed Ti6Al4V alloy parts (Levkulich et al., 2019). The choice of a constant laser scanning strategy throughout the build promotes anisotropy that could also lead to build up of residual stress.

Residual stress is a function of material properties, sample geometry, use of support structures, powder layer thickness, pre-heating of the powders, the LED, number of lasers and scanning strategy. Residual stress could cause many deleterious effects such as distortion of the part, crack growth, interlayer de-bonding, delamination, inferior mechanical properties and decrease in fatigue life (Zou et al., 2020). Hence, it is important to control the residual stress and distortion of the part produced by the LPBF process. Support structures could help reduce distortion of the fabricated parts. Heat treatment is generally recommended before the removal of the part from the build plate to reduce the deformation and cracking of the parts. Residual stress is found to be compressive in the bulk interior while it is tensile at the top, bottom and side surfaces of the part (Li et al., 2018). The peak value of residual stress is identified at the final deposited layer. The use of higher laser power, lower scan speed, higher layer thickness, shorter scan lengths, decreasing the laser striking width, pre-heating the powder bed, adopting a suitable scanning strategy, reducing substrate overhang, decreasing build plan area, rescanning each deposited layer and stress relief post-heat treatment are some of the recommended ways to reduce the residual stress (Li et al., 2018; Levkulich et al., 2019). Adopting a suitable scan strategy is imperative to reduce the residual stress (Zou et al., 2020). An island scanning strategy (chessboard pattern) is recommended to achieve a lower residual stress.

10.1.6 FATIGUE STRENGTH

Fatigue failure is a major concern in biomedical implants as it would lead to implant loosening, stress-shielding, and reduces the implant life. Surface roughness, tensile residual stress, pores and lack-of-fusion (LOF) voids generated during the LPBF process could cause a deleterious influence on fatigue strength, particularly under high cycle conditions (Yadollahi and Shamsaei, 2017; Pegues et al., 2018; Fatemi et al., 2019; Vilardell et al. 2020). As a result, the fatigue limit of LPBF processed Ti6Al4V alloy parts (~110 MPa) is much lesser than its wrought counterpart (550–750 MPa) (Pegues et al., 2018). According to Vilardell et al. (2018), the fatigue limit could be improved if the prior β grain boundaries could act as barriers and deflect the cracks. Crack initiation, rather than crack propagation plays a critical role in determining the fatigue life. The crack initiation is highly sensitive to surface roughness. For parts with a higher surface roughness, the valleys could act as notches with high stress levels promoting initiation of cracks. The LOF voids with irregular shape and sharp corners could cause a greater stress concentration than the spherical shaped gas pores. The fatigue behavior is largely influenced by the geometry of the part than surface area (Pegues et al., 2018). HIP treatment at 920 °C and 100 MPa for 2 h eliminates tensile residual stress and induces compressive residual stress, which would increase the fatigue life of LPBF processed Ti6Al4V alloy parts (Benedetti et al., 2017). HIP is also effective in reducing the pores and LOF voids. Although HIP treatment is capable of eliminating tensile residual stress and reducing the pores and LOF voids, the fatigue life could not be improved if the surface roughness is not reduced. Although shallow machining reduces the surface roughness, it could bring the near surface pores to the surface, which could act as crack initiator, resulting in decreased fatigue strength (Yadollahi and Shamsaei, 2017).

10.1.7 CORROSION BEHAVIOUR

Dai et al. (2016) have evaluated the corrosion resistance of LPBF processed Ti6Al4V alloy and identified the existence of a variation in corrosion rate among the XY-plane (build plane) and XZ-plane (build direction plane) both in 3.5 wt. % NaCl as well as in 1 M HCl. The amount of acicular α' martensite phase and β phase in XY-plane are 88.1% and 11.9%, respectively, whereas they are 95.0% and 5.0%, respectively in the XZ-plane. The "high energy state" and metastable nature of α' martensitic phase has been suggested as the main reason for the inferior corrosion resistance of the XZ-plane (Dai et al., 2016, 2017). The corrosion behavior of LPBF processed CP-Ti and Ti6Al4V alloy in Hank's solution is compared by Xiao et al. (2019). LPBF processed CP-Ti with a single α-Ti phase enables the development of a compact and stable passive film in Hank's solution, as evidenced by the wider passive potential range and a lower passive current density. On the contrary, the presence of a larger amount of acicular α′ martensite phase along with a smaller fraction of the β phase weakens the passive film stability, leading to the development of numerous galvanic cells and increase of the corrosion rate of Ti6Al4V alloy. Zhang et al. (2020) have compared the corrosion behavior of Ti6Al4Al alloy part fabricated by the LPBF process with wrought alloy in artificial saliva having varying concentrations of NaF (0.0020 M, 0.080 M and 0.10 M) and pH (pH: 2, 4 and 6). Irrespective of the concentration of NaF and pH in artificial saliva, the corrosion resistance of the LPBF processed Ti-6Al-4Al alloy part is found to be inferior when compared to the wrought alloy because of the existence of α′ martensite phase and defects in the passive film. The formation of acicular α′ martensitic phase is identified as the main reason for the poor corrosion resistance of the Ti6Al4V alloy produced by wire arc additive manufacturing, rather than those offered by the wrought alloy (Wu et al., 2018). Seo and Lee (2019) have also suggested that the martensitic α′ phase is accountable for the inferior uniform and pitting corrosion resistance of directed energy deposited Ti6Al4V alloy. Using a microdroplet cell, they have identified that the dark grains of the alloy, which is rich in α′ phase, exhibit poor resistance to pitting corrosion.

Dai et al. (2017) have studied how heat treatment could influence the corrosion resistance of LPBF processed Ti6Al4V alloy. During heat treatment, the acicular α′ martensite phase is transformed to a plate-shaped α phase and a lamellar α + β mixture. However, heat treatment also leads to an increase in grain size. After heat treatment at 850 °C, the width of the plate-shaped α phase is increased to 1.14 ± 0.28 μm while it is increased further to 5.71 ± 1.43 μm when heat treated beyond Tβ at 1000 °C. The increase in grain size following heat treatment has deleteriously influenced the corrosion resistance. In spite of an increase in amount of the β phase, the corrosion resistance is dominated by the increase in grain size. Hemmasian Ettefagh et al. (2019) have suggested that post-heat treatment of LPBF processed Ti6Al4V alloy at 800 °C for 2 h could increase the corrosion resistance comparable to the wrought alloy due to stress relief of the martensitic phase and formation of the β phase. Heat treatment at 850 °C and 1000 °C followed by furnace cooling, completely transform the α′ phase to the α + β phases that offers a better corrosion resistance for Ti6Al4V alloy than those of the as-built alloy (Seo and Lee, 2020).

Porosity is an important factor in determining the corrosion behavior of Ti6Al4V alloy parts produced by LPBF. In general, the higher the porosity, the lower is the degree of corrosion protection. Surface roughness is another important factor that determines the corrosion behavior of Ti6Al4V alloy parts. The higher surface roughness of the as-built alloy parts could deleteriously influence the corrosion resistance. An increase in surface roughness increases the surface area and decreases the electron work function. A higher surface roughness also affects the development of a good quality passive film. Chemical etching using a mixture of nitric acid, hydrofluoric acid and demineralized water is effective in removing the partially melted and unmelted particles present on the top layer of Ti6Al4V alloy scaffolds fabricated by the LPBF process. However, it decreases the corrosion resistance of the Ti6Al4V alloy scaffolds due to the formation of a highly adherent fluorine-containing corrosion product layer on the surface of the alloy (Fojt et al., 2020). This inference cautions the use of chemical etching treatment to remove the unmelted particles as well as to improve the surface finish. The higher surface roughness makes LPBF processed Ti6Al4V alloy more susceptible to stress corrosion cracking (SCC) than its wrought counterpart in both Ringer's solution as well as in 3.5% NaCl (Roach et al., 2018). The increase in O_2 content of the powders with re-use is also a concern as it would make the Ti6Al4V alloy part highly susceptible for SCC. Hence, the concentration of O_2 in the powder feed stock should be properly controlled and the surface finish should be improved to minimize the chance of occurrence of SCC in biomedical implants.

The texture of Ti6Al4V alloy manufactured by EBPBF is weak since the transformation of β → α phase occurs by a diffusion controlled process and there is no fixed orientation between the α and β phases (Formanoir et al., 2016). On the contrary, during LPBF of Ti6Al4V alloy, the diffusionless transformation of the β → α′ martensitic phase leads to the development of a stronger texture between the parent β phase and martensitic α′ phase. Chiu et al. (2018) have compared the corrosion and tribocorrosion behavior of LPBF processed Ti6Al4V alloy parts and the wrought alloy in Ringer's solution. The electrochemical corrosion behavior is similar for both the wrought Ti6Al4V alloy as well as those fabricated by LPBF process. However, at potentials above +2.0 V, a two-fold increase in current is observed for the LPBF processed Ti6Al4V alloy parts. The rapid cooling enables the formation of α′ martensitic phase with refined nanoscale grain sizes, which increase the hardness and offer a better wear resistance for the LPBF processed Ti6Al4V alloy parts, which is evidenced by a decrease in the width of the wear track. Under tribocorrosion conditions, the material loss for three different LPBF processed Ti6Al4V alloy samples is 6.90, 5.40 and 5.22 μg, which is relatively lesser than the material loss of 15.96 μg for the wrought alloy. The higher hardness of LPBF processed alloy is accountable for the lower material loss. However, the corrosion-to-wear ratio (C/W ratio) for the three LPBF processed Ti6Al4V alloy samples are 0.28, 0.29, 0.32, which is higher than a C/W ratio of 0.19 for the wrought alloy. This inference indicates that the passive film generated on Ti6Al4V alloy parts produced by the LPBF process is not highly protective under tribocorrosion conditions.

10.1.8 TRIBOLOGICAL BEHAVIOUR

The poor wear resistance is one of the major limitations of utilizing Ti and Ti alloys as orthopaedic implants. Hence, it would be of interest to know whether fabrication of Ti/Ti alloy parts by the LPBF process could improve the tribological properties. The rapid cooling during LPBF of Ti6Al4V alloy refines the prior β grains and enables the development of acicular α′ martensite phase, which enhances the hardness. The residual stress induced during the process also contributes to a higher hardness and elastic modulus. Many researchers have compared the tribological behavior of Ti6Al4V alloys fabricated by the LPBF process with those prepared by the conventional route, but their findings were contradictory. Bartolomeu et al. (2017) have observed an enhancement in wear resistance of LPBF processed Ti6Al4V alloy parts compared to those prepared by the conventional route under dry sliding conditions. This is due to the higher hardness of the former and the wear volume is inversely proportional to the hardness in accordance with Archard's law. In contrast, Palanisamy et al. (2018) have observed no dependence of wear volume on hardness under dry sliding in Hank's solution. The lack of dependence of wear volume on hardness can be explained in terms of an inverse relationship between strength and ductility of the alloy. Lower strength and lower hardness leads to a larger extent of deformation. Higher ductility allows a higher extent of deformation without more amount of fracture and wear loss. Achieving ultrafine microstructure with superior hardness is likely to increase the wear resistance of LPBF processed Ti6Al4V alloy parts (Li et al., 2019).

10.2 POST-TREATMENTS

10.2.1 POST-HEAT TREATMENT

As already mentioned in Section 10.1.3, the fast rate of cooling has promoted the formation of an acicular α′ martensitic structure along with columnar prior β grains. In spite of the higher hardness, UTS and YS, the inhomogeneity in the microstructure, high residual stress and low ductility limit the utility of as-built Ti6Al4V alloy parts. Hence, post HT is essential to achieve better homogeneity in microstructure, to decrease the residual stress and to improve ductility in spite of a moderate loss in strength (Wu et al., 2016; Zhang et al., 2018a, 2019). Residual stress is a serious concern in parts fabricated by LPBF. The built-up of residual stress in Ti6Al4V alloy parts is evidenced by a larger full width at half maximum (FWHM) of the XRD peak at 40.2° 2θ (Tsai et al., 2019). The residual stress should be relieved to achieve improved mechanical properties through appropriate stress-relief treatments. Annealing at 600 °C eliminates the residual stress, as evidenced by a

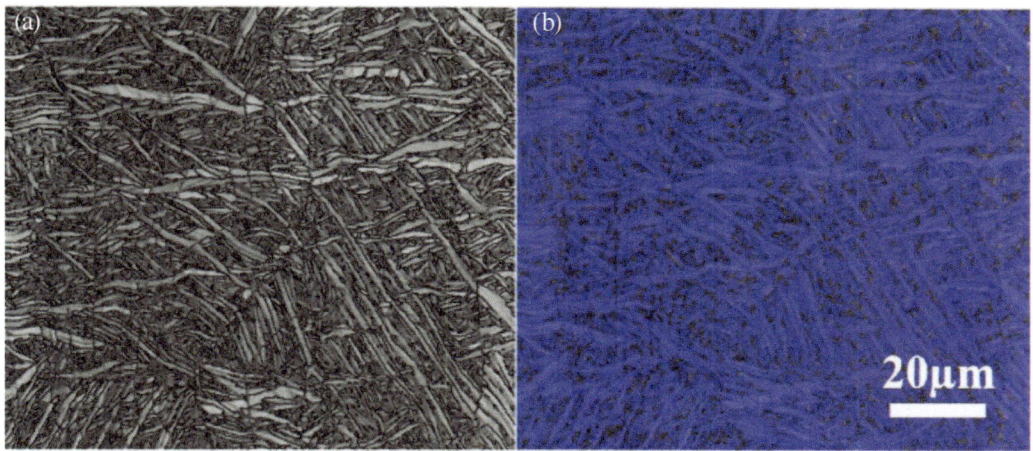

FIGURE 10.2   (a) EBSD band contrast and (b) phase contrast micrographs of stress relieved Ti6Al4V alloy sample fabricated by LPBF process (α/α′-phase: blue color; β-phase: red color).

Source: Reprinted from Vilardell et al. (2019) with permission from Elsevier.

decrease in the FWHM of 40.2° 2θ peak. No significant change in microstructure, YS and UTS of the LPBF processed Ti6Al4V alloy parts could be observed before and after stress-relief treatment. However, the elongation is increased. Stress-relief treatment at 650 °C for 3 h leads to coarsening of acicular α′ needles and promotes precipitation between needles (Figure 10.2) (Vilardell et al., 2019). It is believed that a good mix of strength and ductility could not be achieved only through stress-relieving treatments (Li et al., 2021). The microstructure of LPBF processed Ti6Al4V alloy parts after HT is largely a function of temperature, time and rate of cooling (Zhang et al., 2018a). HT enables a gradual transformation of the metastable acicular α′ martensite phase to a mixture of α and β phases and the extent of phase transformation is a function of temperature and time. The α′ martensite phase slows down/inhibits the growth of the α and β phases (Zhang et al., 2018a).

At temperatures ≤600 °C, only a partial decomposition of the martensite phase is observed. With a further increase in temperature, the acicular α′ martensite starts to decompose leading to the formation of α platelets and β phase. With an increase in temperature from 600 °C to 750 °C, the volume fraction of the α platelets is decreased with a concomitant increase in the volume fraction of the β phase (Figure 10.3(a)) (Tsai et al., 2019). An increase in HT temperature promotes diffusion of vanadium into the β phase, leading to an increase in its lattice parameter. For a given temperature, the fraction of the β phase is increased with treatment time (Figure 10.3(b)). The α′ martensite phase is completely decomposed at temperatures >800 °C (Wu et al., 2016). According to Khorasani et al. (2019), HT at 800 °C for 6 h eliminates high density dislocations and twins. Supertransus HT (>995 °C) leads to a complete breakdown of the original long columnar β grains, promotes extensive grain growth, resulting in the generation of equiaxed β grains along with newly formed weave-type acicular α′ martensite phase with higher hardness (Vrancken et al., 2012; Wu et al., 2016). However, such a condition would promote coarsening of the microstructure and decrease the YS and UTS along with a small improvement in ductility.

In contrast, subtransus HT (< 995 °C) clearly reveals the β grain boundaries and provides a moderate improvement in ductility while maintaining a higher strength of Ti6Al4V alloy (Zhang et al., 2018a). Li et al. (2021) have recommended a multi-stage heat-treatment (MSHT) process, which facilitates decomposition of the α′ martensite phase to α + β phases, globularization of lamellar α and formation of nearly equiaxed α grain structure. The MSHT route offered an excellent improvement in ductility (elongation: 21.8%) with moderate UTS. After HT, no significant change in the density

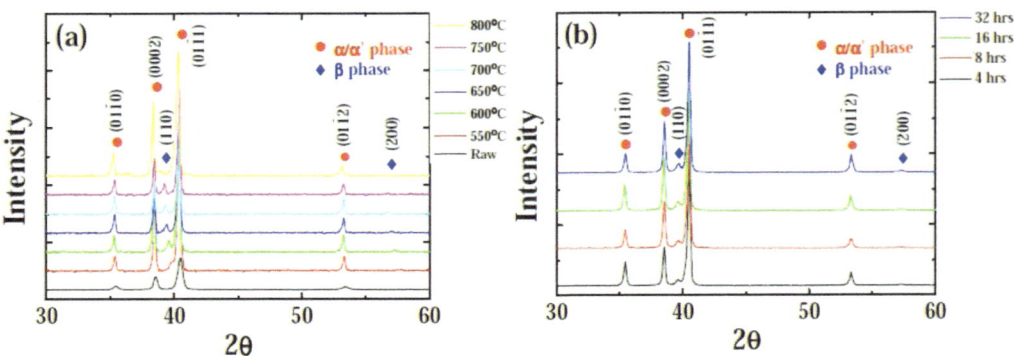

FIGURE 10.3 XRD patterns of (a) as-printed Ti6Al4V alloy part and after heat treated at 550 to 800 °C for 24 h and (b) after heat treated at 600 °C for 4, 8, 16 and 32 h.

Source: Reprinted from Tsai et al. (2020), with permission from Elsevier.

and porosity of the Ti6Al4V alloy parts could be detected (Tsai et al., 2019). However, HT could cause a considerable change in the mechanical properties. Annealing at 600 °C decreased the UTS (from 1362 to 1180 MPa) and YS (from 1311 to 1115 MPa) but increased the elongation from 4.1% to 6.1%. The transformation of α' to α + β phase, large decrease in dislocations and disappearance of twin structures improves elongation (Tsai et al., 2019). Coarsening of needles and relief of residual stresses during HT improves the fatigue life and crack growth resistance (Vilardell et al., 2019).

HIP has been suggested as an effective post-treatment option. HIP at β-transus temperature enables conversion of the acicular α' martensite to a bimodal α/β microstructure. By a careful choice of higher temperature and longer duration of time, it would be possible to modify the prior-β grain structure and remove the anisotropy. HIP is effective in eliminating the pores (Gangireddy et al., 2018). However, the existence of internal pores after HIP has also been identified (Zhang et al., 2019). LOF pores have an irregular shape with sharp edges. HT alone could provide a conditioning effect on pores facilitating a "rounding-off" effect on the LOF pores. However, HIP enables the formation of a highly uniform Widmanstätten structure (Zhang et al., 2019). When compared to conventional HT, HIP at 920 °C for 2 h at 100 MPa has markedly improved the fatigue life of Ti6Al4V ally parts fabricated by the LPBF process.

10.2.2 POST SURFACE FINISHING TREATMENTS

10.2.2.1 Milling, Grinding, Blasting and Electrode Discharge Machining

The primary purpose of surface finishing post-treatments is to decrease the surface roughness and to remove the unmelted particles of as-built Ti6Al4V alloy parts. Milling and grinding are efficient in removing the furrows and surface irregularities, thus decreasing the surface roughness with good dimensional accuracy. Both of them induce compressive residual stress, which is beneficial in terms of improving the fatigue strength. However, their inability to remove internal cavities makes them unsuitable for parts with complex geometries. Blasting is suggested as a suitable surface finishing post-treatment. Blasting assumes significance for industrial applications due to its shorter treatment time and better accessibility of complex shaped parts. The particle bombardment enables plastic deformation on the surface and near vicinity, which often creates upper-surface pitting. The extent of decrease in surface roughness is only marginal due to the sharp-edged blasting media. Blasting is not capable of completely removing the partially molten powder particles. A combination of post-heat treatment followed by sand blasting is found to be effective in removing nearly all lightly bonded metal particles from the as-built parts. Incorporation of the blasting media in the bulk material is also a problem during blasting. Wire electrode discharge machining (EDM) is effective in removing partially melted powder particles. However, EDM induces tensile residual stress and microcracks, which would affect the fatigue life. Surface contamination from the electrode and generation of heat affected zone (HAZ) are other major issues in using EDM.

10.2.2.2 Chemical Polishing

Chemical polishing (CP) using acid mixtures is effective in removing the partially melted and unmelted particles and improving the surface finish. The effectiveness of CP, however, depends on the type of acids and their concentration as well as the interaction between the acidic solutions and the surface of the part (Wysocki et al., 2019). Mixtures of HCl and HNO_3 are found to be ineffective. HF based mixtures are effective. However, the difficulty in handling HF, hydrogen embrittlement and health and safety issues are the major problems in using HF based mixtures for CP. Higher mass loss, maintaining dimensional tolerance, preventing excessive chemical attack and pitting still remains unsolved in CP using acid mixtures. A combination of HIP followed by chemical polishing significantly improved the fatigue life of LPBF processed Ti6Al4V alloy parts.

10.2.2.3 Electropolishing

Electropolishing (EP) is effective in removing partially melted powders, laser scanning lines and offers a smooth surface finish (Wu et al., 2019). The polished surface exhibits no deformation and maintains its original lattice structure. The internal surfaces of parts with complex geometries can be effectively polished using EP (Urlea and Brailovski, 2017). EP reduces the depth of the notch and increases the radius of the notch and eliminates the stress concentration points. An increase in current density and treatment time offers a significant reduction in surface roughness. However, the choice of very high current densities could cause over-etching, resulting in the formation of tiny black dots on the surface of the part. EP has little influence on the average Young's modulus and average YS, a slight improvement in average UTS but offers a good improvement in tensile elongation of Ti6Al4V alloy parts. EP enables the formation of a more dense and homogenous passive TiO_2 layer with fewer defects on the surface of the Ti6Al4V alloy, which increases the corrosion resistance of the Ti6Al4V alloy part in SBF. Urlea and Brailovski (2017) have evaluated the effectiveness of EP using a 1:9 volume ratio of perchloric acid (60%) and glacial acetic acid to improve the surface finish of Ti6Al4V alloy parts fabricated at different build angles from 0 to 135°. In spite of a large variation in the average surface roughness (R_a) ranging from 4 μm (0°) to 23 μm (135°), about 92% improvement in surface finish could be achieved using EP. The use of perchloric acid and glacial acetic acid based electrolyte is highly reactive and requires special cooling arrangements to control the temperature. The acid vapors are toxic and warrants fume hood arrangements. Cathodes need to be specially designed to realize uniform current distribution while polishing complex shaped parts. The pores and cracks, which are inherent to the parts fabricated by LPBF could not be completely eliminated by EP. In addition, the extent of compressive residual stress induced during EP is much lower than those induced during milling and grinding.

10.2.2.4 Laser Polishing

Laser polishing (LP) involves melting of a thin layer and subsequently allowing the molten material to flow from peaks to valleys. LP enables a 70 to 80% reduction in surface roughness of the as-built Ti6Al4V alloy parts (Lee et al., 2021). Unlike mechanical, chemical and EP, during LP material is not removed; rather it is relocated between the peaks and valleys (Marimuthu et al., 2015). When performed under optimized conditions, no significant distortion and no noticeable changes in the mechanical properties are observed after LP. However, precise control of the LED is warranted during LP since excess thermal energy could promote surface oxidation and carbonization. In addition, the high heat input could increase the melt pool velocity, causing striations in the polished surface, which is highly undesirable. The choice of low LED and higher laser scanning speed would help to increase the processing rate.

10.2.2.5 Large Pulsed Electron-Beam Irradiation

Large pulsed electron-beam (LPEB) irradiation can also be used to remove partially melted particles, to improve the surface finish, for selective ablation of impurities, for removal of inclusions and for surface purification. LPEB irradiation is effective in reducing the R_a of maraging steel samples fabricated by LPBF process by about 75% (Sankara Narayanan et al., 2020). LPEB irradiation induces tensile residual stress and involves the formation of a HAZ. LPEB irradiation exhibits similarity with LP in terms of melting of the surface layer, generation of a molten pool, flow of the molten metal pool between the peaks and valleys and smoothening of the surface after re-solidification. In terms of process flexibility, achieving good surface finish and dimensional accuracy, both LP and LPEB irradiation are highly comparable. Wire EDM and LPEB irradiation are comparable in terms of their ability to remove partially melted powder particles, generation of HAZ and inducing tensile residual stresses. However, unlike wire EDM, surfaces treated using LPEB irradiation is free of surface contamination and microcracks.

10.2.3 Other Types of Post-Treatments

10.2.3.1 Laser Shock Peening

Laser shock peening (LSP) is utilized for post-treatment of LPBF processed Ti6Al4V alloy parts (Yeo et al., 2020). In spite of its ability to increase the impact toughness, there is considerable decrease in hardness, wear resistance and corrosion resistance. Hence, the use of LSP alone as a post-treatment option is not helpful. However a combination of HT and LSP has been shown to recover the hardness, wear resistance and corrosion resistance. The higher surface roughness and accumulation of tensile residual stress at the surface decreases the fatigue life of Ti6Al4V alloy parts fabricated by LPBF process. LSP is capable of inducing compressive residual stress, which would improve the fatigue life (Yeo et al., 2020; Hackel et al., 2018). Post-heat treatment alone decreases the wear resistance and corrosion resistance. The high compressive residual stress induced by LSP on the HT sample leads to a considerable decrease in wear loss. HT increased the grain size and corrosion rate. LSP of the HT sample provides an improvement in mechanical properties and corrosion resistance following a decrease in grain size and the compressive residual stress induced in the sample (Lan et al., 2020). Post-heat treatment is employed to increase the ductility at the expense of a slight decrease in strength of the Ti6Al4V alloy parts. LSP on HT sample could offer an improvement in both ductility and strength.

10.2.3.2 Surface Mechanical Attrition Treatment

Surface mechanical attrition treatment (SMAT) is used to post-treat Ti6Al4V alloy parts fabricated by the LPBF process (Eyzat et al., 2019). SMAT reduces the R_a of the as-built Ti6Al4V alloy part by about 80%. SMAT enables an increase in hardness, YS and UTS. Its ability to induce compressive residual stress enables a complete reversal of the residual stress at the surface from +1000 MPa (tensile) to −300 MPa (compressive). Zhang et al. (2018b) have explored the use of electrically assisted ultrasonic nanocrystal surface modification (EA-UNSM) to post-treat Ti6Al4V alloy parts fabricated by LPBF process and compared it with the conventional nanocrystal surface modification (UNSM). Ti6Al4V alloy is difficult to deform. The high defect density induced during the fabrication of Ti6Al4V alloy parts by LPBF process and the presence of porosity make the alloy much harder to deform. The extent of deformation caused by UNSM is low. In contrast, during EA-UNSM, resistive heating from the electric current increases the temperature and plasticity of the sample, thus making it easy to deform, besides closing the pores. As the rough peaks flow into the valleys, the surface finish is improved. EA-UNSM also increases the surface hardness. The combination of decrease in number of pores, improvement in surface finish and hardness could enhance the fatigue properties.

10.3 SUMMARY

- Higher surface roughness of the as-fabricated Ti6Al4V alloy parts is a major concern. The roughness of downward-facing surface is usually higher than the upward-facing surface. The higher surface roughness could cause a deleterious influence on the mechanical properties of the LPBF processed Ti6Al4V alloy parts, particularly the fatigue behavior. Adopting suitable post surface treatments is highly warranted to reduce the surface roughness.

- The rapid cooling facilitates the development of an acicular α' martensitic structure with long columnar prior β grains in the build direction of the LPBF processed Ti6Al4V alloy parts. The needle shaped α' martensite is highly anisotropic, brittle, difficult to deform and accommodate strains evenly. The inhomogeneous microstructure of as-built Ti6Al4V alloy parts is highly undesirable. The formation of an acicular α' martensitic structure would enable an increase in strength. Nevertheless, it decreases the ductility of the alloy, which is critical for implants.

- Adopting suitable post-heat treatment is essential to transform the deleterious α' martensite phase to bimodal α + β phase. The α' martensite phase, generated due to rapid cooling, increased the hardness, YS and UTS. However, the ductility of Ti6Al4V alloy parts fabricated by LPBF process is ~5%, which is lesser than 8% as specified in ASTM F136–13. The ductility of as-built Ti6Al4V alloy parts should be increased by suitable post-heat treatment and HIP in spite of encountering a moderate loss in strength. The development of large thermal gradients, rapid cooling and non-uniform shrinkage leads to build-up of residual stress in LPBF processed Ti6Al4V alloy parts.

- An island scanning strategy (chessboard pattern) is recommended to achieve a lower residual stress. Treatments such as LSP, SMAT are capable of decreasing the tensile residual stress and induce compressive residual stress, which is likely to increase the fatigue life of implants.

- Surface roughness, tensile residual stress, pores and LOF voids generated during the LPBF process cause a deleterious influence on fatigue strength, particularly under high cycle conditions. HIP at 920 °C and 100 MPa for 2 h helps to remove the tensile residual stress, induce compressive residual stress, reducing the gas pores and LOF defects, which would increase the fatigue life of LPBF processed Ti6Al4V alloy parts. However, the fatigue life could be improved only if the surface roughness is also reduced.

- The formation of an acicular martensitic α' phase is the prime reason for the inferior corrosion resistance of the LPBF processed Ti6Al4V alloy parts. Heat treatment at 850 °C and 1000 °C followed by furnace cooling, enables a complete transformation of the α' phase to the α phase and β phase and offers a better corrosion resistance for Ti6Al4V alloy than the as-received alloy.

- The higher surface roughness of the as-built alloy parts could deleteriously influence the corrosion resistance and make it more susceptible to SCC. The increase in O_2 content of the powders with re-use could increase the susceptibility of the alloy part for SCC.

- The formation of acicular α' martensite phase, grain refinement and residual stress induced during the process increase the hardness. Achieving ultrafine microstructure with higher hardness is likely to increase the wear resistance of LPBF processed Ti6Al4V alloy parts.

- Higher processing time and difficulty in scaling-up are the major limitations of the LPBF process. With the use of multiple lasers, a considerable savings in processing time can be achieved. However, a large amount of residual stress will be induced in the as-built part, which warrants suitable scanning strategy to reduce the residual stress. Decrease in processing time, possibility of scaling up and decrease in cost will open the gates for the utility of LPBF processed parts in many areas including automobile, aerospace and biomedical.

REFERENCES

Bartolomeu, F., M. Buciumeanu, E. Pinto, N. Alves, F.S. Silva, O. Carvalho and G. Miranda. 2017. Wear behaviour of Ti6Al4V biomedical alloys processed by selective laser melting, hot pressing and conventional casting. *Transactions of Nonferrous Metals Society of China*, 27:829–838.

Benedetti, M., E. Torresani, M. Leoni, V. Fontanari, M. Bandini, C. Pederzolli and C. Potrich. 2017. The effect of post-sintering treatments on the fatigue and biological behavior of Ti6Al4V ELI parts made by selective laser melting. *Journal of the Mechanical Behavior of Biomedical Materials*, 71: 295–306.

Charles, A., A. Elkaseer, L. Thijs and S. G. Scholz. 2020. Dimensional errors due to overhanging features in laser powder bed fusion parts made of Ti6Al4V. *Applied Sciences*, 10(7):2416.

Chen, Z., X. Wu, D. Tomus and C. H. J. Davies. 2018a. Surface roughness of selective laser melted Ti6Al4V alloy parts. *Additive Manufacturing*, 21:91–103.

Chiu, T.-M., M. Mahmoudi, W. Dai, A. Elwany, H. Liang and H. Castaneda. 2018. Corrosion assessment of Ti6Al4V fabricated using laser powder-bed fusion additive manufacturing. *Electrochimica Acta*, 279:143–151.

Dai, N., J. Zhang, Y. Chen and L.-C. Zhang. 2017. Heat treatment degrading the corrosion resistance of selective laser melted Ti6Al4V alloy. *Journal of the Electrochemical Society*, 164(7):C428–C434.

Dai, N., L.-C. Zhang, J. Zhang, Q. Chen, Q and M. Wu. 2016. Corrosion behavior of selective laser melted Ti-6Al-4 V alloy in NaCl solution. *Corrosion Science*, 102:484–489.

Eyzat, Y., M. Chemkhi, Q. Portella, J. Gardan, J. Remond and D. Retraint. 2019. Characterization and mechanical properties of as-built SLM Ti6Al4V subjected to surface mechanical post-treatment. *Procedia CIRP*, 81:1225–1229.

Fatemi, A., R. Molaei, J. Simsiriwong, N. Sanaei, J. Pegues, B. Torries, N. Phan and N. Shamsaei. 2019. Fatigue behaviour of additive manufactured materials: An overview of some recent experimental studies on Ti6Al4V considering various processing and loading direction effects. *Fatigue & Fracture of Engineering Materials & Structures*, 42(5):991–1009.

Fojt, J., Z. Kacenka, E. Jablonska, V. Hybasek and E. Pruchova. 2020. Influence of the surface etching on the corrosion behaviour of a three-dimensional printed Ti-6Al-4V alloy. *Materials and Corrosion*, 71(10):1691–1696.

Formanoir, C.D., S. Michotte, O. Rigo, L. Germain and S. Godet. 2016. Electron beam melted Ti6Al4V: Microstructure, texture and mechanical behavior of the as-built and heat-treated material. *Materials Science and Engineering: A*, 652:105–119.

Gangireddy, S., E. J. Faierson and R. S. Mishra. 2018. Influences of post-processing, location, orientation, and induced porosity on the dynamic compression behavior of Ti-6Al-4V alloy built through additive manufacturing. *Journal of Dynamic Behavior of Materials*, 4:441–451.

Gokuldoss, P. K., S. Kolla and J. Eckert. 2017. Additive manufacturing processes: Selective laser melting, electron beam melting and binder jetting: Selection guidelines. *Materials*, 10(6):672.

Günther, J., S. Leuders, P. Koppa, T. Tröster, S. Henkel, H. Biermann and T. Niendorf. 2018. On the effect of internal channels and surface roughness on the high-cycle fatigue performance of Ti6Al4V processed by SLM. *Materials & Design*, 143:1–11.

Hackel, L., J. R. Rankin, A. Rubenchik, W. E. King and M. Matthews. 2018. Laser peening: A tool for additive manufacturing post-processing. *Additive Manufacturing*, 24:67–75.

Hemmasian Ettefagh, A., C. Zeng, S. Guo and J. Raush. 2019. Corrosion behavior of additively manufactured Ti6Al4V parts and the effect of post annealing. *Additive Manufacturing*, 28:252–258.

Khorasani, A. M., I. Gibson, A. Ghaderi and M. I. Mohammed. 2019. Investigation on the effect of heat treatment and process parameters on the tensile behaviour of SLM Ti6Al4V parts. *International Journal of Advanced Manufacturing Technology*, 101:3183–3197.

Lan, L., R. Xin, X. Jin, S. Gao, B. He, Y. Rong and N. Min. 2020. Effects of laser shock peening on microstructure and properties of Ti-6Al-4V titanium alloy fabricated via selective laser melting. *Materials*, 13(15):3261.

Lee, S., Ahmadi, Z., Pegues, J. W., Mahjouri-Samani, M. and N, Shamsaei. 2021. Laser polishing for improving fatigue performance of additive manufactured Ti-6Al-4V parts. *Optics & Laser Technology*, 134:106639.

Levkulich, N. C., S. L. Semiatin, J. E. Gockel, J. R. Middendorf, A. T. DeWald and N. W. Klingbeil. 2019. The effect of process parameters on residual stress evolution and distortion in the laser powder bed fusion of Ti6Al4V. *Additive Manufacturing*, 28:475–484.

Li, C., Z. Y. Liu, X. Y. Fang and Y. B. Guo. 2018. Residual stress in metal additive manufacturing. *Procedia CIRP*, 71:348–353.

Li, C.-L., J. -K. Hong, P. L. Narayana, S.-W. Choi, S. W. Lee, C. H. Park, J. T. Yeom and Q. Mei. 2021. Realizing superior ductility of selective laser melted Ti6Al4V through a multi-step heat treatment. *Materials Science and Engineering: A*, 799:140367.

Li, H., M. Ramezani and Z. W. Chen. 2019. Dry sliding wear performance and behaviour of powder bed fusion processed Ti-6Al-4V alloy. *Wear*, 440–441:203103.

Liu, S. and Y. C. Shin. 2019. Additive manufacturing of Ti6Al4V alloy: A review. *Materials & Design*, 164:107552.

Marimuthu, S., A. Triantaphyllou, M. Antar, D. Wimpenny, H. Morton and M. Beard. 2015. Laser polishing of selective laser melted parts. *International Journal of Machine Tools and Manufacture*, 95:97–104.

Palanisamy, C., S. Bhero, B. A. Obadele and P. A. Olubambi. 2018. Effect of build direction on the micro-hardness and dry sliding wear behaviour of laser additive manufactured Ti6Al4V, *Materials Today: Proceedings*, 5:397–402.

Pegues, J., M. Roach, R. S. Williamson and N. Shamsaei. 2018. Surface roughness effects on the fatigue strength of additively manufactured Ti-6Al-4V. *International Journal of Fatigue*, 116:543–552.

Roach, M., R. S. Williamson, J. W. Pegues and N. Shamsaei. 2018. A comparison of stress corrosion cracking susceptibility in additively-manufactured and wrought materials for aerospace and biomedical applications. Solid Freeform Fabrication 2018: Proceedings of the 29th Annual International Solid Freeform Fabrication Symposium: An Additive Manufacturing Conference, The Minerals, Metals & Materials Society, Pittsburgh, Pennsylvania, pp. 1410–1419.

Sankara Narayanan, T. S. N., J. Kim, H. E. Jeong and H. W. Park. 2020. Enhancement of the surface properties of selective laser melted maraging steel by large pulsed electron-beam irradiation. *Additive Manufacturing*, 101125.

Seo, D.-I. and J.-B. Lee. 2019. Corrosion characteristics of additive-manufactured Ti6Al4V using micro-droplet cell and critical pitting temperature techniques. *Journal of The Electrochemical Society*, 166(13):C428–C433.

Seo, D.-I. and J.-B. Lee. 2020. Influence of heat treatment parameters on the corrosion resistance of additively manufactured Ti-6Al-4V alloy. *Journal of The Electrochemical Society*, 167:101509.

Tsai, M.-T., Y.-W. Chen, C.-Y. Chao, J. S. C. Jang, C.-C. Tsai, Y.-L. Su and C.-N. Kuo. 2020. Heat-treatment effects on mechanical properties and microstructure evolution of Ti6Al4V alloy fabricated by laser powder bed fusion. *Journal of Alloys and Compounds*, 816:152615.

Urlea, V. and V. Brailovski. 2017. Electropolishing and electropolishing-related allowances for powder bed selectively laser-melted Ti6Al4V alloy parts. *Journal of Materials Processing Technology*, 242:1–11.

Vanmeensel, K., K. Lietaert, B. Vrancken, S. Dadbakhsh, X. Li, J.-P Kruth, P. Krakhmalev, I. Yadroitsev and J. Van Humbeeck. 2018. Additively manufactured metals for medical applications, In *Additive Manufacturing: Materials, Processes, Qualifications and Applications*, ed. J. Zhang and Y.-G. Jung. Butterworth-Heinemann, pp. 261–309.

Vilardell, A. M., G. Fredriksson, F. Cabanettes, A. Sova and P. Krakhmalev. 2020. Surface integrity factors influencing fatigue crack nucleation of laser powder bed fusion Ti6Al4V alloy. *Procedia CIRP*, 94:222–226.

Vilardell, A. M., G. Fredriksson, I. Yadroitsev and P. Krakhmalev. 2019. Fracture mechanisms in the as-built and stress-relieved laser powder bed fusion Ti6Al4V ELI alloy. *Optics & Laser Technology*, 109:608–615.

Vilardell, A. M., P. Krakhmalev, G. Fredriksson, F. Cabanettes, A. Sova, D. Valentin and P. Bertrand. 2018. Influence of surface topography on fatigue behavior of Ti6Al4V alloy by laser powder bed fusion. *Procedia CIRP*, 74:49–52.

Vrancken, B., L. Thijs, J.-P. Kruth and J. Van Humbeeck. 2012. Heat treatment of Ti6Al4V produced by selective laser melting: Microstructure and mechanical properties. *Journal of Alloys and Compounds*, 541:177–185.

Wu, B., Z. Pan, S. Li, D. Cuiuri, D. Ding and H. Li. 2018. The anisotropic corrosion behaviour of wire arc additive manufactured Ti6Al4V alloy in 3.5% NaCl solution. *Corrosion Science*, 137:176–183.

Wu, S. Q., Y. J. Lu, Y. L. Gan, T. T. Huang, C. Q. Zhao, J. J. Lin, S. Guo and J. X. Lin. 2016. Microstructural evolution and microhardness of a selective-laser-melted Ti-6Al-4V alloy after post heat treatments. *Journal of Alloys and Compounds*, 672:643–652.

Wu, Y.-C., C.-N. Kuo, Y.-C. Chung, C.-H. Ng and J. C. Huang. 2019. Effects of electropolishing on mechanical properties and bio-corrosion of Ti6Al4V fabricated by electron beam melting additive manufacturing. *Materials*, 12:1466.

Wysocki, B., J. Idaszek, J. Buhagiar, K. Szlązak, T. Brynk, K. J. Kurzydłowski and W. Święszkowski. 2019. The influence of chemical polishing of titanium scaffolds on their mechanical strength and in-vitro cell response. *Materials Science and Engineering: C*, 95:428–439.

Xiao, Y., N. Dai, Y. Chen, J. Zhang and S.-W. Choi. 2019. On the microstructure and corrosion behaviors of selective laser melted CP-Ti and Ti6Al4V alloy in Hank's artificial body fluid. *Materials Research Express*, 6:126521.

Yadollahi, A. and N. Shamsaei. 2017. Additive manufacturing of fatigue resistant materials: Challenges and opportunities. *International Journal of Fatigue*, 98:14–31.

Yeo, I., S. Bae and A. Amanov. 2020. Effect of laser shock peening on properties of heat-treated Ti-6Al-4V manufactured by laser powder bed fusion. *International Journal of Precision Engineering and Manufacturing-Green Technology*, (article in press).

Zhang, B., W. J. Meng, S. Shao, N. Phan and N. Shamsaei. 2019. Effect of heat treatments on pore morphology and microstructure of laser additive manufactured parts. *Material Design & Processing Communications*, 1(1):e29.

Zhang, H., C. Man, C. Dong, L. Wang, W. Li, D. Kong, L. Wang and X. Wang. 2020. The corrosion behavior of Ti6Al4V fabricated by selective laser melting in the artificial saliva with different fluoride concentrations and pH values. *Corrosion Science*, 109097.

Zhang, H., J. Zhao, J. Liu, H. Qin, Z. Ren, G. L. Doll, Y. Dong and C. Ye. 2018b. The effects of electrically-assisted ultrasonic nanocrystal surface modification on 3D-printed Ti6Al4V alloy. *Additive Manufacturing*, 22:60–68.

Zhang, X.-Y., G. Fang, S. Leeflang, A. J. A. Böttger, A. Zadpoor and J. Zhou. 2018a. Effect of subtransus heat treatment on the microstructure and mechanical properties of additively manufactured Ti6Al4V alloy. *Journal of Alloys and Compounds*, 735:1562–1575.

Zou, S., H. Xiao, F. Ye, Z. Li, W. Tang, F. Zhu, C. Chen and C. Zhu. 2020. Numerical analysis of the effect of the scan strategy on the residual stress in the multi-laser selective laser melting. *Results in Physics*, 16:103005

11 Biomedical Applications of Laser Texturing

Bruno Gago, Antonio Riveiro Rodríguez, Pablo Pou, Mónica Fernández-Arias, Jesús del Val García, Rafael Comesaña Piñeiro, Aida Badaoui, Mohamed Boutinguiza Larosi and Juan Pou Saracho

CONTENTS

11.1 INTRODUCTION

11.1.1 INFLUENCE OF ROUGHNESS AND WETTABILITY ON IMPLANT SUCCESS

Implants are clinical devices made of one or several materials, which are introduced into the body to replace, repair, or augment a damaged tissue or organ. Materials used for these applications are called biomaterials. Not all the biomaterials are integrated in the same way by the biological tissues. Their degree of the bioactivity depends on the type of biomaterial (namely, polymeric, metallic, or ceramic), but for a large extent depends on the surface characteristics of the implant. Surface topography (consisting of form, waviness, and roughness), wettability, chemical composition (including presence of functional groups), surface charge, surface stiffness, or surface energy have been reported to be relevant surface characteristics of implants as they influence the interfacial reactions of biological tissues with the biomaterial. In this chapter, we will focus our attention to only two of these characteristics: surface topography and wettability. As we will see, these are not independent, as wettability depends, to a certain extent, on the roughness.

DOI: 10.1201/9781003173533-11

Surface topography includes the macro- (range of millimeters up to tens of microns), micro- (range of tens of microns up to a few microns), and nanoscale (range of hundreds of nanometers up to a few nanometers) features of the surface implant (Barfeie, Wilson, and Rees 2015). Different techniques (e.g. sandblasting, machining and polishing, plasma spraying, etching, sol-gel coatings, anodization, chemical vapor deposition, etc.) can be used to modify the surface characteristics of implants at these scales. However, in general, neither of them is able to modify simultaneously the surface topographies in the three scales (macro-, micro-, and nanoscale) in a controllable and repro- ducible way. On the contrary, laser texturing, as we will see, is a surface processing technique that can address this problem to a certain extent.

Surface wettability is also a relevant property determining the performance of the implant (Bonn et al. 2009). As the surface of the implant is the first to come into contact with the biological tissues, wettability (in conjunction with other surface properties) will determine the biocompatibility of the biomaterial (Menzies and Jones 2010). This parameter can be defined as the ability of one fluid to spread on, or to adhere to a given surface. It mainly depends on the topography (roughness and waviness) and chemistry of the surface of materials.

The wettability is commonly expressed in terms of the contact angle (CA), that is, the angle formed by the interface liquid-vapor with a solid surface (as depicted in Figure 11.1) (Yuan and Lee 2013). If the affinity of the solid surface to the liquid is high, the liquid easily spreads over the surface and the value of the CA formed by the liquid drop is low, typically lower than 90° (Figure 11.1(a)). Such surfaces are known as hydrophilic when water is the test liquid. On the contrary, if the liquid does not spread, it forms a quasi-spherical drop resting at equilibrium on the solid surface. In this case, the CA is large, typically higher than 90°; these surfaces are called hydrophobic when water is the test liquid.

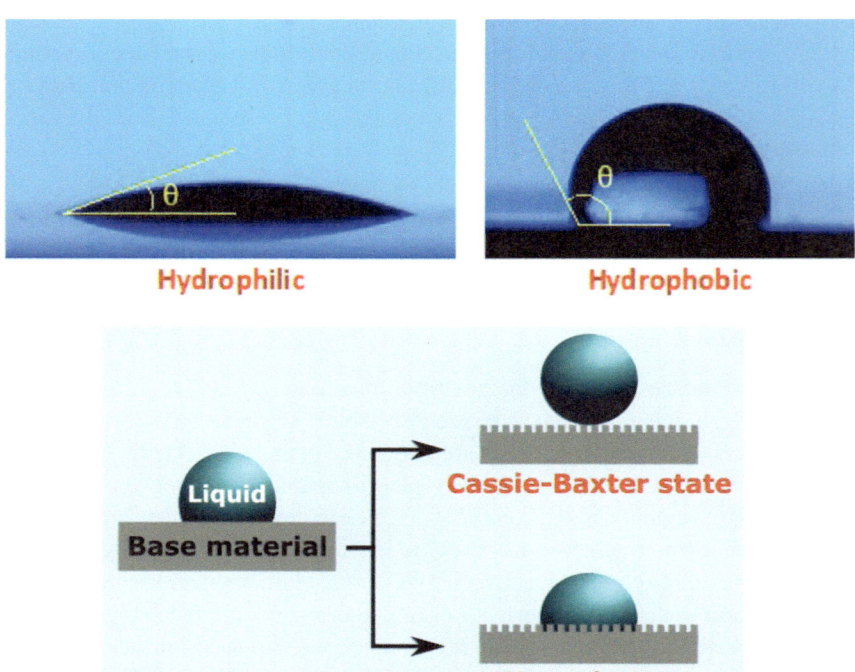

FIGURE 11.1   (a) Contact angle for hydrophilic ($\theta < 90°$) and hydrophobic materials ($\theta > 90°$). (b) Scheme of the influence of the roughness on the wettability under the framework of the Wenzel and Cassie-Baxter models.

The wetting behavior of a surface mainly depends on the chemistry of both liquid and surface. For example, the presence of polar or charged functional groups in a surface increases the adhesive interactions with water, a polar liquid (Fowkes 1964). In fact, the presence of polar hydroxides (OH-groups) is directly related to the water-wettability of glasses (Takeda et al. 1999). Common presence of polar oxides/hydroxides in the outer layers of metals also explains their high wettability. On the contrary, if the surface is covered with non-polar groups, the molecular adhesive interactions with water are decreased, which in turn results into lower wettability. Another factor influencing the wettability of surfaces is the surface topography. The relation of both will be discussed in the next section.

The influence of the wettability on the biocompatibility of biomaterials has been largely studied. In general, it is assumed that the higher the wettability the larger the biocompatibility of the surface; however, this situation is more complex. Cells must attach, grow, migrate, and proliferate on the surface of the implant to guarantee a good integration of implant-living tissue. Then, the cell-surface interactions must predominate. These interactions are largely controlled by proteins such as fibronectin, fibrinogen, and vitronectin that facilitate the adhesion of cells to the surface of the bio-material. In consequence, a good cell adhesion will depend on the adhesion of these serum proteins on the surface of the biomaterial. Tzoneva et al. showed that adsorption of proteins (fibronectin, fibrinogen, and collagen) onto hydrophobic surfaces is lower than on hydrophilic ones; however, some works have demonstrated that highly hydrophilic surfaces do not lead to a larger biocompat-ibility (Tzoneva, Faucheux, and Groth 2007). There exists a range of wettability that improves the adhesion of these proteins and in consequence of the cells. Faucheux et al. showed that adhesion of these proteins to highly hydrophobic or hydrophilic surfaces is limited and the growth of cells on these surfaces is greatly reduced (Faucheux et al. 2004). Arima et al. showed that adhesion of endothelial cells is maximum when the CA is around 35–50° (Arima and Iwata 2007).

11.1.2 Roughness-Wettability Interaction

Surface topography plays a relevant role on the wettability of surfaces. Two main approaches have been developed to explain this relation. Wenzel developed a model predicting that roughness rein-forces the intrinsic wetting behavior of an initially flat surface. According to this model, wettable surfaces become more wettable and non-wettable surfaces become less wettable when roughness is added. This model assumes that the liquid is able to penetrate into the cavities or grooves formed in the surfaces (see Figure 11.3(b)); however, in some other cases, air pockets remain trapped under the drop and the liquid is not able to penetrate into these features. In this situation, the model of Wenzel no longer applies. To address this problem, another approach, called the Cassie-Baxter model, con-siders the surface under the drop as a mixture of two materials, one being the actual solid and the other air (Gennes, Brochard-Wyart, and Quere 2004). Then, the overall wettability is a weighted combination of the individual wettability of each phase (solid or air). According to this model, the addition of a micro- and/or nanofeatures to a flat surface increases the air fraction under the drop and the contact angle increases, independently of the intrinsic wettability of the solid (Figure 11.3). Low wettability states are usually better approximated by the Cassie-Baxter's model if the rough-ness is high, whereas Wenzel's predictions are accurate at low roughness levels.

11.2 LASER TEXTURING FUNDAMENTALS

Laser texturing (also called laser surface texturing—LST, laser structuring, or laser patterning), is based on the direct irradiation of surfaces by (usually, focused) laser beams (see Figure 11.2). In consequence, different micro- and nanofeatures, such as regular or irregular patterns of grooves, bumps, dimples, ripples (LIPSS or laser-induced periodic surface structures), etc., are created on the treated surfaces, as depicted in Figure 11.3. These are produced due to the localized ablation, evaporation, or melting in the surface.

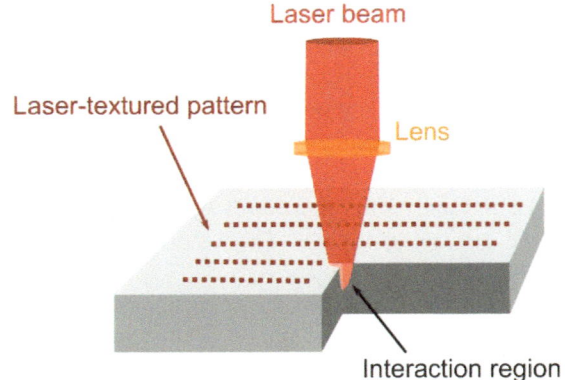

FIGURE 11.2 Schematic of the laser surface texturing process.

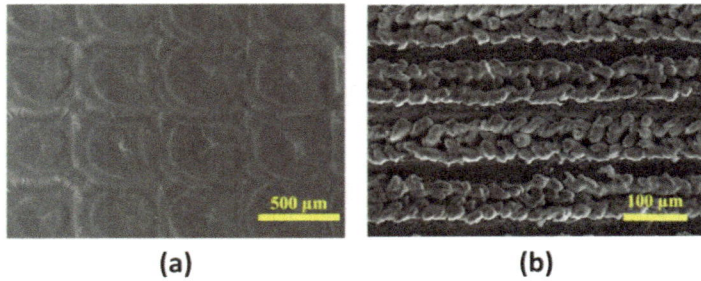

FIGURE 11.3 Typical features produced during laser texturing (a) dimples (produced in titanium with a Nd:YAG laser) and (b) grooves (produced in stainless steel with a frequency-doubled Nd:YVO₄ laser).

11.2.1 Laser Ablation Mechanisms

The structures formed during laser texturing are the consequence of a process known as ablation, a general term referring to the localized material removal in the zone directly irradiated by the laser beam. The mechanisms involved in the process highly depend on the material properties (optical, structural, thermal, etc.), the laser beam characteristics (wavelength, power density, continuous/pulsed regime, etc.), and the processing environment (gas, liquid, vacuum, etc.). Despite a brief explanation given here, we invite the interested reader to refer to more specialized works referred in this topic (Krüger and Kautek 2004; Leitz et al. 2011).

When a material surface is irradiated by a laser beam, the former phenomenon always involves the absorption of light by coupling of the optical energy. This can happen through different mechanisms depending on the material and the irradiation conditions, as excitation of free electrons (metals), phonon or lattice vibrations (insulators), etc. Eventually, the coupled energy is transformed into heat (thermalization), increasing the temperature of the affected material volume. The time required for the thermalization defines the nature of the main ablation mechanism: thermal, photochemical, or photophysical ablation (Bäuerle 2013).

11.2.2 Laser-Induced Periodic Surface Structures (LIPSS)

Under some processing conditions, when a surface is irradiated predominantly with laser pulses in the ps- or fs-regime, nanoscale ripples, also called laser-induced periodic surface structures

(LIPSS) are formed (J. Bonse et al. 2012). These have a spatial wavelength close or smaller than the laser wavelength and originate on all types of materials when they are ablated close to their ablation threshold. The mechanism originating these nanoscale features is still controversial; however, the most accepted explanation is based on the interference of the incident laser light with the light scattered at the surface interface. For more information on this phenomena, we refer to more specialized works (J. Bonse et al. 2012; Jörn Bonse et al. 2017).

11.2.3 LASER BEAM CHARACTERISTICS

Regarding the laser beam characteristics, a wide range of wavelengths going from 10 micron (infrared radiation) to a few hundreds of nanometres (ultraviolet radiation) is available in commercial laser sources. The most suitable radiation depends on the optical absorption properties of the material. In general, the surface quality of the features improves with shorter wavelengths due to the fine focusing (Steen and Mazumder 2010).

Another important laser property is the emission temporal regime, going from continuous-wave emission (CW) to pulsed regime in a wide range of pulse lengths from milli- to femto-second. In general, CW or long millisecond pulses generate large thermal effects on a relatively large depth and a substantial amount of melt, resulting in the formation of debris. Therefore, the generated textures have limited precision and quality, and the bulk properties of the material can be easily altered during processing. Reducing the pulse length also reduces the bulk thermal damage, the amount of melt and debris, and allows the production of finer features with a higher quality. In high thermal diffusivity materials, like metals or some semiconductors, the thermalization is not avoided until the pulse length is reduced to the ultrashort regime, for instance up to pico lengths or femtosecond lengths (Chichkov et al. 1996). Under this particular processing condition, the intensity is high enough to allow the occurrence of multiphoton events. Also bond breaking in insulators is eased if the energy of several photons is simultaneously absorbed (Steen and Mazumder 2010; Bäuerle 2013).

11.2.4 SURFACE CHEMISTRY AND PROCESSING ENVIRONMENT

Direct modification of the surface chemistry with laser irradiation is generally limited to polymers. The relatively low bonding energy of polymer chains makes them sensitive to chemical modifications by photochemical processes during direct laser irradiation.

On the other hand, surface chemistry can also be altered during laser texturing due to thermally driven reactions with the surrounding environment. Typically, laser texturing is performed in atmospheric conditions. Oxidation is a very common result of laser irradiation in air. However, controlling the composition of the processing environment, either gas (Lehr et al. 2014) or liquid (Wang et al. 2020) can be useful to tailor the resulting surface chemistry. Few works have paid attention on the impact of this parameter on the final wettability of biomaterial surfaces.

11.3 LASER SOURCES AND SETUPS FOR SURFACE TEXTURING

11.3.1 LASER SOURCES FOR SURFACE TEXTURING

Among the most used laser sources for laser texturing one finds Nd:YAG (Vilhena et al. 2009), Nd:YVO$_4$ (and its 2nd and 3rd harmonics) (Riveiro et al. 2012), femtosecond (Jörn Bonse et al. 2018), or pulsed fiber laser sources (Demir, Maressa, and Previtali 2013). Other lasers like excimer (Geiger, Popp, and Engel 2002) or CO$_2$ (Riveiro et al. 2020) have also been employed.

Besides, as only a thin layer of material must be treated, sources with a relatively low average power (typically P < 200W) are usually employed for laser texturing. Even then, only a few watts are enough for texturing with very short pulses.

11.3.2 LASER TEXTURING SETUPS

The most common approaches for laser texturing can be divided in two main categories, depending on whether the laser beam is moving relative to the irradiated surface or not.

Regarding stationary laser beam texturing approaches, interference patterning or direct laser interference patterning (DLIP) is one of the most commonly employed techniques. Two or more beams are superposed on the treated surface to generate an interference pattern with a specific distribution of intensity, which is converted into a topography pattern through the ablation mechanisms (Aguilar-Morales et al. 2018). Another approach with a stationary laser is the use of a mask to produce the pattern. The mask can be placed before the focusing lens to partially block the beam or be directly applied to the surface of the material and removed after the treatment (Kumar et al. 2012). These approaches, with a stationary beam, suffer from a lack of flexibility, as time and complex setups are required.

Techniques based on the relative motion between the laser beam and the workpiece usually use a Cartesian axis system (similarly to conventional machining), or a couple of galvanometer mirrors to deflect the beam to the desired location. Compared to the stationary beam approaches, the main advantages of these approaches are the high flexibility in geometries and the high processing speeds achievable, especially with the galvanometer mirrors systems.

11.4 BIOMEDICAL APPLICATIONS OF LASER TEXTURING

11.4.1 TAILORING THE SURFACE OF IMPLANT MATERIALS TO IMPROVE THE BIOCOMPATIBILITY

11.4.1.1 Polymers

Polymers are one of the most widely used materials for biomedical applications. Typical polymers used as implant materials include polyetheretherketone (PEEK), ultra-high-molecular-weight polyethylene (UHMWPE), polypropylene (PP), or acrylic bone cements (PMMA). Surface topography and wettability of polymers can be easily modified by laser texturing (Riveiro et al. 2018). Figure 11.4(a) shows the modification of the topography and wettability in PEEK surfaces after the laser treatment using a Nd:YVO$_4$ laser emitting different wavelengths. Wettability evolves from hydrophilic (when treated with 355 nm laser radiation) up to hydrophobic (when treated with 532 nm laser radiation) (Riveiro et al. 2012). The correlation between, roughness, wettability, and the formation of polar groups in the surfaces of laser-treated PEEK surfaces demonstrated an increased biocompatibility (Zheng et al. 2015).

Other polymers, such as UHMWPE, are largely transparent to visible and near-IR radiation. Therefore, the application of absorbing coatings (such as a black carbon coating) are required to increase the laser absorption (Riveiro et al. 2014). As seen in Figure 3.4(b), the influence of the laser wavelength on the surface wettability is not as big as for PEEK; but some differences are found. This result is a consequence of the lower laser absorption.

Polycarbonate (PC) is an amorphous thermoplastic polymer exhibiting also a low absorption to visible light. It is usually used as in cardiac devices or renal applications. As shown by Ramazani et al., Nd:YAG laser texturing of PC surfaces improves the fibroblast attachment and proliferation (Ramazani et al. 2009).

The modification of the wettability of PP by laser texturing has been also addressed in the literature. Using UV laser radiation was determined as an efficient way to promote the formation of functional groups which increase the wettability of PP surfaces (Riveiro et al. 2016). Their formation explains the increment of the CA after laser texturing. Khaledian et al. showed that cell viability increased on PP surfaces with oriented topographies produced after UV laser treatment (KrF laser, 248 nm) rather than on untreated surfaces (Khaledian, Jiroudhashemi, and Biazar 2017).

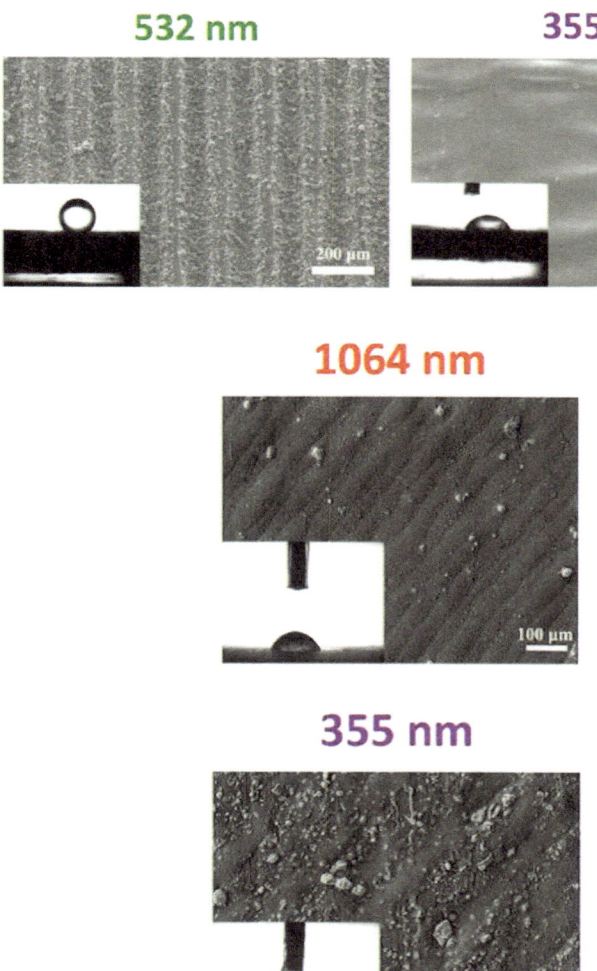

FIGURE 11.4 (a) PEEK—laser texturing allows obtaining hydrophobic surfaces (at 532 nm) or hydrophilic (at 355 nm).

Source: Adapted with permission from Riveiro et al. (2012).

(b) UHMWPE—laser texturing under different wavelength allows increasing wettability by 25% which is beneficial for cell attachment.

Source: Reprinted with permission from Riveiro et al. (2014).

11.4.1.2 Metals

Several metals are widely used as biomaterial in different clinical applications. These materials are easily textured using near-IR or visible (typically green) laser radiation. One of the most common metals used in implant devices is stainless steel. Bizi-Bandoki et al. (Bizi-Bandoki et al. 2011) tailored the wettability and topography of AISI 316L stainless steel using a femtosecond laser ($\lambda = 800$ nm). As demonstrated, micro- and nanoscale features are produced. The wettability of the surface is reduced with the increment of the number of laser pulses, as a Cassie-Baxter state is produced. Similar results were found by Lin et al. (Lin, Cheng, and Ou 2012). Microscale features and nanoscale LIPSS features improved the cell attachment, adhesion, migration, and cell division during in-vitro tests performed on stainless steel (SUS 304) substrates. The same biological results were found on 316L stainless steel by Kenar et al. using mesenchymal stem cells (Kenar et al. 2013). Martinez-Calderon et al. demonstrated that cells tends to align with the LIPSS features produced by femtosecond laser texturing of AISI304 austenitic stainless steel (Martínez-Calderon et al. 2016). On the other hand, femtosecond laser texturing of stainless steel was also demonstrated as an excellent alternative to avoid restenosis in stents as enough endothelialization is observed, while inflammation and fibrosis formation are avoided on the laser-treated surfaces (Oberringer et al. 2013).

In contrast to polymers, the processing atmosphere has some impact on the wettability. Mainly, oxides formed during the processing change the surface chemistry. Pou et al. demonstrated that the wettability of stainless steel is different depending if a reactive or an inert atmosphere is used under identical laser processing conditions. As shown in Figure 11.5, the Nd:YVO_4 laser ($\lambda = 532$ nm) treatment of stainless steel in an oxidizing air atmosphere (Figure 11.5(b)) promotes the superhydrophilic behavior of the surface, while the same treatment in an argon atmosphere (Figure 11.5(c)) promotes a superhydrophobic response (Pou et al. 2019).

Other metals used for biomedical applications, such as titanium alloys, magnesium, or Cr-Co-Mo have also been studied. In this regard, Cunha et al. showed that LIPSS and nanopillar nanotextures produced on the surface of a Ti6Al4V alloy can, potentially, promote and improve the human mesenchymal stem cells (hMSC) differentiation into an osteoblastic lineage (Cunha et al. 2015). In-vivo studies performed by Pető et al. demonstrated that the surface of titanium screws used for dental applications become covered with micron- and submicron features after laser treatment with a Nd:glass laser source (power density of 5×10^7 W/mm²), while the chemical composition remained unaltered (Pető et al. 2002). The in-vivo experiments showed that this treatment improved the osseointegration by 20%. The increased biocompatibility of this kind of nanotextures was also confirmed on other metals. Zhang et al. showed that production of micron-sized line-like patterns or nano-sized LIPSS on the surface of a Mg-Gd-Ca alloy increased the biocompatibility and the mechanical properties (Zhang et al. 2019); moreover, the wear rate and hardness were increased, while the elastic modulus remained unaltered.

11.4.1.3   Ceramics

The application of laser texturing has been explored on different bioceramic materials. The advantage of this surface modification on these materials is the absence of mechanical actions during the processing; this avoids the potential mechanical damage of the implant during the treatment.

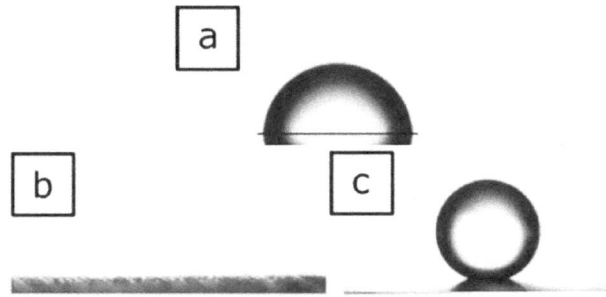

FIGURE 11.5 Comparison of the wettability of (a) base material (AISI 304 SS) with θ = 88 ± 2°, (b) superhydrophilic surface with θ = 0 ± 2° and (c) superhydrophobic surface with θ = 152 ± 4°. Surface (b) was treated in air atmosphere, while surface (c) was obtained in argon atmosphere.

Source: Reprinted with permission from Pou et al. (2019).

However, thermal stresses arising during processing with ms or ns laser sources can produce microcracks that can compromise the long-term mechanical integrity of the implant (Roitero et al. 2017). These microcracks are avoided by the utilization of fs laser sources due to their negligible heat input.

The most suitable laser sources used for these materials include those emitting in the infrared range (mainly, Nd:YAG lasers emitting at 1064 nm), but UV and visible laser sources have also been used. Laser powers used for the treatment were < 30 W in most of the experiments (see Table 11.1).

Zirconia bioceramics are relevant in clinical practice as they are used in implants (e.g. hip joint balls, knee joint components, esthetic crowns, etc.); however, they show a low bioactivity. Typical surface treatments of zirconia implants include etching, sandblasting, polishing, and coating. Conventional surface treatment involves sandblasting followed by etching in order to produce a rough surface on the implant with the aim to improve the bone-implant contact. Moura et al. showed that laser texturing of zirconia produces surfaces with better friction properties that improve the bone-implant contact (Moura et al. 2017). Different authors showed that these kinds of laser-textured

TABLE 11.1

Laser Texturing of Bioceramics and Bioactive Glasses (Y-TZP—Yttria-Stabilized Tetragonal Zirconia Polycrystals; 3Y-TZP—Tetragonal Yttria-Stabilized Tetragonal Zirconia Polycrystals; ATZ—Alumina Toughened Zirconia).

Material	Laser	Pulse Length	Pulse Energy (J)/Laser Power (W)/Fluence (J/cm²)	Geometrical Feature	Remarks	Ref.
Y-TZP	Nd:YAG (1064 nm)		0.9–1.8 W	Line-like structures covered with a submicron roughness	Friction properties (relevant on bone-implant contact) improved after laser treatment	(Moura et al. 2017)
3Y-TZP	Q-switched Nd:YAG laser (532 nm and 355 nm)	10 ns	4 J/cm² for 532 nm, 3.5 J/cm² for 355 nm	Line-like structures	Microcracks detected during processing with ns laser pulses	(Roitero et al. 2017)
Y-TZP	Picosecond laser (355 nm)	10 ps	20–200 J/cm²	Line-like and grid structures	Depending on the pattern produced on the surface the contact angle varies from 10° (superhydrophilic) to 133° (hydrophobic)	(Yan et al. 2018)
3Y-TZP	Nd:YAG (1064 nm)	35 ns	3 W	Line-like and micropillar structures	Osteoblast viability, proliferation, and differentiation improved on laser-textured surfaces	(Fernandes et al. 2020)
Al_2O_3/ZrO_2 composite	Nd:YVO$_4$ (1064 nm)	10 ns	800 µJ	Grid structure	Hierarchical surface composed of a micro-topography and a laser-induced roughness	(Baino et al. 2019)
ATZ	Yb:KYW (1030 nm)	560 fs	875 µJ	Line-like structure covered with LIPSS	Surface textures induce a faster osteogenic differentiation of hMSC; they also modulate the cell attachment, alignment, and proliferation	(Carvalho et al. 2018)

surfaces provided the required cues for the increased osteogenic differentiation of hMSC (Elena Sima et al. 2020; Fernandes et al. 2020). Similar findings were found for other bioceramics, such as in alumina-containing bioceramics. In these, Carvalho et al. showed that surface features produced by femtosecond laser texturing on alumina toughened zirconia are able to modulate the cell attachment, alignment, and proliferation of hMSC. These also showed a faster and greater osteogenic differentiation (Carvalho et al. 2018).

11.4.2 Tailoring the Surface of Implant Materials to Avoid Bacteria and Biofilm Formation

Infections caused by the contamination of the surgical site and surrounding areas are the most common cause of implant failure. This is a major clinical problem aggravated by the increased resistance of some bacteria to antibiotics (Campoccia, Montanaro, and Arciola 2006).

Once the bacteria become adhered to the surface of an implant, they stay firmly attached (using specialized tail-like structures) and form colonies which secret an extracellular matrix forming a biofilm layer; this protects the bacteria from chemicals, antibiotics, or the immune system. In this regard, tailoring the surface of implant materials to obtain a better and faster osseontegration, while showing antibacterial properties to avoid bacteria attachment and proliferation, would be an ideal scenario.

Current techniques to avoid bacteria attachment and biofilm formation involve the reduction of the adhesion of bacteria combined with certain bactericidal effects produced by the own surface or by chemicals (e.g. antibiotics, silver coatings, etc.) (Orapiriyakul et al. 2018). Material properties affecting biofilm formation include chemical (surface chemistry), physical (surface charge, wettability, topography, viscoelasticity or stiffness) and biological properties (e.g. presence of proteins in the surface) (Song, Koo, and Ren 2015; Dou et al. 2015; Lee et al. 2021). While the influence of the wettability is recognized, it is still controversial if hydrophobic or hydrophilic surfaces are better to reduce the bacterial adhesion. On the contrary, tailoring the surface topography of the implant has been recently recognized as a promising strategy to avoid the failure of implants by biofilm formation. While the influence of the microtopography is controversial as, in general, microtopographies reduce bacteria adhesion but bactericidal effects are not found, nanotropographical features show clear bactericidal effects by damaging the bacteria membrane (Lee et al. 2021). In this regard, laser texturing arises as an excellent candidate to produce micro- and nanoscale topographies with reduced bacteria adhesion or bactericidal effects. In Table 11.2, we have summarized selected works in this regard. As can be deduced from the analysis of these works, the bactericidal effect is evident after patterning different materials (polymers, metals, or glasses) with different laser sources and processing parameters.

Most of the works have explored the use of femtosecond, nanosecond, or picosecond laser sources. Different surface patterns have been produced, as observed in Table 11.2, by scanning laser beams under programmed trajectories or by the interference of two laser beams (DLIP technique). On the other hand, the pulse energy required to produce these topographies is in the range of μJ to mJ, with fluences reaching hundreds of J/cm^2.

Several works showed that nanoscale features clearly reduce the bacterial adhesion, independently of the surface pattern (Cunha et al. 2016; Doll et al. 2016; Du et al. 2020). Different explanations for the bactericidal effect of laser-treated surfaces have been proposed, mainly dependent on the type of topography and nature of the material. In this way, Truong et al. hypothesized that air trapped in femtosecond laser-textured titanium surfaces induced a lower bacterial adhesion (Truong et al. 2012). Surface roughness is hypothesized to contribute to the bactericidal action due to the reduced bonding opportunities for the bacteria, consequence of the large number of peaks on the generated surface which avoids a continuous contact (Shaikh et al. 2018).

TABLE 11.2

Main Research Works Dealing with Bacterial Adhesion in Laser-Textured Samples (Using Laser Texturing—LST—or Direct Laser Interference Patterning—DLIP).

Material	Laser (Wavelength)/ Texturing Technique	Pulse Length	Pulse Energy/ Laser Power/ Fluence	Geometrical Feature	Bacteria	Remarks	Ref.
Polystyrene	Nd:YAG (266 nm)/DLIP			Line-like, pillar-like, lamella (or grid)-like structures	S. aureus	Lamella-like surface reduced bacteria adhesion (in contrast to line- and pillar-like structures). Reduced bacteria adhesion and biofilm formation confirmed in lamella-like structures by in-vivo tests.	(Valle et al. 2015)
Titanium pure (Grade 2)	Ti:Sapphire (800 nm)/LST	sub-30 fs	1 mJ	Lotus-like surface (microscale features decorated with nanoscale features)	S. aureus, S. epidermidis, P. maritimus	Bacteria tend to accumulate on the crevices formed between the microscale features. Air-trapped by the surface features inhibit contact of bacteria with Ti surface.	(Truong et al. 2012)
Titanium pure (Grade 2)	Yb:KYW (1030 nm)/LST	500 fs		LIPSS (periodic ripples), nanopillars	S. aureus	Average fraction of surface covered with bacteria is ~7% for laser-treated samples, while reaching ~25% for polished surfaces. Nanoripples and nanopillars show similar reductions in bacteria adhesion.	(Cunha et al. 2016)
Titanium pure (Grade 4)	Ti:Sapphire (800 nm)/LST	30 fs		Diamond-like (Sharklet™) pattern design, grooved-like and grid structures	S. aureus	Significant reduction of bacteria coverage in laser-treated samples (only ~40% of surface covered with bacteria after 24 h, compared to ~50% for untreated surfaces). Influence of topography on bacteria adhesion not relevant.	(Doll et al. 2016)
45S5 bioactive glass	Ti:Sapphire (800 nm)/LST	45 fs	0.94–1.0 J/cm^2	Line-like structures	S. aureus, P. aeruginosa, E. coli.	Adhesion of S. aureus minimal while no adhesion detected for P. aeruginosa and E. coli after 24 h.	(Shaikh et al. 2018)
Zr-bulk metallic glasses	Yb:KGW (1030 nm)/LST	200 fs	23–230 J/cm^2	LIPSS (periodic ripples), nanoparticle structure	E. coli, S. aureus	LIPSS substantially reduced bacterial adhesion compared to nanoparticle covered surface.	(Du et al. 2020)

11.5   CONCLUSIONS

In this chapter, we have demonstrated the ability and potential of laser texturing to improve and promote the biocompatibility of biomaterials. Moreover, the potential of this technique to avoid bacteria and biofilm formation has been also introduced. This demonstrates the great advantages of this surface processing technique for its introduction in the biomedical industry; however, despite the results being promising, more research is needed to discern the role of the wettability and nano-topography on the biocompatibility and antibacterial action; the influence of features in this range has provided excellent performance from the point of view of biocompatibility and bactericidal action. Moreover, the most suitable processing parameters for each particular material with the aim to improve the biocompatibility should be determined while guaranteeing, at the same time, an optimum antibacterial action. Finally, the integration of these processes on current manufacturing chains, although it is foreseen not to be problematic, must be also considered and studied.

11.6   ACKNOWLEDGMENTS

The authors acknowledge the financial support of the European Union program Bluehuman (EAPA_151/2016 Interreg Atlantic Area), Government of Spain [RTI2018–095490-J-I00 (MCIU/AEI/FEDER, UE, FPU16/05492)], and Xunta de Galicia (ED431C 2019/23, ED481D 2017/010, ED481B 2016/047–0).

REFERENCES

Aguilar-Morales, Alfredo I., Sabri Alamri, Tim Kunze, and Andrés Fabián Lasagni. 2018. "Influence of Processing Parameters on Surface Texture Homogeneity Using Direct Laser Interference Patterning". *Optics & Laser Technology* 107: 216–227.

Arima, Yusuke, and Hiroo Iwata. 2007. "Effect of wettability and surface functional groups on protein adsorption and cell adhesion using well-defined mixed self-assembled monolayers". *Biomaterials* 28 (20): 3074–3082.

Baino, Francesco, Maria Angeles Montealegre, Joaquim Minguella-Canela, and Chiara Vitale-Brovarone. 2019. "Laser Surface Texturing of Alumina/Zirconia Composite Ceramics for Potential Use in Hip Joint Prosthesis". *Coatings* 9 (6): 369.

Barfeie, A., J. Wilson, and J. Rees. 2015. "Implant Surface Characteristics and Their Effect on Osseointegration". *British Dental Journal* 218 (5): E9.

Bäuerle, Dieter. 2013. *Laser Processing and Chemistry*. 3rd ed. Springer Science & Business Media, Berlin, Heidelberg.

Bizi-Bandoki, P., S. Benayoun, S. Valette, B. Beaugiraud, and E. Audouard. 2011. "Modifications of Roughness and Wettability Properties of Metals Induced by Femtosecond Laser Treatment". *Applied Surface Science* 257 (12): 5213–5218.

Bonn, Daniel, Jens Eggers, Joseph Indekeu, Jacques Meunier, and Etienne Rolley. 2009. "Wetting and Spreading". *Reviews of Modern Physics* 81 (2): 739–805.

Bonse, J., J. Krüger, S. Höhm, and A. Rosenfeld. 2012. "Femtosecond Laser-Induced Periodic Surface Structures". *Journal of Laser Applications* 24 (4): 042006.

Bonse, Jörn, Sandra Höhm, Sabrina V. Kirner, Arkadi Rosenfeld, and Jörg Krüger. 2017. "Laser-Induced Periodic Surface Structures: A Scientific Evergreen". *IEEE Journal of Selected Topics in Quantum Electronics* 23 (3).

Bonse, Jörn, Sabrina V. Kirner, Michael Griepentrog, Dirk Spaltmann, and Jörg Krüger. 2018. "Femtosecond Laser Texturing of Surfaces for Tribological Applications". *Materials* 11 (5): 801.

Campoccia, Davide, Lucio Montanaro, and Carla Renata Arciola. 2006. "The Significance of Infection Related to Orthopedic Devices and Issues of Antibiotic Resistance". *Biomaterials* 27 (11): 2331–2339.

Carvalho, Angela, Liliana Cangueiro, Vítor Oliveira, Rui Vilar, Maria H. Fernandes, and Fernando J. Monteiro. 2018. "Femtosecond Laser Microstructured Alumina Toughened Zirconia: A New Strategy to Improve Osteogenic Differentiation of HMSCs". *Applied Surface Science* 435 (marzo): 1237–1245.

Chichkov, Boris N., C. Momma, Stefan Nolte, F. Von Alvensleben, and A. Tünnermann. 1996. "Femtosecond, Picosecond and Nanosecond Laser Ablation of Solids". *Applied Physics A* 63 (2): 109–115.

Cunha, Alexandre, Anne-Marie Elie, Laurent Plawinski, Ana Paula Serro, Ana Maria Botelho do Rego, Amélia Almeida, Maria C. Urdaci, Marie-Christine Durrieu, and Rui Vilar. 2016. "Femtosecond Laser Surface Texturing of Titanium as a Method to Reduce the Adhesion of Staphylococcus Aureus and Biofilm Formation". *Applied Surface Science* 360 (enero): 485–493.

Cunha, Alexandre, Omar Farouk Zouani, Laurent Plawinski, Ana Maria Botelho do Rego, Amélia Almeida, Rui Vilar, and Marie-Christine Durrieu. 2015. "Human Mesenchymal Stem Cell Behavior on Femtosecond Laser-Textured Ti-6Al-4V Surfaces". *Nanomedicine* 10 (5): 725–739.

Demir, A. G., P. Maressa, and B. Previtali. 2013. "Fibre Laser Texturing for Surface Functionalization". *Physics Procedia, Lasers in Manufacturing (LiM 2013)*, 41: 759–768.

Doll, Katharina, Elena Fadeeva, Nico S. Stumpp, Sebastian Grade, Boris N. Chichkov, and Meike Stiesch. 2016. "Reduced Bacterial Adhesion on Titanium Surfaces Micro-Structured by Ultra-Short Pulsed Laser Ablation". *Bio Nano Materials* 17 (1–2): 53–57.

Dou, Xiao-Qiu, Di Zhang, Chuanliang Feng, and Lei Jiang. 2015. "Bioinspired Hierarchical Surface Structures with Tunable Wettability for Regulating Bacteria Adhesion". *ACS Nano* 9 (11): 10664–10672.

Du, Cezhi, Chengyong Wang, Tao Zhang, Xin Yi, Jianyi Liang, and Hongjian Wang. 2020. "Reduced Bacterial Adhesion on Zirconium-Based Bulk Metallic Glasses by Femtosecond Laser Nanostructuring". *Proceedings of the Institution of Mechanical Engineers, Part H: Journal of Engineering in Medicine* 234 (4): 387–397.

Elena Sima, Livia, Anca Bonciu, Madalina Baciu, Iulia Anghel, Luminita Nicoleta Dumitrescu, Laurentiu Rusen, and Valentina Dinca. 2020. "Bioinstructive Micro-Nanotextured Zirconia Ceramic Interfaces for Guiding and Stimulating an Osteogenic Response In Vitro". *Nanomaterials* 10 (12): 2465.

Faucheux, N., R. Schweiss, K. Lützow, C. Werner, and T. Groth. 2004. "Self-Assembled Monolayers with Different Terminating Groups as Model Substrates for Cell Adhesion Studies". *Biomaterials* 25 (14): 2721–2730.

Fernandes, Beatriz Ferreira, Mariana Brito da Cruz, Joana Faria Marques, Sara Madeira, Óscar Carvalho, Filipe Samuel Silva, António Duarte Sola Pereira da Mata, and João Manuel Mendez Caramês. 2020. "Laser Nd:YAG Patterning Enhance Human Osteoblast Behavior on Zirconia Implants". *Lasers in Medical Science* 35 (9): 2039–2048.

Fowkes, Frederick M. 1964. "Attractive Forces at Interfaces". *Industrial & Engineering Chemistry* 56 (12): 40–52.

Geiger, M., U. Popp, and U. Engel. 2002. "Excimer Laser Micro Texturing of Cold Forging Tool Surfaces: Influence on Tool Life". *CIRP Annals* 51 (1): 231–234.

Gennes, Pierre-Gilles de, Francoise Brochard-Wyart, and David Quere. 2004. *Capillarity and Wetting Phenomena: Drops, Bubbles, Pearls, Waves.* Springer-Verlag, New York.

Kenar, Halime, Erhan Akman, Elif Kacar, Arif Demir, Haiwoong Park, Hashim Abdul-Khaliq, Cenk Aktas, and Erdal Karaoz. 2013. "Femtosecond Laser Treatment of 316L Improves Its Surface Nanoroughness and Carbon Content and Promotes Osseointegration: An in Vitro Evaluation". *Colloids and Surfaces B: Biointerfaces* 108: 305–312.

Khaledian, Mohammad, Faeze Jiroudhashemi, and Esmaeil Biazar. 2017. "Chitosan- and Polypropylene-Oriented Surface Modification Using Excimer Laser and Their Biocompatibility Study". *Artificial Cells, Nanomedicine, and Biotechnology* 45 (1): 135–138.

Krüger, Jörg, and Wolfgang Kautek. 2004. "Ultrashort Pulse Laser Interaction with Dielectrics and Polymers". In *Polymers and Light*, edited by Thomas. K. Lippert, 247–290. Advances in Polymer Science. Springer Verlag, Berlin, Heidelberg.

Kumar, K., K. Lee, J. Nogami, P. Herman, and N. Kherani. 2012. "Ultrafast Laser Direct Hard-Mask Writing for High Performance Inverted-Pyramidal Texturing of Silicon". *2012 38th IEEE Photovoltaic Specialists Conference* (Austin, TX, USA), 002182-002185.

Lee, Sang Won, K. Scott Phillips, Huan Gu, Mehdi Kazemzadeh-Narbat, and Dacheng Ren. 2021. "How Microbes Read the Map: Effects of Implant Topography on Bacterial Adhesion and Biofilm Formation". *Biomaterials* 268: 120595.

Lehr, Jorge, Fabrizio de Marchi, Luke Matus, Jennifer MacLeod, Federico Rosei, and Anne-Marie Kietzig. 2014. "The Influence of the Gas Environment on Morphology and Chemical Composition of Surfaces Micro-Machined with a Femtosecond Laser". *Applied Surface Science* 320: 455–465.

Leitz, Karl-Heinz, Benjamin Redlingshöfer, Yvonne Reg, Andreas Otto, and Michael Schmidt. 2011. "Metal Ablation with Short and Ultrashort Laser Pulses". *Physics Procedia, Lasers in Manufacturing 2011: Proceedings of the Sixth International WLT Conference on Lasers in Manufacturing* 12: 230–238.

Lin, C. Y., C. W. Cheng, and K. L. Ou. 2012. "Micro/Nano-Structuring of Medical Stainless Steel Using Femtosecond Laser Pulses". *Physics Procedia, Laser Assisted Net Shape Engineering 7* (LANE 2012), 39: 661–668.

Martínez-Calderon, M., M. Manso-Silván, A. Rodríguez, M. Gómez-Aranzadi, J. P. García-Ruiz, S. M. Olaizola, and R. J. Martín-Palma. 2016. "Surface Micro- and Nano-Texturing of Stainless Steel by Femtosecond Laser for the Control of Cell Migration". *Scientific Reports* 6 (1): 36296.

Menzies, Kara L., and Lyndon Jones. 2010. "The Impact of Contact Angle on the Biocompatibility of Biomaterials". *Optometry and Vision Science* 87 (6): 387–399.

Moura, C. G., R. Pereira, M. Buciumeanu, O. Carvalho, F. Bartolomeu, R. Nascimento, and F. S. Silva. 2017. "Effect of Laser Surface Texturing on Primary Stability and Surface Properties of Zirconia Implants". *Ceramics International* 43 (17): 15227–15236.

Oberringer, Martin, Erhan Akman, Juseok Lee, Wolfgang Metzger, Cagri Kaan Akkan, Elif Kacar, Arif Demir, et al. 2013. "Reduced Myofibroblast Differentiation on Femtosecond Laser Treated 316LS Stainless Steel". *Materials Science and Engineering: C* 33 (2): 901–908.

Orapiriyakul, Wich, Peter S Young, Laila Damiati, and Penelope M Tsimbouri. 2018. "Antibacterial Surface Modification of Titanium Implants in Orthopaedics". *Journal of Tissue Engineering* 9: 2041731418789838.

Pető, G., A. Karacs, Z. Pászti, L. Guczi, T. Divinyi, and A. Joób. 2002. "Surface Treatment of Screw Shaped Titanium Dental Implants by High Intensity Laser Pulses". *Applied Surface Science* 186 (1): 7–13.

Pou, P., J. del Val, A. Riveiro, R. Comesaña, F. Arias-González, F. Lusquiños, M. Bountinguiza, F. Quintero, and J. Pou. 2019. "Laser Texturing of Stainless Steel under Different Processing Atmospheres: From Superhydrophilic to Superhydrophobic Surfaces". *Applied Surface Science* 475: 896–905.

Ramazani S. A., Ahmad, Seyyed Abbas Mousavi, Ehsan Seyedjafari, Reza Poursalehi, Shohreh Sareh, Kaveh Silakhori, Ali Akbar Poorfatollah, and Amir Nasser Shamkhali. 2009. "Polycarbonate Surface Cell's Adhesion Examination after Nd:YAG Laser Irradiation". *Materials Science and Engineering: C* 29 (4): 1491–1497.

Riveiro, A., T. Abalde, P. Pou, R. Soto, J. del Val, R. Comesaña, A. Badaoui, M. Boutinguiza, and J. Pou. 2020. "Influence of Laser Texturing on the Wettability of PTFE". *Applied Surface Science* 515: 145984.

Riveiro, A., A. L. B. Maçon, J. del Val, R. Comesaña, and J. Pou. 2018. "Laser Surface Texturing of Polymers for Biomedical Applications". *Frontiers in Physics* 6: 16.

Riveiro, A., R. Soto, R. Comesaña, M. Boutinguiza, J. del Val, F. Quintero, F. Lusquiños, and J. Pou. 2012. "Laser Surface Modification of PEEK". *Applied Surface Science* 258 (23): 9437–9442.

Riveiro, A., R. Soto, J. del Val, R. Comesaña, M. Boutinguiza, F. Quintero, F. Lusquiños, and J. Pou. 2014. "Laser Surface Modification of Ultra-High-Molecular-Weight Polyethylene (UHMWPE) for Biomedical Applications". *Applied Surface Science* 302: 236–242.

Riveiro, A., R. Soto, J. del Val, R. Comesaña, M. Boutinguiza, F. Quintero, F. Lusquiños, and J. Pou. 2016. "Texturing of Polypropylene (PP) with Nanosecond Lasers". *Applied Surface Science* 374: 379–386.

Roitero, E., F. Lasserre, J. J. Roa, M. Anglada, F. Mücklich, and E. Jiménez-Piqué. 2017. "Nanosecond-Laser Patterning of 3Y-TZP: Damage and Microstructural Changes". *Journal of the European Ceramic Society* 37 (15): 4876–4887.

Shaikh, Shazia, Deepti Singh, Mahesh Subramanian, Sunita Kedia, Anil Kumar Singh, Kulwant Singh, Nidhi Gupta, and Sucharita Sinha. 2018. "Femtosecond Laser Induced Surface Modification for Prevention of Bacterial Adhesion on 45S5 Bioactive Glass". *Journal of Non-Crystalline Solids* 482: 63–72.

Song, F., H. Koo, and D. Ren. 2015. "Effects of Material Properties on Bacterial Adhesion and Biofilm Formation". *Journal of Dental Research* 94 (8): 1027–1034.

Steen, William M., and Jyotirmoy Mazumder. 2010. *Laser Material Processing*. 4th ed. Springer, London.

Takeda, Satoshi, Kiyoshi Yamamoto, Yuki Hayasaka, and Kiyoshi Matsumoto. 1999. "Surface OH Group Governing Wettability of Commercial Glasses". *Journal of Non-Crystalline Solids* 249 (1): 41–46.

Truong, V. K., H. K. Webb, E. Fadeeva, B. N. Chichkov, A. H. F. Wu, R. Lamb, J. Y. Wang, R. J. Crawford, and E. P. Ivanova. 2012. "Air-Directed Attachment of Coccoid Bacteria to the Surface of Superhydrophobic Lotus-Like Titanium". *Biofouling* 28 (6): 539–550.

Tzoneva, R., N. Faucheux, and T. Groth. 2007. "Wettability of Substrata Controls Cell-Substrate and Cell-Cell Adhesions". *Biochimica et Biophysica Acta (BBA): General Subjects* 1770 (11): 1538–1547.

Valle, Jaione, Saioa Burgui, Denise Langheinrich, Carmen Gil, Cristina Solano, Alejandro Toledo-Arana, Ralf Helbig, Andrés Lasagni, and Iñigo Lasa. 2015. "Evaluation of Surface Microtopography Engineered by Direct Laser Interference for Bacterial Anti-Biofouling". *Macromolecular Bioscience* 15 (8): 1060–1069.

Vilhena, L. M., M. Sedlaček, B. Podgornik, J. Vižintin, A. Babnik, and J. Možina. 2009. "Surface Texturing by Pulsed Nd:YAG Laser". *Tribology International* 42 (10): 1496–1504.

Wang, Yutong, Xiaoyan Zhao, Changjun Ke, Jin Yu, and Ran Wang. 2020. "Nanosecond Laser Fabrication of Superhydrophobic Ti6Al4V Surfaces Assisted with Different Liquids". *Colloid and Interface Science Communications* 35: 100256.

Yan, Tianyang, Lingfei Ji, Jian Li, Pengxiang Zhao, and Xuemei Ma. 2018. "Tailoring Surface Wettability of TZP Bioceramics by UV Picosecond Laser Micro-Fabrication". *Applied Physics A* 124 (2): 97.

Yuan, Yuehua, and T. Randall Lee. 2013. "Contact Angle and Wetting Properties". In *En Surface Science Techniques*, edited by Gianangelo Bracco and Bodil Holst, 3–34. Springer Series in Surface Sciences. Springer, Berlin, Heidelberg.

Zhang, Jiaru, Yingchun Guan, Wenting Lin, and Xuenan Gu. 2019. "Enhanced Mechanical Properties and Biocompatibility of Mg-Gd-Ca Alloy by Laser Surface Processing". *Surface and Coatings Technology* 362: 176–184.

Zheng, Yanyan, Chengdong Xiong, Zhecun Wang, Xiaoyu Li, and Lifang Zhang. 2015. "A Combination of CO2 Laser and Plasma Surface Modification of Poly(Etheretherketone) to Enhance Osteoblast Response". *Applied Surface Science* 344: 79–88.

12 Laser Processing and On-Line Monitoring for Biomedical Applications

Guoqing Hu, Xuan Wang, Jingwen He, Jie Yang and Feng Zhao

CONTENTS

INTRODUCTION

12.1 LASERS AND LASER PROCESSING APPLICATIONS

The laser is an acronym for "Light Amplification by Stimulated Emission of Radiation". The mechanism of laser emission can be traced back to the hypothesis put forward by Einstein in 1917 when he explained the blackbody radiation law. A laser consists of a special gain medium, a mechanism to energize it, and a resonant cavity such as a pair of mirrors. Different from the other light sources, it owns unique characteristics of monochromaticity, directionality, coherence, and high laser fluence. Thus, laser techniques have attracted increasing interest since the 1960s (Maiman 2018). With the further development of chirp pulse amplification (Strickland 2019) and coherent beam combination (Goodno et al. 2006), the exploitation of lasers with versatile characteristics, including wavelength band, pulse energy, and beam quality, has accelerated laser processing, which built the foundation of practical applications. The typical lasers for laser processing are listed as shown in Table 12.1.

Because of the advantages of simplicity, eco-friendliness, reproducibility, wide practicality, and easy integration, laser processing has attracted increasing research interest. The typical laser processing techniques include laser cutting and drilling, laser polishing, laser cleaning, and micro- and nano-processing, which can be applied in biomedical applications such as laser surgery (Liu et al. 2021; Jivrajka et al. 2012; Li et al. 2014; Liang et al. 2009; Williams et al. 2015; Sacks et al. 2000; Grewal et al. 2015; Augello et al. 2018; Bernal et al. 2018; Levesque

DOI: 10.1201/9781003173533-12

167

TABLE 12.1
Typical Lasers for Laser Processing.

Types	Gain medium	Wavelength
Solid-State Laser	Ruby (Edmonds et al. 2013)	694 nm
	Nd:YAG (Andrea et al. 2017)	1064, 532, 355, 266 nm
	Er:YAG (Aljdaimi et al. 2018)	2.9 µm
	Yb:YAG (Cao et al. 2010)	1030 nm
	Nd:YVO$_4$ (Xu et al. 2012)	1064 nm
	Ti:sapphire (Jeong et al. 2019)	840–1100 nm
Gas Laser	He-Ne (Tan et al. 2020)	351, 364, 458, 488, 514 nm
	CO (Kitamura et al. 2007)	6–8 µm
	CO$_2$ (Boyd et al. 2015)	10.6, 9.6 µm
	Argon ion (Wang et al. 2014)	351, 364, 458, 488, 514 nm
	HeCd (Christof et al. 2018)	442 nm
	XeF (Yanlong et al. 2019)	351–353 nm
	XeCl (Chelnokov et al. 2008)	308 nm
	KrF (Li, Dou, et al. 2020)	248 nm
	F$_2$ (Yoshida et al. 2019)	157 nm
	Krypton (Cui et al. 2014)	476, 528, 568, 647 nm
	N$_2$ (Uno et al. 2008)	337 nm
	Copper-vapor (Li, Li, et al. 2020)	510–578 nm
	H$_2$O (Mathieu et al. 2019)	1180 nm
Semiconductor Laser	GaAs (Schroder et al. 2002)	840 nm
	AlGaAs (Zhang et al. 2019)	670–850 nm
	InGaAlP (Ohtsu et al. 1990)	630–680 nm
	InGaN (Fibrich et al. 2021)	400–490 nm
	GaInAsP (Lei et al. 2006)	0.98, 1.3, 1.48, 1.55 µm
Fiber Laser	Yb-doped fiber laser (Pan et al. 2015)	1030–1080 nm
	Er-doped fiber laser (Li, et al. 2020)	1550 nm
	Tm-doped fiber laser (Zhang et al. 2020)	1600–2000 nm
Other	Chemical laser (Hui et al. 2018)	2.7 µm
	Dye laser (Pramanik et al. 2020)	Tunable 570–650 nm
	Free electron laser (Ko et al. 2021)	0.08–200 nm

et al. 2017; Yumoto et al. 2018; Jiaru et al. 2019; Hu et al. 2018; Lu et al. 2019) and surface microstructure fabrication of biomaterials (Slepicka et al. 2017; Liu et al. 2013; Wang, Li, et al. 2015; Wang et al. 2013). Laser cutting and drilling can process almost any materials, especially the hard and high-strength materials such as bone materials (Gautam et al. 2018). The extremely high-fluence laser beam is focused and absorbed to melt and vaporize the material for kerf and hole formations. Besides, the short and ultrashort pulse sources can be adopted to further suppress the thermal effect in the machining process (Zhao et al. 2020; Barsch et al. 2003). Laser polishing is a noncontact, pollution-free, and high-efficiency tool. When the laser fluence is higher than the threshold of the target materials, the irradiated area rapidly melts and the liquid becomes horizontal due to the surface tension and gravity. After removing the laser beam, the surface solidifies and the roughness decreases (Wang, Morrow, et al. 2015). Laser cleaning is adopted to remove the surface pollutions of biomaterials. The high-fluence laser beam irradiates the surface of the workpiece to make the pollutions vaporize and burst for pollution removal (Siano et al. 2010). In addition, the laser processing technique is used for the efficient fabrication of micro- and nano-structure without any additional mask. Such structures can be applied for functional surfaces of biomaterials such as the micro- and nano-structures for cell culture (Anselme et al. 2011; Huang et al. 2020).

12.2   LASER PROCESSING FOR BIOMEDICAL APPLICATIONS

As an advanced manufacturing tool, laser processing techniques have been gradually applied in biomedical applications such as laser surgery and laser modification. Ablation, cutting, and drilling are some of the most classic operations in surgery, and laser processing provides a precise and damage-free tool. Therefore, laser processing techniques are applied in laser surgery. The typical applications of laser surgery include orthopedic surgery (Augello et al. 2018; Bernal et al. 2018; Levesque et al. 2017; Yumoto et al. 2018), eye surgery (Liu et al. 2021; Jivrajka et al. 2012; Li et al. 2014; Liang et al. 2009; Williams et al. 2015; Sacks et al. 2000; Grewal et al. 2015), and so on. Recently, laser modification of biomaterials for cell research has also attracted increasing interest in biomedical fields. In comparison with the traditional surface-modification techniques such as electron beam lithography, electrospinning, photolithography, and chemical patterning, the unique advantages of laser modification techniques are low cost with excellent uniformity and reproducibility, high enhancement factor, non-pollution, and capability to tailor topography at the micro- or nanoscales (Jiaru et al. 2019; Hu et al. 2018).

12.2.1   LASER SURGERY

Due to the unique characteristics of laser processing, the applications of laser surgery technique have been widely reported while the degrees of development of different applications are different. For example, the commercial techniques and setups of laser surgery have already appeared in eye surgery (Grewal et al. 2015). However, laser orthopedic surgery is on the way to the large-scale commercial level (Lo et al. 2012; Kawata et al. 2001; Jianjun 2004). Although the applications of laser surgery are not limited in eye surgery and orthopedic surgery, these two surgeries are described as the most typical examples.

12.2.1.1   Laser Orthopedic Surgery

With social development, lifestyle changes, and aging of the world's populations, the incidences of orthopedics-related diseases are increasing year by year (Xiangyang 2016). According to the statistics from "Chinese Surgery Yearbook", there are over 20 million patients with orthopedic trauma in China each year, of which 79.35% require surgical treatment (Jing et al. 2016). Processing operations such as bone drilling are the most commonly used surgical procedures in orthopedic surgery. They are widely applied in craniotomy (Vitek et al. 2010), dental surgery (Duperron et al. 2019), laminectomy (Ito et al. 2009), stapedotomy (MacDougall et al. 2016), total knee arthroplasty (Mihalko et al. 2012), fracture treatment (Martins et al. 2011), and so on. Due to the mutual extrusion between the tool and the bone tissue during bone processing, it is risky to damage the bone tissue and its surrounding soft tissues. Meanwhile, the accumulated thermal effect can easily carbonize the bone tissue, and even damage the spinal cord and nerve roots around the bone tissue, resulting in limb dysfunction. Therefore, damage suppression of bone processing is the prerequisite basis for orthopedic surgery (Abbasi et al. 2018). To meet the requirements of precise and damage-free orthopedic surgery, the widely adopted solution is a robot-assisted orthopedic surgical technique based on the improved mechanical processing system and priori model (Liao et al. 2017; Changshu et al. 2014; Changshu et al. 2015). Currently, in comparison with the orthopedic surgical robots with a positioning accuracy of 1 mm from the companies such as Mazor Robotics Ltd., MAKO Surgical, and RoboDoc, TiRobot with the positioning accuracy of 0.8 mm and image distortion rate of 1.49%, developed by Beijing Jishuitan Hospital and TIN AVI Co. Ltd., has been obtained. Nevertheless, the processing subsystem is still based on traditional mechanical processing.

To further eliminate the damage risks, laser processing techniques provide an attractive alternative choice (Lo et al. 2012; Kawata et al. 2001; Jianjun 2004). In comparison with mechanical processing, the noncontact processing and high precision of laser processing could be beneficial for bone processing without mechanical damage (Papadaki et al. 2007). In some special cases, the growth of the bone cells and the postoperative recovery of bone tissues could also be promoted under laser irradiation (Blaskovic et al. 2016; Davoudi et al. 2018). Since the water and hydroxyapatite in the bone tissues have strong absorption of mid-infrared laser, the bone materials could be stripped due to the thermomechanical effect. Therefore, mid-infrared lasers such as Er:YAG laser (Augello et al. 2018; Bernal et al. 2018), CO_2 laser, and Cr:CdSe laser (Yumoto et al. 2018) were first proposed for bone processing. However, since the removal mechanism is mainly based on the thermomechanical effect, the cooling parameters should be carefully optimized to suppress the thermal damages, especially for the tissues surrounded with nervous tissue. On the other hand, when the laser irradiates the materials, the electron-phonon coupling time scale is in the order of several to tens of picoseconds while those of thermal diffusion and material melting are tens to hundreds of picoseconds. Besides the time scale of surface ablation formation is hundreds of picoseconds to nanoseconds. Therefore, if the pulsewidths of ultrashort pulses are picoseconds and femtoseconds, the irradiation of ultrashort pulses will finish before the thermal diffusion start. Namely, femtosecond laser processing could be a cold processing tool, which could be beneficial for thermal damage suppression (Lo et al. 2012). Besides, the beam diameter of the femtosecond laser could be focused to the micrometer level and the multiphoton absorption dominates in the femtosecond laser irradiation to achieve the spatial resolution of subdiffractional limit. Therefore, up to nm-level precision and processing of almost any materials can be achieved (Kawata et al. 2001). Then, the femtosecond laser bone processing was demonstrated (Huang et al. 2015; Hu et al. 2019; Song et al. 2020). As shown in Figure 12.1, the femtosecond laser bone processing system generally consists of the laser processing subsystem and monitoring subsystem (Huang et al. 2015), which has already shown high potential in bone processing applications such as craniotomy (Vitek et al. 2010), dental surgery (Duperron et al. 2019), total knee arthroplasty (Mihalko et al. 2012), and stapedotomy (MacDougall et al. 2016).

12.2.1.2 Laser Eye Surgery

Refractive error, cataracts, and glaucoma are the most common ophthalmic diseases of all of humankind (Liu et al. 2021; Jivrajka et al. 2012; Li et al. 2014; Liang et al. 2009; Williams et al. 2015; Sacks et al. 2000; Grewal et al. 2015). As one of the most classic refractive errors, myopia affects ~30% of the whole world population and it's believed that it will reach approximately 50% by 2050 (Ruiz-Pomeda et al. 2020). Similarly, the incidence of the other three types of refractive

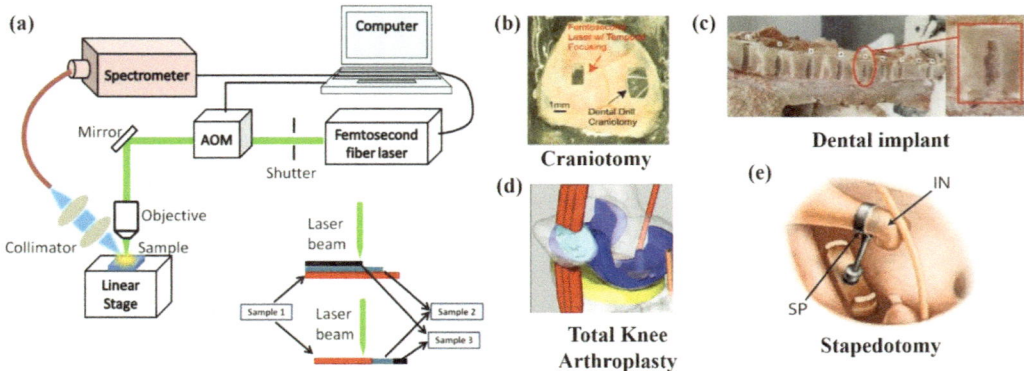

FIGURE 12.1 (a) Sketch of laser bone processing system (Huang et al. 2015) and its applications in (b) craniotomy (Vitek et al. 2010), (c) dental implant (Duperron et al. 2019), (d) total knee arthroplasty (Mihalko et al. 2012), and (e) stapedotomy.

Source: MacDougall et al. (2016).

error (i.e. hyperopia, astigmatism, and presbyopia) is also not optimistic. The cataract is an ophthalmic disease with the highest rate of blindness, while surgery is the most effective method to heal it (Grewal et al. 2015; Nagy et al. 2009). Glaucoma is also the primary cause of blindness in the world (Sacks et al. 2000). Therefore, except for effective precautions, damage-free, high-efficiency, and high-precision treatment is extremely urgent.

Due to the unique advantages of laser surgery as demonstrated before, it provides an attractive alternative treatment tool. Besides, excimer laser photons have high energy exceeding the bond energy of many polymers. Its energy is sufficient to break molecular bonds of materials for material removal without heating. Thus, it is most suitable for the organic materials such as the components of the eye due to their weak bonds. Actually, research on the applications of laser eye surgery in ophthalmic diseases has lasted for tens of years (Alio 2014; McAlinden 2012; Mosquera et al. 2014). The first excimer laser surgery, executed already over 31 years ago, was a historical breakthrough (McDonald 1990). As demonstrated previously, the femtosecond laser processing could be a cold processing tool to effectively suppress thermal damage (Lo et al. 2012). In 1998, Kurtz et al. proposed the first experimental verification of femtosecond laser-assisted eye surgery (Kurtz et al. 1998). In 2000, Sacks et al. demonstrated the femtosecond laser cutting of the sclera in glaucoma surgery (Sacks et al. 2000). At that time, over five million people all over the world had become blind due to complications arising from glaucoma. In 2009, femtosecond laser-assisted cataract surgery was reported (Nagy et al. 2009). Nowadays, the systems of femtosecond laser-assisted cataract surgery (Grewal et al. 2015) and laser surgery of glaucoma and refractive error have been commercially produced for practical applications. The diagram of the optical and mechanical interface between the laser system and the eye is depicted as shown in Figure 12.2. For safety, an on-line monitoring subsystem (i.e. optical coherence tomography (OCT)) is also adopted for imaging. Nevertheless, higher precision and efficiency, lower cost, and better postoperative recovery are still a continuous pursuit.

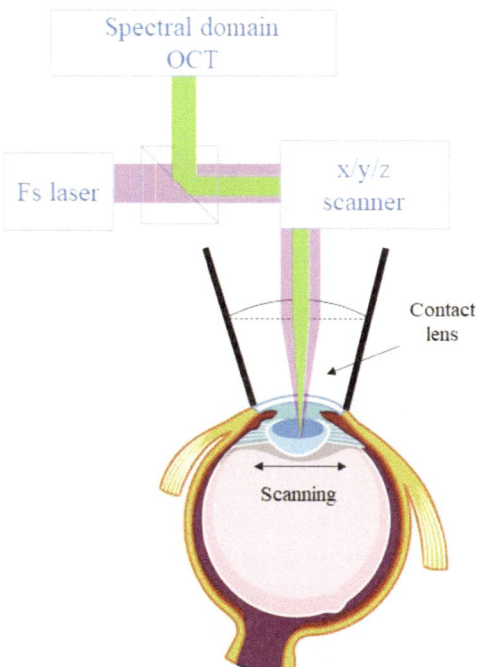

FIGURE 12.2 Diagram of the optical and mechanical interface.

12.2.1.3 Other Laser Surgeries

Actually, besides laser eye surgery and orthopedic surgery, laser surgery is also adopted in other applications such as removing spots. Similar to the earlier demonstration, characteristics of high precision, low damage, easy integration, and so on are preferred. With the further development of laser techniques and cost reduction, it is believed that laser surgery could become more widespread.

12.2.2 LASER MODIFICATION OF BIOMATERIALS FOR CELL RESEARCH

Biomaterials constructed for mimicking of extracellular matrix (ECM) environment which support cell culture are an important part of the wide endeavor in advancing medical research. The preparation of more relevant models with various structures to study cell behavior is an important issue (Slepicka et al. 2017; Liu et al. 2013; Wang, Li, et al. 2015; Wang et al. 2013). Many approaches are used to modify the structure of materials to produce order or disorder patterns or scaffolds from microscale to nanoscale (Anselme et al. 2011; Huang et al. 2020). Lasers are highly advantageous instruments for materials engineering as they may be used for the processing of a broad range of materials, including metals, ceramics, polymers, and natural biomaterials (Skoog et al. 2013; Rebollar et al. 2015). By selectively ablating specific regions of biomaterials, laser processing could be used to generate microscale to sub-microscale structures from one dimension to three dimensions. Laser processing could topographically functionalize biomaterial surfaces, which is beneficial for optimizing cell responses and enhancing biocompatibility (Fadeeva et al. 2014). The detailed advantages of laser modified surfaces of different biomaterials are listed in Table 12.2. In summary, special structures such as laser induced periodic surface structure (LIPSS), and finely designed micropatterns which couldn't be achieved by traditional methods can be obtained by laser processing. The technique enlarges the research areas of cell and materials interactions.

TABLE 12.2
Microscale Modification of Biomaterials with Laser Process Technique.

Structures	Materials	Advantages	References
1D micropatterned lines	PVA	Generates simpler and more representative system for study cell migration.	(Doyle et al. 2009)
2D micropatterns	DNA and proteins	Maintains properties and biological functions of biomoleculars.	(Zergioti et al. 2005)
2D micropatterns	Fibronectin protein, bovine serum albumin	Precisely controlled cells adhere to target area.	(Da Sie et al. 2015)
3D micropatterned structure on surface on biomaterials	Fibronectin protein, bovine serum albumin	3D structures offer great potential for simulating real in vivo environments.	(Da Sie et al. 2015) (Jonušauskas et al. 2019)
3D surface topography with controlled geometry parameters	ORMOCER® US-S4	Precise fabrication of 3D topography with various geometry parameters is a powerful technique for measuring contractile forces of single cells and understanding interaction of cells to extracellular matrix.	(Jeon et al. 2010) (Jeon et al. 2011)
3D microgrooves and osteon mimicking structure	VT6 titanium	Develops surfaces more beneficial for stem cells differentiation.	(Veiko et al. 2020)
3D scaffolds	Acrylated poly(ethylene glycol)	Varies the arrangement of cells, and emulates the 3D assembly of natural tissues.	(Ovsianikov et al. 2010)

Structures	Materials	Advantages	References
3D microfluidic structures	SU-8 photoresist, Foturan glass	1. The surface quality of glass could be improved for integration of polymer microstructures. Microchannels with arbitrary cross-sectional shapes could be created. 2. True 3D microchannels different from previous 2D planar channels could be achieved, which provides a more biomimetic 3D environment.	(Wu et al. 2014) (Sima et al. 2017) (Sima et al. 2018)
3D sub-microscale LIPSS	Polystyrene	Compared to spin-coated polystyrene, hydrophilic laser-irradiated polystyrene accelerates C6 cells to attach and adhere, inhibits cells from migrating, and promotes cell proliferation.	(Hwang et al. 2008)
	Ti, 15(Ti/Zr)/Si	Induces a high degree of cell orientation and proliferation.	(Oya et al. 2012) (Petrović et al. 2020)
	PET	Varies different types of structures by regulating parameters of laser beams. The influence of different surface properties on cell behavior could be evaluated.	(Rebollar et al. 2014)
Regular sub-microscale pits	Quartz	Special arrangement of submicro-pits could generate cell repellent surfaces.	(Jeon et al. 2015)
Scaffolds with microscale configuration	PEG-DA/PETA, Fibronectin	Full control of cell adhesion and cell shape in the three dimensions.	(Klein et al. 2011)

12.3　ON-LINE MONITORING OF LASER PROCESSING

With the development of modern applications, the requirements of laser processing with complex processing environments and objects, large dynamic range, high precision and efficiency, and damage-free operations are becoming more and more urgent (Sony et al. 2019). The traditional off-line measurement (i.e. measuring the processing quality after processing) significantly hinders the practical applications of laser processing. The on-line monitoring and real-time optimization of laser processing parameters are important solutions (Huang et al. 2015). Meanwhile, a variety of acoustic, optical, temperature, and image signals are generated during the process of laser irradiation on the materials (Ding et al. 2018). The typical signals include the plasma signals corresponding to the chemical components (Skruibis et al. 2019); harmonic signals generated during the interaction between laser and highly nonlinear materials (Hu et al. 2019); fluorescence and Raman signals when a laser irradiates certain materials (Moretti et al. 2019; Allen et al. 2017); reflected light signals related to the surface roughness, chemical components, and material damage (Moretti et al. 2019; Lee, Watkins, et al. 2001; Lee et al. 2000; Lee, and Steen 2001; Marimuthu et al. 2010); acoustic signals induced by changes in regional structure and volume during laser irradiation (Orzi et al. 2013; Gu et al. 1996; Hong et al. 1997); the temperature signals related to thermal effect (Marshall et al. 2016); and so on. According to the mapping relationship among the monitoring signals, processing parameters, processing status, and target properties, several thresholds, including generation threshold of monitoring signals, ablation threshold, switch threshold, and damage threshold, are first proposed. Correspondingly, processing parameters including pulse energy, laser fluence, focus, scanning speed, overlapping, cooling rate, cooling medium are adjusted according to the indicators of the thresholds. Here, the unique characteristics of on-line monitoring techniques of laser processing are respectively described here. The typical monitoring signals during laser processing and their characteristics are listed as shown in Table 12.3.

TABLE 12.3
Typical Monitoring Signals during Laser Processing.

Signal Types	Monitoring Signals	Monitoring Equipment	Monitoring Targets	Advantages	Limitations
Optical signals	Plasma signals	Optical spectrum analyzer	Element type and content (Cabalín et al. 1998), thermal effect (Deng et al. 2006), focus adjustment (Skruibis et al. 2019; Diego-Vallejo et al. 2013)	High precision and accuracy, immunity to environmental disturbance, wide range of application scenarios	Short working distance Low efficiency in identifying slight defect, accuracy depending on the plume behavior
	Reflected light signals	Optical spectrum analyzer, photodiode	Degree of cleaning, damage, chroma	Long working distance, immunity to the environmental disturbance, simple setup	Low resolution, difficulty in phase analysis Limited application scenarios, specific materials and measurement setup
	Nonlinear optical signals	Spectrometer	Material type, damage	Damage-free signals, additional signals for material analysis	
Acoustic signals	Sound and ultrasonic waves	Piezoelectric ceramic, microphone detection	Ablative thresholds, removal efficiency and matrix damage, defect identification and control	Simple measurement setup, fast response time	Sensitivity to the environmental disturbance
Image signals	Image	CCD camera, CMOS camera, optical coherence tomography	Surface morphology, roughness and size, plasma plume	Wide range of application scenarios, simple operation	Low sampling speed, high computing demands, invalidation in high-brightness conditions
Temperature signal	Temperature	Infrared camera, pyrometer, thermocouple	Defect, microstructure, microstructure, molten pool morphology	Wide range of application scenarios	Relatively slow response time, indirect measurement of material property

12.3.1 ON-LINE MONITORING BASED ON OPTICAL SIGNALS

Optical signals exist in almost any laser processing. Spectra, intensity and intensity ratios of optical signals are generally adopted to reveal the types, concentration, and distribution of elements (Vadillo et al. 2004), the real-time adjustment of focus (Skruibis et al. 2019; Diego-Vallejo et al. 2013), monitoring of plasma temperature and electron density (Deng et al. 2006; Kong et al. 2012), surface topology (Moretti et al. 2019; Lee, Watkins, et al. 2001; Lee et al. 2000; Lee, and Steen 2001; Marimuthu et al. 2010), and so on. In 1998, Cabalín et al. proposed the thresholds of plasma generation, which built the foundation of plasma monitoring (Cabalín et al. 1998). Subsequently, the plasma monitoring was first applied in laser welding by Connolly et al. (Connolly et al. 2000). In 2002, Hong et al. further studied the plasma length under different cavity pressure, which revealed the generation threshold and intensity change under different laser fluence (Hong et al. 2002). Then, plasma spectroscopy was proposed for laser processing. Except for the composition analysis, Deng et al. demonstrated that the characteristic spectral intensity of element, plasma temperature, and shape were directly related to the thermal effect during processing, which could be used for the monitoring of thermal effect (Deng et al. 2006). Furthermore, in 2013, Diego et al. proposed the adjustment method of focus based on the intensity of characteristic lines of plasma spectra (Diego-Vallejo et al. 2013). Based on this research, plasma monitoring started to be widely applied. Meanwhile, as a non-destructive monitoring method over a relatively long distance, reflection spectroscopy can effectively monitor the cleanliness, damages, chroma, and compositional changes of the material surface by measuring spectral integral power of the reflected light signal in a specific band, position, and intensity of characteristic spectral peaks and bands (Moretti et al. 2019; Lee, Watkins, et al. 2001; Lee et al. 2000; Lee, and Steen 2001; Marimuthu et al. 2010). In 2017, Allen et al. first demonstrated the on-line monitoring based on Raman signals, which provided a new on-line monitoring method based on the nonlinear optical signals (Allen et al. 2017). The nonlinear optical signals excited under certain conditions, including the harmonic signals (Song et al. 2020), the fluorescence signals (Moretti et al. 2019), and the Raman signals (Allen et al. 2017), can also provide additional methods of spectral measurements. For example, Song et al. proposed a smart femtosecond laser bone processing system integrated with the second-harmonic-generation (SHG) green positioning and spectral monitoring functions (Song et al. 2020). Since the main inorganic component of bone is hydroxyapatite with high second-order susceptibility, SHG green signal centered at half wavelength of the laser source was first generated at lower fluence and provided a direct-viewing and label-free tool for initial positioning. With the increase of laser fluence, the color of excited optical signals continuously turned from green to red and the spectral intensity ratio of the green and other visible components continuously changed due to accelerated removal of hydroxyapatite and enhanced ablation process. Besides, the abnormal enhancement of the red component ranged from 700 to 800 nm could be observed when various degrees of carbonization occurred. Hence, the spectral intensity ratio of the red component and the rest of the visible spectral component were also calculated to distinguish the thermal damage. With real-time feedback, control of color evolution and the previously mentioned two spectral ratios, the focal length, water cooling, and laser fluence could be effectively optimized. Although their application scenarios are limited, they provide a new monitoring method for component analysis, focus, and material damage.

12.3.2 On-Line Monitoring Based on Photoacoustic Signals

The photoacoustic phenomenon was found in the 1880s. In 1996, Lu et al. first demonstrated the on-line monitoring of laser processing based on acoustic signals (Lu et al. 1996). When the laser with sufficient laser fluence irradiates the materials, shock waves containing sound and ultrasonic waves are excited. Some photon energy is absorbed and transferred to the lattice. The instantaneous increase of temperature can induce the thermoelastic effect, which can excite a shock wave inside the material. In addition, the materials evaporate rapidly and local particle density increases in the region where the laser interacts with the matter. The increase of local pressure can also induce the shock wave. Besides, the resulting slag splashes outward at high speed, causing a strong recoiling force inside the materials (Schmid 2006).

Since acoustic signals can be directly measured by a microphone or piezoelectric ceramic, it is a simple monitoring tool with high-speed response time. The frequency and amplitude of specific acoustic and ultrasonic signals are generally adopted as the indicators of processing thresholds. Although it has been applied to realize real-time monitoring of the ablation threshold (Hong et al. 1997), removal efficiency, matrix damage (Tserevelakis et al. 2019), and defect recognition (Ye, Fuh, et al. 2018; Ye, Hong, et al. 2018), it is easily affected by environmental noise, which hinders the further widespread applications (Ding et al. 2018; Orzi et al. 2013).

12.3.3 On-Line Monitoring Based on Image Signals

The on-line monitoring based on image signals was immediately proposed when the laser processing techniques started to be applied. In 1995, Akira et al. proposed the on-line monitoring of material damage during laser processing by using laser speckle images (Akira et al. 1995). Since the information such as the surface topography, roughness, size, and plasma plume can be directly observed with CCD or CMOS camera, OCT technique (Palanker et al. 2010), on-line monitoring of images are suitable for almost all kinds of laser processing techniques, including laser cutting and drilling (Wen et al. 2012; Douplik et al. 2010), laser cleaning (Kane et al. 2003), micro- and nano-structure fabrication (Zeng et al. 2017). Nevertheless, in a situation with high brightness, it will not work. Besides, the measurement and analysis of image characteristics such as gray scale and texture need to match the image processing algorithms, such as Gaussian filtering, edge enhancement, tracking algorithm, and so on (Jinqiang et al. 2011). Recently, by introducing artificial intelligence algorithms such as steady-state and dynamic neural network algorithms, the processing efficiency and speed have been improved (Luo et al. 2015; Tan et al. 2013). However, the relatively long processing time is still needed, which hinders the on-line monitoring and real-time feedback for high-speed and efficient laser processing.

12.3.4 ON-LINE MONITORING BASED ON TEMPERATURE SIGNALS

As one of the most classic effects during processing, the thermal effect is directly related to the processing status and quality. The temperature information is the key indicator of this monitoring method. In 1997, Schanwald et al. first proposed the on-line monitoring of laser processing based on temperature signals (Schanwald 1997). The temperature signals during laser processing can be measured using the infrared camera, pyrometer, and thermocouple. The transient and dynamic variance of temperature and its spatial and temporal distribution of pool, steam, and plasma can be measured and analyzed. Then, the defects (Schwerdtfeger et al. 2012), microstructure and molten pool morphology, and so on could be deduced indirectly. It is suitable for laser processing closely related to the thermal effect such as laser cladding (Marshall et al. 2016) and laser welding (Fanrong et al. 2012). Nevertheless, it is still an indirect monitoring method based on temperature information, and the implementation of more accurate monitoring still requires the combination of the chemical composition and physical parameters of materials (Pregowski et al. 2003).

12.3.5 ON-LINE MONITORING BASED ON OTHER SIGNALS

Besides the earlier mentioned signals, on-line monitoring methods based on some special signals for certain situations have also been reported. For example, Song et al. utilized air free particle counters for measuring particles ejected during laser cleaning to monitor the cleaning thresholds, efficiency, and cleanliness (Song et al. 2003). However, injection particles tend to contaminate peripheral optical equipment and damage operators' health. Although some other signals are demonstrated for on-line monitoring, they are not as widespread as previous signals and can only be implemented for some specific scenarios and applications. Nevertheless, it could still be an innovative idea for on-line monitoring.

12.3.6 PROSPECT OF ON-LINE MONITORING OF LASER PROCESSING

In order to monitor and control the laser processing more comprehensively and precisely, the collaborative monitoring of multiple signals is proposed. In 2002, Hong et al. first proposed the composite monitoring of acoustic, optical, and electrical signals to optimize the laser ablation process. The microphone, phototube, metal probe, spectrometer, and high-speed camera are adopted to measure acoustic, optical, and electrical signals in laser ablation (Hong et al. 2002). Song et al. proposed the collaborative monitoring of the spectral and temperature signals to ensure damage-free bone drilling (Song et al. 2020). Although the collaborative monitoring of optical signals and other signals can complement each other, more measuring equipment is additionally introduced. It will limit the practical applications and popularization of this method in terms of cost, volume, and system complexity. Therefore, it is necessary to carry out a series of research studies from the aspects of system cost, volume, system complexity, on-line monitoring algorithm, computation, and response time.

In addition, on-line monitoring is to build the mapping relationship among monitoring signals, laser processing parameters, processing status, and processing quality. Therefore, abundant experiments are required to establish a relationship. On the other hand, artificial intelligence algorithms such as neural networks (Barletta et al. 2006), support vector machines (SVM) (Lee et al. 2020), and deep belief networks (Ye, Hong, et al. 2018) are effective ways for data preprocessing, denoising, and feature extraction. These will be beneficial for high-speed and high-resolution on-line monitoring with a little calculation. Besides, self-study, self-diagnosis, and self-optimization after training will accelerate the intelligence of on-line monitoring and high-efficiency laser processing.

12.4 SUMMARY

In summary, due to the unique characteristics of laser processing, it has shown high potential in biomedical applications such as laser surgery and laser modification. Laser orthopedic surgery and eye surgery are depicted as the typical tools of laser surgeries. Besides, laser modification of biomaterials is widely considered to be an effective way to control cell behavior of surfaces due to its advantages of non-pollution, low cost with high reproducibility and uniformity, high enhancement factor and capability to tailor topography at the micro- and nanoscales. Moreover, the on-line monitoring and real-time optimization of laser processing parameters could further promote the practical biomedical applications requiring large dynamic range, high efficiency, and high precision. The characteristics of on-line monitoring based on the optical, acoustic, image, temperature, and some other signals are discussed in detail here. Besides, with the further combination of collaborative monitoring and artificial intelligence, the intellectualization and high efficiency of laser processing and its on-line monitoring will be promoted.

12.5 ACKNOWLEDGMENTS

This work was supported in part by National Natural Science Foundation of China (62105038, 61903042, and 62005020), Beijing Natural Science Foundation (3204047), China Postdoctoral Science Foundation (2019M650423).

REFERENCES

Abbasi, H., G. Rauter, R. Guzman, P.C. Cattin, and A. Zam. 2018. Laser-induced breakdown spectroscopy as a potential tool for autocarbonization detection in laserosteotomy. *Journal of Biomedical Optics* 23:071206.

Akira, K., K. Mitsuo, and N. Ichiro. 1995. Damage monitoring of metal materials by laser speckle assisted by image processing techniques: Relationship between distribution of laser speckle and surface properties *Jsme International Journal. ser. a Mechanics & Material Engineering* 38:249–257.

Alio, J. 2014. Refractive surgery today: Is there innovation or stagnation? *J. Eye Vision* 1:4.

Aljdaimi, A., H. Devlin, M. Dickinson, A. Alfutimie, and C. Mao. 2018. Effect of 2.94 m Er: YAG laser on the chemical composition of hard tissues. *Microscopy Research & Technique* 81(8):1–10.

Allen, F.I., E. Kim, N.C. Andresen, C.P. Grigoropoulos, and A.M. Minor. 2017. In situ TEM Raman spectroscopy and laser-based materials modification. *Ultramicroscopy* 178:33–37.

Andrea, A., A. Benedicenti, S Ravera, et al. 2017. Short-pulse neodymium:yttrium-aluminium garnet (Nd:YAG 1064nm) laser irradiation photobiomodulates mitochondria activity and cellular multiplication of Paramecium primaurelia (Protozoa). *European Journal of Protistology* 61(Pt A):294–304.

Anselme, K., and M. Bigerelle. 2011. Role of materials surface topography on mammalian cell response. *International Materials Reviews* 56:243–266.

Augello, M., W. Deibel, K. Nuss, P. Cattin, and P. Jurgens. 2018. Comparative microstructural analysis of bone osteotomies after cutting by computer-assisted robot-guided laser osteotome and piezoelectric osteotome: An in vivo animal study. *Lasers in Medical Science* 33:1471–1478.

Barletta, M., and A. Gisario. 2006. An application of neural network solutions to laser assisted paint stripping process of hybrid epoxy-polyester coatings on aluminum substrates. *Surface & Coatings Technology* 200:6678–6689.

Barsch, N., K. Korber, A. Ostendorf, and K.H. Tonshoff. 2003. Ablation and cutting of planar silicon devices using femtosecond laser pulses. *Applied Physics A Materials Science & Processing* 77:237–242.

Bernal, L.M.B., I.T. Schmidt, N. Vulin, J. Widmer, J.G. Snedeker, et al. 2018. Optimizing controlled laser cutting of hard tissue (bone). *At-Automatisierungstechnik* 66:1072–1082.

Blaskovic, M., D. Gabric, N.J. Coleman, I.J. Slipper, M. Mladenov, et al. 2016. Bone healing following different types of osteotomy: Scanning electron microscopy (SEM) and three-dimensional SEM analyses. *Microscopy and Microanalysis* 22:1170–1178.

Boyd, K., S. Rees, N. Simakov, J.M.O. Daniel, R. Swain, et al. 2015. High precision 9.6 μm CO_2 laser end-face processing of optical fibres. *Optics Express* 23:15065–15071.

Cabalín, L.M., and J.J. Laserna. 1998. Experimental determination of laser induced breakdown thresholds of metals under nanosecond Q-switched laser operation. *Spectrochimica Acta Part B: Atomic Spectroscopy* 53:723–730.

Cao, H., H. Peng, M. Zhang, Y. Chen, and H. Tan. 2010. Laser diode array (LDA) end-pumped multi-watt Yb:YAG 1030 nm laser. *Optica Applicata* 40:653.

Ci-Ling, P., Z., Alexey, C. Lin, Y.J. You, 2015. Progress in short-pulse Yb-doped fiber oscillators and amplifiers. *Current Trends of Optics and Photonics* 129: 61.

Changshu, L., L. Qigui, H. Donghong, L. zheng, S. Yunlong, et al. 2015. Testing and research of drilling feed force on fresh corpse femoral. *Basic Research of Digital Medicine* 10:57–61.

Changshu, L., B. Yuzhe, K. Xiangxue, C. Lan, L. Jianyi, et al. 2014. Testing of drilling feed force on fresh porcine femur. *Journal of Medical Biomechanics* 29:560–566.

Chelnokov, E., L. Soustov, N. Sapogova, M. Ostrovsky, and N. Bityurin. 2008. Nonreciprocal XeCl laser-induced aggregation of beta-crystallins in water solution. *Optics Express* 16:18798–18803.

Christof, J., E. Hadj, and G.B. Julian. 2018. A new photometric ozone reference in the Huggins bands: The absolute ozone absorption cross section at the 325nm HeCd laser wavelength. *Atmospheric Measurement Techniques* 11:1707–1723.

Connolly, J.O., G.J. Beirne, G.M. O'Connor, T.J. Glynn, and A.J. Conneely. 2000. Optical monitoring of laser generated plasma during laser welding. *Laser Plasma Generation and Diagnostics* 3935:132–138.

Cui, J., Y. Liu, J. Zhang, and H. Yan. 2014. An experimental study on choroidal neovascularization induced by Krypton laser in rat model. *Photomedicine & Laser Surgery* 32:30.

Da Sie, Y., Y.-C. Li, N.-S. Chang, P.J. Campagnola, and S.-J. Chen. 2015. Fabrication of three-dimensional multi-protein microstructures for cell migration and adhesion enhancement. *Biomedical Optics Express* 6:480–490.

Davoudi, A., M. Amrolahi, and H. Khaki. 2018. Effects of laser therapy on patients who underwent rapid maxillary expansion: A systematic review. *Lasers in Medical Science* 33:1387–1395.

Deng, Y.Z., H.Y. Zheng, V.M. Murukeshan, and W. Zhou. 2006. Analysis of optical emission towards optimisation of femtosecond laser processing. *Journal of Laser Micro Nanoengineering* 1:136–141.

Diego-Vallejo, D., D. Ashkenasi, and H.J. Eichler. 2013. Monitoring of focus position during laser processing based on plasma emission. In *Lasers in Manufacturing*, edited by C. Emmelmann, M.F. Zaeh, T. Graf and M. Schmidt, 904–911. Amsterdam: Elsevier Science Bv.

Ding, Y., Y. Xue, J. Pang, L. Yang, and M. Hong. 2018. Advances in in-situ monitoring technology for laser processing. *SCIENTIA SINICA Physica, Mechanica & Astronomica* 49:60–78.

Douplik, A., A. Zam, R. Hohenstein, A. Kalitzeos, E. Nkenke, et al. 2010. Limitations of cancer margin delineation by means of autofluorescence imaging under conditions of laser surgery. *Journal of Innovation in Optical Health Science* 3:45–51.

Doyle, A.D., F.W. Wang, K. Matsumoto, and K.M. Yamada. 2009. One-dimensional topography underlies three-dimensional fibrillar cell migration. *Journal of Cell Biology* 184:481–490.

Duperron, M., K. Grygoryev, G. Nunan, C. Eason, R. Burke, et al. 2019. Diffuse reflectance spectroscopy-enhanced drill for bone boundary detection. *Biomedical Optics Express* 10:961–977.

Edmonds, A.M., M.A. Sobhan, V.K.A. Sreenivasan, E.A. Grebenik, J.R. Rabeau, et al. 2013. Nano-ruby: A promising fluorescent probe for background-free cellular imaging. *Particle & Particle Systems Characterization* 30:506–513.

Fadeeva, E., A. Deiwick, B. Chichkov, and S. Schlie-Wolter. 2014. Impact of laser-structured biomaterial interfaces on guided cell responses. *Interface Focus* 4:20130048.

Fanrong, K., and, J. Ma, et al. 2012. Real-time monitoring of laser welding of galvanized high strength steel in lap joint configuration. *Optics and Laser Technology* 44:2186–2196.

Fibrich, M., J. Ulc, R. Vejkar, and H. Jelínková. 2021. Continuous-wave efficient cyan-blue Pr:YAlO3 laser pumped by InGaN laser diode. *Applied Physics B* 127:1–6.

Gautam, G.D., and A.K. Pandey. 2018. Pulsed Nd:YAG laser beam drilling: A review. *Optics and Laser Technology* 100:183–215.

Goodno, G.D., H. Komine, S.J. McNaught, S.B. Weiss, S. Redmond, et al. 2006. Coherent combination of high-power, zigzag slab lasers. *Optics Letters* 31:1247–1249.

Grewal, D., T. Schultz, S. Basti, and H. Dick. 2015. Femtosecond laser assisted cataract surgery: Current status and future directions. *Survey of Ophthalmology* 61:00154-X.

Gu, H., R.E. Mueller, and W. Duley. 1996. Acoustic monitoring of modulated laser beam processing of metals. In *Lasers as tools for manufacturing of durable goods and microelectronics*, edited by J.J. Dubowski, J. Mazumder, L.R. Migliore, C.S. Roychoudhuri and R.D. Schaeffer, Vol. 2703. Bellingham: Spie-Int Soc Optical Engineering.

Hong, M.H., Y.F. Lu, and T.C. Chong. 2002. Diagnostics and real-time monitoring of pulsed laser ablation. In *Second international symposium on laser precision microfabrication*, edited by I. Miyamoto, Y.F. Lu, K. Sugioka and J.J. Dubowski, 51–54. Bellingham: Spie-Int Soc Optical Engineering.

Hong, M.H., Y.F. Lu, W.D. Dong, D.M. Liu, and T.S. Low. 1997. Audible acoustic wave real-time monitoring in laser processing of microelectronic materials. *Microelectronic Packaging and Laser Processing*, edited by Y.K. Swee, H.Y. Zheng and R.T. Chen, Vol. 3184. Bellingham: Spie-Int Soc Optical Engineering.

Hu, G., K. Guan, L. Lu, J. Zhang, N. Lu, et al. 2018. Engineered functional surfaces by laser microprocessing for biomedical applications. *Engineering* 4:822–830.

Hu, G., Y. Song, Z. Zheng, and Y. Guan. 2019. Femtosecond laser bone drilling with the second-harmonic-generation green positioning and on-line spectral monitoring. Frontiers in Optics + Laser Science APS/DLS, Washington, DC, 2019/09/15.

Huang, H., L.M. Yang, S. Bai, and J. Liu. 2015. Smart surgical tool. *Journal of Biomedical Optics* 20:7.

Huang, W., C. Li, L. Gao, Y. Zhang, Y. Wang, et al. 2020. Emerging black phosphorus analogue nanomaterials for high-performance device applications. *Journal of Materials Chemistry C* 8:1172–1197.

Hui, L., S. Jia, T. Zhao, and H. Ying. 2018. Skeletal and reduced chemical mechanism for hydrogen fluoride chemical laser. *Journal of Mathematical Chemistry* 56:1–16.

Hwang, D.J., N. Misra, C.P. Grigoropoulos, A.M. Minor, and S.S. Mao. 2008. In situ monitoring of laser cleaning by coupling a pulsed laser beam with a scanning electron microscope. *Applied Physics A Materials Science & Processing* 91:219–222.

Ito, K., S. Ishizaka, T. Sasaki, T. Miyahara, T. Horiuchi, et al. 2009. Safe and minimally invasive laminoplastic laminotomy using an ultrasonic bone curette for spinal surgery: Technical note. *Surgical Neurology* 72:470–475.

Jeon, H., H. Hidai, D.J. Hwang, K.E. Healy, and C.P. Grigoropoulos. 2010. The effect of microscale anisotropic cross patterns on fibroblast migration. *Biomaterials* 31:4286–4295.

Jeon, H., E. Kim, and C. Grigoropoulos. 2011. Measurement of contractile forces generated by individual fibroblasts on self-standing fiber scaffolds. *Biomedical Microdevices* 13:107–115.

Jeon, H., S. Koo, W.M. Reese, P. Loskill, C.P. Grigoropoulos, et al. 2015. Directing cell migration and organization via nanocrater-patterned cell-repellent interfaces. *Nat Mater* 14:918–923.

Jeong, J., S. Cho, S. Hwang, B. Lee, and T.J. Yu. 2019. Modeling and analysis of high-power Ti:sapphire laser amplifiers: A review. *Applied Sciences* 9:2396.

Jianjun, Y. 2004. Femtosecond laser "cold" micro-machining and its advanced applications. *Laser & Optoelectronics Progress* 41:42–52.

Jiaru, Z., H. Guoqing, L. Libin, G. Yingchun, and M.H. Hong. 2019. Enhancing protein fluorescence detection through hierarchical biometallic surface structuring. *Optics Letters* 44:339–342.

Jinqiang, G., Q. Guoliang, Y. Jialin, H. Jianguo, Z. Tao, et al. 2011. Image processing of weld pool and keyhole in Nd:YAG laser welding based on edge predicting. *China Welding* 20:67–70.

Jivrajka, R.V., M.C. Shammas, and H.J. Shammas. 2012. Improving the second-eye refractive error in patients undergoing bilateral sequential cataract surgery. *Ophthalmology* 119:1097–1101.

Jonušauskas, L., D. Gailevicius, S. Rekštytė, T. Baldacchini, and S. Juodkazis. 2019. Mesoscale laser 3D printing. *Optics Express* 27:15205.

Kane, D.M., A.J. Fernandes, and R.P. Mildren. 2003. Optical microscopy imaging and image-analysis issues in laser cleaning. *Applied Physics A Materials Science & Processing* 77:847–853.

Kawata, S., H.B. Sun, T. Tanaka, and K. Takada. 2001. Finer features for functional microdevices. *Nature* 412:697–698.

Kitamura, R., L. Pilon, and M. Jonasz. 2007. Optical constants of silica glass from extreme ultraviolet to far infrared at near room temperature. *Appl Opt* 46:8118–8133.

Klein, F., B. Richter, T. Striebel, C.M. Franz, G. von Freymann, et al. 2011. Two-component polymer scaffolds for controlled three-dimensional cell culture. *Adv Mater* 23:1341–1345.

Ko, J.H., H.S. Chi, I. Nam, D. Na, and H.S. Kang. 2021. Two-dimensional tilt control of electron bunch for X-ray free electron laser. *Nuclear Instruments & Methods in Physics Research* 986:164726.

Kong, F., J. Ma, B. Carlson, and R. Kovacevic. 2012. Real-time monitoring of laser welding of galvanized high strength steel in lap joint configuration. *Optics & Laser Technology* 44:2186–2196.

Kurtz, R.M., C. Horvath, H.H. Liu, R.R. Krueger, and T. Juhasz. 1998. Lamellar refractive surgery with scanned intrastromal picosecond and femtosecond laser pulses in animal eyes. *Journal of Refractive Surgery* 14:541–548.

Lee, J.M., and W.M. Steen. 2001. In-process surface monitoring for laser cleaning processes using a chromatic modulation technique. *International Journal of Advanced Manufacturing Technology* 17:281–287.

Lee, J.M., W.M. Steen, and K.G. Watkins. 2000. Chromatic surface monitoring and diagnostic system for laser cleaning process. In *Icaleo*, edited by P. Christensen, Vol. 87. Orlando: Laser Inst America.

Lee, J.M., K.G. Watkins, and W.M. Steen. 2001. In-process chromatic monitoring in the laser cleaning of marble. *Journal of Laser Applications* 13:19–25.

Lee, S.H., J. Mazumder, J. Park, and S. Kim. 2020. Ranked feature-based laser material processing monitoring and defect diagnosis using k-NN and SVM. *Journal of Manufacturing Processes* 55:307–316.

Lei, P., C. Yang, M. Wu, M. Wu, K.Y. Cheng, et al. 2006. Effects of n-type modulation-doping barriers and a linear graded-composition GaInAsP intermediate layer on the 1.3 μm AlGaInAs/AlGaInAs strain-compensated multiple-quantum-well laser diodes. *Journal of Vacuum Science & Technology B* 24:623.

Levesque, L., and A. Robaczewski. 2017. Very accurate temperature control of bones by a CO2 laser for medical applications. *Applied Optics* 56:3923–3928.

Li, N., W.Y. Zhang, J. Zhang, M. Guo, and Z.X. Guo. 2020. Mode-locked Er-doped fiber laser based on non-linear multimode interference. *Laser Physics Letters* 17:085105.

Li, X., X.A. Dou, H. Zhu, Y. Hu, and X. Wang. 2020. Nanosecond laser-induced surface damage and its mechanism of CaF2 optical window at 248nm KrF excimer laser. *Scientific Reports* 10:5550.

Li, Y., M. Li, Y. Utaka, C. Yang, and M. Wang. 2020. Effect of copper surface modification applied by combined modification of metal vapor vacuum arc ion implantation and laser texturing on anti-frosting property. *Energy & Buildings* 223.

Li, Z.J., K.K. Xu, S.B. Wu, J. Lv, D. Jin, et al. 2014. Population-based survey of refractive error among school-aged children in rural northern China: The Heilongjiang eye study. *Clinical and Experimental Ophthalmology* 42:379–384.

Liang, Y.B., T.Y. Wong, L.P. Sun, Q.S. Tao, J.J. Wang, et al. 2009. Refractive errors in a rural Chinese adult population the Handan eye study. *Ophthalmology* 116:2119–2127.

Liao, Z., D.A. Axinte, and G. Dong. 2017. A novel cutting tool design to avoid surface damage in bone machining. *International Journal of Machine Tools and Manufacture* 116:52–59.

Liu, M.H., K.H. Nan, and Y.J. Chen. 2021. The progress in thermogels based on synthetic polymers for treating ophthalmic diseases. *Acta Polymerica Sinica* 52:47–60.

Liu, X., D. Han, Z. Sun, C. Zeng, H. Lu, et al. 2013. Versatile multi-wavelength ultrafast fiber laser mode-locked by carbon nanotubes. *Scientific Reports* 3:2718.

Lo, D.D., M.A. Mackanos, M.T. Chung, J.S. Hyun, D.T. Montoro, et al. 2012. Femtosecond plasma mediated laser ablation has advantages over mechanical osteotomy of cranial bone. *Lasers Surg Med* 44:805–814.

Lu, L.B., J.R. Zhang, L.S. Jiao, and Y.C. Guan. 2019. Large-scale fabrication of nanostructure on bio-metallic substrate for surface enhanced Raman and fluorescence scattering. *Nanomaterials* 9:14.

Lu, Y.F., M.H. Hong, S.J. Chua, B.S. Teo, and T.S. Low. 1996. Audible acoustic wave emission in excimer laser interaction with materials. *Journal of Applied Physics* 79:2186–2191.

Luo, M.S.Y., and Y. Shin. 2015. Estimation of keyhole geometry and prediction of welding defects during laser welding based on a vision system and a radial basis function neural network. *International Journal of Advanced Manufacturing Technology* 81:263–276.

MacDougall, D., J. Farrell, J. Brown, M. Bance, and R. Adamson. 2016. Long-range, wide-field swept-source optical coherence tomography with GPU accelerated digital lock-in doppler vibrography for real-time, in vivo middle ear diagnostics. *Biomedical Optics Express* 7:4621–4635.

Maiman, T.H. 2018. *The laser inventor: Memoirs of Theodore H. Maiman.* Gewerbestrasse: Springer Nature.

Marimuthu, S., A.M. Kamara, D. Whitehead, P. Mativenga, and L. Li. 2010. Laser removal of TiN coatings from WC micro-tools and in-process monitoring. *Optics and Laser Technology* 42:1233–1239.

Marshall, G.J., W.J. Young, S.M. Thompson, N. Shamsaei, S.R. Daniewicz, et al. 2016. Understanding the microstructure formation of Ti-6Al-4V during direct laser deposition via in-situ thermal monitoring. *JOM* 68:778–790.

Martins, G.L., E. Puricelli, C.E. Baraldi, and D. Ponzoni. 2011. Bone healing after bur and Er:YAG laser ostectomies. *Int. J. Oral Maxillofac. Surg.* 69:1214–1220.

Mathieu, O., L.T. Pinzon, T.M. Atherley, C.R. Mulvihill, I. Schoel, et al. 2019. Experimental study of ethanol oxidation behind reflected shock waves: Ignition delay time and H_2O laser-absorption measurements. *Combustion & Flame* 208:313–326.

McAlinden, C. 2012. Corneal refractive surgery: Past to present. *Clinical & Experimental Optometry* 95:386–398.

McDonald, M. 1990. Central photorefractive keratectomy for myopia. *Archives of Ophthalmology* 108:799.

Mihalko, W.M., and J.L. Williams. 2012. Total knee arthroplasty kinematics may be assessed using computer modeling: A feasibility study. *Orthopedics* 35:40–44.

Moretti, P., M. Iwanicka, K. Melessanaki, E. Dimitroulaki, O. Kokkinaki, et al. 2019. Laser cleaning of paintings: in situ optimization of operative parameters through non-invasive assessment by Optical Coherence Tomography (OCT), reflection FT-IR spectroscopy and Laser Induced Fluorescence spectroscopy (LIF). *Heritage Science* 7:12.

Mosquera, S.A., and J.L. Alió. 2014. Presbyopic correction on the cornea. *Eye and Vision* 1:5.

Nagy, Z., A. Takacs, T. Filkorn, and M. Sarayba. 2009. Initial clinical evaluation of an intraocular femtosecond laser in cataract surgery. *Journal of Refractive Surgery* 25:1053–1060.

Ohtsu, M., H. Suzuki, K. Nemoto, and Y. Teramachi. 1990. Narrow-linewidth tunable visible InGaAlP laser, application to spectral measurements of lithium, and power amplification. *Japanese Journal of Applied Physics* 29:L1463–L1465.

Orzi, D.J.O., F.C. Alvira, and G.M. Bilmes. 2013. Determination of femtosecond ablation thresholds by using Laser Ablation Induced Photoacoustics (LAIP). *Applied Physics A Materials Science & Processing* 110:735–739.

Ovsianikov, A., M. Gruene, M. Pflaum, L. Koch, F. Maiorana, et al. 2010. Laser printing of cells into 3D scaffolds. *Biofabrication* 2:014104.

Oya, K., S. Aoki, K. Shimomura, N. Sugita, K. Suzuki, et al. 2012. Morphological observations of mesenchymal stem cell adhesion to a nanoperiodic-structured titanium surface patterned using femtosecond laser processing. *Japanese Journal of Applied Physics* 51:125203.

Palanker, D.V., M.S. Blumenkranz, D. Andersen, M. Wiltberger, G. Marcellino, et al. 2010. Femtosecond laser-assisted cataract surgery with integrated optical coherence tomography. *Sci Transl Med* 2:58–85.

Papadaki, M., A. Doukas, W.A. Farinelli, L. Kaban, and M. Troulis. 2007. Vertical ramus osteotomy with Er: YAG laser: A feasibility study. *International Journal of Oral and Maxillofacial Surgery* 36:1193–1197.

Petrović, S., D. Peruško, A. Mimidis, P. Kavatzikidou, J. Kovač, et al. 2020. Response of NIH 3T3 fibroblast cells on laser-induced periodic surface structures on a 15×(Ti/Zr)/Si multilayer system. *Nanomaterials* 10:2531.

Pramanik, A., S. Biswas, P. Kumbhakar, and P. Kumbhakar. 2020. External feedback assisted reduction of the lasing threshold of a continuous wave random laser in a dye doped polymer film and demonstration of speckle free imaging—ScienceDirect. *Journal of Luminescence* 230:117720.

Pregowski, P., J. Marczak, and A. Koss. 2003. Study of thermal effects on artwork surfaces cleaned with laser ablation method. Proceedings of SPIE the International Society for Optical Engineering.

Rebollar, E., M. Castillejo, and T.A. Ezquerra. 2015. Laser induced periodic surface structures on polymer films: From fundamentals to applications. *European Polymer Journal* 73:162–174.

Rebollar, E., S. Pérez, M. Hernández, C. Domingo, M. Martín, et al. 2014. Physicochemical modifications accompanying UV laser induced surface structures on poly(ethylene terephthalate) and their effect on adhesion of mesenchymal cells. *Physical Chemistry Chemical Physics* 16:17551–17559.

Ruiz-Pomeda, A., and C. Villa-Collar. 2020. Slowing the progression of myopia in children with the MiSight contact lens: A narrative review of the evidence. *Ophthalmology and Therapy* 9:783–795.

Sacks, Z.S., R.M. Kurtz, G. Mourou, and T. Juhasz. 2000. Subsurface femtosecond photodisruption for glaucoma surgery. Conference on Lasers and Electro-Optics, San Francisco, CA, 2000/05/07.

Schanwald, L.P. 1997. Two thermal monitors for high power laser processing. *Metal Powder Report* 52.

Schmid, T. 2006. Photoacoustic spectroscopy for process analysis. *Anal Bioanal Chem* 384:1071–1086.

Schroder, S., and H. Grothe. 2002. Submilliampere operation of selectively oxidised GaAs-QW vertical cavity lasers emitting at 840nm. *Electronics Letters* 32:348.

Schwerdtfeger, J., F. Singer Robert, and C. Körner. 2012. In situ flaw detection by IR-imaging during electron beam melting. *Rapid Prototyping Journal* 18:259–263.

Siano, S., and R. Salimbeni. 2010. Advances in laser cleaning of artwork and objects of historical interest: The optimized pulse duration approach. *Accounts of Chemical Research* 43:739–750.

Sima, F., H. Kawano, A. Miyawaki, L. Kelemen, P. Ormos, et al. 2018. 3D Biomimetic chips for cancer cell migration in nanometer-sized spaces using "ship-in-a-bottle" femtosecond laser processing. *ACS Applied Bio Materials* 1:1667–1676.

Sima, F., D. Serien, D. Wu, J. Xu, H. Kawano, et al. 2017. Micro and nano-biomimetic structures for cell migration study fabricated by hybrid subtractive and additive 3D femtosecond laser processing. Vol. 10092, *SPIE LASE*: SPIE.

Skoog, S., and R. Narayan. 2013. Laser processing of biomaterials and cells. In *Encyclopedia of Biophysics*, edited by Gordon C.K. Roberts, 1226–1233. Berlin, Heidelberg: Springer Berlin Heidelberg.

Skruibis, J., O. Balachninaite, S. Butkus, V. Vaicaitis, and V. Sirutkaitis. 2019. Multiple-pulse Laser-induced breakdown spectroscopy for monitoring the femtosecond laser micromachining process of glass. *Optics and Laser Technology* 111:295–302.

Slepicka, P., J. Siegel, O. Lyutakov, N. Kasálková, Z. Kolská, et al. 2017. Polymer nanostructures for bioapplications induced by laser treatment. *Biotechnology Advances* 36:30163–30165.

Song, W.D., M.H. Hong, S.H. Lee, Y. Lu, and T.C. Chong. 2003. Real-time monitoring of laser cleaning by an airborne particle counter. *Applied Surface Science* 208:306–310.

Song, Y., G. Hu, Z. Zhang, and Y. Guan. 2020. Real-time spectral response guided smart femtosecond laser bone drilling. *Optics and Lasers in Engineering* 128:106017.

Sony, S., S. Laventure, and A. Sadhu. 2019. A literature review of next-generation smart sensing technology in structural health monitoring. *Structural Control & Health Monitoring* 26:22.

Strickland, D. 2019. Nobel lecture: Generating high-intensity ultrashort optical pulses. *Reviews of Modern Physics* 91:7.

Tan, W.D., N.S. Bailey, and Y.C. Shin. 2013. Investigation of keyhole plume and molten pool based on a three-dimensional dynamic model with sharp interface formulation. *Journal of Physics D-Applied Physics* 46:12.

Tan, Z., X. Zhang, J. Liu, and B. Zhang. 2020. Performance test of an internally modulated He-Ne laser based on optical tunneling effect. *Optics & Laser Technology* 127:106154.

Tserevelakis, G.J., J.S. Pozo-Antonio, P. Siozos, T. Rivas, P. Pouli, et al. 2019. On-line photoacoustic monitoring of laser cleaning on stone: Evaluation of cleaning effectiveness and detection of potential damage to the substrate. *Journal of Cultural Heritage* 35:108–115.

Uno, K., K. Nakamura, T. Goto, and T. Jitsuno. 2008. Longitudinally excited N2 lasers without high-voltage switches. *Review of Scientific Instruments* 79:944.

Vadillo, J.M., and J.J. Laserna. 2004. Laser-induced plasma spectrometry: Truly a surface analytical tool. *Spectrochimica Acta Part B-Atomic Spectroscopy* 59:147–161.

Veiko, V.P., Y.Y. Karlagina, E.E. Egorova, E.A. Zernitskaya, D.S. Kuznetsova, et al. 2020. In vitro investigation of laser-induced microgrooves on titanium surface. *Journal of Physics: Conference Series* 1571:012010.

Vitek, D.N., D.E. Adams, A. Johnson, P.S. Tsai, S. Backus, et al. 2010. Temporally focused femtosecond laser pulses for low numerical aperture micromachining through optically transparent materials. *Optics Express* 18:18086–18094.

Wang, Q., J.D. Morrow, C. Ma, N.A. Duffie, and F.E. Pfefferkorn. 2015. Surface prediction model for thermocapillary regime pulsed laser micro polishing of metals. *Journal of Manufacturing Processes* 20:340–348.

Wang, X., S. Li, C. Yan, P. Liu, and J. Ding. 2015. Fabrication of RGD micro/nanopattern and corresponding study of stem cell differentiation. *Nano Lett* 15:1457–1467.

Wang, X., K. Ye, Z. Li, C. Yan, and J. Ding. 2013. Adhesion, proliferation, and differentiation of mesenchymal stem cells on RGD nanopatterns of varied nanospacings. *Organogenesis* 9:280–286.

Wang, X.Y., B.J. Shen, L.H. Jin, X.L. Zhao, H.Y. Wang, et al. 2014. Excess heat measurement and transmutation study of Pd wires after lasers stimulation in a D2 gas-loading system. *Advanced Materials Research* 977:300–303.

Wen, P., Y. Zhang, and W. Chen. 2012. Quality detection and control during laser cutting progress with coaxial visual monitoring. *Journal of Laser Applications* 24:032006.

Williams, K.M., V.J.M. Verhoeven, P. Cumberland, G. Bertelsen, C. Wolfram, et al. 2015. Prevalence of refractive error in Europe: The European Eye Epidemiology (E-3) consortium. *European Journal of Epidemiology* 30:305–315.

Wu, D., S.Z. Wu, J. Xu, L.G. Niu, K. Midorikawa, et al. 2014. Hybrid femtosecond laser microfabrication to achieve true 3D glass/polymer composite biochips with multiscale features and high performance: The concept of ship-in-a-bottle biochip. *Laser & Photonics Reviews* 8:458–467.

Xiangyang, L. 2016. Current situation and thinking of orthopaedic rehabilitation. *Rehabilitation Medicine* 26:1–4.

Xu, L., H. Zhang, J. He, X. Yu, L. Cui, et al. 2012. Double-end-pumped Nd:YVO4 slab laser at 1064 nm. *Applied Optics* 51:2012–2014.

Yanlong, S., F. Zhu, Y. Li, et al. 2019. High energy closed-loop cycle narrow linewidth optically pumped XeF(C-A) blue laser at a repetition rate of 10 Hz. *Optics Express* 27:2258–2267.

Ye, D., G.S. Hong, Y. Zhang, K. Zhu, and J.Y.H. Fuh. 2018. Defect detection in selective laser melting technology by acoustic signals with deep belief networks. *The International Journal of Advanced Manufacturing Technology* 96:2791–2801.

Ye, D.S., Y.H.J. Fuh, Y.J. Zhang, G.S. Hong, K.P. Zhu, et al. 2018. Defects recognition in selective laser melting with acoustic signals by SVM based on feature reduction. 2018 3rd International Conference on Advanced Materials Research and Manufacturing Technologies, Iop Publishing Ltd., Bristol.

Yoshida, T., and M. Okoshi. 2019. A resist-less patterning method of Al thin film on polycarbonate by F 2 laser irradiation. *Surfaces and Interfaces* 17:100373–100373.

Yumoto, M., N. Saito, T. Lin, R. Kawamura, A. Aoki, et al. 2018. High-energy, nanosecond pulsed Cr:CdSe laser with a 2.25–3.08 um tuning range for laser biomaterial processing. *Biomedical Optics Express* 9:5645–5653.

Zaiping, J. 2016. *Chinese yearbook of surgery*. Shanghai:Shanghai Publisher of Science and Technology.

Zeng, Y., J. Xu, D. Kang, S. Zhuo, X. Zhu, et al. 2017. Microstructural imaging of human esophagus using multiphoton microscopy. International Conference on Photonics and Imaging in Biology and Medicine, Suzhou, 2017/09/26.

Zergioti, I., A. Karaiskou, D.G. Papazoglou, C. Fotakis, M. Kapsetaki, et al. 2005. Femtosecond laser micro-printing of biomaterials. *Applied Physics Letters* 86:163902.

Zhang, L., J. Zhang, Q. Sheng, S. Sun, and J. Yao. 2020. Efficient multi-watt 1720 nm ring-cavity Tm-doped fiber laser. *Optics Express* 28:37910.

Zhang, P., C. Liu, M. Xiang, X. Ma, and W. Guo. 2019. 850 nm GaAs/AlGaAs DFB lasers with shallow surface gratings and oxide aperture. *Optics Express* 27:31225.

Zhao, W.Q., X.W. Shen, H.D. Liu, L.Z. Wang, and H.T. Jiang. 2020. Effect of high repetition rate on dimension and morphology of micro-hole drilled in metals by picosecond ultra-short pulse laser. *Optics and Lasers in Engineering* 124:8.

13 Laser Assisted Production of Calcium Phosphate Nanoparticles from Marine Origin

*Mónica Fernández-Arias, Mohamed Boutinguiza Larosi,
Jesús del Val García, Antonio Riveiro Rodríguez,
Rafael Comesaña Piñeiro, Fernando Lusquiños
Rodríguez and Juan Pou Saracho*

CONTENTS

13.1   INTRODUCTION

Nanometric materials have different physical, chemical and biological properties from the same bulk materials or materials at molecular or atomic scale. This peculiarity related to manipulating matter at nanoscale, opens a wide range of applications in different branches of technology, medicine and science.

In the particular case of calcium phosphates (CaP), and more specially hydroxyapatite (HA) with a chemical formula $(Ca_{10}(PO_4)_6(OH)_2)$ and a Ca/P molar ratio of 1.67, shows great similarity of chemical composition with the inorganic part of the hard tissues of vertebrates. For this reason it is widely used in areas of orthopedics, dentistry, maxillofacial surgery, etc. with excellent results of biocompatibility, osteoconductivity and osteoinductivity (Dorozhkin 2013). However, not only the composition is behind the success of calcium phosphates in biomedical applications, but also their size.

Calcium phosphate particles, with their micrometric size, have surface areas with a range from 2 to 5 m^2/g, which are quite lower than those of nanometric particles of calcium phosphate (around 100 m^2/g) (Padilla et al. 2008). Furthermore lower bone resorption has been observed in micrometric crystals of synthetic hydroxyapatite (HA) than that attributed to the mineral part of the bone (Kalita et al. 2007).

Submicron-sized hydroxyapatite (HA) (with grain size around 180 nm) has lower surface roughness than nanometric HA (those of 67 nm), while the contact angle of the nanometric HA is significantly lower than that of the submicron one (Webster et al. 2000). In brief, differences in size lead to different properties, and therefore, to a different behavior in a given application. So, surface roughness is known to stimulate the function of osteoblasts, while the presence of porosity favors the osteoinduction (Sato et al. 2004). On the other hand, the sintering of tricalcium phosphate (β-TCP) as a nanometric powder at low temperature, gives better results in terms of mechanical properties, leading to a higher densification than when the β-TCP is used in the form of micrometric grains. The histological results of bone substitution, performed with HA-based biocomposites, point in the same direction.

Other characteristics related to the chemical composition, such as the Ca/P molar ratio have also a great influence. Although synthetic and biological HA have the same composition, the latter has a better metabolic activity and better dynamic response to the surrounding environment than the first one (Best et al. 2008). The solubility and bioactivity of HA are closely related to its crystallographic structure and Ca/P molar ratio. The lower the Ca/P ratio, the higher the solubility and the acidity, so Ca/P ratio is a determining factor in the biocompatibility of HA and it should be as close as possible to 1.67. Biogenic HA has disordered nanostructures and a non-stoichiometric composition, in addition to a lower number of hydroxyl groups, whose content in cortical bone is around 20% lower than synthetic HA.

In brief, natural or biogenic HA differs from the synthetic one by its morphology, size and the way in which it is ordered within the biological system. Taking into account that materials at nanometric scale have different chemical and physical properties than those of the bulk material, it is expected that biogenic HA presents better results in biomedical applications than synthetic ones. Furthermore, natural HA obtained from animal bones has the advantage of preserving some of the inherent properties of the precursor material, as its structure and composition (Herliansyah et al. 2009). In addition, fish bones, especially those obtained from species that are traded filleted and headless, may be a cheap source of natural HA that would contribute to a better use of the available resources.

In recent years, great efforts have been made by researchers to develop synthetic HA with similar properties to their corresponding biogenic ones. Although there are different techniques for producing calcium phosphate nanoparticles, such as solid state reaction method (Rhee 2002), coprecipitation method (Siddharthan et al. 2005), sol-gel method (Vijayalakshmi et al. 2006), hydrothermal processing (Sadat-Shojai et al. 2012), etc., most of these methods are often complex, time consuming, as well as relatively expensive.

In the present work, we have used the laser ablation technique of fish bones targets in de-ionized water to produce CaP nanoparticles. This method presents several advantages such as direct formation of nanoparticles in solution, the absence of contamination, easiness of preparation, low costs of processing, etc.

13.2 BIOMEDICAL APPLICATIONS OF CALCIUM PHOSPHATE NANOPARTICLES

One of the most relevant applications of calcium phosphates is as supply material for bone repairs and filling of bone defects. The use of implants in industrialized societies increases as the life expectancy of the population increases, and the bone graft is the most common type of tissue transplantation after blood tissue transplantation (Shegarfi et al. 2009). As an example, to illustrate the magnitude of the use of implants, approximately 600,000 bone graft procedures are performed each year in the United States, and about 2.2 million of such procedures are performed worldwide annually (Wang et al. 2014). In most cases, these implants are metallic, with titanium and its alloys the most commonly used material due to their biocompatibility. Over the years, different types of coatings have been tested to improve the osteointegration of implants made from titanium and its alloys. Among all these coating materials, calcium phosphates, and more specifically HA, have demonstrated a great capacity to improve the interaction between implant and bone tissue.

On the other hand, metallic implants often present complications due to the presence of infections, caused by microorganisms such as bacteria and antibiotic resistant fungi that adhere to the implant surface, forming a layer known in the scientific literature as "biofilm", which surrounds the implant as a foreign body (Kiedrowski et al. 2011). This often leads to the failure of implants and in some cases, can lead to complications or even to the death of the patient (Stevens et al. 2009), together with the consequent enormous expense of the infection treatment. For this reason, implants are usually coated with calcium phosphate to improve their functionalities.

Considering that, the similarity of coating properties to those of the natural bone is crucial in preventing the rejection of the implant by the human body. In order to recover the functionality of damaged tissues due to disease or aging, tissue engineering aims to restore such functionality by means of the design and construction of scaffolds from biomaterials that favor the adhesion, proliferation and differentiation of bone cells.

Although there are several methods and techniques of preparing a structure for the purpose of being used in bones (Kundu et al. 2013), the presence of calcium phosphate nanoparticles is essential to reach as far as possible in the regeneration of the bone. In this sense, the structure of the spongy part of the bone or of the trabecular bone is easier to reproduce than the compact part, due to the complex structure of the compact part of bone. The current tissue engineering techniques, such as rapid prototyping, are able to develop structures that quite correctly mimic the spongy part of the bone with the presence of pores (Wang et al. 2012). The design of bone scaffolds as well as the pore size and porosity are extremely important. Therefore, these techniques currently come together with the finite element method to improve the results, and to get a controlled and well-defined pore size distribution (see Figure 13.1).

However, an ideal tissue scaffold should have not only a similar porous structure to that of human bone to facilitate the transport of nutrients, tissue growth, improve the conformation of typical adhesive proteins and accelerate cell attachment, but also should be strong enough, biocompatible and bioresorbable (Cai et al. 2012).

The use of nanometric calcium phosphates has been studied also to promote the remineralization of damaged enamel and consequently its regeneration. The use of this nanometric material in

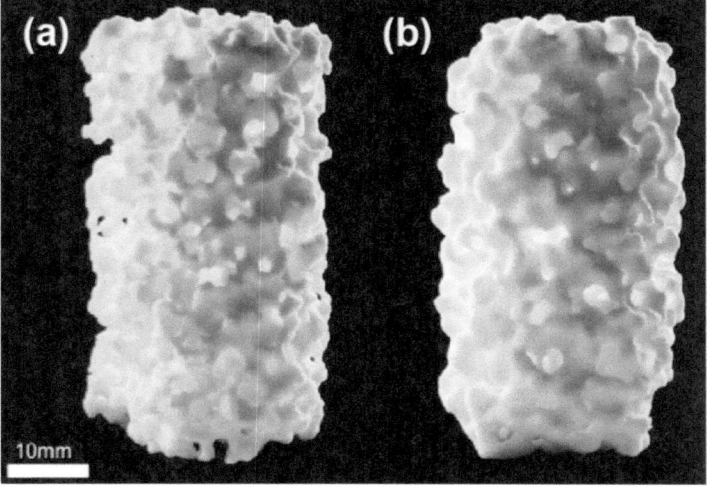

FIGURE 13.1 Two manufactured microscale structure scaffolds utilizing multi-resolution modelling approach: (a) high resolution (24 μm) and (b) low resolution (48 μm).

Source: Reprinted with permission from Elsevier; Podshivalov et al. (2013).

toothpastes promotes the remineralization of the demineralized enamel (Dorozhkin 2013; Tschoppe et al. 2011), reducing dentin hypersensitivity (Gopinath et al. 2015), together with the whitening effect which can be obtained (Niwa et al. 2001).

Another application of calcium phosphate nanoparticles is as a drug delivery system, since they are the most natural option for releasing drugs in bone therapies. The materials manufactured from nanoparticles have large specific surfaces and pore volume, which enable transporting a large concentration of drugs. They are one of the safest and most biocompatible nanomaterials tested to date (Singh et al. 2009) and they present also good adsorption capacity to biomolecules (Saito et al. 2013). Calcium phosphate nanoparticles constitute ideal vehicles to transport active principles since their porous nature does not change with the pH and does not present swelling. On the other hand, the different varieties of stoichiometry of the nanometric calcium phosphates allow different kinetics of drug release, which makes it possible to arrange the time of drug release from hours to months (Loomba et al. 2015).

In addition to releasing drugs, calcium phosphate nanoparticles may also be used to perform other functions simultaneously, such as their use as contrast agents for the detection of diseases, in therapeutic properties, as bactericidal agents, etc. (Uskoković et al. 2013). Figure 13.2 illustrates the functionalization of a nanoparticle to perform several simultaneous tasks (Sanvicens et al. 2008).

Transfection is another promising biomedical application of the calcium phosphate nanoparticles. This method allows to control intracellular processes such as induction or inhibition of protein expression (Kovtun et al. 2009), where the size of the nanoparticle plays an important role. The inhibition of protein or gene silencing, also known as RNA interference (RNAi), is the introduction of small interfering RNA into the cytoplasm of the cells that specifically turns off the production of proteins, a powerful tool to inhibit a specific gene function for disease treatment (Sokolova et al. 2014).

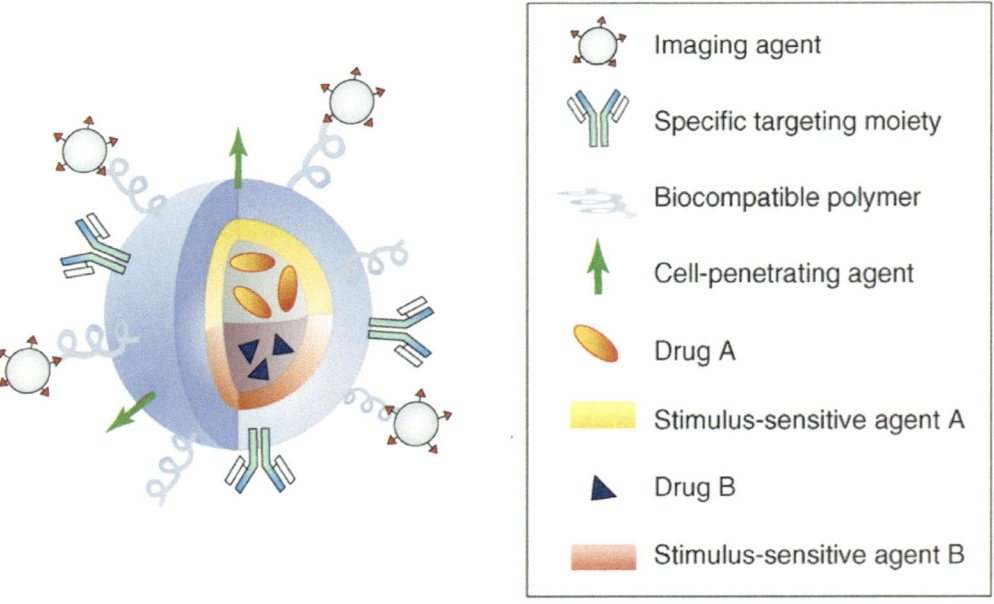

FIGURE 13.2   Multifunctional nanoparticle for both imaging and drug delivery.

Source: With permission from Elsevier; Sanvicens et al. (2008).

13.3 CALCIUM PHOSPHATE NANOPARTICLES FROM FISH BONE

For the aforementioned applications, especially those related to problem solving in bones or teeth, the similarity of calcium phosphate nanoparticles to those forming part of human bone is essential for success. This resemblance includes structure, morphology, size of crystals, crystalline orientation, etc.

One option could be the use of nanoparticles from biological origin, since they retain inherent properties of their biogenic origin, such as crystal size, composition, degree of crystallinity, etc. Among the different calcium phosphate compositions, hydroxyapatite shows high similarities to the mineral part of bone, since human and animal bones are basically composed of hydroxyapatite (Sobczak et al. 2009).

Hydroxyapatite can be obtained from different materials such as algae (Walsh et al. 2008), egg shells, seashells, corals and animal bones (Laonapakul 2015). Nevertheless there are incipient lines of research focused on management and use of by-products from fishing and shelling activities (M. Boutinguiza et al. 2011). On one hand, this can contribute to solve the problem of residues; on the other hand this can give added value to the by-products, which can be used for new applications. Taking into account that in the fish processing industry, many species are traded headless and filleted; discarding the bony part of the fish, which is mainly composed of calcium phosphate, bone fish can be used as a source CaP. Intensifying research on the use of by-products from fishing activities would mean reducing organic wastes, as well as relieving the pressure on marine resources by obtaining a higher yield from them, which implies new ways of industrialization and creation of employment.

But, why is the nanometer-scale so important in bone? Several studies have shown that natural nanocomposites possess a mechanical structure in which the nanometric sizes of the minerals fulfil the function of ensuring the optimum strength and the maximum tolerance to defects (Gao et al. 2003; Gupta et al. 2006). Bones can be considered as an assembly of different levels of seven hierarchical structural units from the macroscale to the nanoscale through units of micrometric size (Cui et al. 2007) as shown in Figure 13.3.

Another key function of nanometric calcium phosphates for organisms is to provide calcium and orthophosphate ions, which are necessary for a wide variety of metabolic functions (Dorozhkin 2009), where orthophosphate and calcium ions are supplied or consumed in a continuous process of resorption and formation of nanometric HA by the osteoclasts and osteoblasts respectively (Dorozhkin 2011).

The methods and techniques of nanoparticle synthesis enable controlling their size and morphology depending on the parameters and conditions used. In brief, nanoparticle production techniques can be classified into two large groups. The "Bottom-Up methods" are basically chemical methods that start by assembling small units to obtain nanometer-scale material. Within this group, there are several techniques and methods for the synthesis and production of calcium phosphate nanoparticles, such as, the sol-gel method, hydrothermal synthesis or microwave irradiation. The "Top-Down methods" group includes physical methods, which start from a target of macroscopic material to reduce it to nanometric portions by structural decomposition. This approach includes grinding, thermal evaporation, vapor phase deposition, laser ablation, etc.

Although each method presents its own pros and cons regarding the use of nanoparticles in the biomedical field, obtaining nanoparticles without harmful contamination is very important. In this regard, laser ablation makes it possible to obtain pure nanoparticles directly, without any additional reagent. Moreover, relevant features such as size or shape of nanoparticles can be easily controlled by adjusting the laser parameters.

Within the framework of the work carried out in several research projects (IBEROMARE, PROTEUS and MARMED) whose main aim was revalorization of by-products or wastes from fishing activities, we have used the laser as a tool to obtain calcium phosphates nanoparticles from fishbone as precursor material. For this purpose, as targets, we have used fish bones of different varieties of fish

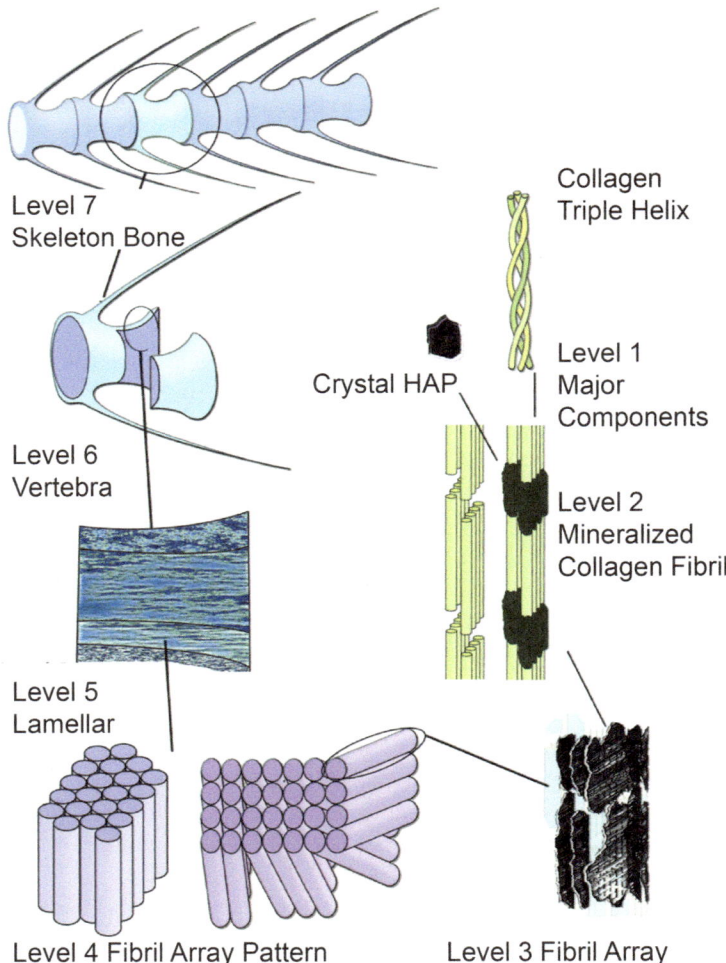

FIGURE 13.3 The seven hierarchical levels of organization of the zebrafish skeletal bone.

Source: Modified with permission from Elsevier; Cui et al. (2007).

previously treated to remove organic matter and preserve only the mineral part. The targets were irradiated with several laser sources, with different wavelengths and under different conditions to obtain calcium phosphate nanoparticles following different processes.

One of the strategies involved was the fracture of micrometric particles by laser. Here we report the results obtained with two different lasers operating at 1064 and 1075 nm wavelength respectively. The first one consisted in a pulsed Nd:YAG laser with a pulse duration of f 1.0 ms and an energy per pulse of 2.0 to 8.0 mJ, while the other laser source consisted in an Ytterbium doped fiber laser operating in continuous mode with an irradiance of approximately 10^6 W/cm^2.

For this purpose, particles obtained from swordfish skeletons previously calcined at 950 °C to remove organic matter, were milled by means of ball milling to achieve a powder composed by micrometric particles. The morphology of these particles is shown in Figure 13.4(a).

The XRD and XRF analysis performed on the resulting material showed that it is mainly composed of HA, although it also revealed the presence of Na, Mg, Cl, K and Sr as can be seen from Table 13.1 and Figure 13.4(b).

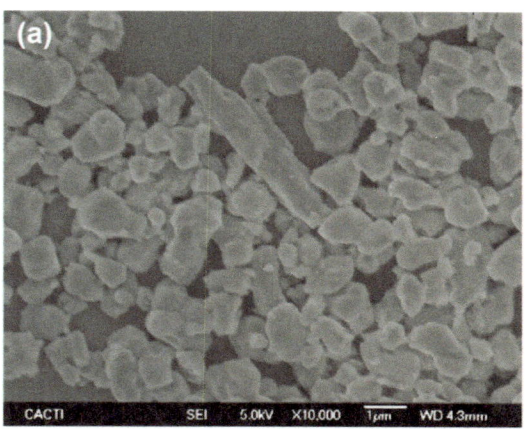

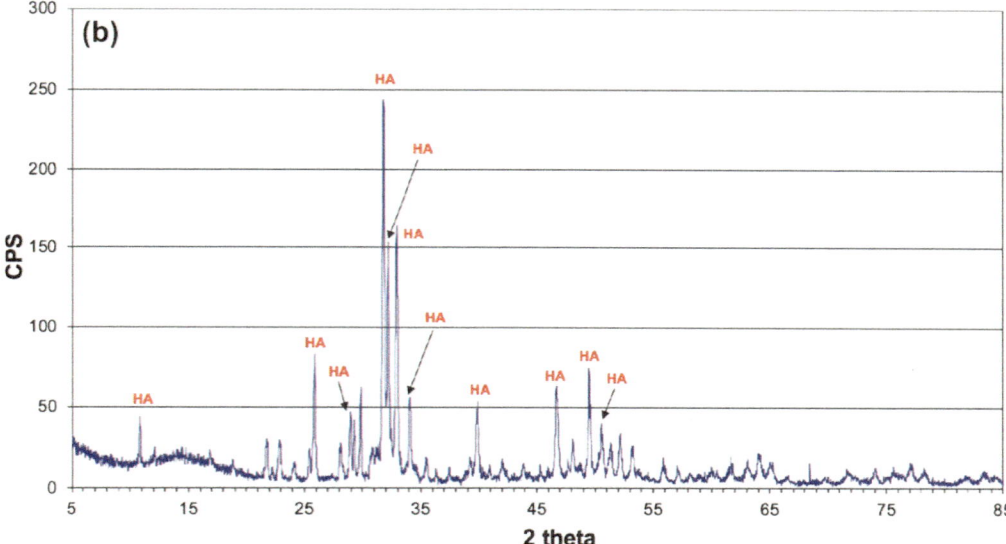

FIGURE 13.4   (a) SEM image showing the starting particles from fish bones before irradiation. (b) XRD pattern of calcined fish bones powder.

Source: With permission from Elsevier; Boutinguiza et al. (2009).

TABLE 13.1
Composition of Precursor Material Measured by XRF.

Element	O	Ca	P	Na	Mg	Cl	K	Sr
wt%	42,94	34,26	22,1	0,71	0,57	0,33	0,1	0,099

Source: With permission from Elsevier; Boutinguiza et al. (2009).

With the resulting particles, a solution in distilled water of pH 8 was prepared to maintain the stability of HA. The obtained solution was kept in sonication to avoid agglomeration of the precursor particles and finally, it was subjected to the irradiation of a laser beam.

When the laser beam impinges on precursor particles, they suffer a rapid rise of temperature, which can lead to fracture of the material, its melting and/or evaporation, depending on processing parameters. In this case, the accumulation of successive pulses delivered on particles involves a photo-thermal process. This thermal confinement leads to the increase of energy in the form of heat in the material (HA), which cannot be dissipated rapidly due to the low thermal conductivity of the target. This produces local superheat that results in thermal stress fracture and fusion of the precursor particles. Particles of nanometric size fractured by this mechanism together with smaller ones, whose rounded form indicates that they have been formed by fusion of the starting material and rapid solidification, are produced as shown in the micrograph of Figure 13.5(a).

With this technique and using a pulsed laser, calcium phosphate nanoparticles of a few nanometers have been obtained, as can be seen in Figure 13.5(b), showing high resolution transmission electron microscopy of several nanoparticles, which exhibit clear patterns of their crystal lattice together with the corresponding Fast Fourier Transform (FFT). The measurements of their interplanar distances permitted the indexation of the obtained particles as crystalline nanoparticles of

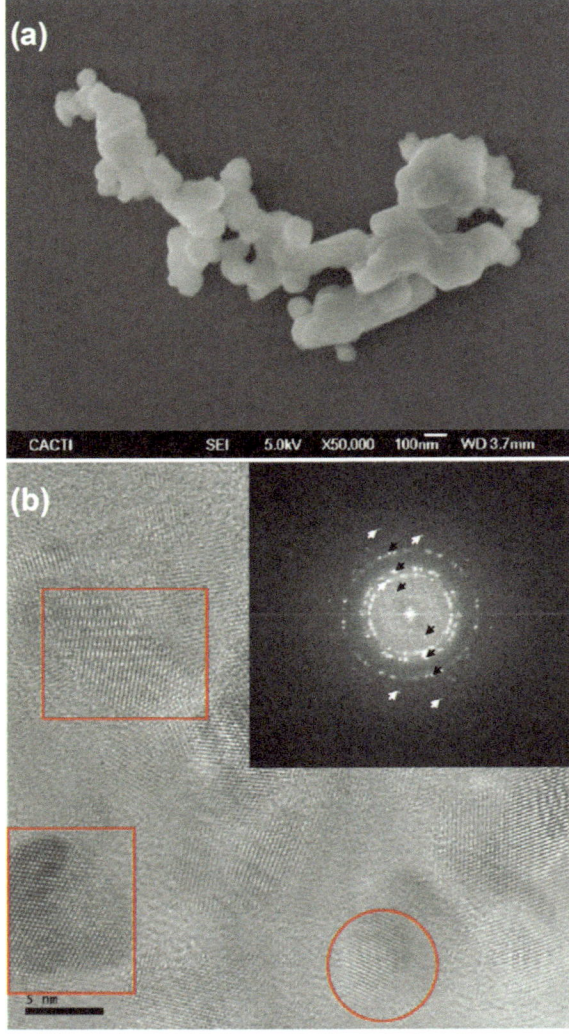

FIGURE 13.5 (a) Appearance of fragmented particles obtained by cw laser. Irradiance: 6×10^5 W/cm^2. (b) HRTEM of a group of crystalline nanoparticles obtained. Some particles smaller than 10 nm are indicated. The insert corresponds to the Fast Fourier Transform (FFT) and it allows to identify the presence of TCP (white arrows) and HA (black arrows).

Source: With permission from Elsevier; Boutinguiza et al. (2009).

HA with others of tri-calcium phosphate (TCP). In some applications such as bone regeneration, the combination of HA and TCP is very interesting. HA has excellent biocompatibility, promoting direct adhesion and proliferation of osteoblasts and additionally, it is very stable in body fluids. Nevertheless TCP has a too high dissolution rate compared to the rate of bone bonds formation. For this reason, it is particularly interesting to use a mixture of both materials with a nanometric size, so that bone tissues can grow occupying the holes left by the TCP after its rapid dissolution (Epple et al. 2010).

Regarding this technique, there are crucial parameters in the formation of nanoparticles, such as the pulse width, pulse duration, power density or irradiance of the laser beam on the particles. When the irradiance value increases to 2×10^6 W/cm^2, nanoparticles with spherical morphology are obtained, accompanied by occasional presence of spherical particles with submicronic size. The formation of this kind of particles is probably due to a rapid melting, which can cause explosive ejection of drops. Material can condense forming particles by the quenching on the liquid environment together with the presence of chunks fractured from starting material (M. Boutinguiza et al. 2011). The analyses carried out on several groups of these particles revealed that their fundamental composition is Ca and P, but not crystalline. Their almost spherical morphology suggests they are formed from molten material due to the high temperatures reached at the interaction zone between the laser beam and the target. Surface tension plays a fundamental role in conferring them such shape, while its rapid cooling favored by the temperature of the surrounding aqueous medium contributes to make them amorphous. In terms of composition, the calculations of the Ca/P ratio performed on groups of particles provided a value of 1.82, somewhat lower than the value measured for precursor particles of 1.86. This higher value for HA of biological origin than for synthetic HA can be explained by the presence of carbonates substituting phosphate ions. The biogenic HA is normally carbonated and less crystalline than the synthetic one. This is due to the incorporation of certain ions as Mg^{2+}, K^+, Na^+, etc. in very small amounts, which not only affect its crystallinity, but also modify its surface charge, solubility and other properties that determine its behaviour in the presence of body fluids, being essential in the metabolism of the bone. The decrease in the Ca/P ratio observed in the nanoparticles obtained by laser is in accordance with the partial transformation of the HA in TCP. This transformation normally occurs at temperatures above 1000 °C, which can be reached and exceeded by the particles when the interaction time with the laser beam is relatively high (about ms). This particular composition is in accordance with other results reported in the scientific literature (Shi et al. 2005). Several studies carried out by different researchers point out that HA of animal origin is biologically more active than synthetic HA because it preserves part of the original properties, inherent to the mineral part of the bone, such as crystallinity, nanometric size of crystals, as well as the presence of minor elements in its composition such as Na^+, Mg^{2+}, K^+, etc. The size and shape of crystals are of seminal importance in the biomechanical properties of bones, due to the preferential orientation of the HA crystals in collagen, which determines the bone strength as well as the bioactivity or flexibility of the structure.

Finally, considering the ability of HA to store heavy metals by ion exchange or adsorption of heavy ions on the crystalline surface (Moriguchi et al. 2008), it is important to study the content of heavy metals and toxic elements in the inorganic part of the fish bones used as the source of calcium phosphate. In the analyzed samples, Pb, Cd and Hg concentrations are below the limit stablished by the American Society for Testing and Materials (ASTM) for the inorganic part of bone in implant surgeries (ASTM 1999). The results are in agreement with those obtained in the MTT Cell Proliferation Assay, which show that the HA obtained is not cytotoxic (Boutinguiza et al. 2011).

13.4   CHALLENGES AND FUTURE PROSPECTS

Nowadays, there are several lines of research focused on reducing the environmental impact of fishing activities and generating high added value products from fishing by-products, especially in the fields of biomedicine, pharmacy or cosmetics.

Hydroxyapatite, like other calcium phosphates, has been widely used in several medical applications such as implant coatings or bone defect fillers, due to its similarity with the mineral part of human bone in terms of chemical composition, which makes it compatible with living tissues. In this sense, calcium phosphate can be considered the best bone graft substitutes because they promote rapid bone formation on their surface, and may assure bone healing within a year (Habraken et al. 2016). But due to the magnificent improvements achieved in the field of biomedical applications, the first commercial CaP bone grafts, launched 40 years ago, are currently regarded as "old biomaterials" or even as an "obsolete" research topic (Habraken et al. 2016). Today, research studies in the field of tissue engineering are aimed at getting functionalized bone grafts, where in addition to repairing damaged bone, the implant incorporates active principles that can be released to solve other musculoskeletal diseases and disorders (Saber-Samandari et al. 2017). That is to combine the function of drug release with that of bone regeneration, taking advantage of the beneficial properties of hydroxyapatite.

Although this chapter has exposed the study carried out to obtain nanoparticles from fish bones, as an alternative of a better use of the resources and to provide added value products from waste material, new perspectives show the possibility of obtaining hydroxyapatite from sea snails (Gunduz et al. 2014) or fish scales (Pon-On et al. 2016).

On the other hand, calcium phosphates possess good biocompatibility and biodegradability, and in addition they are not affected by microbiological degradation (Epple et al. 2010). But the process of transfection through this material is not really efficient due to its physical instability and weak electropositivity (Xu et al. 2016). Current efforts in research activity are intensified to optimize the method using lipids and polymer coatings. Recent experiments, such as tests carried out in cancer therapy by PEGylated calcium phosphate hybrid micelles, have shown their efficacy (Pittella et al. 2014).

In order to satisfy clinical needs, new promising developments as nanostructured CaP composites in tissue engineering (Wang et al. 2014) or negatively charged calcium phosphate nanoparticles as disease-specific biomarkers and key regulators of cardiac dysfunction (Di Mauro et al. 2016) are being investigated.

All of these new developments could take advantage of the benefits of using calcium phosphate nanoparticles from marine origin synthesized by laser.

13.5 CONCLUSIONS

Nanotechnology permits designing and engineering materials with new properties obtainable only at nanoscale. At nanosize scale, materials have unique physical, chemical and biological properties, which differ from both properties of the bulk precursor material and the atoms and molecules of the same material. Natural nanometric calcium phosphates derived from fish bones have excellent mechanical, physical and chemical properties, similar to those of bones and dental tissue. Due to these unique properties of nanometric calcium phosphates such as high surface area, biocompatibility, morphology and size similar to the mineral constituents of human bone, CaP nanoparticles can be used in a wide range of applications, such as bone regeneration, pharmacology, etc.

A natural source of nanometric calcium phosphates are fish bones, which are available as a by-product of fishing activity. There are different methods and techniques for obtaining nanometric calcium phosphate from fish bones. The results of the research carried out to obtain CaP nanoparticles from fish bones by laser ablation, show that fish bones can be a sustainable source of HA. This can contribute to reduce the negative impact of residues from fishing activities and to obtain new added value products from by-products. In several works, we have used the laser as a tool to obtain calcium phosphate nanoparticles from fish bones as precursor material. The results showed that CaP nanoparticles, with the composition and properties similar to that of the mineral part of fish bones, can be obtained.

13.6 ACKNOWLEDGEMENTS

This work was partially supported by the EU research project Bluehuman (EAPA_151/2016 Interreg Atlantic Area), Government of Spain (MAT2015–71459-C2-P), Xunta de Galicia (ED431C 2019/23, ED481D 2017/010, ED481B 2016/047–0).

REFERENCES

Best, S. M., Porter, A. E., Thian, E. S., et al. 2008. Bioceramics: Past, present and for the future. *Journal of the European Ceramic Society* 28(7):1319–1327.

Boutinguiza, M., Comesaña, R., Lusquiños, F., et al. 2011. Production of nanoparticles from natural hydroxylapatite by laser ablation. *Nanoscale Research Letters* 6(1):255.

Boutinguiza, M., Lusquiños, F., Riveiro, A., et al. 2009. Hydroxylapatite nanoparticles obtained by fiber laser-induced fracture. *Applied Surface Science* 255(10):5382–5385.

Boutinguiza, M., Pou, J., Comesaña, R., et al. 2011. Biological hydroxyapatite obtained from fish bones. *Materials Science and Engineering C* 32:478–486.

Boutinguiza, M., Pou, J., Lusquiños, F., et al. 2011. Laser-assisted production of tricalcium phosphate nanoparticles from biological and synthetic hydroxyapatite in aqueous medium. *Applied Surface Science* 257(12):5195–5199.

Cai, S., Xi, J., and Chua, C. K. 2012. A novel bone scaffold design approach based on shape function and all-hexahedral mesh refinement. *Methods in Molecular Biology* 868:45–55.

Cui, F. Z., Li, Y., and Ge, J. 2007. Self-assembly of mineralized collagen composites. *Materials Science and Engineering R: Reports* 57(1–6):1–27.

Di Mauro, V., Iafisco, M., Salvarani, N., et al. 2016. Bioinspired negatively charged calcium phosphate nanocarriers for cardiac delivery of MicroRNAs. *Nanomedicine* 11(8):891–906.

Dorozhkin, S. V. 2009. Calcium orthophosphates in nature, biology and medicine. *Materials* 2:399–498.

Dorozhkin, S. V. 2011. Calcium orthophosphates. *Biomatter* 1:121–164.

Dorozhkin, S. V. 2013. Calcium orthophosphates in dentistry. *Journal of Materials Science: Materials in Medicine* 24(6):1335–1363.

Epple, M., Ganesan, K., Heumann, R., et al. 2010. Application of calcium phosphate nanoparticles in biomedicine. *Journal of Materials Chemistry* 20(1):18–23.

Gao, H., Ji, B., Jäger, I. L., et al. 2003. Materials become insensitive to flaws at nanoscale: Lessons from nature. *Proceedings of the National Academy of Sciences of the United States of America* 100(10):5597–5600.

Gopinath, N. M., John, J., Nagappan, N., et al. 2015. Evaluation of dentifrice containing nano-hydroxyapatite for dentinal hypersensitivity: A randomized controlled trial. *Journal of International Oral Health: JIOH* 7(8):118–122.

Gunduz, O., Sahin, Y. M., Agathopoulos, S., et al. 2014. A new method for fabrication of nanohydroxyapatite and TCP from the sea snail Cerithium vulgatum. *Journal of Nanomaterials* 2014:1–6.

Gupta, H. S., Seto, J., Wagermaier, W., Zaslansky, P., et al. 2006. Cooperative deformation of mineral and collagen in bone at the nanoscale. *Proceedings of the National Academy of Sciences* 103(47):17741–17746.

Habraken, W., Habibovic, P., Epple, M., et al. 2016. Calcium phosphates in biomedical applications: Materials for the future? *Materials Today* 19(2):69–87.

Herliansyah, M. K., Hamdi, M., Ide-Ektessabi, A., et al. 2009. The influence of sintering temperature on the properties of compacted bovine hydroxyapatite. *Materials Science and Engineering C* 29(5):1674–1680.

Kalita, S. J., Bhardwaj, A., and Bhatt, H. A. 2007. Nanocrystalline calcium phosphate ceramics in biomedical engineering. *Materials Science and Engineering C* 27(3):441–449.

Kiedrowski, M. R., and Horswill, A. R. 2011. New approaches for treating staphylococcal biofilm infections. *Annals of the New York Academy of Sciences* 1241(1):104–121.

Kovtun, A., Heumann, R., and Epple, M. 2009. Calcium phosphate nanoparticles for the transfection of cells. *Bio-Medical Materials and Engineering* 19(2):241–247.

Kundu, J., Pati, F., Shim, J.-H., et al. 2013. Rapid prototyping technology for bone regeneration. In *Rapid Prototyping of Biomaterials: Principles and Applications*. Elsevier, pp. 254–284.

Laonapakul, T. 2015. Synthesis of hydroxyapatite from biogenic wastes. *Kku Engineering Journal* 42(3):269–275.

Loomba, L., and Sekhon, B. S. 2015. Calcium phosphate nanoparticles and their biomedical potential. *Journal of Nanomaterials & Molecular Nanotechnology* 4(1):1–12.

Moriguchi, T., Nakagawa, S., and Kaji, F. 2008. Reaction of Ca-deficient hydroxyapatite with heavy metal ions along with metal substitution. *Phosphorus Research Bulletin* 22:54–60.

Niwa, M., Sato, T., Li, W., et al. 2001. Polishing and whitening properties of toothpaste containing hydroxyapatite. *Journal of Materials Science: Materials in Medicine* 12(3):277–281.

Padilla, S., Izquierdo-Barba, I., and Vallet-Regí, M. 2008. High specific surface area in nanometric carbonated hydroxyapatite. *Chemistry of Materials* 20(19):5942–5944.

Pittella, F., Cabral, H., Maeda, Y., et al. 2014. Systemic siRNA delivery to a spontaneous pancreatic tumor model in transgenic mice by PEGylated calcium phosphate hybrid micelles. *Journal of Controlled Release* 178(1):18–24.

Podshivalov, L., Gomes, Z.M., Zocca, A., et al. 2013. Design, analysis and additive manufacturing of porous structures for biocompatible micro-scale scaffolds. *Procedia CIRP* 5:247–252.

Pon-On, W., Suntornsaratoon, P., Charoenphandhu, N., et al. 2016. Hydroxyapatite from fish scale for potential use as bone scaffold or regenerative material. *Materials Science and Engineering C* 62:183–189.

Rhee, S. H. 2002. Synthesis of hydroxyapatite via mechanochemical treatment. *Biomaterials* 23(4):1147–1152.

Saber-Samandari, S., and Saber-Samandari, S. 2017. Biocompatible nanocomposite scaffolds based on copolymer-grafted chitosan for bone tissue engineering with drug delivery capability. *Materials Science and Engineering C* 75:721–732.

Sadat-Shojai, M., Khorasani, M. T., and Jamshidi, A. 2012. Hydrothermal processing of hydroxyapatite nanoparticles: A Taguchi experimental design approach. *Journal of Crystal Growth* 361(1):73–84.

Saito, M., Kurosawa, Y., and Okuyama, T. 2013. Scanning electron microscopy-based approach to understand the mechanism underlying the adhesion of dengue viruses on ceramic hydroxyapatite columns. *PLoS One* 8(1).

Sanvicens, N., and Marco, M. P. 2008. Multifunctional nanoparticles—properties and prospects for their use in human medicine. *Trends in Biotechnology* 26(8):425–433.

Sato, M., and Webster, T. J. 2004. Nanobiotechnology: Implications for the future of nanotechnology in orthopedic applications. *Expert Review of Medical Devices* 1(1):105–114.

Shegarfi, H., and Reikeras, O. 2009. Review article: Bone transplantation and immune response. *Journal of Orthopaedic Surgery* 17(2):206–211.

Shi, J., Klocke, A., Zhang, M., et al. 2005. Thermally-induced structural modification of dental enamel apatite: Decomposition and transformation of carbonate groups. *European Journal of Mineralogy* 17:769–775.

Siddharthan, A., Seshadri, S. K., and Kumar, T. S. S. 2005. Rapid synthesis of calcium deficient hydroxyapatite nanoparticles by microwave irradiation. *Trends Biomaterials and Artificial Organs* 18(2):110–113.

Singh, N., Manshian, B., Jenkins, G. J. S., et al. 2009. NanoGenotoxicology: The DNA damaging potential of engineered nanomaterials. *Biomaterials* 30(23–24):3891–3914.

Sobczak, A., Kida, A., Kowalski, Z., et al. 2009. Evaluation of the biomedical properties of hydroxyapatite obtained from bone waste. *Polish Journal of Chemical Technology* 11(1):37–43.

Sokolova, V., and Epple, M. 2014. Bioceramic nanoparticles for tissue engineering and drug delivery. In *Tissue Engineering Using Ceramics and Polymers*, Elsevier, 2nd Edition, pp. 633–647.

Stevens, K. N. J., Crespo-Biel, O., van den Bosch, E. E. M., et al. 2009. The relationship between the antimicrobial effect of catheter coatings containing silver nanoparticles and the coagulation of contacting blood. *Biomaterials* 30(22):3682–3690.

Tschoppe, P., Zandim, D. L., Martus, P., et al. 2011. Enamel and dentine remineralization by nano-hydroxyapatite toothpastes. *Journal of Dentistry* 39(6):430–437.

Uskoković, V., and Desai, T. A. 2013. Phase composition control of calcium phosphate nanoparticles for tunable drug delivery kinetics and treatment of osteomyelitis. *J. Biomed Mater Res A* 101(5):1427–1436.

Vijayalakshmi, U., and Rajeswari, S. 2006. Preparation and characterization of microcrystalline hydroxyapatite using sol gel method. *Trends in Biomaterials and Artificial Organs* 19(2):57–62.

Walsh, P. J., Buchanan, F. J., Dring, M., et al. 2008. Low-pressure synthesis and characterisation of hydroxyapatite derived from mineralise red algae. *Chemical Engineering Journal* 137(1):173–179.

Wang, P., Zhao, L., Liu, J., et al. 2014. Bone tissue engineering via nanostructured calcium phosphate biomaterials and stem cells. *Bone Research* 14017.

Wang, Z., Tang, Z., Qing, F., et al. 2012. Applications of calcium phosphate nanoparticles in porous hard tissue. *Nano Brief Reports and Reviews* 7(4):1–18.

Webster, T. J., Ergun, C., Doremus, R. H., et al. 2000. Specific proteins mediate enhanced osteoblast adhesion on nanophase ceramics. *Journal of Biomedical Materials Research* 51(3):475–483.

Xu, X., Li, Z., Zhao, X., et al. 2016. Calcium phosphate nanoparticles-based systems for siRNA delivery. *Regenerative Biomaterials* 3.

14 Surface Modifications of Biometals

G. Keerthiga, V.S. Simi and L. Mohan

CONTENTS

14.1 INTRODUCTION

Biomaterials are engineered devices or implants that are introduced into the biological system for assisting, augmenting, replacing a normal bodily function or sometimes in theranostics (Shi 2005). Annual increase in ageing population, trauma and improved functional demands has led to scientists delving for more efficient and safer natural or synthetic sourced biomaterials. As such, these biomaterials are chosen and modified to fit in for a desired application. The implants or device introduced into the functional system to replace or subordinate tissue is expected to be non-toxic, non-cancerous, mechanically ample, synthetically latent and stable (Patel and Gohil 2012). Depending upon the stay, permanent and temporary fixating devices are designed using the most common metallic substrates made of stainless steel, cobalt-chromium, titanium, and magnesium alloys. Biocompatibility is the fundamental requirement for any biomaterial as it dictates the ability of the material to generate appropriate host responses that befits the acceptance or rejection of the biomaterial (Anderson 2001). Periphery of the biomaterial comes in immediate contact with the physiological environment that aid in protein adsorption and cell adhesion. Although bulk properties determine the biological and mechanical properties of the biomaterial, the top surface governs the biomaterial-tissue interactions that are proposed to occur in narrowed precinct of less than 1 nm. The structure and chemistry of the top facade is necessarily constituted by the atoms or molecules immediately below. Surface atomic molecules are partly reactive to the environment as it lapses in the fringe of the biomaterial.

14.1.1 NEED FOR SURFACE MODIFICATIONS

Broadly, biomaterials are classified into ceramics, glasses, metallics, polymeric and composite systems. Any biomaterial introduced to the physiological environment networks with body fluids that contain electrically, mechanically and chemically active molecules with immediate biomaterial-biological interface formation.

The history of metallic biomaterial in dental application using gold dates back to 500 BC. Reports on using noble metals for structural dentistry since the fifteenth century to current research in biometallics have made metallic biomaterial indispensable in clinical and other applications. Some of the metallic components in clinical application are tabulated in Table 14.1.

DOI: 10.1201/9781003173533-14

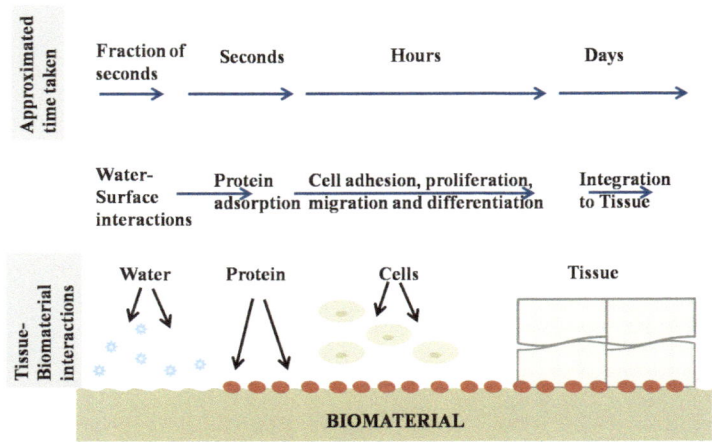

FIGURE 14.1 Various tissue-biomaterial interactions that occur at the biomaterial-tissue interface in the physiological system in different time scales.

Source: Redrawn from Nouri and Wen (2015).

TABLE 14.1
Metallic Biomaterials in Clinical Applications.

S. No	Biomaterial Used	Clinical Application	Example	Reference
1.	Ferritic steel	Surgical instruments	Guide pins, fasteners.	(Brar et al. 2009)
2.	Austenitic steel	Short-term implants, hip replacements, other non-implantable medical devices	Guide pins, canulae, hypodermic needles, dental impression trays, steam sterilizers, thoracic retractors.	(Talha, Behera, and Sinha 2013)
3.	Martensitic steel	Dental and surgical instruments	Dental curettes, explorers, root elevators and scalers, burs and chisels, bone curettes, chisel and gouges, scalpels, forceps, retractors.	(Disegi and Eschbach 2000)
4.	Titanium and its alloys	Permanent load bearing implants and circulatory devices	Artificial bones, joints, dental implants, screws and staples for spinal surgery. Prosthetic heart valves, pacemaker cases.	(Kaur and Singh 2019)
5.	Magnesium	Resorbable bone implants	Magnesium wires.	(Witte 2015)
6.	Cobalt-chromium and its alloys	Dental, orthopedic, maxillofacial implant components	Partial dentures, fracture mending plates and screws, hip and knee prostheses.	(Eliaz 2019)

Surface modifications are considered to be inexpensive and non-tedious as opposed to inventing novel materials. While retaining the bulk physical property of the biomaterial, tissue biocompatibility, mechanical and chemical durability can be improved with surface modifications. The main notion of introducing surface modification is to enhance wear and corrosion resistance, anti-bacterial property, biocompatibility and cell adhesion (especially bone in-growth) by better integration and biomolecular interactions. Surface treatments to modify surface chemistry, energy, topography and charge by retaining the bulk property can be achieved by one or combinatorial methods.

Considering the spotlight of this chapter on the surface modifications and coatings introduced onto the metallic implant materials, there is no universal technique applicable to all types of biomaterials. This chapter briefly discusses selective surface coatings and modification techniques and the principle behind the same. These surface modification and coatings in turn influence the surface properties of the biomaterial that is debriefed further.

14.1.2 SURFACE PROPERTIES

Medical devices or implants that are introduced into the body tissues are subjected to dynamic courses of events that can strongly affect their durability and lifespan. As mentioned earlier, body fluid is an aqueous corrosive medium of which water molecules are the first to be in contact with the biomaterial's surface. The surface characteristics of the biomaterial influence the binding of water molecules and subsequent protein/other molecular adsorption. Initial binding of water molecules occurs in nanoseconds and depending upon the anatomical location, the implant material will be exposed to various environments such as its adjoining hard and soft tissue exposure in orthopedic applications; in cardiovascular implants, which encounter soft tissue and blood components; and in the orthodontic environment that poses drastic components. Figure 14.1 briefly sketches the outline of events that occur in-vivo after biomaterial introduction.

Biomaterial introduction in a host causes an inflammatory host response as the human body considers it as a foreign material. After a series of immunological events, fibrous encapsulation and in some cases bacterial adhesion were reported at the bio-interfacial layer. Surface properties of the biomaterials such as surface chemistry, topography and roughness, surface energy, wettability play a vital role in determining the initial response occurring at the material-host interface. Accordingly, a suitable surface modification/coating choice is made for tweaking surface protein adsorption, thrombogenicity, control of cell adhesion, growth and differentiation, inflection of fibrous encapsulation, etc.

14.1.2.1 Surface Chemistry

Surface chemistry of biomaterials influence and participate in the cascade of events that occur after introduction of the implant/device into the physiological system. Depending on the functional groups extended on the biomaterial surface, protein adsorption, denaturation and epitope exposure are determined. To nullify or lessen the adverse cellular events when biomaterial is introduced into the microenvironment, efforts to engineer an ideal protein repelling surface are worked upon. This is achieved by introducing a uniform protein-resistant functional group that is neutral and hydrophilic with the presence of hydrogen bond acceptors. Tailoring the surface functional group affects protein adsorption and interactions that occur successively and can control biocompatibility. Computational stimulation studies to comprehend the effect of hydroxyl groups tethered to the surface facilitating fibrinogen adsorption were reported (Benesch et al. 2001). Reports on abundant fibrinogen adsorption in methyl and amine functional surfaces through hydrophobic and hydrogen bonding respectively serve as a proof that surface chemistry deeply affects the cascade of molecular signals and protein binding (Martins, Ratner, and Barbosa 2003). Negatively charged surfaces such as carboxyl groups showed little or no fibrinogen adhered as it was not adroit to outdo water molecule interactions. In-vivo investigations and the influence of mixed functionalities on the surface in biomaterial-surface interactions are to be widely explored. Besides influencing protein adsorption, the surface chemistry of the biomaterial is also known to participate in both classical and alternate complement immune pathways. These direct and activate phagocytes, attachment and activation of leukocytes. Radio frequency glow discharge polymer technique to know the influence of increased surface functionality density on protein adsorption and epitope exposure is also reported (Kamath et al. 2008). Understanding the impact of surface chemistry in-vivo is very influential in tailoring the specific requirements and betterment of medical devices such as implantable sensors, drug release devices, etc. The surface chemistry of the implant material can be modified by applying a suitable coating. The addition of ions by modulating the architecture of titania and deposition of other material distinguished into organic or mineral layers with a suitable deposition technique to regulate surface chemistry, to increase implant functionality were also studied extensively (Mahajan and Sidhu 2018).

14.1.2.2   Surface Topography and Roughness

Surface topography has a significant influence in cell-cell signaling and the implant-cell adhesion as it determines cell morphology, self-orientation, proliferation and differentiation. All implant materials for bioactive applications are anticipated to have complex topographical features, especially in bone related structures as they expose additional surface area for protein interactions. This is validated in resorption surfaces of existing bone by osteoclasts by development of new bone matrix deposition that consequently becomes inter-digitated and interlocked. Porous, textured, rough structures accelerate cell attachment and the extracellular matrix (ECM) formation. On the other hand, smooth surface (i.e. endothelium-mimicking surface) is used in blood contact biomaterials. Surface modifications to alter the topography are aimed at the creation of 3D structures in the form of gratings, columns, microgrooves, dots, pits, micropores and nanopores. Typically characterized by a succession of peaks and valleys, surface topography is quantified by 2D and 3D parameters.

Surface roughness is categorized into three main catalogues depending upon the scale of irregularities as macroroughness (100 μm-millimeters) related directly to the implant geometry, microroughness (100 nm–100 μm) and nano roughness (< 100 nm) introduced into the surface by various methods like acid etching, plasma spraying, anodization and blasting. These topographies determine size and shape of the cell adhered to the surface as they exhibit unique persuasion on cellular response. Nanostructures are preferred to smaller cells closer to the dimensions of the proteins and adapt to such as in human vein endothelial cells' adhesion and growth while higher topographies influence larger cells such as neurons and osteoblasts. The phenomenon in in-vitro observation of cells arranging themselves along the axis of the groove in the anisotropic groove topography for further development is called 'contact guidance'. It is also seen when only a single side of the surface is polished and edges of the structure act as a guide, giving clear directions for cell orientation. Despite isotropic and anisotropic structure difference, flat titanium substrate shows that osteoblasts maintain their round shape whereas the cells on the nanotube arrays become elongated with increased extensions and filopodia (Subramanian, Tran, and Nguyen 2012). Further, cells that can sense stiff, solid surface promote attachment and distribution, while soft flexible surfaces hinder the same.

Alterations to surface topography affect surface chemistry as well as they are interrelated to each other. Homogenous non-alloyed substrates that are further modified by micro- and nano-patterning that do not use strong amounts of energy are effectively used to understand the means of topography in cell response. This however does not reflect the real-time implant applications. For discriminating relative effect of surface chemistry and topography at the micrometer scale, research on depositing a thin layer of another material to minimize the effect of surface chemistry is being explored. Further, surface topography modifications, such as creating porous structures in the bulk or the surface of metallic biomaterials that minimize stress shielding effects and prevent loosening of implants, provide better mechanical stability.

14.1.2.3   Surface Energy and Wettability

Surface properties are interrelated to each other as the surface of the material is the terminal extension of 3D molecular structures. Zooming into the atomic scale, these surface molecules are unterminated chemical bonds that are more loosely bound than bulk material. The surface energy of the atoms in the outer surface has higher energy that readily reacts with the available molecules of air or water to achieve a more favorable energy state. Thus, it is necessary to consider the surface chemistry which is quite distinguished from the bulk chemistry as when bulk chemistry is introduced to the functional environment, it undergoes ready reactions with tissues and cellular components to attain a lower favorable energy state.

Surface wettability is nearly related to surface energy of the biomaterial. The former restricts the capacity of the surface interaction with liquids based on the hydrophilicity and hydrophobicity of the biomaterial surface, while the latter relates to surface interface with different range of materials. Surface wettability is an important criterion in evaluation of biocompatibility as the surface controls protein adsorption and surfeit of successive cell cascade. Liquids (e.g. water) have a colossal capacity of bonding, and materials with high surface energy readily form interfacial interactions thereby favoring hydrophilic surfaces. Hydrophobic surfaces like polystyrene or polytetrafluoroethylene (PTFE) have lower energies that do not favor spontaneous liquid reactions. Protein adsorption is reported quite contrary to the liquid surface wettability. Low energy hydrophobic surfaces show increased protein adsorption while very high hydrophobic surfaces suppress protein adsorption and favor blood biocompatibility. Various hydrophilic biomaterial surfaces defy protein adsorption and reportedly reduce platelet adhesion and thrombus formation.

Investigations to explain high affinity protein adsorption on hydrophobic surfaces are being carried out. It is allegedly reasoned to the lower water heat of adsorption on hydrophobic surfaces and due to interactions between the internal hydrophobic protein domains and the surface hydrophobicity, which cause protein unfolding. These interactions lead to protein denaturation and increased internal protein entropy. It has been noted that a hydrophobic surface greatly favors protein conformational changes than a hydrophilic surface, which can be elucidated by the structuration of the proteins. Proteins are generalized to have hydrophobic residues well concealed in the core and their hydrophilic charged and polar molecules flaunting along the external lining accessible to the surface interactions. Attributable to the thermodynamic net force, hydrophobic patches on the protein's amphiphilic surface when interacting with a hydrophobic surface, aid in protein adsorption. The protein then tends to unfold and spread its hydrophobic core across the surface. Alternatively, a hydrophilic surface influences the adsorbed protein orientation as it rallies with the surface charged polar functional groups of the protein molecule, preserving its core structure.

Expressed as a measure of contact/wetting angle to determine the average wettability of the surface, a zero degree corresponds to complete wetting (i.e. hydrophilic surface and angles recorded greater than 80°–90° consign to a hydrophobic surface). Surface nature is interrelated to surface energies and distribution of charges which is a corollary of surface chemistry, surface topography and crystallinity. Generalizing cell adhesion based on the surface nature is not possible as osteoblasts adhesion decreases along the contact angles of 0° to 106° and fibroblast adhesion increases along 60°–80°. The influence of surface roughness and energy on cell adhesion studies suggests that for high surface energy biomaterials (i.e. metals), surface roughness parameter consideration is secondary on modulating cell adhesion.

14.1.2.4 Surface Charge

Solid-solution interfaces in the metallic biomaterials tether various biological electrostatic interactions that modulate protein adsorption and cell adhesion. It is prominent that on a negatively charged surface at low ionic strength, cell adhesion is very low or negligible (Trommler, Gingell, and Wolf 1985). On the other hand, at high ionic strength, electrostatic repulsion is trifling and van der Waals attraction governs favoring cell attachment. High affinity protein adsorption is also seen with surface-oriented carboxyl groups. Protein adsorption is also dependent on shared electrostatic interactions between the surface and protein charges. Electrochemical series with respect to standard hydrogen, highly electropositive metals like Ag and Au readily adsorb proteins in-vitro and in-vivo whereas naturally passivating materials are difficult to arrive at conclusion despite their position in the electrochemical series. The pH of the human physiological system is 7.35–7.45 with circulating proteins with an isoelectric point less than 7. Considering our prime interest proteins, fibronectin and the bone morphogenetic protein (BMP) with isoelectric points approximately of 5.8 and 4.8 respectively, it demands a more positively charged surface to facilitate cell adhesion and better osseointegration. A Ti surface after UV irradiation becomes a positively charged surface that favors electrostatic attraction that leads to protein adsorption and consequently osteoblast adhesion.

14.2 SURFACE COATINGS AND MODIFICATIONS

We have discussed briefly the effect of various surface properties and their role in biocompatibility modulation. The main notion of various processes for introducing surface modifications is to enable bio-interfacial reactions in the peri-implant region of the host tissue. The peri-implant region refers to the immediate region to which the surface of the implant is exposed when introduced into the biological environment. Broadly the existing methodologies for metallic implants surface altera-tions are classified into surface chemistry modifications and topographical treatments. As discussed previously, topographical treatments on the metallic substrates inevitably alter the surface chem-istry of the biomaterial. For industrial applications, a combinatorial methodology to enhance host cellular response is preferred.

The various techniques used for surface alterations categorized into surface modifications and coatings are depicted in Figure 14.2. It is fair to note that surface coatings are also classified as surface modifications; it is also necessary to distinguish other physiochemical, electrochemical and biochemical methods from the coating process. Considering the span of this chapter, some of the exemplary approaches used for surface coatings and modifications of metallic biomaterials in spe-cific applications are discussed next.

14.2.1 SURFACE COATINGS

Surface coatings are considered to be applications of thin sheathing of materials to facilitate better metallic implant integration and acceptance in the host system. Bioactive ceramics like bioglass, sintered hydroxyapatite (HA) form spontaneous bone-like apatite growth when introduced into the living body for orthopedic and dental applications. The formed apatite layer is similar to bone mineral composition and structure, and hence facilitates easier bonding of the implant and host tis-sue. Osteoblasts also preferentially attach on this apatite layer, proliferate and differentiate to yield additional apatite and collagen. Extending ceramics into load bearing bone implants is not possible as they possess high elastic modulus and low fracture toughness, wherein metallic biomaterials such as Titanium (Ti) and its alloys (especially commercially pure and β alloys) pitch into the picture. Introducing surface coatings ideally of 10–15 Å thickness, bearing in mind distinguishable bulk and surface properties, is preferred. However, to ensure uniformity, functionality and durability, thicker layers of 10–100 nm are practiced. It is vital that the coating is consistent, uniform and intact along

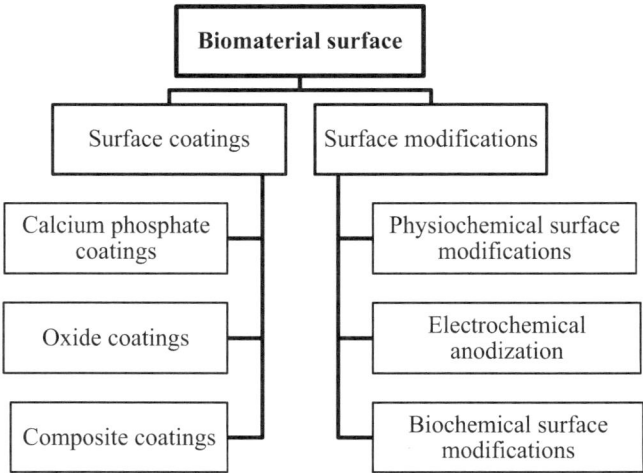

FIGURE 14.2 Classification of different methods of surface coatings and modifications introduced into the biomaterial surface.

the surface of the biomaterial. Also, it is necessary that the coatings do not introduce any impurities as they can adversely affect interfacial interactions and biocompatibility. Substrate-coating adhesion is also a primary concern as implant failure may be due to the release of ions and coating debris in the human body, which in turn causes fallouts evident by tampering biocompatibility and bio-functionality of the biomaterials.

14.2.1.1 Calcium Phosphate (Ca-P) Coatings

Hard tissue engineering with improved bio-conductivity through the formation of the bioactive layer is the main goal in providing coatings on metallic surfaces. Bone mineral composites are approximately 70% equivalent to the non-stoichiometric hydroxyapatite (HA) (HA) $(Ca_{10}(OH)_2(PO_4)_6)$, a resorbable material in the biotic system. It is more relevant to provide HA coatings on the metallic surface more generically referred to as Ca-P, as they are not recognized as foreign by the host system and aid in faster healing and better integration (Duan and Wang 2006).

Most frequently used techniques to coat bone like Ca-P on the metal substrate are plasma spraying, sol-gel, laser methods, biomimetic coatings, radio frequency sputtering, electrophoretic deposition, physical vapor deposition, electrochemical cathodic deposition, glow discharge plasma. What follows next is a brief discussion on plasma spraying, a widely used technique to coat HA on Ti and its alloys.

The plasma spraying process involves multifaceted thermal changes between the substrate, powder particles and the plasma zone that offer quick coating adhesion. Particles infused into the plasma jet endure a high heating rate in a few seconds. Some particles may achieve melting or boiling, depending upon their size and stay in the plasma zone. Droplets propel and impact the surface forming a film of several tenths of microns with a mean roughness of a few micrometers. Variation in coating density as a result of this coating process, accumulation of particulates at the rough surface, poor coating-metal adhesion, microcracks on the coating facade and propensity to delaminate are a few of the very common limitations seen in plasma spraying. Despite these shortcomings, the plasma spraying process is widely accepted for clinical applications. Limited to surface coating of thermally stable molecules such as HA, it is highly impossible to immobilize bone morphogenetic proteins and promote bone healing. Also, only substrates that are in-line to the plasma spraying are preferred to achieve uniform coating rather than complex, porous metallic shapes in orthopedic applications.

Plasma spraying and sol-gel coating are primary methods for bioactive coating on bioglass. Glass ceramics such as Bioglass and Cerabone essentially contain glass phase and hard ceramic microstructures that underline the material. Glass ceramics are composed of SiO_2, CaO, P_2O_5 with trace amounts of MgO, Na_2O or CaF_2. Commonly used methods to apply glasses as coating are glazing or enamelling using glass frits, rapid-immersion coating and flame spray coating.

14.2.1.2 Oxide Coatings

Biomaterials that are spontaneously passivating in nature such as surgical grade stainless steels (SS), cobalt-chromium (Co-Cr) alloys, Ti and its alloys form a fine and stable oxide layer when exposed to aqueous environments or atmospheric oxygen. This few nanometer thick native oxide film acts as a barrier between the substrate and the environment, which is also reasoned to increase biocompatibility (inertness) of the biomaterial. The protective nature of the oxide layer depends on the solidity of the oxide layer against dissolution and ion diffusion through the layer. This in turn is regulated by numerous factors such as oxide composition, homogeneity, microstructure and defect concentration, which are interrelated with each other. Increasing the oxide layer thickness not only affects surface roughness, wettability and surface energy, but also is known to augment long-term biocorrosion resistance in Ti and increase fibrinogen adsorption in tantalum substrate. Some metals such as vanadium (V) and aluminium (Al) exert toxicity in its oxide form, which is to be consciously replaced by strontium (Sr), niobium (Nb) and zirconium (Zr) to bypass adverse cellular response in orthopedic applications. Titanium is one of the widely investigated metals in designing oxide coatings as it provides a hard, adherent protective layer. Depending upon the choice of methods (dip coating, plasma spraying or microsphere precipitation) and the conditions involved in the process, oxide layer thickness varies from several nanometers to micrometers. Oxide coatings can also be used to increase adhesion of HA and metallic surface. Microstructure cracks along the coated layer are also subjugated and completely covered by the successive HA layer. Further, an oxide layer can also be used to create other surface microstructures by electrochemical anodization (see Section 14.2.2.2), and a detailed review by Losic et al. can give readers more insight on the same.

14.2.1.3 Composite Coatings

Composite coatings are an emerging arena for surface modifications as improved functionality is difficult to attain using a single biomaterial. For improved functionality and many other requirements, a composite coating on the biomaterial that gratifies the physical, mechanical and biological properties covering the core bulk structure is being explored. Titanium/porous titanium have better mechanical properties and Ti/ceramic composite features special physical properties that make suitable lead for pacemaker candidature. In the course of developing metallic composites, parallel ceramic/ceramic and ceramic/polymer coatings were also explored. Bio-mimicking coatings take inspiration from the naturally present human tissues and aim to reconstruct the same. In case of natural bone, biopolymers such as collagen, chitosan, gelatin, silk fibroin, poly (lacide-co-glycolide) are used along with inorganic Ca-P molecules to increase the fracture toughness of the material surface through coatings. Coating dicalcium phosphate dehydrate (DCPD) and polycaprolactone (PCL) on Mg-Zn alloys exhibited amended corrosion resistance quantified by observable increase in corrosion potential, reduced corrosion current and decreased hydrogen release.

14.2.2 Surface Modifications

The various surface modification methods discussed in this chapter are aimed to attain better biocompatibility. However, for better understanding and classification purposes, surface topography altering methodologies that influence other surface parameters should be focused. Conventionally, surface modifications were grouped into two major classes, namely physiochemical and biochemical treatments. Electrochemical techniques were also grouped under the former, while it can be deemed as a separate class of surface modifications to be introduced on the biomaterials.

14.2.2.1 Physiochemical Surface Modifications

Multiple physiochemical modifications have emerged to tailor the metallic surface to improve adhesion and stability of biological molecules and regulate molecular cascade and induce biological response. Machining, polishing, grit-blasting, acid etching, alkali etching are a few of the common processes involved in inexpensive physiochemical surface modification methods. Electrostatic interactions, van der Waals force, hydrophobic interaction and hydrogen bonding are major chemical interactions that offer rationale for the foresaid surface modification techniques.

Grit-blasting is an erosion process to create porous seams on the implant structure through collision with microscopic particles. Depending upon the dimensions of the particle, the effect is down to micrometer scale with a mean roughness index less than 1μm. Chemical etching uses acid or alkali environment for dissolution of a native oxide layer of the metal. Major shortcomings of the grit-blasting and chemical etching processes are that they are completely random with no definitive uniformity and distribution of nanostructures on the implant biomaterial. These surfaces are further used for plasma spraying and deposition or used as a starting material for other electrochemical processes (Nouri and Wen 2015).

Ion beam assisted deposition (IBAD) is a non-equilibrium process that uses ionized particles bombarded on specific metallic surfaces. When the ions impact the metallic surface, attachment, sputtering and embedding proportional to the ion's energy is observed. Accurate dose and depth control, versatility of the ions, reliability and low temperature processing are a few of the advantages that IBAD offers; however, narrow penetration depth and relatively high cost refrains common usage of this process. Laser modifications are much the preferred mode for introducing surface modifications on opaque structures. The electromagnetic radiation is precisely focused to tailor only the surface properties not influencing the bulk characteristics. Remote non-contact and easily automated processing in a regulated thermal penetration and profile can modify surface hardness, surface alloying and/or surface melting. It is also used in the surface coating process, wherein the thickness is tuned by increasing laser power, powder feed rate and laser scan speed.

14.2.2.2 Electrochemical Surface Modifications

Electrochemical methods are a variety of physiochemical surface modifications that are performed by immersing the metallic implant substrate and counter electrode in a suitable electrolyte connecting to an electric circuit pole. Anodization, electropolishing and electro-erosion are the common methods of introducing surface modifications of which anodization will be debriefed, considering its simplicity and easy commercialization of the formation of different nano- and microstructures on the implant surface.

Self-aligned perpendicular nanostructures are grown on metallic substrate that overrule delamination and fracture possibility in coatings (Figure 14.3). Among other methods such as template assisted methods, hydrothermal, sol-gel techniques for the synthesis of titania nanostructures, anodization is the most preferred for its cost-effective, easy industrialization and tuneable geometry at nanoscale. There are a number of factors that influence the geometry and dimensions of the nanostructure fabricated on the titanium substrate such as anodization time (Mohan et al. 2020), voltage (Lockman et al. 2011), electrolyte composition (Simi and Rajendran 2017), pH (Chin et al. 2011) and temperature (Mohan et al. 2020). The choice of alloy composition and phase structure of titanium persuade the formation of nanostructures and studies on influence of electrolyte composition in morphology and degree of nanostructure were observed (Simi and Rajendran 2017).

Three competing chemical reactions that dictate the formation of nanostructures are field assisted oxidation at surface, field assisted dissolution of titanium metal to form titanium oxide and chemical dissolution of the formed oxide layer into definite nanostructures due to the etching of fluorides in the electrolyte (Regonini et al. 2013). Surface functionalization of the fabricated nanostructures to achieve better osseointegration, mechanical properties and cell adhesion can also serve as an add-on. Hanks' immersion studies on TNTs fabricated by L. Mohan et al. on Ti-6Al-4V substrate showed apatite formation (Figure 14.4) that increases with prolonged incubation in simulated body fluid (SBF). This can facilitate better osseointegration and minimize inflammatory and/or implant rejection.

Addition of inorganic elements such as zinc, strontium, silver and zirconium facilitate long term anti-bacterial activity and osteogenesis induction (Indira, Kamachi Mudali, and Rajendran 2014; Mohan et al. 2013). Incorporation of polypyrrole into titania nanotubes to form PPy/TNTA hybrid structures by electro-deposition is also considered to be an alternative in orthopedic implants (Simi et al. 2018). Other methodologies to improve surface functionalization on the anodized titanium surface, like plasma nitriding (Mohan, Anandan, and Rajendran 2015a), cathodic arc vapor deposition (Mohan et al. 2015), were also investigated to enhance the mechanical parameters of the versatile titania nanotubes (TNT) structures. Loading of bioactive or drug molecules of interest onto the TNTs have made Ti an impeccable biomaterial for localized drug delivery systems (LDDS). Biopolymer coating to control the drug release in the physiological system is also explored in designing drug releasing implants (DRIs) (Mohan, Anandan, and Rajendran 2016).

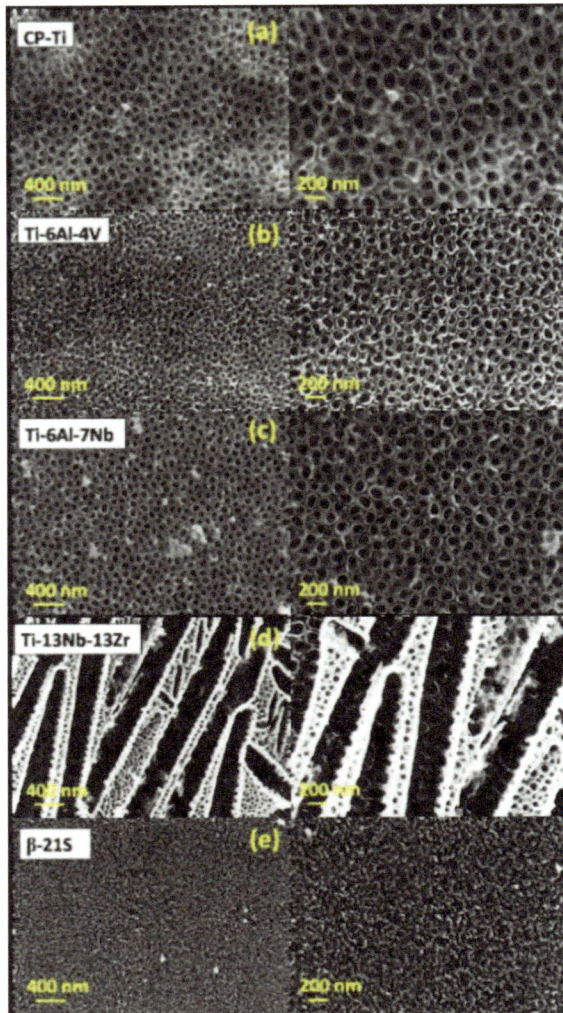

FIGURE 14.3 Surface morphology obtained by Field Emission Scanning Electron Microscope (FESEM) of different Ti and its alloys anodized for 1 h at 25°C (a) CP Ti, (b) Ti-6Al-4V, (c) Ti-6Al-7Nb, (d) Ti-13Nb-13Zr and (e) β-21s.

Source: Reprinted with permission from Mohan et al. (2020).

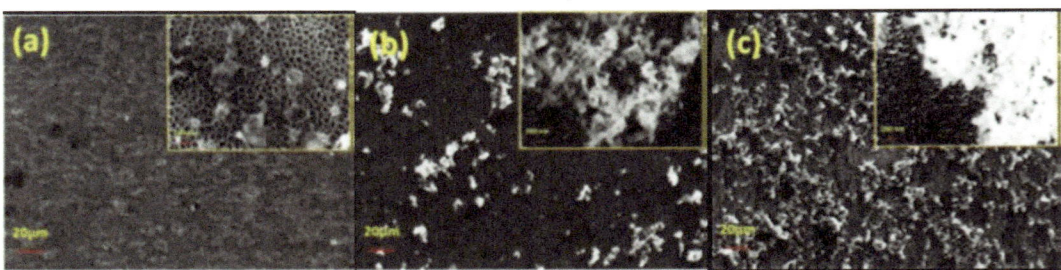

FIGURE 14.4 FESEM images of anodized Ti-6Al-4V with surface nanotubes immersed in Hanks' solution for (a) 1 day, (b) after 7 days bare substrate (Ti-6Al-4V), (c) nanotubes after 7 days.

Source: Reprinted with permission from Mohan, Anandan and Rajendran (2015b).

14.2.2.3 Biochemical Surface Modifications

Biochemical surface modifications are a recent advancement that integrates surface modification and coating process using hybrid biomaterials to immobilize bioactive molecules (proteins, enzymes and growth factors) (García 2011). Covalent adhesion of these bioactive molecules is difficult to adsorb and entrap on a metallic surface to ensure better integration and elicit favorable biological response (Wise, Mithieux, and Weiss 2009). The metallic surface is subjected to silanization process in which the surface has self-assembled reactive functional groups. The bioactive molecules are conjugated into the surface by retorting to the reactive groups. Most commonly, silane and thiols with different functional group ends are introduced into the metallic surface and the bioactive molecule of interest is complimentarily reactive to the same. Peptide molecules are ensured to have Arg-Gly-Asp (RGD) as it is essential for cell attachment cascade, depending on the application. In vascular applications, anti-thrombotic molecules are immobilized using specific functional groups to achieve blood compatibility.

14.3 APPLICATIONS AND ADVANCEMENTS

Biomimetic coating is a method for bioactive deposition on synthetic materials under physiological conditions (temperature 37° C and pH 7.4). Biomaterials made of ceramics, metals and polymers are soaked in the simulated body fluid (SBF) containing nearly equal the saturated ion composition of human body fluid. Apatite nuclei readily form on the surface favorably functionalized and continue to develop, guzzling the calcium and phosphate ions in the SBF. Essential prerequisites for this biomimetic coating are the existence of functional groups (hydroxyl, carboxyl, phosphate) on the surface supporting induction of apatite nucleation and increase in supersaturation of apatite surrounding the material surface. Combinatorial alkaline and heat treatment of thin titanate layers form a bone-like apatite layer of up to 10–15 µm thickness that aids in increased strength of the bone-implant interface. Tuneable coating thickness is achieved by modulating the saturation of Ca-P and soaking time in the physiological solution, volume of the SBF used. This method of coating is advantageous over other conventional methods as it involves formation of resorbable bioactive crystals on all forms of biomaterials. It is also non-specific to the shape and geometry of the biomaterial as long as the prerequisites are satisfied. Further, this process takes place at physiological parameters enabling provision for incorporating thermally unstable compounds (for example bone morphogenetic proteins (BMP), antibiotics, adhesion molecules) that can boost clinical performance of metallic biomaterials.

Additive manufacturing (AM) is one of the emergent fields that have reached the level of building various machineries since printing of the materials was conceptualized. Different AM techniques are available with convincing part quality and reliability for aerospace and the United States Food and Drug Administration (US-FDA). It is envisioned to create and solve for complex geometries while retaining the required mechanical properties. Extrusion techniques, like fused deposition modelling (FDM), powder fed selective laser sintering (SLS) and electron beam melting (EBM), etc., suffer from resolution limitation in introducing surface modification on the biomaterial surface. Further, homogeneity of the inlet is maintained, refraining from creating composites in a definitive printing time and causing difficulties in the scaling-up process. In these aspects, layer by layer printing and atomic/molecular layer deposition (ALD/MLD) offers controlled surface deposition allowing adherence of molecules to the surface exposed for a non-line-of-sight process. AM is a relatively new arena that has been able to fancy bioprinting in creating humanized tissues or organoids as a model for tissue engineering study or even to a near replacement of an organ. A trade-off between the resolution and the viscosity of the printing material is made; however, the end products, despite the choice of technique, are subjected to post processing. This is attributed to be the main reason that obstructs the applicability of AM as it can alter the modified surface when processed further.

Advancements occur as we speak about techniques such as continuous liquid interface production (CLIP), 3D inkjet printing, laser-induced forward transfer (LIFT) to improvise resolution and automation such that the bulk material along with spatially controlled surface micro/nano topographies can be developed. In-vitro and in-vivo validation of the implant materials and medical devices developed by various AM techniques are available, suggesting the future progression in real-time applications.

Metallic biomaterials are predominantly found in biomedical devices, surgical tools and implant design. Surface modifications in load bearing metals can alter characteristics close to biological requirements. These techniques are targeted for smoother integration of a synthetic external component into the physiological system and have advanced further in loading and release of bioactive molecules of interest.

14.4 CONCLUSIONS

Biomedical implants are inevitably designated to replace/modify/support the natural component of the human body. Biocompatibility and mechanical properties are vital; however, favoring longer stay and aiding augmented healing is an add-on. Further, it is very clear that the surface of the biomaterial is the first encounter of the physiological system and it is the bio-interfacial interactions that decide the cascade of further cellular and molecular events.

This chapter has discussed various surface modifications and coating methods in brief to give a clear picture on the mechanism and application involved. It is noteworthy that the surface properties are as crucial as the bulk properties of a biomaterial. Different techniques and their concepts were discussed and this is necessary to understand to make a sound choice in choosing the appropriate method for a particular biomaterial alloy. Surface treatments and modifications aid in better post-operative performances, healing and better integration of the synthetic or hybrid materials in the host environment, leading to development of new medical devices and improvised functionalization of the existing biomaterial. It is more fitting to carefully design the choice of surface modification and the steps involved such that it is insensitive to smaller environmental condition changes, relatively easy in industrializing and commercializing the end material.

REFERENCES

Anderson, J. M. 2001. "Biological Responses to Materials." *Annual Review of Materials Science*. https://doi.org/10.1146/annurev.matsci.31.1.81.

Benesch, Johan, Sofia Svedhem, Stefan C. T. Svensson, Ramunas Valiokas, Bo Liedberg, and Pentti Tengvall. 2001. "Protein Adsorption to Oligo(Ethylene Glycol) Self-Assembled Monolayers: Experiments with Fibrinogen, Heparinized Plasma, and Serum." *Journal of Biomaterials Science, Polymer Edition*. https://doi.org/10.1163/156856201316883421.

Brar, Harpreet S., Manu O. Platt, Malisa Sarntinoranont, Peter I. Martin, and Michele V. Manuel. 2009. "Magnesium as a Biodegradable and Bioabsorbable Material for Medical Implants." *JOM*. https://doi.org/10.1007/s11837-009-0129-0.

Chin, Lim Ying, Zulkarnain Zainal, Mohd Zobir Hussein, and Tan Wee Tee. 2011. "Fabrication of Highly Ordered TiO 2 Nanotubes from Fluoride Containing Aqueous Electrolyte by Anodic Oxidation and Their Photoelectrochemical Response." *Journal of Nanoscience and Nanotechnology*, 11: 4900–4909. https://doi.org/10.1166/jnn.2011.4108.

Disegi, J. A., and L. Eschbach. 2000. "Stainless Steel in Bone Surgery." *Injury*. https://doi.org/10.1016/S0020-1383(00)80015-7.

Duan, Ke, and Rizhi Wang. 2006. "Surface Modifications of Bone Implants through Wet Chemistry." *Journal of Materials Chemistry* 16 (24): 2309–2321. https://doi.org/10.1039/b517634d.

Eliaz, Noam. 2019. "Corrosion of Metallic Biomaterials: A Review." *Materials*. https://doi.org/10.3390/ma12030407.

García, Andrés J. 2011. "Surface Modification of Biomaterials." In *Principles of Regenerative Medicine*. https://doi.org/10.1016/B978-0-12-381422-7.10036-7.

Indira, K., U. Kamachi Mudali, and N. Rajendran. 2014. "In Vitro Bioactivity and Corrosion Resistance of Zr Incorporated TiO 2 Nanotube Arrays for Orthopaedic Applications." *Applied Surface Science* 316 (1): 264–275. https://doi.org/10.1016/j.apsusc.2014.08.001.

Kamath, Shwetha, Dhiman Bhattacharyya, Chandana Padukudru, Richard B. Timmons, and Liping Tang. 2008. "Surface Chemistry Influences Implant-Mediated Host Tissue Responses." *Journal of Biomedical Materials Research: Part A.* https://doi.org/10.1002/jbm.a.31649.

Kaur, Manmeet, and K. Singh. 2019. "Review on Titanium and Titanium Based Alloys as Biomaterials for Orthopaedic Applications." *Materials Science and Engineering C.* Elsevier Ltd. https://doi.org/10.1016/j.msec.2019.04.064.

Lockman, Zainovia, Syahriza Ismail, Khairunisak Abdul Razak, and Lim Shu Lee. 2011. "Effect of Anodisation Parameters on the Formation of Porous Anodic Oxide on Ti, Zr and W." In *IOP Conference Series: Materials Science and Engineering.* https://doi.org/10.1088/1757-899X/18/5/052004.

Mahajan, A., and S. S. Sidhu. 2018. "Surface Modification of Metallic Biomaterials for Enhanced Functionality: A Review." *Materials Technology.* https://doi.org/10.1080/10667857.2017.1377971.

Martins, Ma Cristina L., Buddy D. Ratner, and Mário A. Barbosa. 2003. "Protein Adsorption on Mixtures of Hydroxyl- and Methyl-Terminated Alkanethiols Self-Assembled Monolayers." *Journal of Biomedical Materials Research: Part A.* https://doi.org/10.1002/jbm.a.10096.

Mohan, L., C. Anandan, and N. Rajendran. 2015a. "Effect of Plasma Nitriding on Structure and Biocompatibility of Self-Organised TiO2 Nanotubes on Ti-6Al-7Nb." *RSC Advances.* https://doi.org/10.1039/c5ra05818j.

Mohan, L., C. Anandan, and N. Rajendran. 2015b. "Electrochemical Behaviour and Bioactivity of Self-Organized TiO2 Nanotube Arrays on Ti-6Al-4V in Hanks' Solution for Biomedical Applications." *Electrochimica Acta* 155: 411–420. https://doi.org/10.1016/j.electacta.2014.12.032.

Mohan, L., C. Anandan, and N. Rajendran. 2016. "Drug Release Characteristics of Quercetin-Loaded TiO2 Nanotubes Coated with Chitosan." *International Journal of Biological Macromolecules* 93: 1633–1638. https://doi.org/10.1016/j.ijbiomac.2016.04.034.

Mohan, L., C. Dennis, N. Padmapriya, C. Anandan, and N. Rajendran. 2020. "Effect of Electrolyte Temperature and Anodization Time on Formation of TiO2 Nanotubes for Biomedical Applications." *Materials Today Communications*, no. February: 101103. https://doi.org/10.1016/j.mtcomm.2020.101103.

Mohan, L., P. Dilli Babu, Prateek Kumar, and C. Anandan. 2013. "Influence of Zirconium Doping on the Growth of Apatite and Corrosion Behavior of DLC-Coated Titanium Alloy Ti-13Nb-13Zr." *Surface and Interface Analysis* 45 (11): 1785–1791. https://doi.org/10.1002/sia.5323.

Mohan, L., S. Viswanathan, C. Anandan, and N. Rajendran. 2015. "Corrosion Behaviour of Tetrahedral Amorphous Carbon (Ta-C) Filled Titania Nano Tubes." *RSC Advances* 5 (113): 93131–93138. https://doi.org/10.1039/c5ra19625f.

Nouri, A., and C. Wen. 2015. "Introduction to Surface Coating and Modification for Metallic Biomaterials." In *Surface Coating and Modification of Metallic Biomaterials.* https://doi.org/10.1016/B978-1-78242-303-4.00001-6.

Patel, N. R., and P. P. Gohil. 2012. "A Review on Biomaterials: Scope, Applications & Human Anatomy Significance." *International Journal of Emerging Technology and Advanced Engineering* 2 (4): 91–101.

Regonini, D., C. R. Bowen, A. Jaroenworaluck, and R. Stevens. 2013. "A Review of Growth Mechanism, Structure and Crystallinity of Anodized TiO2 Nanotubes." In *Materials Science and Engineering R: Reports.* Elsevier Ltd. https://doi.org/10.1016/j.mser.2013.10.001.

Shi, Donglu. 2005. *Introduction to Biomaterials.* https://doi.org/10.1142/6002.

Simi, V. S., and N. Rajendran. 2017. "Influence of Tunable Diameter on the Electrochemical Behavior and Antibacterial Activity of Titania Nanotube Arrays for Biomedical Applications." *Materials Characterization* 129 (July): 67–79. https://doi.org/10.1016/j.matchar.2017.04.019.

Simi, V. S., Aishwarya Satish, Purna Sai Korrapati, and N. Rajendran. 2018. "In-Vitro Biocompatibility and Corrosion Resistance of Electrochemically Assembled PPy/TNTA Hybrid Material for Biomedical Applications." *Applied Surface Science* 445 (July): 320–334. https://doi.org/10.1016/j.apsusc.2018.03.151.

Subramanian, K., D. Tran, and K. T. Nguyen. 2012. "Cellular Responses to Nanoscale Surface Modifications of Titanium Implants for Dentistry and Bone Tissue Engineering Applications." In *Emerging Nanotechnologies in Dentistry.* https://doi.org/10.1016/B978-1-4557-7862-1.00008-0.

Talha, Mohd, C. K. Behera, and O. P. Sinha. 2013. "A Review on Nickel-Free Nitrogen Containing Austenitic Stainless Steels for Biomedical Applications." *Materials Science and Engineering C.* https://doi.org/10.1016/j.msec.2013.06.002.

Trommler, A., D. Gingell, and H. Wolf. 1985. "Red Blood Cells Experience Electrostatic Repulsion But Make Molecular Adhesions with Glass." *Biophysical Journal*. https://doi.org/10.1016/S0006-3495(85)83842-X.

Wise, Steven G., Suzanne M. Mithieux, and Anthony S. Weiss. 2009. "Engineered Tropoelastin and Elastin-Based Biomaterials." *Advances in Protein Chemistry and Structural Biology*. https://doi.org/10.1016/s1876-1623(08)78001-5.

Witte, Frank. 2015. "Reprint of: The History of Biodegradable Magnesium Implants: A Review." *Acta Biomaterialia*. https://doi.org/10.1016/j.actbio.2015.07.017.

15 Surface Treatments of Load Bearing Bio-implant Materials

K. Saranya, P. Agilan, M. Kalaiyarasan and N. Rajendran

CONTENTS

15.1 INTRODUCTION

Biomaterials for medical implants comprise metals, polymers and ceramics, which are employed as a single material or as parts of material compounds inserted into the living tissue to fulfil the functionality for an extended period. Based on the requisite at the implementing site concerning the functionality, biocompatibility and viability of the implant in the body, the materials are chosen to perform the specified application (Moseke and Gbureck 2019). Among the available materials, the proportion of medical materials feasible for application is less. Hence, surface modifications are preferably done to cope with the constraints and improve the morphological as well as the chemical composition of the coating to support its usage in biological applications. The parameters that need to be focused on designing a surface coating for biomaterial application are biocompatibility, corrosion resistance and surface properties.

Biocompatibility is defined as "the ability of a material to perform with an appropriate host response in a specific application" (Marin et al. 2020). The materials employed in the fabrication of biomaterials should be devoid of toxicity and allergic reactions. It should not cause any physical irritation or carcinogenic effect in *in-vivo* condition. The fate of the implant material is relying on the reactions occurring between the host and the implant. The success of the implant lies in the ability to support cell survival and functions after implantation. The different kinds of tissues have various kinds of chemical signals; hence it is very hard to consider a material is compatible with all the body needs (Putra et al. 2020).

The corrosion behavior of the biomedical materials in the body environment is governed by pH, temperature, oxygen saturation level and the ions present in the biological fluids. The physiological temperature is almost constant throughout the lifetime of a person, whereas the pH varies from acidic to alkaline (4.0 to 9.0) due to infections, recovery phase, acidosis, etc. The insertion of an implant in the body can drop the pH up to 4.0 and lasts at a related pH for several weeks, leading to severe corrosion at the local sites of the implant. Additionally, the liberation of hydrogen peroxide is seen in *in-vivo* conditions during the inflammatory responses. The local pH change affects the corrosion rate of the implant by forming galvanic cells on the surface thereby affecting the integrity and viability of the implant material.

The oxygen saturation level and ionic composition of the bodily fluids play the most important role in corrosion behavior. The chloride concentrations are less and bicarbonate levels are high in blood; hence, limiting the aggressiveness of corrosion. Metals like Mg form a passive surface with carbonates that minimize the corrosion rate (Eliaz 2019). The corrosion performance of the metal in the simulated body fluid (SBF) solution at atmospheric condition is different from that of the body condition. The oxygen saturation levels of SBF at the atmospheric condition is high, hence the surface passivation of the implant is efficient whereas the implant in the body is exposed to minimum levels of free oxygen as it is bound to hemoglobin.

The corrosion behavior of the material in the body condition also depends on the proteins and cells that affect the anodic/cathodic reactions. The proteins influence the corrosion behavior: (a) the removal of metal ions from the implant surface, thereby destabilizing the electrical double layer causing the dissolution of metal. (b) It modifies the electrode potential because of its electron carrying capabilities. (c) The adsorption of protein molecules prevents oxygen entry into the implant surface leading to the breakage of the passive layer. (d) The protein molecules adhered to the surface prevent the cathodic reaction and fasten the corrosion rate. The interplay of the cells and implant surface is very specific to the type of material. Cell adherence to the implant layer acts as a physical barrier and prevents corrosion. The oxidizing agents and enzymes released by the cells may aggravate the corrosion (Virtanen 2012).

Surface characteristics are the very important parameters that determine the biological cascade reaction. The surface properties including topography, stiffness, chemical constituents, roughness, interfacial free energy, charge, wettability, etc. determine the fate of the implant. The topographical patterns and roughness have a direct impact on the cytoskeleton and protein orientations. The cell alignment was in an anisotropic pattern on an anisotropic patterned topography. Nevertheless, the isotropic topographies show enhanced cellular responses through cell signaling pathways. Among the various topographic parameters, the micro- and nano-scale patterns play a significant role in governing cellular functions. The micro-scale patterns have an impact on the cell morphology whereas the nano-scale pattern controls the physicochemical sensing pathways. The micro-environment and physical forces adjacent to the cells can play a crucial role in the implantation. At the microlevel, the cells sense and utilize mechanical stimuli like roughness into biological responses that influence the integration of the implant. When cells come across stiff implant material, the focal adhesions get activated and transport the signals to the cytoskeleton thereby the implant establishes the link with the cells (Rahmati et al. 2020).

15.1.1 Surface Treatments on Metals

Surface treatments on biomaterials have helped to improve the biological responses of biomaterials through changes in a material's surface properties via. surface energy, topography and charge. The metallic material Ti was employed for long-term application because of its high corrosion resistance and bio-inertness. The biodegradable metals like Mg and Zn are preferably studied as it is the essential micro-nutrient of the body and can avoid revision surgery. However, surface treatments are essential to improve the bioactivity and corrosion resistance of these materials (Duan and Wang 2006). Surface treatments are classified as follows:

1. Addition of desirable materials to the biomaterial surface
2. Converting the prevailing surface into more appropriate composition and/or topographies
3. Excision of material from the surface to develop required topographies.

These surface modifications can be achieved on the biomaterial's surface (metals, ceramics and polymers) using various surface modification techniques (Figure 15.1). Surface modifications should offer distinctive properties of interaction with biomolecules or cells in the biological environment. The physical, chemical and biological properties of the biomaterial can be altered by various surface treatments. The key motive for the surface treatment is to retard the corrosion, antimicrobial property and induce the interaction at bone-implant interfaces (Niinomi 2019).

15.1.2 Conversion Coating

The process of converting the native surface oxides to new oxides/compounds with the aid of chemicals to achieve the desirable properties is called conversion coating. The modified layer is anticipated to provide better corrosion resistance and satisfy the requisites of biocompatibility (Saranya et al. 2019). Conversion coating is a strong adhesive coating that is formed by wetting the metal in an electrolyte with external power (electrochemical conversion or anodization) or without external power (chemical conversion). The superficial layer of the substrate interacts with the chemical bath solution forming a consistent coating layer and an intermittent chemical bonding layer (Chen et al. 2015).

15.1.2.1 Electrochemical Anodization

Electrochemical anodization is a simple electrochemical process that enables the production of protective dense oxide film on the metal surface with amendable surface nanostructure or microstructure and crystal structure (Wang et al. 2016). Among the various surface modification techniques, the electrochemical anodization technique is one of the most favorable techniques to fabricate nanostructures on metal surfaces, which is capable of forming a fine-tuning thick oxide film, feature size, nano-topography and chemistry. It was used to fabricate dense, stable and passive oxide layers on valve metals such as titanium, aluminium, zirconium, tantalum and hafnium. Generally, the electrochemical anodization process is the oxidation of metals using various fluoride-comprising organic or inorganic electrolytes, namely, H_2SO_4/HF, H_3PO_4/NaF, glycerol/HF, citric acid/HF, etc., leading to the formation of nanotubular and nanoporous structures due to the presence of fluoride ions. The geometry, diameter and length of the nanotubes are strongly influenced by electrolyte, voltage, anodization time and temperature. Organic (or neutral) electrolytes are viscous in nature and form well-organized and longer length nanotubes. In the case of inorganic (or aqueous) electrolytes, larger diameter nanotubes with shorter lengths were formed (Indira et al. 2015).

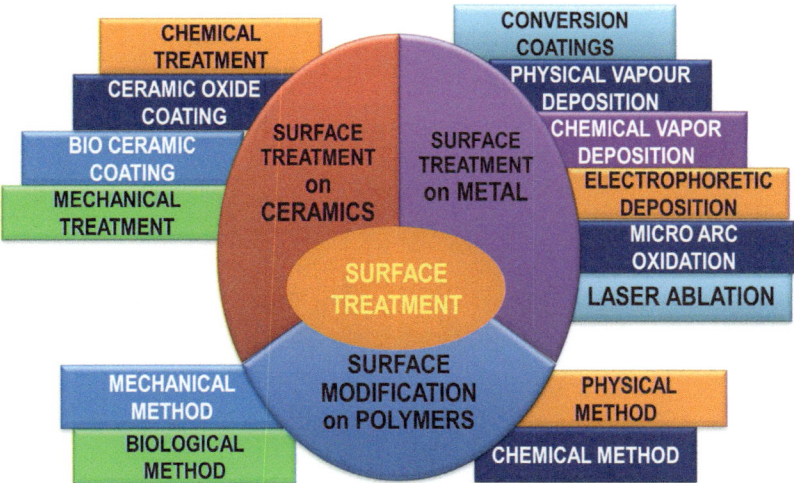

FIGURE 15.1 Surface treatment techniques on biomaterials.

The fabrication of nanostructures with different architectures and properties on metal surfaces with the thick oxide layer ranging from 5 to 25 μm can be obtained. The passive anodic layer has good adherence with the base metal substrate that makes them durable for a long period with enhanced corrosion resistance. Among the valve metals, titanium and its alloys are widely employed for biomedical applications. Nano-scale structured artificial implant materials simulate extracellular matrix (ECM) and interact better with cell physiology; thereby they could resolve the obstacles related to the orthopedic implant materials (Souza et al. 2018).

The nanostructures on the Ti surface can control the protein adsorption, cell responses, bone-tissue contact and enhance osseointegration. The anodizing electrolyte comprises F- and OH- ions that get absorbed on the TiO_2 nanotube surface, making it a negatively charged surface. Hence, the positively charged proteins easily attach to the nanotubes, which facilities cell attachment (Awad et al. 2017). Besides, the TiO_2 nanostructures on the Ti surface match with the nano-scale regime of the natural bone thereby enhancing cell adhesion and proliferation accompanied by improved corrosion resistance because of the oxide film (Perumal et al. 2021). Ercan et al. (2010) investigated the effect of the anodic TiO_2 nanostructures (40–60 nm) on bone-cell formation, osteoblasts adhesion and bacterial adhesion. The nanotubes significantly improved osteoblast cell proliferation and adhesion compared to the titanium substrates. Moreover, calcium phosphate deposition on anodized substrates was higher compared to that of the titanium substrates.

The growth process of self-ordered TiO_2 nanotubes arrays (TNTA) was the competition between the field-assisted anodic oxidation and TiO_2 layer dissolution in the electrochemical conditions. The surface properties of the titanium changed remarkably due to the formation of the thick oxide film and the simultaneous H_2 evolution at the cathode. The fluoride ions influence the TNTA formation and had the capability to combine with TiO_2 to form a water-soluble hexafluorotitanate complex ion $[TiF_6]^{2-}$. Moreover, as the fluoride ions are smaller in size, they can move into the growing TiO_2 lattice and disintegrate the oxide layer, thereby resulting in the increased thickness of the tubular oxide layer. The anodic layer consists of Ti^{4+}, which reacts with the fluoride ions to form $[TiF_6]^{2-}$ (Perumal et al. 2020).

The surface morphology and properties of TiO_2 nanotubes were governed by anodization parameters like voltage, electrolyte concentration, electrolyte temperature, anodization time and electrode distance (Agilan and Rajendran 2018). The movement of ions and the diameter of the nanotubes are influenced by the applied voltage (Figure 15.2). In organic electrolytes, the applied voltage ranges from 10 to 60 V whereas in inorganic electrolytes, the applied voltage ranges from 5 to 30 V for the formation of well-organized nanostructures (Regonini et al. 2009).

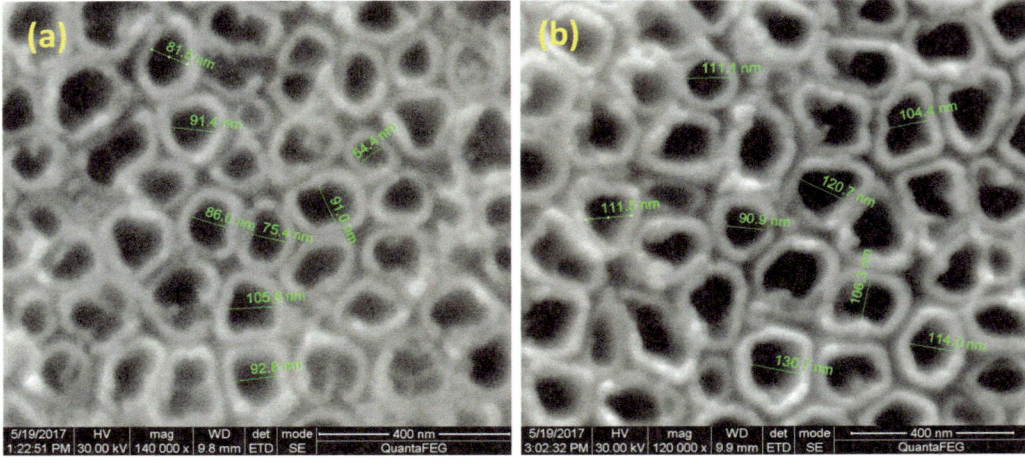

FIGURE 15.2   Nanotubes on Cp-Ti at (a) 20 V and (b) 30 V in inorganic electrolyte.

Inorganic electrolytes are aggressive in nature, which helps the nanotube growth on the surface of titanium metal within a few minutes after the onset of the anodization process. The time range of anodization typically ranges from 30 to 120 min for structural rearrangement and increases the degree of self-ordering. In the case of organic electrolytes, the nanotube growth process is slower compared to the inorganic electrolyte. The rate of dissolution of nanotubes in the organic electrolyte is comparatively less than the inorganic electrolyte. These phenomena in organic and inorganic electrolytes explain the concentration of movement of ions between electrodes (Cai et al. 2006). Electrolyte temperature is one of the factors that influence nanotube growth. At lower temperatures (< 2 °C) the nanotube growth is inhibited in the inorganic electrolyte. The temperature ranging from 0 °C and 40 °C is the most favorable condition for the nanotube formation in organic electrolyte. Annealing temperature influences the crystallinity and hydrophilicity of the nanotubes that supports improved corrosion resistance and enhanced cell interactions. The annealing temperature from 200–450 °C forms the anatase phase and above 450 °C the rutile phase is formed (Prida et al. 2007). Table 15.1 overviews the cellular responses on the nanostructures of the metal implants.

TABLE 15.1
Cellular Responses on the Nanostructures of the Metal Implants

Metallic Biomaterials	Electrolyte	Applied Voltage	Anodization Time	Annealing Temp and Time	Nanotube Diameter	Nanotube Length	Cellular Response
CP-Ti (Perumal et al. 2021)	H_2SO_4 + HF	20 V	1 hr	500° C, 2 hr	~ 90 nm	~ 250 nm	Better MG63 osteoblast cell adhesion and proliferation.
Ti foil (Yu et al. 2010)	H_3PO_4 + HF	15 V	3 hr	450° C, 3 hr	~ 70 nm	-	Good MC3T3-E1 pre-osteoblast cell response.
Ti-6Al-4V (Mohan et al. 2015)	H_2SO_4 + HF	20 V	30 min	450° C, 3 hr	~80 nm	~250 nm	Good osteoblast cell attachment.
Ti-30Ta (Capellato et al. 2012)	H_2SO_4 + HF + 5% DMSO	35 V	40 min	530° C, 3 hr	~ 80–100 nm	~ 100 nm	Well adhered fibroblast and differentiated.
Zirconium foil (Frandsen et al. 2011)	1st step $(NH_4)_2 SO_4$ + NH_4F 2nd step	20 V 20 V	30 min 15 min	300 ° C, 6hr	40 nm	10 μm	MC3T3-E1 cells are well attached with a concentrated calcified extracellular matrix.
Ti-50Zr (Vardaki et al. 2019)	1st step Glycerol + H_2O + NH_4F 2nd step	55 V 75 V	4 hr 1 hr	450° C, 1 hr	~ 160 nm	~ 9.4 μm	Showed HGF-1 gingival fibroblast cell viability and cellular response.
AZ31 Magnesium alloy (Saranya et al. 2021)	Hexafluorotitanic acid + HF	80 V	1 hr	-	-	-	Enhanced ALP activity, osteocalcin levels along with antibacterial activity.
Tantalum foil (Uslu et al. 2021)	H_2SO_4	10 to 55 V	1 min Nanotube		30–140 nm	0.5–8.5 μm	Higher protein absorption with better cell adhesion.
	H_2SO_4/HF + DMSO	10–35 V	10–30 min Nanodimple				
	H_2SO_4 + NH_4F	20 V	1–2.5 h Nanocoral		120 nm		

15.1.2.2 Chemical Conversion Coating

The chemical conversion coatings are usually done on reactive metals like Mg and Zn dipped in acidic bath solution (via. hydrofluoric acid, phosphoric acid, selenious acid, etc.) that initiates the reaction. When the metal interacts with the acidic solution, the metal ions eventually come out vigorously leading to the elevation of metal ions concentration at the metal/solution interface. The bath solution penetrates the dissolving metal oxide layer, thereby facilitating the formation of the conversion layer and the OCP vs. the time curve is shown in Figure 15.3 (Liu et al. 2015). The eminence of the coating is influenced by different parameters including immersion time, the concentration of the chemicals, co-additives, bath temperature, agitation conditions and post-treatment (Saji 2019). Different morphological structures like small flakes, leaf-like, flower-like morphology, etc. are usually observed in chemical conversion coating that supports the apatite deposition and cell adhesion. The topographical changes with various structures are often found in alloys. The α and β phases act as micro-anodes and micro-cathodes leading to the nucleation of the formation at the β phase (Zhou et al. 2008).

Amidst different surface coating and treatment techniques, the chemical conversion coatings have gained greater importance because of their ease of preparation, less operating cost (low-temperature reactions, cheaper chemicals, less energy consumption, speedy reactions) and adherent coatings. The chemical conversion coating is segregated into organic, inorganic and composite coating. Table 15.2 summarizes the formulated conversion coatings that support the biomedical application.

15.1.3 ELECTROPHORETIC DEPOSITION

The electrophoretic deposition (EPD) process is associated with the migration and accumulation of charged particles by an applied electric field onto a conductive electrode to fabricate films and coatings on materials. EPD facilitates the fabrication of a variety of structures with different

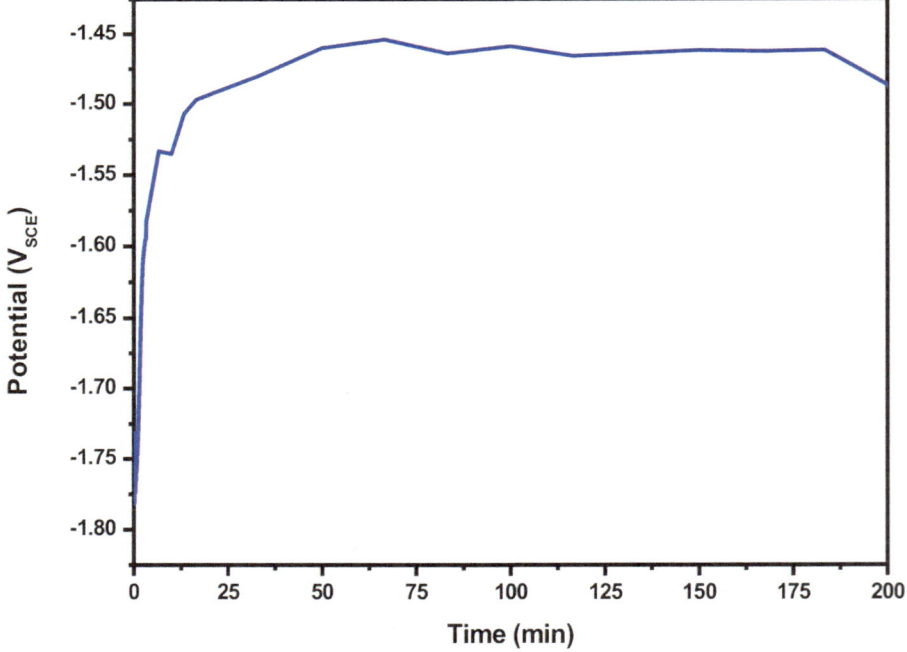

FIGURE 15.3 OCP vs time curve during the selenium conversion coating on AZ31 Mg alloy.

TABLE 15.2
Conversion Coating Employed for Biomedical Application.

Type	Material Used	Electrolyte	Biomedical Application
		Inorganic Conversion Coating	
SrP (Chen et al. 2014)	Pure-Mg	0.1 M $Sr(NO_3)_2$ and 0.06 M $NH_4H_2PO_4$	Supports the differentiation of human MSC in the formation of bones.
Fluoride (Saranya et al. 2019)	AZ31 Mg alloy	HF	Supports cell adhesion. Lowered degradation rate in dynamic electrochemical studies.
Phosphate (Su et al. 2019)	Pure-Zn	0.07 M $Zn(NO_3)_2$ and 0.15 M H_3PO_4	Considerably enhanced hemocompatibility, cytocompatibility and antibacterial performance.
Carbonate (Prabhu et al. 2020)	Mg- 4 wt% Zn	Saturated $NaHCO_3$ solution	Supports apatite formation.
Fe-Phosphate (Yin et al. 2020)	AZ31 Mg alloy	$FeSO_4$, H_3PO_4, Na_3PO_4	Improved degradation rate. Better hemocompatibility.
		Organic Conversion Coating	
Epigallocatechin Gallate (EGCG) (Zhang et al. 2015)	MgZnMn	EGCG in Tris buffer	EGCG-metal complex reduced the corrosion density.
CA/PEI (Zhang et al. 2018)	MgZnMn	CA (catechol) and PEI (polyethyleneimine) in Tris buffer	Improved hemocompatibility. Obvious HUVEC proliferation and inhibits the proliferation of HUASMC.
		Composite Conversion Coating	
Phosphate-GO (Zhang et al. 2020)	Pure Zn	0.07 M $Zn(NO_3)_2$ and 0.15 M H_3PO_4 at a pH of 2.5 along with graphene oxide	Slows down corrosion resistance.
Ag/HA (Hu et al. 2017)	2Zn-1Mn-0.5Ca	$NaSiO_3$, $Ca(NO_3)_2$, $AgNO_3$	Excellent coating adhesion with good antibacterial activity.
PA/Ca^{2+} (Liu et al. 2019)	Mg-Sr	Phytic acid solution and calcium nitrate	Osteoinductive with improved cell adhesion.

compositions, complex shapes, advanced materials with nano- and micro-sized, thin and thick films, and also form porous scaffolds with highly compact coatings. The EPD process can be widely used for the deposition of colloidal particles of metals, ceramics, polymers and composite materials. Mostly, it has been used for depositing bioactive ceramics on metallic implant materials. Bioactive nanostructures of titania, alumina, zirconia and hydroxyapatite coatings on metal substrates were produced by EPD (Abbass et al. 2021).

15.1.4 LASER ABLATION OR LASER SURFACE TEXTURING

Laser sources are used to tailor the favorable surface properties on the metal surfaces without changing the fundamental properties of a material. Laser ablation is a widely used technique to remove or restructure a portion of the material to create a micro- or nano-patterned material using a laser beam. Laser treatment is a quick and precise process for the formation of the ideal structures in the biomedical field (Shivakoti et al. 2021). Pulse frequency, intensity of the lamp and scan speed play a substantial role that influences the surface roughness of the implant (Mohammad et al. 2016). Laser ablation fabricates the micro-textured grooves which are 30–50 μm in width and 15–50 μm deep on ZrO_2, increasing the surface wettability (Liu et al. 2017). Roy et al. (2008) reported tri-calcium phosphate (TCP) coated Cp-Ti by laser treatment enhanced the bone cell interaction with bio-implant material. Yu et al. (2018) demonstrated that the micro-texture formed by picosecond laser ablation on Ti6Al4V alloy exhibited good biofunctionalization and better mesenchymal stem cell adhesion on the micro-groove of the Ti surface. Ti and Mg alloys engineered with functional bioactive materials by laser micro-processing method showed improved cell attachment and better corrosion resistance (Hu et al. 2018). Pereira et al. (2020) studied that laser ablation is a suitable process for the fabrication of hydrophilic surfaces on ceramics. The laser textured Ti_6Al_4V and 316L SS coated with hydroxyapatite offered more corrosion resistance and good cell interaction (Stango et al. 2018).

15.1.5 MICRO ARC OXIDATION

Micro Arc Oxidation (MAO) is an attractive surface engineering process for metallic materials like Mg, Ti. Generally, it is an electrochemical oxidation method that forms a dense and well-adhered oxide layer on Mg and Ti alloys. The coating thickness of ~100 μm can be achieved. The formed dense oxide coatings offer enhanced corrosion-resistant and mechanical properties (Rúa et al. 2019). It is akin to the electrochemical anodization process, but it requires high potentials. The topography and phase transitions of the oxide layer were changed by the applied potential. The thickness and roughness of the oxide layer were enhanced with the increase in potential. The MAO treated Cp-Ti (Grade 2) brings about positive change in the physical and chemical properties related to the biomedical application (Koshuro et al. 2018). The porous nano-crystalline, titania film formed on a titanium substrate by the MAO method showed induced apatite formation and exhibited better bioactivity (Li et al. 2019). Butt et al. (2020) demonstrated MAO treated AZ31 Mg alloy with PLA displayed controlled degradation in Hank's solution with higher tensile strength.

15.2 SURFACE TREATMENT OF CERAMIC OXIDE COATING

Several clinical failures of dentistry (e.g. teeth root fracture) are frequently associated with cementation flaws. To enhance the bonding strength between the ceramic and luting agent, various surface treatments are employed.

Acid etching is the most effective treatment for achieving high bonding strength. After etching with HF, it enhances the magnitude of ceramic surface roughness and enhances the interaction of ceramic-resin cement (Posritong et al., 2013). The acid etching removes the smooth glassiness and leads to a microporous surface with high energy that induces better mechanical locking between the tissue and implant. Xie et al. (2013) have discussed that the bond strength increased between zirconia and resin with no effect on mechanical properties.

Coupling agents play a vital role in the adhesion of silica and resin in ceramic repair and cementation in dentistry application. A silane coupling agent is the effective bonding in silica-based restorative materials including acid etched to resin composite luting cement. Silane coupling agents are organic-inorganic bifunctional groups at the end of the molecules. The general formula of silane bifunctional is $L-(CH_2)_k-Si-(OR)_3$. Where L is an organofunctional group (epoxy, methacrylate acrylate), OR is alkoxy group and $-(CH_2)_k-$ is a linker group of the organofunctional group and Si atom. The bi-functional groups at the end create a siloxane network with the hydroxyl (-OH) of the Si in the ceramic surface and copolymerize with the resin matrix of composites. The silane group enhances the wettability by increasing surface energy; thereby the hydrophobic resin matrix can interact with hydrophilic ceramic surfaces (Matinlinna et al. 2018). Silane coupling has proven to increase the bond strength in ceramic composite during the luting procedure in repairing chipped ceramic restoration. Researchers have reported the efficiency of the silane coupling agent in increasing the bond strength of the fiber and ceramic composites (Bona et al. 2000). Silane coupling agent was effective in improving adhesion and bond strength between ceramic and resin. Park and Jin (2001) studied the increase in interfacial bonding between the C=O and $-NH_2$.

15.3 SURFACE MODIFICATION ON POLYMERS

The distinctive properties of polymers including viscosity, malleability, moldability and their mechanical strength make them an exquisite material in the biomedical field. They can perform well in various biological requisites, which cover knee prostheses with high mechanical strength, heart values with flexibility and durability to sustain the repeated cycle of blood pressure, oxygenator membranes with air permeability and so on. To induce the biological responses and biocompatibility of the polymeric materials, the surface properties such as roughness, adhesion, adsorption of molecules, wettability, ionic charge, pH, etc. are tailored. To facilitate understanding of the significance of surface modification techniques, they are categorized into four types (physical, chemical, mechanical and biological) and are discussed in the following sections.

15.3.1 Chemical Method

Chemical modification of the polymeric topography was done by introducing new molecules or ions by chemically grafting the functional groups (acetylation, fluorination) or by altering the prevailing functional groups (oxidation, reduction). With specific chemical reactions, stronger interacting molecules can be grafted onto the surface. Hydrolysis and aminolysis are the two wet techniques used for the grafting of polymers. Hydrolysis is a simple pH-dependent process that may lead to the degradation of the polymeric surface whereas aminolysis produces organic salt that gets degraded at elevated temperatures.

The radiation-induced surface modification is a cost-efficient, controllable and well-established technique. Among all the radiation methods, the most widely used ones are gamma, UV and electron beam radiation. The local surface chemistry is modified alone and bulk properties are retained. The process consists of four basic steps: (a) the initiation of free radicals formation, (b) chain propagation, (c) chain transfer and (d) termination. Two parameters that govern the rate of the reaction are the confined spatial effect due to the free-ends of the polymeric chain tethering that prevents the growth of the nearby chain and the persistent free radical effect maintains irreversible dynamic equilibrium. The photo-induced grafting of PDMS with various polymers including acrylic acid (AA) and acrylamide (AM) enables the application-specific functionalized surface (Sun et al. 2020). Polymer grafted PEEK under the UV radiation of 15 min had increased the protein adsorption and cell adhesion (Ishihara et al. 2020).

15.3.2 Physical Method

A variety of physical techniques are widely employed to enhance the addition of oxygen-containing molecules rather than modifying the polymeric surface with chemical agents that impart the polymer surface morphology and are discussed next.

15.3.2.1 Plasma-Assisted Coating Methods

Plasma treatment is the widely preferred technique to specifically modify the physical and chemical properties of the polymeric surface. Plasma treatments can introduce functional groups (oxygen, amine, carboxyl), graft polymerization, crosslink formation, along with cleaning and etching to improve sterilization efficiency and biocompatibility (Petlin et al. 2017). Based on the outcome of the plasma reaction, it can be classified into three types:

- *Plasma Polymerization* is defined as "the formation of polymeric materials under the influence of plasma" (Yasuda 2012). The plasma polymerization involves high energy electrons and UV radiation. It is a two-step process that occurs simultaneously: (a) the removal of surface molecules bombarded with ions from the plasma and the ionization of the precursor molecules produce reactive free radicals by breaking bonds. (b) The grafting and crosslinking on the polymer surface can be done on the free radicals that act as anchor sites to enhance the coating stability. The gaseous molecules including Ar, He, O_2, N_2, NH_3 and CF_4 are employed to achieve the desired surface properties (Aziz et al. 2018).

- *Plasma Treatment* can be easily applied on thin polymeric films and nanofibers to enhance biocompatibility. The cold plasma treatments are widely preferred as it employs AC source with the diminution of temperature up to 150 °C. Many biologically significant electrospun polymers like PCL, PLA exhibit a hydrophobic nature. The post-plasma treatment with oxygen-rich and amine-rich functionalities not only increases the wettability of the polymer but also enhances the adhesion, growth, differentiation and spreading of cells. The post-plasma-treated nanofibrous mats were tested with osteoblasts, myoblasts, fibroblasts and stem cells and showed an improvement in biocompatibility and no separate sterilization is required (Baradaran et al. 2015).
- *Plasma Etching and Cleaning*: reactive-ion plasma etching is a technique that aids in producing specific bioengineered materials. The distinctive nanopatterns and biofunctionalization can be produced on the polymeric surface (biochip) to aid the detection of the diseases. The bladder cancer diagnosis is made possible with the selective capture of cancer cells from the urine (Macgregor et al. 2017). Nanostructured PCL film was fabricated on the silicon wafer with the plasma etching to biomimetic surface that supported the high cellular adhesion performance. Then, PCL film was removed from the silicon wafer and it was free-standing (Limongi et al. 2017).

15.3.3 MECHANICAL METHOD

The surface of the polymers is tailored with the mechanical method using micromanipulation and roughening. The cleaning along with the texturing of the polymeric surface can be done effectively using the mechanical method.

The surface modification of the polymers by mechanical roughening aids the topographic alteration from microporous to porous. Govindarajan and Shandas (2014) employed alumina slurry particles on polymer surfaces to enhance the surface roughness, thereby improving the endothelial cells' attachment without altering the chemical constituents of the polymer.

The physical interaction of the polymer surface with the mechanical process is employed in micromanipulation. The microscope was used to position them and to carve the nano and micro level structures with high precision. Scanning tunnelling microscope and atomic force microscope are preferred to manipulate the surface. Scanning probe-assisted patterning with the heating has provoked remarkable freedom in the surface patterning (Ciftci et al. 2019).

15.3.4 BIOLOGICAL METHOD

The immobilization of the biological molecules on the surface of the polymer aids the biocompatibility, hemocompatibility and vascularization with the specific cellular response at the required site. The frequently employed biomolecules are carbohydrates, proteins (collagen, gelatin, RGD peptides), lipids, synthetic chemicals (ascorbic acid) and growth factors (stromal-derived growth factor and erythropoietin). The non-toxic and biomimicking nature prevents adverse reactions like inflammatory reactions, etc. that are caused by synthetic polymers. Moreover, the biomolecules promote the cell adhesion, proliferation, differentiation and organization of ECM that support faster tissue regeneration and wound healing.

15.3.4.1 Protein Coating

Altering the superficial layer of the polymer with protein provokes cellular interaction by mediating biomolecular recognition which can be attained by chemical or physical methods with ECM proteins or peptides. The amino acids and the long chain of ECM proteins act as a signalling domain and interact with cell membrane receptors. The proteins are employed for surface modification which initiates the series of reactions that control cell behavior. The protein adsorption on the surface may cause unwanted reactions like coagulation. To avoid unnecessary reactions and encourage the absorption of specific proteins, surface passivation is done.

15.3.4.2 Carbohydrate Coating

Among the different types of carbohydrate, heparin and chitosan are widely employed as a surface coating to improve biocompatibility, tissue engineering, antimicrobial activity and to ensure drug-controlled release. Heparin is an anticoagulant, widely employed to enhance hemocompatibility. The heparin and heparin-mimicking functionalized polymers are used in blood cleansing, artificial organs and other clinical medical devices. The heparinization of the polymer is done with electrostatic interaction between the positively charged polymer and the negatively charged heparin to enhance the stability (He et al. 2019). Chitosan is a preferred coating material on the polymer to enhance the antimicrobial property and prevent clot formation. The immobilization of the chitosan is done by the creation of NH_2 active sites by aminolysis with 1,6-hexane diamine on the polyester substrate, proceeded by chemical grafting of chitosan with the crosslinking agents (Jeznach et al. 2019).

15.3.4.3 Physical Adsorption and Self-Crosslinking of Biomolecules

Physical adsorption is the effortless method in which the polymer is incubated with biomolecules solution. Here, the biomolecules bind to the polymeric surface using van der Waal forces, electrostatic forces, hydrophobic interactions and hydrogen bonds. Two main parameters influence the adsorption of the biomolecule on the surface: (a) physicochemical properties of the biomolecule, liquid medium and polymeric surface properties; (b) parameters employed to characterize biomolecule adsorption (surface-adsorbed amount, conformational change, aggregation state and state reversibility) (Migliorini et al. 2018). The physical adsorption technique does not require coupling reagents, seldom activation or chemical modification. Physical "entrapment" systems include (a) microcapsules, (b) hydrogels and (c) physical mixtures such as matrix drug delivery systems. It is a simple and cost-effective method with a small conformational change of the biomolecules. However, the interaction is weak between the polymer and the biomolecules; therefore long-term stable material cannot be prepared.

15.4 SUMMARY

Surface treatment on biomaterials is a vast subject that has an immense propensity for enhancing the implant efficiency and viability in a human body. It has a great purpose in the medical field with many challenges that have to be addressed. The detailed study on literature emphasizes that the surface treatments aid in the betterment of biological and blood compatibility, sterility, surface properties, mechanical properties, corrosion resistance, etc. However, the surface treatments improve only certain properties. For instance, oxidation enhances cell adhesion and apatite formation, ion implantation aids corrosion resistance and chemical conversion coating supports antimicrobial and cell adhesion. Also, only certain techniques can be scaled up. The techniques like MAO and anodization can't be scaled up and non-uniform shapes can't be coated because uniform coating can't be formed on the entire surface. The techniques like micromanipulation require huge efforts and time. The laser ablation technique can be scaled up easily and used for uniform coating even in irregular shapes. Also, they can be used for coating various materials, namely, bioglass, polymers, composites, etc. on the metallic implant materials. However, it is too costly and requires extensive work for colloidal solution preparation and very stringent conditions. On the whole, all the techniques have one or more disadvantages. On taking into accounts the merits and demerits, laser ablation can be one of the viable techniques to treat the implant surface. Ceramic materials like CaP compounds and alumina are treated with acids and base to improve biocompatibility. The introduction of the hydroxyl groups on the ceramic surface supports the ALP activity. Moreover, there is no change in the mechanical properties of the ceramics employed in the hard tissue replacement.

The surface treatment on polymers improves the physicochemical properties and biocompatibility and thereby acts as a carrier in bone-tissue engineering. Chemical modification, plasma treatment and ion implantation on polymers are widely preferred. The chemical modifications on the polymers improve the biological functionalities. Nevertheless, during the modification, biosafety and green synthesis should be considered as many of the chemical reactions involve toxic chemicals, which are harsh on the environment. The grafting and crosslinking of the polymers with plasma treatment precisely controls and modulates the material properties, thereby exploring the complete potential of the polymer.

15.5 FUTURE SCOPE

Many of the surface modification techniques are still in the laboratory stage and the few that have been employed in the industries are not very efficient in preparing biofunctional coatings. This opens up a new venture on the optimization of surface treatment techniques with a multiple stage process that can form multi-functional coatings to aid biocompatibility and antimicrobial activity.

REFERENCES

Abbass, M.K. Khadhim, M.J. Jasim, A.N. Issa, M.J. 2021. A study the effect of porosity of bio-active ceramic hydroxyapatite coated by electrophoretic deposition on the Ti6Al4V alloy substrate. *Journal of Physics: Conference Series* 1773: 012035.

Agilan, P. Rajendran, N. 2018. In-vitro bioactivity and electrochemical behavior of polyaniline encapsulated titania nanotube arrays for biomedical applications. *Applied Surface Science* 439: 66–74.

Awad, N.K. Edwards, S.L. Morsi, Y.S. 2017. A review of TiO_2 NTs on Ti metal: Electrochemical synthesis, functionalization and potential use as bone implants. *Materials Science and Engineering: C* 76: 1401–1412.

Aziz, G. Ghobeira, R. Morent, R. De Geyter, N. 2018. Plasma polymerization for tissue engineering purposes. *Recent research in polymerization* 69–93, IntechOpen Limited.

Baradaran, R.A. Biazar, E. HeidarI-keshel, S. 2015. Cellular response of stem cells on nanofibrous scaffold for ocular surface bioengineering. *Asaio Journal* 61 5: 605–612.

Bona, A.D. Anusavice, K.J. Shen, C. 2000. Microtensile strength of composite bonded to hot-pressed ceramics. *Journal of Adhesive Dentistry* 2 4: 305–313.

Butt, M.S. Maqbool, A. Saleem, M. Umer, M.A. Javaid, F. Malik, R.A. Hussain, M.A. Rehman, Z. 2020. Revealing the effects of microarc oxidation on the mechanical and degradation properties of Mg-based biodegradable composites. *ACS Omega* 5 23: 13694–13702.

Cai, Q. Yang, L. Yu, Y. 2006. Investigations on the self-organized growth of TiO_2 nanotube arrays by anodic oxidization. *Thin Solid Films* 515: 1802–1806.

Capellato, P. Smith, B.S. Popat, K.C. Claro, A.P. 2012. Fibroblast functionality on novel Ti30Ta nanotube array. *Materials Science and Engineering: C* 32: 2060–2067.

Chen, X.B. Chong, K. Abbott, T. Birbilis, N. Easton, M. 2015. Biocompatible strontium-phosphate and manganese-phosphate conversion coatings for magnesium and its alloys. *Surface modification of magnesium and its alloys for biomedical applications*, ed. T.S.N. Sankara Narayanan, I.S. Park, and M.H. Lee, 407–432. Woodhead Publishing.

Chen, X.B. Nisbet, D.R. Li, R.W. Smith, P. Abbott, T.B. Easton, M.A. 2014. Controlling initial biodegradation of magnesium by a biocompatible strontium phosphate conversion coating. *Acta Biomaterialia* 10: 1463–1474.

Ciftci, H.T. Van, L.P. Koopmans, B. Kurnosikov, O. 2019. Polymer patterning with self-heating atomic force microscope probes. *The Journal of Physical Chemistry A* 123 37: 8036–8042.

Duan, K. Wang, R. 2006. Surface modifications of bone implants through wet chemistry. *Journal of Materials Chemistry* 16: 2309–2321.

Eliaz, N. 2019. Corrosion of metallic biomaterials: A review. *Materials* 12 3: 407.

Ercan, B. Webster, T.J. 2010. The effect of biphasic electrical stimulation on osteoblast function at anodized nanotubular titanium surfaces. *Biomaterials* 31: 3684–3693.

Frandsen, C.J. Brammer, K.S. Noh, K. Connelly, L.S. Oh, S. Chen, L.H. 2011. Zirconium oxide nanotube surface prompts increased osteoblast functionality and mineralization. *Materials Science and Engineering: C* 31: 1716–1722.

Govindarajan, T. Shandas, R. 2014. A survey of surface modification techniques for next-generation shape memory polymer stent devices. *Polymers* 6 9: 2309–2331.

He, C. Ji, H. Qian, Y. Wang, Q. Liu, X. Zhao, W. Zhao, C. 2019. Heparin-based and heparin-inspired hydrogels: size-effect, gelation and biomedical applications. *Journal of Materials Chemistry B* 7 8: 1186–1208.

Hu, G. Guan, K. Lu, L. Zhang, J. Lu, N. Guan, Y. 2018. Engineered functional surfaces by laser microprocessing for biomedical applications. *Engineering* 4 6: 822–830.

Hu, G. Zeng, L. Du, H. Fu, X. Jin, X. Deng, M. Zhao, Y. Liu, X. 2017. The formation mechanism and bio-corrosion properties of Ag/HA composite conversion coating on the extruded Mg-2Zn-1Mn-0.5 Ca alloy for bone implant application. *Surface and Coatings Technology* 325: 127–135.

Indira, K. Mudali, U.K. Nishimura, T. Rajendran, N. 2015. A review on TiO$_2$ nanotubes: Influence of anodization parameters, formation mechanism, properties, corrosion behavior, and biomedical applications. *Journal of Bio-and Tribo-Corrosion* 1: 1–22.

Ishihara, K. Yanokuchi, S. Fukazawa, K. Inoue, Y. 2020. Photoinduced self-initiated graft polymerization of methacrylate monomers on poly (ether ketone) substrates and surface parameters for controlling cell adhesion. *Polymer Journal* 52: 731–741.

Jeznach, O. Kolbuk, D. Sajkiewicz, P. 2019. Aminolysis of various aliphatic polyesters in a form of nanofibers and films. *Polymers* 11 10: 1669.

Koshuro, V. Fomin, A. Rodionov, I. 2018. Composition, structure and mechanical properties of metal oxide coatings produced on titanium using plasma spraying and modified by micro-arc oxidation. *Ceramics International* 44 11:12593–12599.

Li, B. Xia, X. Guo, M. Jiang, Y. Li, Y. Zhang, Z. Liu, S. Li, H. Liang, C. Wang, H. 2019. Biological and antibacterial properties of the micro-nanostructured hydroxyapatite/chitosan coating on titanium. *Scientific Reports* 9 1: 1–10.

Limongi, T. Tirinato, L. Pagliari, F. Giugni, A. Allione, M. Perozziello, G. Candeloro, P. Di Fabrizio, E. 2017. Fabrication and applications of micro/nanostructured devices for tissue engineering. *Nano-Micro Letters* 9 1: 1–13.

Liu, B. Zhang, X. Xiao, G. Lu, Y. 2015. Phosphate chemical conversion coatings on metallic substrates for biomedical application: A review. *Materials Science and Engineering: C* 47: 97–104.

Liu, L. Yang, Q. Huang, L. Liu, X. Liang, Y. Cui, Z. Yang, X. Zhu, S. Li, Z. Zheng, Y. 2019. The effects of a phytic acid/calcium ion conversion coating on the corrosion behavior and osteoinductivity of a magnesium-strontium alloy. *Applied Surface Science* 484: 511–523.

Liu, Y. Liu, L. Deng, J. Meng, R. Zou, X. Wu, F. 2017. Fabrication of micro-scale textured grooves on green ZrO$_2$ ceramics by pulsed laser ablation. *Ceramics International* 43 8: 6519–6531.

Macgregor, R.M. McNicholas, K. Ostrikov, K. Li, J. Michael, M. Gleadle, J.M. Vasilev, K. 2017. A platform for selective immuno-capture of cancer cells from urine. *Biosensors and Bioelectronics* 96: 373–380.

Marin, E. Boschetto, F. Pezzotti, G. 2020. Biomaterials and biocompatibility: An historical overview. *Journal of Biomedical Materials Research Part A* 108 8: 1617–1633.

Matinlinna, J.P. Lung, C.Y.K. Tsoi, J.K.H. 2018. Silane adhesion mechanism in dental applications and surface treatments: A review. *Dental Materials* 34 1: 13–28.

Migliorini, E. Weidenhaupt, M. Picart, C. 2018. Practical guide to characterize biomolecule adsorption on solid surfaces. *Biointerphases* 13 6: 06D303.

Mohammad, A. Mohammed, M.K. Alahmari, A.M. 2016. Effect of laser ablation parameters on surface improvement of electron beam melted parts. *The International Journal of Advanced Manufacturing Technology* 87 1: 1033–1044.

Mohan, L. Anandan, C. Rajendran, N. 2015. Electrochemical behaviour and bioactivity of self-organized TiO$_2$ nanotube arrays on Ti-6Al-4V in Hanks' solution for biomedical applications. *Electrochimica Acta* 155: 411–420.

Moseke, C. Gbureck, U. 2019. Surface treatment. *Metals for biomedical devices*, ed. M. Niinomi, 355–367. Woodhead Publishing.

Niinomi M. 2019. *Metals for biomedical devices.* Woodhead Publishing.

Park, S.J. Jin, J.S. 2001. Effect of silane coupling agent on interphase and performance of glass fibers/unsaturated polyester composites. *Journal of Colloid and Interface Science* 242 1: 174–179.

Pereira, R. Moura, C. Henriques, B. Chevalier, J. Silva, F. Fredel, M. 2020. Influence of laser texturing on surface features, mechanical properties and low-temperature xdegradation behavior of 3Y-TZP. *Ceramics International* 46 3: 3502–3512.

Perumal, A. Kannan, S. Nallaiyan, R. 2021. Silver nanoparticles incorporated polyaniline on TiO_2 nanotube arrays: A nanocomposite platform to enhance the biocompatibility and antibiofilm. *Surfaces and Interfaces* 22: 100892.

Perumal, A. Kanumuri, R. Rayala, S.K. Nallaiyan, R. 2020. Fabrication of bioactive corrosion-resistant polyaniline/TiO_2 nanotubes nanocomposite and their application in orthopedics. *Journal of Materials Science* 55: 15602–15620.

Petlin, D. Tverdokhlebov, S. Anissimov, Y. 2017. Plasma treatment as an efficient tool for controlled drug release from polymeric materials: A review. *Journal of Controlled Release* 266: 57–74.

Posritong, S., Borges, A. L. S., Chu, T.-M. G., Eckert, G. J., Bottino, M. A., & Bottino, M. C. (2013). The impact of hydrofluoric acid etching followed by unfilled resin on the biaxial strength of a glass-ceramic. *Dental materials, 29*(11), e281–e290.

Prabhu, D.B. Gopalakrishnan, P. Ravi, K. 2020. Morphological studies on the development of chemical conversion coating on surface of Mg-4Zn alloy and its corrosion and bio mineralisation behaviour in simulated body fluid. *Journal of Alloys and Compounds* 812: 152146.

Prida, V. Manova, E. Vega, V. Hernandez-Velez, M. Aranda, P. Pirota, K. 2007. Temperature influence on the anodic growth of self-aligned Titanium dioxide nanotube arrays. *Journal of Magnetism and Magnetic Materials* 316: 110–113.

Putra, N. Mirzaali, M. Apachitei, I. Zhou, J. Zadpoor, A. 2020. Multi-material additive manufacturing technologies for Ti-, Mg-, and Fe-based biomaterials for bone substitution. *Acta Biomaterialia* 109: 1–20.

Rahmati, M. Silva, E.A. Reseland, J.E. Heyward, C.A. Haugen, H.J. 2020. Biological responses to physico-chemical properties of biomaterial surface. *Chemical Society Reviews* 49: 5178–5224.

Regonini, D. Satka, A. Allsopp, D. Jaroenworaluck, A. Stevens, R. Bowen, C. 2009. Anodised titania nanotubes prepared in a glycerol/NaF electrolyte. *Journal of Nanoscience and Nanotechnology* 9: 4410–4416.

Roy, M. Krishna, B.V. Bandyopadhyay, A. Bose, S. 2008. Laser processing of bioactive tricalcium phosphate coating on titanium for load-bearing implants. *Acta Biomaterialia* 4 2: 324–333.

Rúa, J. Zuleta, A. Ramírez, J. Fernández-Morales, P. 2019. Micro-arc oxidation coating on porous magnesium foam and its potential biomedical applications. *Surface and Coatings Technology* 360: 213–221.

Saji, V.S. 2019. Organic conversion coatings for magnesium and its alloys. *Journal of Industrial and Engineering Chemistry* 75: 20–37.

Saranya, K. Bhuvaneswari, S. Chatterjee, S. Rajendran, N. 2021. Titanate incorporated anodized coating on magnesium alloy for corrosion protection, antibacterial responses and osteogenic enhancement. *Journal of Magnesium and Alloys.* Doi: 10.1016/j.jma.2020.11.011

Saranya, K. Kalaiyarasan, M. Chatterjee, S. Rajendran, N. 2019. Dynamic electrochemical impedance study of fluoride conversion coating on AZ31 magnesium alloy to improve bio-adaptability for orthopedic application. *Materials and Corrosion* 70: 698–710.

Saranya, K. Kalaiyarasan, M. Rajendran, N. 2019. Selenium conversion coating on AZ31 Mg alloy: A solution for improved corrosion rate and enhanced bio-adaptability. *Surface and Coatings Technology* 378:124902.

Shivakoti, I. Kibria, G. Cep, R. Pradhan, B.B. Sharma, A. 2021. Laser surface texturing for biomedical applications: A review. *Coatings* 11 2: 124.

Souza, J.C. Bins-Ely, L. Sordi, M. Magini, R. Aparicio, C. Shokuhfar, T. 2018. Nanostructured surfaces of cranio-maxillofacial and dental implants. *Nanostructured biomaterials for cranio-maxillofacial and oral applications*, ed. J.C.M. Souza, D. Hotza, B. Henriques, and A.R. Boccaccini, 13–40. Elsevier.

Stango, S.A.X. Karthick, D. Swaroop, S. Mudali, U.K. Vijayalakshmi, U. 2018. Development of hydroxyapatite coatings on laser textured 316 LSS and Ti-6Al-4V and its electrochemical behavior in SBF solution for orthopedic applications. *Ceramics International* 44 3: 3149–3160.

Su, Y. Wang, K. Gao, J. Yang, Y. Qin, Y.X. Zheng, Y. 2019. Enhanced cytocompatibility and antibacterial property of zinc phosphate coating on biodegradable zinc materials. *Acta Biomaterialia* 98: 174–185.

Sun, W. Liu, W. Wu, Z. Chen, H. 2020. Chemical surface modification of polymeric biomaterials for biomedical applications. *Macromolecular Rapid Communications* 41 8: 1900430.

Uslu, E. Mimiroglu, D. Ercan, B. 2021. Nanofeature size and morphology of tantalum oxide surfaces control osteoblast functions. *ACS Applied Bio Materials* 4: 780–794.

Vardaki, M. Mohajernia, S. Pantazi, A. Nica, I.C. Enachescu, M. Mazare, A. 2019. Post treatments effect on TiZr nanostructures fabricated via anodizing. *Journal of Materials Research and Technology* 8: 5802–5812.

Virtanen, S. 2012. Degradation of titanium and its alloys. *Degradation of implant materials*, ed. N. Eliaz, 29–55. Springer Publishing.

Wang, X. Xu, S. Zhou, S. Xu, W. Leary, M. Choong, P. 2016. Topological design and additive manufacturing of porous metals for bone scaffolds and orthopaedic implants: A review. *Biomaterials* 83:127–141.

Xie, H. Chen, C. Dai, W. Chen, G. Zhang, F. 2013. In vitro short-term bonding performance of zirconia treated with hot acid etching and primer conditioning etching and primer conditioning. *Dental Materials Journal* 32 6: 928–938.

Yasuda H.K. 2012. *Plasma polymerization*. Academic Press, Inc.

Yin, Z.Z. Huang, W. Song, X. Zhang, Q. Zeng, R.C. 2020. Self-catalytic degradation of iron-bearing chemical conversion coating on magnesium alloys: Influence of Fe content. *Frontiers of Materials Science* 14: 296–313.

Yu, Wq. Jiang, Xq. Zhang, Fq. Xu, L. 2010. The effect of anatase TiO_2 nanotube layers on MC3T3-E1 pre-osteoblast adhesion, proliferation, and differentiation. *Journal of Biomedical Materials Research Part A* 94: 1012–1022.

Yu, Z. Yang, G. Zhang, W. Hu, J. 2018. Investigating the effect of picosecond laser texturing on microstructure and biofunctionalization of titanium alloy. *Journal of Materials Processing Technology* 255: 129–136.

Zhang, H. Luo, R. Li, W. Wang, J. Maitz, M.F. Wang, J. 2015. Epigallocatechin gallate (EGCG) induced chemical conversion coatings for corrosion protection of biomedical MgZnMn alloys. *Corrosion Science* 94: 305–315.

Zhang, H. Xie, L. Shen, X. Shang, T. Luo, R. Li, X. You, T. Wang, J. Huang, N. Wang, Y. 2018. Catechol/poly-ethyleneimine conversion coating with enhanced corrosion protection of magnesium alloys: Potential applications for vascular implants. *Journal of Materials Chemistry B* 6 43: 6936–6949.

Zhang, L. Tong, X. Lin, J. Li, J. Wen, C. 2020. Enhanced corrosion resistance via phosphate conversion coating on pure Zn for medical applications. *Corrosion Science* 169:108602.

Zhou, W. Shan, D. Han, E.H. Ke, W. 2008. Structure and formation mechanism of phosphate conversion coating on die-cast AZ91D magnesium alloy. *Corrosion Science* 50: 329–337.

16 Advance Surface Treatments for Enhancing the Biocompatibility of Biomaterials

Abhinay Thakur and Ashish Kumar

CONTENTS

16.1 INTRODUCTION

Biomaterials are synthetic or natural substances capable of precisely replacing the improper biological structure of manufactured implants or biomaterials to restore shape and function. Biomaterials help to improve the quality of life and sustainability of human beings, as well as to meet the requirements of a growing population. Biomaterials are extensively utilized in a variety of medical applications, including cardiology, gynecology, orthopedics, dentistry, plastic surgery, wound healing, and synthetic biology. As the demand for biomaterials rises with the aging population and the desire to protect health and well-being, the global biomaterials market is expected to surpass around USD 390.92 billion by 2027, and growing at noteworthy CAGR of 15.2% over the forecast period 2020–2027. Metals and bioceramics form the major and vast parts of biomaterials as well as medical devices and implants. However, implant and prosthesis design are a challenging phase as its intended components must emulate the structure and properties of biological tissue while also some specific requirements such as biomaterials must be non-allergenic, non-toxic, non-chromogenic, non-carcinogenic, and non-antigenic. Additionally, biomaterials must have optimized properties such as ductility, elasticity, yield strain, resilience, and less manufacturing cost.

16.1.1 BIOCOMPATIBILITY OF METALS

A key component of a biomaterial is biocompatibility. In a particular application, a biocompatible material induces a significant host reaction (i.e., minimal disruption of normal body function). Thus, when it is placed in vivo, the substance does not induce a chromogenic, toxic, or allergic inflammatory reaction. The biocompatibility of a substance is determined by two main factors: the host reactions to the material and the degradation of the material in the body's environment.

DOI: 10.1201/9781003173533-16

In metals, biocompatibility requires the adoption by the underlying tissues and by the system wherein an artificial implant is placed. Metal implants should not infuriate the surrounding structures, do not trigger an unnecessary allergic reaction, do not induce allergic and immunological reactions, and do not induce cancer. As most metallic system applications are for structural implants, the biocompatibility of metals is of significant concern as metals can corrode in the in vivo environment. Metallic implant corrosion can impact the neighboring tissues in three different ways: (a) electric current could affect the cell behavior, (b) undesirable corrosion process may alter the chemical environment, and (c) metal ions may influence the growth of cells. Metallic implant corrosion affects the surrounding tissues as well as the implant itself. It emits chemicals that are hazardous to human organs and degrades the mechanical properties of the implant. As a result, the corrosion resistance of a metallic implant is a significant aspect of its biocompatibility. However, metals that can decay are recommended for temporary implants in specific situations, but definitely without disregarding the biocompatibility criterion. In Figure 16.1, two eventual toxic pathways by metal ions discharged into internal fluids as a result of corrosion have been shown.

16.2 METALS AND ALLOYS AS BIOMATERIALS

There are a broad variety of biomedical uses for metals and alloys, including fracture fixing products, craniofacial plates, external splints and screws, cardiovascular surgery (e.g. balloon catheters, artificial heart sections, and valve substitutes), dental amalgams, and so on [1]. The predominant metals and their alloys currently being utilized as biomaterials such as titanium and its alloys are heavily used in the spinal fixation devices, fracture fixation devices, maxillofacial surgery, dental implants, artificial heart valves, cardiac pacemakers, and high-speed blood centrifugal components. Ni-Ti Alloys (Nitinol) are used in applications involving kink resistance, durability, compress susceptibility under high mechanical stress and abrasion ratios. Dental amalgams are among the ancient compounds used throughout oral healthcare. For rear teeth, they are the most valuable therapeutic materials, produced by mixing liquid Hg (45–55 wt. %) and Sn, Ag, and Cu powder with low quantities of Pd, Se, Zn. Owing to its light weight and easy biodegradability, magnesium (Mg) is well considered. Mg is available in the bone instinctively. In a comparison to pure Mg, magnesium alloys showed excellent corrosion characteristics. Even then, the release of metal ions may induce hazardous, inflammatory, and carcinogenic effects in the body; thereby the degradation products of such alloys should be thoroughly investigated. Stainless steel has excellent strength, chemical

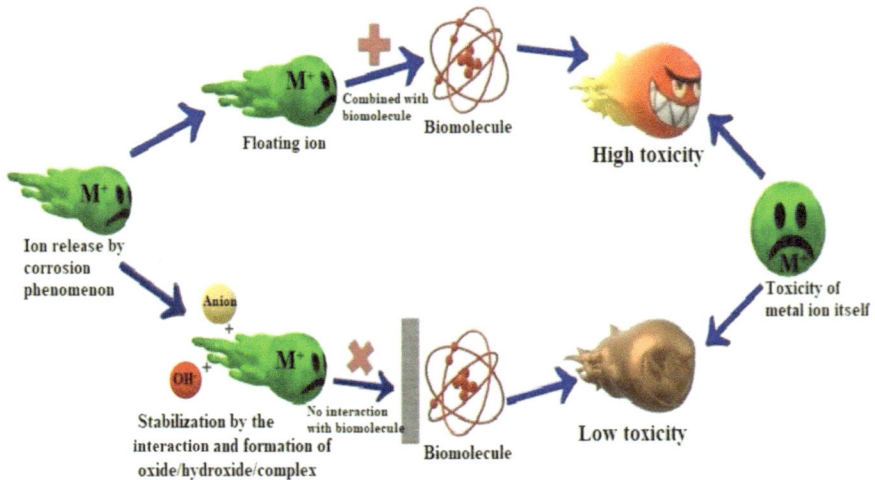

FIGURE 16.1   Two eventual toxic pathways by metal ions released into body fluids due to corrosion.

stability, and corrosion resistance due to the availability of chromium (Cr), which forms a thin oxide film that prevents oxidation (Cr_2O_3). Tantalum is being used due to its low cytotoxic effects and high corrosion resistance, owing to the stable Ta_2O_5 oxide produced mostly on the surface of a material, which makes tantalum (Ta) among the most biocompatible metals. Tantalum is used in artificial hips, knees, and other joints. Bioactive ceramics implants induce a specific biological activity that contributes to the formation of a chemical bond among the material interface and the surrounding tissue. Few compounds have demonstrated the ability to form bioactive tissue fixations with variable intercellular zone thicknesses and potentials, as well as distinct binding methods and frequencies. They develop a hydroxyl carbonate apatite layer over their surface that is similar to the chemical and physical process of bone mineralization.

16.3 SURFACE TREATMENT ON IMPLANT MATERIALS

One significant consideration regarding recent advances in metallic biomaterials is surface treatment or surface modification. Surface morphology is an important implant feature since it affects the ability of cells to adhere to the solid substrate and protects it against corrosion, mechanical stress, biochemical impact, and extends its lifelong working efficiency [2]. In the case of chemical modification, the purpose of the modification is to provide the metal surface with a specific biological approach and to enhance the efficacy of bio-molecules. Various such surface treatments will be addressed in brief in this section.

16.3.1 SURFACE TREATMENTS/MODIFICATIONS ON DENTAL IMPLANTS

16.3.1.1 Surface Treatments on Titanium Based Dental Implants

Titanium includes certain characteristics of dentistry, including resistance to corrosion, excellent biocompatibility, relatively high strength and formability. The significant outcome, such as improved bone-to-implant interaction, could be achieved by modifying or changing the surfaces' texture to achieve the roughness of Ti implants.

16.3.1.1.1 *Physical Modifications of Titanium (Ti) Implants*

Physical modification allows sustaining the morphology of the surface and roughness of most of the Ti implants and is useful for the osseointegration phenomenon. The most widely used physical approach to alter the surface is the sandblasting technique. Microroughness is induced on the surface of the implant by a frequent airstream of accelerated particles like silicon dioxide (SiO_2), aluminium oxide (Al_2O_3), titanium dioxide (TiO_2), or HA powders under adequate pressure. The titanium surface can be designed with the attributes of both topography and wettability by this modification process, which results in cell proliferation, higher cell attachment, and osteoblast cell differentiation characteristics. When the substrate compositions and surface topography of a titanium implant for osseointegration were examined, they [3] introduced a physical surface modification to enhance osseous healing. For a period of 12 weeks, 48 zirconia and titanium implants were inserted into the tibias of 12 mini pigs. The zirconia and titanium surfaces had direct bone contact, as per histological findings. Throughout the healing cycle, the bone impact contact (BIC) increased by more than 65%, based on the conclusion of researchers on their study using various techniques.

16.3.1.1.2 *Chemical Treatment of Titanium Implants*

Chemical modification of a surface refers to the modification of the implant surface's structure and environment as a result of chemical adsorption or interaction between the titanium metal surface and the changed surface. One of the most practiced procedures for the chemical treatment of titanium implants is anodization. It utilizes the traditional electrochemical conversion method, in which the substrate product produces a layer with a thickness of 5–200 m. During anodization, the metal acting as anode is converted into an oxide film having decorative and corrosion protective characteristics. The approach has the advantage of being able to develop nano-structured oxide films from the aqueous/non-aqueous solutions, as well as a nano-tubular porous network on the substrate under certain optimized conditions. The osteoconductivity of hydrophilic microstructured titanium implants containing phosphate ion chemistry was studied using the anodization method [4]. The findings demonstrate that by getting high and strong bone apposition, the microstructured phosphate ion-incorporated Ti oxide surface formed by hydrothermal treatment and anodization of H_3PO_4 and extreme thermal treatment was effective in improving the osteoconductivity of Ti implants.

16.3.1.1.3 Zirconia Implant Surface Modifications

The rapidly increased aesthetic demand for zirconia as an implant biomedical metal in the dental field is owed to its white color and its appealing potent properties, such as high biocompatibility, high thermal conductivity, and low modulus of elasticity. It is also possible to enhance the mechanical stability of zirconia by incorporating tetragonal yttrium polycrystals. Although, there are still some disputes over zirconia implant osseointegration due to an inadequacy of long-term clinical research. Several physicochemical methods, such as acid etching, grit blasting, machining, laser treatment and ultraviolet light treatment, have been used to enhance the effectiveness of the zirconia surface [5]. These techniques were designed to improve zirconia implant osseointegration and surface morphology. In a sheep pelvis prototype and over the duration of two, four, and eight weeks, six different types of dental implants for osseointegration were assessed [6]. A sandblasted and acid-etched Ti surface was employed as a reference. The implant that had been chemically treated was either plasma-anodized or calcium phosphate-coated. Both titanium implants had the same BIC at two weeks (57–61%); zirconia was better (77%). The most significant increase in BIC occurred around 2 to 4 weeks. When compared to the reference implant, the biochemically coated implants (78–79%) and the calcium phosphate coating (83%) had equivalent results after 8 weeks (80%).

16.3.1.1.4 HA Layer and Nanocomposite Coatings on Dental Implants

The HA coated implants are among the most prevalent alternatives among the commercially available. It is a benign and natural source of calcium phosphate, having non-immunogenic and non-inflammatory chemical properties that mineralize the organic matrix and strengthen it. It is made up of ions that are found naturally in the body and possesses superior osteoconductive and osteointegration capabilities. Several ion-substituted HA, as well as diverse bio-functions, paved the path for implant development. Whereas, ion-substituted HA coatings have been shown to have a cytotoxic effect, reducing the proliferation phenomenon and differentiation of the cells adhering on the coating surface of the implant, thus significantly improving the cell attachment. By laser processing, a unique nanocomposite coating of Ag-HA on a biomedical substrate Ti6Al4V was developed [7]. EDS, SEM, and XRD analysis confirmed the development of Ag-HA coating, approximately 200 nm bounded to the substrate. The average particle size of the hydroxyapatite (HA) powders used in this study were approximately 100 nm, whereas the Ag nanoparticles formed were 25–35 nm in size. To increase adhesion with preplaced powder, the Ti6Al4V sample was polished using 60-grit sandpapers to provide a rippled surface. The specimens were ultrasonically cleaned in deionized water, wiped using acetone while rippling/roughening. The Ag-HA composite pastes with 0 wt. % Ag, 1 wt. % Ag, 2 wt. % Ag, and 5 wt. % Ag was formed by blending the Ag-HA powder with PVA (polyvinyl alcohol), which functioned as an adhesive. A brush was utilized to apply the Ag-HA paste on the Ti6Al4V samples. An antibacterial assay concluded that most of the Ag-HA nanocomposite layers are capable of killing bacteria effectively; however, an increased Ag content could decrease the coating cytocompatibility. Ion release testing has shown that as the Ag concentration in such coatings exceeds 5%, cell viability decreases as the amount of Ag leached out increases.

16.3.1.2 Polyetheretherketone (PEEK) as an Alternative to Titanium in Dental Applications

PEEK is a synthetic polymeric organic material that was developed in 1978. PEEK is an old member of the poly-aryl-ether-ketone polymer group with a high steady-state temperature of above 300ºC. It offers excellent chemical resistance, mechanical strength, and biocompatibility. It is a tooth-colored substance that has recently been employed as a dental implant material in instances where aesthetics is important. PEEK is also bio inert in nature, with no osseoconductive qualities embedded. The carbon fiber reinforced PEEK (CFR-PEEK) and FDM-printed pure PEEK composites were effectively manufactured using FDM and mechanically evaluated [8]. The sample surfaces were polished and sandblasted to investigate surface roughness and topography influenced cell adhesion and cytotoxicity. Both CFR-PEEK and PEEK materials exhibited acceptable biocompatibility with and without surface treatment and modification. As a result, the FDM-printed CFR-PEEK composite appears to be an effective candidate to be utilized in the tissue engineering applications and bone grafting.

16.3.2 CALCIUM PHOSPHATE COATING METHODS

Calcium phosphate (CaP) has intrinsic biocompatibility as the key inorganic component in bone tissue when used as biomaterials in the human body. Various techniques included under the plasma assisted surface treatment are used to achieve physical and chemical deposition on biomaterials, and some of them are discussed next.

16.3.2.1 Plasma Spray

Plasma is a compressed ionized gas with a very large range of temperature and pressure that can be regulated. By separating electrons from their parent atoms or molecules that are still in an ionized state, plasma generation is achieved. Usually, the manufacture and maintenance of low-temperature plasmas for technical applications are accomplished by the application of an electric field to a neutral gaseous precursor. Plasma spray techniques start with the use of a solid metal wire or solid molten powder and discharge liquid metal droplets onto the substrate through the air. The solid is placed in a system that warms it, separates it into miniscule droplets, and moves the droplets to a molten state on the substrate. The wire or powder is then placed in the funnel, which subsequently directs a stream of liquid metal particles on the substrate. The "plasma" spray utilizes an electric arc as a source of heat to initially melt the wire or powder into the compressed airstream. The gap between the molten metal's source and the ground determines the type of coating that is generated. Most substrates have good adhesion, which is always ensured by washing and roughening the surface. If properly designed, this inorganic coating could attain high mechanical strength to serve as a bearing surface.

The effect of electrical parameters and solution pH on HAP coatings synthesized with the help of plasma-assisted electrophoresis was studied [9]. Conventional unbalanced AC and DC power supply were utilized to bias the Ti alloy over the metal surface. Furthermore, the coated TiO_2 layers on the HA layer have a polycrystalline structure that was nano-scaled (10–20 nm). At the interface between both the substrate and the TiO_2 layer, a thick, persistent crystalline titania layer (10 nm thick) was formed, that further improved the substrate's corrosion resistance.

16.3.2.2 Ion Beam-Assisted Deposition (IBAD)

IBAD is a vacuum deposition surface modification technique that integrates PVD and ion implantation techniques. It is also known as "ion beam enhanced deposition" (IBED). In the IBAD, mostly during deposition, an ion beam penetration is constant during the operation to clear the substrate surface after deposition and regulation of film depositing properties [10]. IBAD has the advantage of being able to produce a progressive transition layer between the substrate and the deposited film, as well as substrate material and depositing a film, allowing the coating to adapt tightly to the surface. Figure 16.2 illustrates the IBD technique; bombardment of the vapor and ion stream on the substrate surface create a defensive layer against corrosion and wear [11].

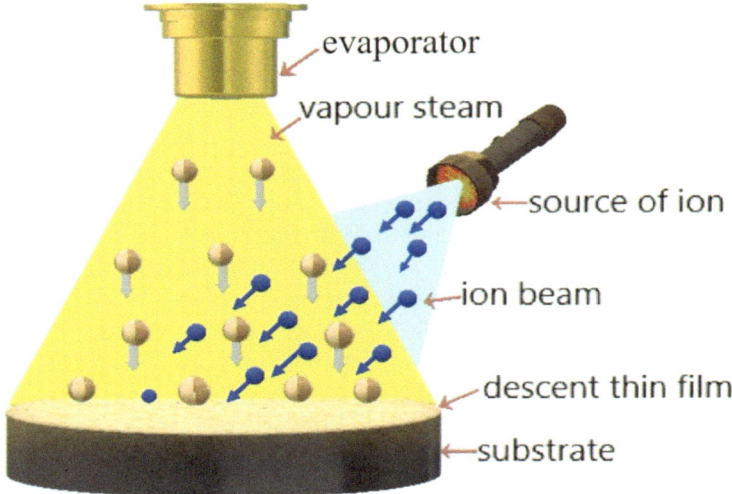

FIGURE 16.2 IBD technique; bombardment of the vapor and ion stream on the substrate surface to create defensive layer against corrosion and wear.

Source: Figure adapted from [11].

Ion-beam-based methods like ion implantation and IBAD provide advantageous surface layers with favorable properties without altering bulk characteristics [12]. They reported that ion implantation was effective in biomaterials modifications, including artificial joint component wear resistance, anticoagulability, improved permeability, and reduced biofouling of medical devices. Biocompatible diamond-like C-N films and carbon coatings, antimicrobial coatings, and adhesive coatings have all been developed using this technique.

16.3.2.3 Electrophoretic Deposition

It is a simple process of colloidal processing that utilizes the electrophoresis mechanism to transfer charged particles embedded within an electric field in a solution to deposit them on a substrate in an orderly direction to create thin and sometimes thick film coating bodies. Here, Figure 16.3 shows the EPD cell illustrating positively charged suspended particles moving towards the negative electrodes [13].

The microstructure of nanocrystalline TiO_2 films formed by electrophoretic deposition on Ti-6Al-7Nb alloy was studied [14]. To boost the coating's adhesion ability, it was subjected to a post-heat treatment at 850°C. A nanocrystalline structure was observed in the TiO_2 film. SAED and XRD tests revealed anatase and rutile phases in the as-deposited nanocrystalline TiO_2 films. Micromechanical properties such as microhardness and Young's modulus were calculated. TiO_2 films have a low hardness and Young's modulus, as per the findings obtained through the study.

16.3.2.3.1 Dip Coating

The dip-coating method comprises of four stages: immersion of the substrate into the sol solution, withdrawal of the substrate, drainage and withdrawal of sample followed by sufficient evaporation and drying after and during withdrawal of the sample. Thermal treatment is also used after drying to finish the dehydration process or to eradicate any remaining organic substances. The sintering heat treatment is utilized at the coating substrate interface for achieving the effective and reliable coating structure, adhesive strength, and cohesive strength. Figure 16.4 illustrates the stages involved in the dip coating process before the final sintering procedure [15].

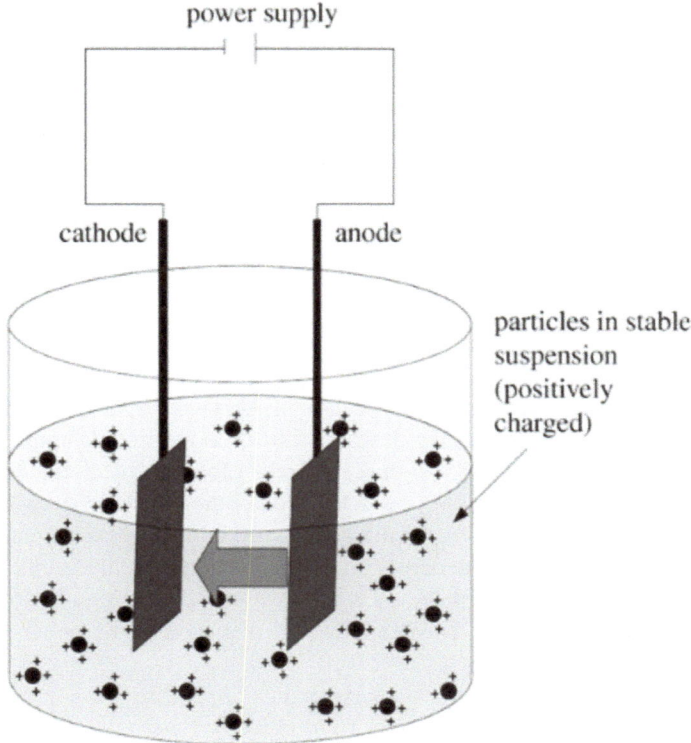

FIGURE 16.3 EPD cell illustrating positively charged suspended particles moving towards the negative electrodes.

Source: Figure adapted from [13].

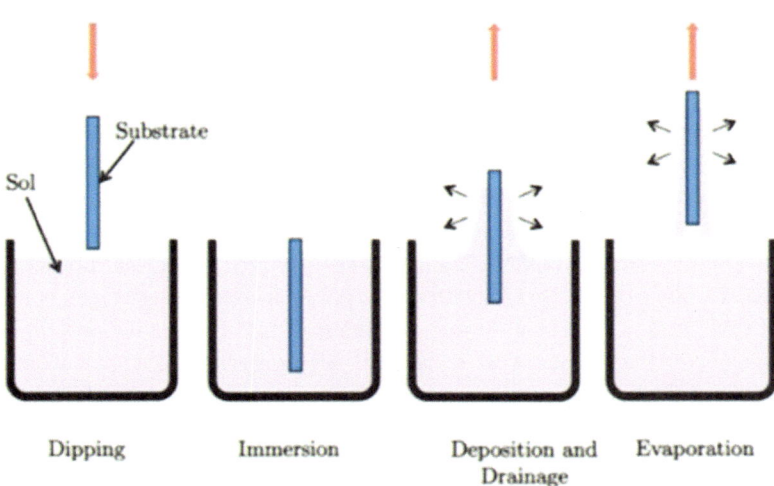

FIGURE 16.4 Illustration of the stages involved in the dip coating process before the final sintering procedure.

Source: Figure adapted from [15].

Another study investigated the effects of calcium hydroxyapatite dip coating on Ti-6Al-4V substrates [16]. For sample characterization, X-ray diffractometry and scanning electron microscopy were used. Poly(ethylene glycol), gelatin/glycerol were used as organic additives in the solutions. The dip-coating procedure was carried out using specific equipment that could keep withdrawal and dipping rates constant. The HA dip-coated Ti-6Al-4V strips were calcined in a nitrogen-gas environment at about 840°C. The developed HA coatings exhibited extremely permeable property, having bonding strengths of above 30 MPa.

16.3.2.4 Electrochemical Deposition (ED)

ED is a non-substrate-containing coating process widely utilized for the formation of coating for biodegradable magnesium and its alloys. Although inorganic phase deposition is primarily concerned with management, the coating formed can be achieved through both pure deposition and conversion at the interface. It can produce coherent coatings by regulating the required chemical composition and thickness of the coating. In the case of AZ91D magnesium alloy, the reduction in the rate of bio-corrosion was monitored [17], where they used the ED technique to cover the alloy surface with zirconia, a biocompatible ceramic. After that, the effects of changing ED parameters like period time, current density, and ZrO_2 particle concentration on the coating characteristics like adhesion, morphology, and thickness were studied. In vitro corrosion experiments in SBF solution revealed that the coating decreased the corrosion current density of alloy from 18.4 to 12 A/cm^2 in the maintained environment. Furthermore, EDs findings of a coated sample following a week of dipping in SBF solution and heating to 37°C indicated that the coating enhanced hydroxyapatite development on the surface.

16.3.3 Biofilm Formation on Implants

A biofilm is an organized community of bacteria that covers themselves in a matrix of exopolysaccharides, resulting in firm implant adhesion. Intercellular bacterial contact, survival during nutrient starvation and phenotypic variation triggering gene expression and biofilm adaptation are some of the main attributes of biofilms.

16.3.3.1 Surface Modification with Cellulose and Its Derivatives

Cellulose and its derivatives are considered among the most common natural occurring biopolymers that are used for numerous biomedical applications. A protective coating for magnesium-based implants have been recently employed by some researchers [18]. Organosoluble cellulose acetate was deposited over Mg implants using the spin-coated process, and its capability to regulate the pH and mitigate corrosion of the material in the environment was investigated.

The use of zwitterionic polymers for the surface modifications of cellulose membranes was studied [19] to increase resistance towards platelet adhesion and protein adsorption. Three zwitterionic polymers were grafted from cellulose membranes: poly(2-(methacryloyloxyethyl) ethyl-dimethyl-(3-sulfopropyl)-ammonium), poly(N,N-dimethyl-N-(p-vinylbenyl)-N-(3-sulfopropyl) ammonium), and poly (CMs). The hydrophilicity of zwitterionic polymer-modified surfaces is higher than that of the original CM surface. The in vitro measurements of this cumulative protein adsorption and platelet adherence over the surface were measured. All zwitterionic surfaces were found to have outstanding platelet adhesion resistance and enhanced resistance to nonspecific protein adsorption.

16.3.3.2 Surface Modification with Chitosan

Chitosan is a natural amino polysaccharide composed of (1/4)-linked D-glucosamine residues and N-acetyl-glucosamine groups. The cationic chitosan is bioactive, eco-friendly, and has beneficial biological properties. Along with its natural abundance, chitosan is a potential prospect for a variety of biomedical uses, such as a corrosion-resistant coating for biomedical implants. Another study

[20] used an electrophoretic deposition process to form inorganic-organic composite coatings to be employed for biomedical applications. A wet chemical method used to make needle-shaped hydroxyapatite (HA) particles, was later used to make chitosan-HA layers. The c-axis of the HA particles in the chitosan matrix tended to be parallel to the substrate surface. The surface modification of these coatings was achieved using composite chitosan-heparin layers. The findings open the path for the electrophoretic manufacturing of novel blood-compatible FGM coatings for biomedical applications such as implants.

16.3.4   Other Coating and Treatments Techniques

16.3.4.1   Thermal Spray Technique

The premise of this process is that the coating substances are melted (or partly melted) by way of heating in a gaseous environment and the melted substances are projected at a high speed onto the surface to be coated. Usually, the coating materials are in powder form or (less often) wire.

The coating material, in the powder form or wire, is injected into the plasma jet during deposition of the coating by thermal spray, where it is melted and transmitted to the surface of a substrate. Thermal energy supplies heating to melt the powder whereas the kinetic energy function is twofold: to supply the particles with deformation energy after the substrate effect and to produce heat due to non-elastic impacts. The particles partially melt just after powder infusion into flame or plasma jet and impact mostly on substance being coated, "splat" configuration. The particles reinforce the base material and transfer heat to it. A comparison of HAP coatings produced by cold spray with conventional thermal spray techniques was demonstrated in vitro [21]. Primary human osteoblasts were tested in vitro after 1, 7, 14 days of cell culture on the substrates. MTS and LIVE/DEAD assays were used to determine cell viability, alkaline phosphatase quantification was used to determine cell differentiation, and SEM was employed to investigate the cell morphology. HVOF HA coatings allowed cells to adhere well, but didn't produce prolonged filopodia similar cells on APS HA coatings. HA coatings with increased crystallinity demonstrated increased cell proliferation and differentiation following 14 days of cell culture. Surface micro-features, as well as moderate surface wettability, favored cell attachment, while HA crystallinity and crystal size imparted a strong effect on cell proliferation and differentiation.

16.3.4.2   Spin Coating

A method of coating a substance on a flat surface is a spin coating. Here, a substrate on a rotating device is first placed. In a solution, the coating material is dissolved and added to the surface of substrate such that the whole substrate is macroscopically coated. Most of the material is spun off to the side when the spinner is turned on. The thickness of the film depends on a combination of both the spinning speed and the viscosity of the solution applied. With a reduction in spinning speed and an increase in the solution viscosity, the film thickness increases. By utilizing polyproylene microtitre plates, the preparation of metal nitrate (Ag, Cu, Zn) doped methyltriethoxysilane (MTEOS) coatings as well as the gradual evaluation of their antibacterial activity were studied [22]. Different volumes of liquid sol-gel were applied to microtitre plate wells and cured under various conditions. Thermogravimetric analysis and visual inspection were utilized in order to examine various curing parameters. Using a volume of 200 ul, the optimum curing conditions were calculated to be 50–70°C. Microtitre plates allowed a wide range of sol-gel coatings to be tested for antibacterial activity towards several bacteria in a moderately period.

16.3.4.3   Photon Irradiation

In certain particular applications, surface alteration by UV and infrared (IR) lasers could be used with one key benefit, such as the possibility of treating very small and localized surface areas. The procedure depends on intense bursts of light to produce an instant increase in pressure to break bonds at the sample surface. The procedure can be employed in biological tissues, ceramic, composites, crystal, glasses, paper, and polymers with outstanding effect because there is surplus heat passed to the ambient medium. To modify the surface of starch-based biomaterials, oxygen plasma or UV irradiation techniques were used [23]. Using a direct contact method, the influence of different surface characteristics on osteoblast-like cell adhesion was examined. A higher number of cells adhered to the treated surfaces after shorter culture cycles. It is impossible to establish a direct link among the amplitude of each surface characteristic and cell behavior. The roughness induced by the surface treatments was at the bio mimicking nano-scale, stimulating cell activity.

16.3.4.4 Ion Implantation

This technique requires a phase wherein the subjected surface is accelerated and impregnated with ions. Without having any electrolytic solutions, the ion implantation technique integrates ions into the substrate. Only a thin layer of the subjected surface could be developed by this process; typically the corrosion resistance is indeed not sufficient in the lengthy period. Based on the implanted components, the ion implantation for Mg modification has been classified into three forms: gas ion implantation, metal ion implantation, and dual ion implantation. For the surface modification and to improve the corrosion resistance, Mg and its alloy substrates were incorporated with the organic elements such as nitrogen, carbon dioxide, and oxygen gases. M^+ implantation incorporates metal components into the magnesium and its alloy substrate to make surface alloys, whereas gas ion implantation forms N_2 or O_2-rich coating/layer on the substrate. The few metallic elements that are frequently implanted onto the magnesium alloys are Zn, Al, Ti, Zr, and Ta. The modified metal ion implantation layer improves the corrosion resistance. The impact of ion implantation on the mechanical and physical properties of Ti-Si-N multifunctional coatings towards biomedical applications was examined [24]. The researchers used the CAVD method to deposit multifunctional Ti-Si-N coatings with the intention of examining their structural, physical, chemical, and mechanical properties. Through high-intensity ion implantation, copper ions with a dosage $D = 2 \times 10^{17}$ ions/ cm^2 and an energy $E = 60$ keV were used to modify Ti-Si-N coatings. The findings revealed that ion implantation affected the bulk density, strength, and Young's modulus of Ti-Si-N coatings.

16.3.4.5 Physical Vapor Deposition (PVD)

PVD is among the best coating techniques used commercially, where a defensive film is applied on the surface of the implant. Owning to its ability to develop thick dense nano-structured coatings, this approach is often utilized to deposit a strong protective coating on the surface of the implant/ material to enhance the wear and corrosion resistance. The PVD method create layers on the substrate that are dense and strongly adherent. A sol-gel was accompanied by physical vapor deposition technique to render Ag:TiO_2 nanocomposite thin films [25]. Glancing angle X-ray diffractometer and UV-Vis spectroscopy were utilized to analyze the surface, microstructure, and plasmonic properties of nanocomposite films. Based on microstructural analysis, Ag nanoparticles were imbedded in a TiO_2 matrix containing heterogeneous portions of anatase and rutile. According to the Scherrer formula, the overall crystallite size of Ag nanoparticles was 23 nm. Optical tests confirmed the production of Ag nanoparticles with confirming an SPR-induced absorption pattern in the visible spectrum. The intensity of this band increased as the deposition time was increased.

16.4 PERSPECTIVES AND CHALLENGES

Biomaterials and implant materials have corrosion risks in their pristine state which makes them not an ideal source for further in vivo use and as implants. In order to improve its biocompatibility, we have shown some of the major biomaterial surface modification techniques such as laser and plasma therapy, ion implantation, and PVD, as well as various others that play an important role. Changes in the physicochemical properties, chemical composition, morphology, and biocompatibility of a range of these metallic implant substrates may be enhanced by surface modification. Several researchers such as Ashish Kumar et al. have performed the utilization of organic compounds as corrosion inhibitors for several metals such as mild steel, stainless steel which are considered potential materials for implanted biomedical materials and application. Various organic compounds studied by them such as acarbose [26], analgin [27], phenylephrine [28], 1-methyl- 3-propylimidazolium iodide [29] on the corrosion inhibition on steel in acidic media revealed a maximum corrosion inhibition efficiency of > 95%. This type of inhibitor strategies attributing to the adsorption of the molecule on the substrate by the formation of a defensive layer against corrosive species could be used for extended use in the implant materials' protection toward corrosion. We agree that there are still some major challenges, primarily focused on the development of new raw materials (polymers, metals, and others) that can be further processed and new surface treatment techniques based on different sources of plasma, laser, or ion beams in conjunction with various grafting processes. New materials with high potentials, such as tissue replacements, organ replacements, biosensors, antibacterial materials, or materials for unique applications may be synergized with newly established materials in conjunction with improved surface modification and grafting techniques.

16.5   CONCLUSION

Metallic biomaterials impart today's class of materials with the widest usage across biomedical devices and implants in humans. These biomaterials include stainless steel, titanium (its alloys), memory alloy of nitinol form, dental amalgams. Corrosion acts as a key element in the selection and design of metals and alloys for use in vitro. Corrosion resistance of a biomaterial is a key factor that affects its biocompatibility by influencing its efficiency, lifespan, and physical properties. Even at low corrosion rates in comparison to the implant's physical output, the ion toxic effect of corrosion in the body can cause hypersensitivity and cancer. From a corrosion viewpoint, dissolved O_2, chloride, and pH levels are the most critical attributes. Different body sections have varying pH values and concentrations of oxygen. Besides that, due to its enhancement in corrosion resistance behavior and bioactivity, the development of a range of coating techniques in biomaterials has received in-depth analysis. Dip, spray, or spin-coating are often used for traditional functionalization and can be compliant with metal substrates. To prevent metal-related toxicity and prevent corrosion adverse effects in the body, metals must be cautiously chosen. There are several other variables to consider, like the impact of a pH shift on corresponding tissue and metabolic function in the intervening phases. As per several analyses, a desirable candidate to be used as a biodegradable implant could be magnesium alloys coated with bioactive nano-structure glass. Additionally, a brief and intensive clinical investigation is also required to evaluate the performance of various coatings and analyze the effectiveness of emerging novel implant coatings. Furthermore, more research should be conducted to determine whether conventional implant surface treatments and coatings can offer consistent therapeutic results, particularly in the areas of osseointegration stability, mobility, infection prevention, inflammation, and structural failure.

REFERENCES

1. Mehdipour M, Afshar A, (2012) A study of the electrophoretic deposition of bioactive glass-chitosan composite coating. *Ceram Int* 38: 471–476.
2. Singh J, Wolfe DE, (2005) Nano and macro-structured component fabrication by electron beam-physical vapor deposition (EB-PVD). *J Mater Sci* 40: 1–26.
3. Depprich R, Zipprich H, Ommerborn M, Naujoks C, Wiesmann HP, Kiattavorncharoen S, Lauer HC, Meyer U, Kübler NR, Handschel J, (2008) Osseointegration of zirconia implants compared with titanium: An in vivo study. *Head Face Med* 4: 1–8.
4. Park JW, Jang JH, Lee CS, Hanawa T, (2009) Osteoconductivity of hydrophilic microstructured titanium implants with phosphate ion chemistry. *Acta Biomater* 5: 2311–2321.

5. Li B, Fintan T, (2020) Racing for the surface. *Racing Surf.* doi: 10.1007/978-3-030-34475-7
6. Langhoff JD, Voelter K, Scharnweber D, Schnabelrauch M, Schlottig F, Hefti T, Kalchofner K, Nuss K, von Rechenberg B, (2008) Comparison of chemically and pharmaceutically modified titanium and zirconia implant surfaces in dentistry: A study in sheep. *Int J Oral Maxillofac Surg* 37: 1125–1132.
7. Liu X, Man HC, (2017) Laser fabrication of Ag-HA nanocomposites on Ti6Al4V implant for enhancing bioactivity and antibacterial capability. *Mater Sci Eng C* 70: 1–8.
8. Han X, Yang D, Yang C, Spintzyk S, Scheideler L, Li P, Li D, Geis-Gerstorfer J, Rupp F, (2019) Carbon fiber reinforced PEEK composites based on 3D-printing technology for orthopedic and dental applications. *J Clin Med* 8: 240.
9. Nie X, Leyland A, Matthews A, Jiang JC, Meletis EI, (2001) Effects of solution pH and electrical parameters on hydroxyapatite coatings deposited by a plasma-assisted electrophoresis technique. *J Biomed Mater Res* 57: 612–618.
10. Polo TOB, da Silva WP, Momesso GAC, Lima-Neto TJ, Barbosa S, Cordeiro JM, Hassumi JS, da Cruz NC, Okamoto R, Barão VAR, Faverani LP, (2020) Plasma electrolytic oxidation as a feasible surface treatment for biomedical applications: An in vivo study. *Sci Rep* 10: 1–11.
11. Qiu ZY, Chen C, Wang XM, Lee IS, (2014) Advances in the surface modification techniques of bone-related implants for last 10 years. *Regen Biomater* 1: 67–79.
12. Cui FZ, Luo ZS, (1999) Biomaterials modification by ion-beam processing. *Surf Coatings Technol* 112: 278–285.
13. Boccaccini AR, Keim S, Ma R, Li Y, Zhitomirsky I, (2010) Electrophoretic deposition of biomaterials. *J R Soc Interface.* doi: 10.1098/rsif.2010.0156.focus
14. Moskalewicz T, Czyrska-Filemonowicz A, Boccaccini AR, (2007) Microstructure of nanocrystalline TiO2 films produced by electrophoretic deposition on Ti-6Al-7Nb alloy. *Surf Coatings Technol* 201: 7467–7471.
15. Aymerich M, Gómez-Varela AI, Álvarez E, Flores-Arias MT, (2016) Study of different sol-gel coatings to enhance the lifetime of PDMS devices: Evaluation of their biocompatibility. *Materials (Basel).* doi: 10.3390/ma9090728
16. Mavis B, Taş AC, (2000) Dip coating of calcium hydroxyapatite on Ti-6Al-4V substrates. *J Am Ceram Soc* 83: 989–991.
17. Amiri H, Mohammadi I, Afshar A, (2017) Electrophoretic deposition of nano-zirconia coating on AZ91D magnesium alloy for bio-corrosion control purposes. *Surf Coatings Technol* 311: 182–190.
18. Jorfi M, Foster EJ, (2015) Recent advances in nanocellulose for biomedical applications. *J Appl Polym Sci* 132: 1–19.
19. Liu PS, Chen Q, Wu SS, Shen J, Lin SC, (2010) Surface modification of cellulose membranes with zwitterionic polymers for resistance to protein adsorption and platelet adhesion. *J Memb Sci* 350: 387–394.
20. Sun F, Pang X, Zhitomirsky I, (2009) Electrophoretic deposition of composite hydroxyapatite-chitosan-heparin coatings. *J Mater Process Technol* 209: 1597–1606.
21. Vilardell AM, Cinca N, Garcia-Giralt N, Dosta S, Cano IG, Nogués X, Guilemany JM, (2020) In-vitro comparison of hydroxyapatite coatings obtained by cold spray and conventional thermal spray technologies. *Mater Sci Eng C* 107: 110306.
22. Jaiswal S, McHale P, Duffy B, (2012) Preparation and rapid analysis of antibacterial silver, copper and zinc doped sol-gel surfaces. *Colloids Surfaces B Biointerfaces* 94: 170–176.
23. Pashkuleva I, Marques AP, Vaz F, Reis RL, (2010) Surface modification of starch based biomaterials by oxygen plasma or UV-irradiation. *J Mater Sci Mater Med* 21: 21–32.
24. Shypylenko A, Pshyk AV, Grześkowiak B, Medjanik K, Peplinska B, Oyoshi K, Pogrebnjak A, Jurga S, Coy E, (2016) Effect of ion implantation on the physical and mechanical properties of Ti-Si-N multifunctional coatings for biomedical applications. *Mater Des* 110: 821–829.
25. Kumar M, Parashar KK, Tandi SK, Kumar T, Agarwal DC, Pathak A, (2013) Fabrication of Ag:TiO2 nanocomposite thin films by sol-gel followed by electron beam physical vapour deposition technique. *J Spectrosc.* doi: 10.1155/2013/491716
26. Bashir S, Thakur A, Lgaz H, Chung I-M, Kumar A, (2020) Corrosion inhibition performance of acarbose on mild steel corrosion in acidic medium: An experimental and computational study. *Arab J Sci Eng.* https://doi.org/10.1007/s13369-020-04514-6

27. Bashir S, Sharma V, Lgaz H, Chung I-M, Singh A, Kumar A, (2018) The inhibition action of analgin on the corrosion of mild steel in acidic medium: A combined theoretical and experimental approach. *J Mol Liq* 263: 454–462.
28. Bashir S, Thakur A, Lgaz H, Chung I-M, Kumar A, (2019) Computational and experimental studies on Phenylephrine as anti-corrosion substance of mild steel in acidic medium. *J Mol Liq* 293: 111539.
29. Parveen G, Bashir S, Thakur A, Saha SK, Banerjee P, Kumar A, (2020) Experimental and computational studies of imidazolium based ionic liquid 1-methyl- 3-propylimidazolium iodide on mild steel corrosion in acidic solution Experimental and computational studies of imidazolium based ionic liquid 1-methyl- 3-propylimidazolium. *Mater Res Express* 7: 016510.

17 Surface Nanostructuring of Ti Based Alloys for Biomedical Applications
Role of Surface Mechanical Attrition Treatment

T. Balusamy, Sivakumar Bose, T.S.N. Sankara Narayanan and K. Ravichandran

CONTENTS

17.1 INTRODUCTION

Ti and its alloys have gained widespread acceptance for biomedical applications because of their high strength-to-weight ratio, low modulus, good mechanical properties, low corrosion rate, and good biocompatibility. However, the lack of bioactivity and inferior tribological properties are the major limitations for their use in load bearing implants, particularly for hip and knee joints (Zhang and Chen, 2019). Nano/ultra-fine grained materials prepared by severe plastic deformation (SPD) approaches like accumulative roll bonding, multiple compression, equal channel angular pressing, cold rolling, and so on, improve the previously mentioned properties required for biomedical uses (Ratna Sunil et al., 2016). Conversely, the ductility of materials induced by SPD is a major concern. In view, S^2PD assumes significance as they are capable of imparting desirable surface properties such as better mechanical and tribological properties, better strength/ductility combination, better fatigue strength and electrochemical properties of metallic implants. As the material failures occur at the surface and propagate through a matrix, the surface modification would be a good approach to increase the properties and performance of metallic biomaterials. In this regard, SMAT is a versatile S^2PD process that produces the nanometer-sized grains on the surface with the gradient microstructure by plastic deformation

DOI: 10.1201/9781003173533-17

without affecting the surface chemistry. Moreover, surfaces can be modified by altering the chemistry of a surface using plating/deposition or thermochemical diffusion coatings that are bounded with surface contamination, porosity, and uncertainty in obtaining a strong bonding at their interface. In addition, mismatch in hardness and higher thermal expansion coefficient of surface coatings applied for load bearing application may chip off during the service. SMAT is capable of generating nanostructure at the surface on metals/alloys during severe plastic deformation. Hence, limitations experienced by surface coatings via changing the surface chemistry can be nullified, particularly for load bearing implant application. Various types of metals and alloys (ferrous and non-ferrous) have been performed to understand the influence of SMAT on their properties including microstructural, mechanical, tribological, and electrochemical properties as well as biocompatibility (Alikhani Chamgordani et al., 2018; Du et al., 2019; Wei et al., 2018; Mani Prabu Chandra et al., 2020; Singh et al., 2020; Duan et al., 2020; Gao et al., 2020; Wu et al., 2020). In addition, improving the surface reactivity through formation of nanostructure at the surface has a beneficial influence on the diffusion and chemical conversion coatings, which are the added advantages of SMAT (Grosdidier and Novelli, 2019; Balusamy et al., 2013b; Kavitha et al., 2013, 2014; Olugbade and Lu, 2020; Li and Wang, 2018).

17.2 SURFACE MECHANICAL ATTRITION TREATMENT

The involvement of large numbers of dislocation movements following impingement of balls (typically 1–8 mm Ø balls) with high strain rate during SMAT on the materials surface leads to a NC microstructure from the coarse one. The nanocrystallization is attributed to the slip or twinning deformation mechanism depending upon the stacking fault energy and crystal structure of the materials as well as the deformation conditions imposed (Lu and Lu, 1999, 2004; Azadmanjiri et al., 2015). Bulk deformation via equal channel angular pressing, etc., requires large amounts of plastic working energy as compared to S^2PD methods. Moreover, titanium and titanium alloys possess high strength and it is very difficult to process them with bulk deformation. SMAT is quite a versatile, one-step, high throughput method and it can be processed with relatively low cost—added advantages of SMAT.

SMAT is an advanced version of shot peening (SP). The SMAT is similar to that of SP in which both processes are involved in the repeated impingement on the sample surface with spherical shots. However, the notable difference exists between them such as shot size, impact velocity, changes in kinetic energy, strain, and strain rate. This leads to a change in thickness of the NC surface layer and work hardened layer. Besides, the residual compressive stress induced by SMAT is much deeper as compared to SP (Jelliti et al., 2013).

The schematic set-up and SMAT process is explained well by Lu and Lu (1999, 2004) and Balusamy et al. (2010). The ball diameter, type and number of balls used, vibration amplitude, and treatment time are the major process variables. The frequency of vibration ranges from 50 Hz–20 kHz, while within the vibrating chamber, the spherical balls will be placed at the bottom surface while the material to be modified will be fixed at the upper part (at top). During SMAT the vibrating chamber is sealed and evacuated. The continuous multidirectional impact of the spherical balls (typically 1–8 mm Ø) induces plastic deformation that enables the development of a NC surface and a work hardened layer. Due to repeated impacts, the coarse grains are transformed into nanosized grains on the surface, having the gradient microstructure towards matrix. SMAT induces the compressive residual stress at the surface. The NC surface, along with the work hardened layer and compressive residual stress imparted during SMAT, improves the mechanical, tribological, and fatigue performance, etc. (Alikhani Chamgordani et al., 2018). The extent of deformation depends on the strain and its rate, which is influenced by the type and size of the balls, SMAT time, and the type of materials to be modified and the presence of alloying elements.

17.3 CHANGE IN PROPERTIES OF SMATED TI AND ITS ALLOYS

17.3.1 MICRO-STRUCTURAL CHARACTERIZATION

The SEM images of cross-sectioned surfaces of SMAT treated pure Ti (using 3 mm Ø balls (stainless steel) at 20 kHz for 10 to 60 min) are shown in Figure 17.1. SMAT of Ti under such conditions has led to the formation of graded microstructure with nano grains on the Ti surface. With the increase of SMAT duration (10 to 60 min), the extent of deformation is enhanced (Zhu et al., 2004), leading to the emergence of mechanical twins (Figure 17.1(a)). The extent of formation of the mechanical twins is increased with time (Figures 17.1(a)–17.1(d)) that produces a gradient microstructure from the top surface to the bulk (with nano grains at the top surface, submicron-sized grains at the sub-surface and coarse grains in the bulk, respectively). Jamesh et al. (2013) have also shown that after SMAT (conditions: 8 mm Ø Al_2O_3 balls at 50 Hz) of Ti, the extent of deformation (from 35 to 84 µm) of Ti is increased with the increase of SMAT duration (15 to 45 min), resulting in the development of a graded layer structure. Zhu et al. (2004) have explained the mechanistic information about the formation of gradient microstructure on Ti subjected to SMAT process. They predicted four different stages are (a) initiation of twins' formation and their intersections, (b) low angle disoriented lamellae formation with high density of dislocations, (c) partitioning of microbands and

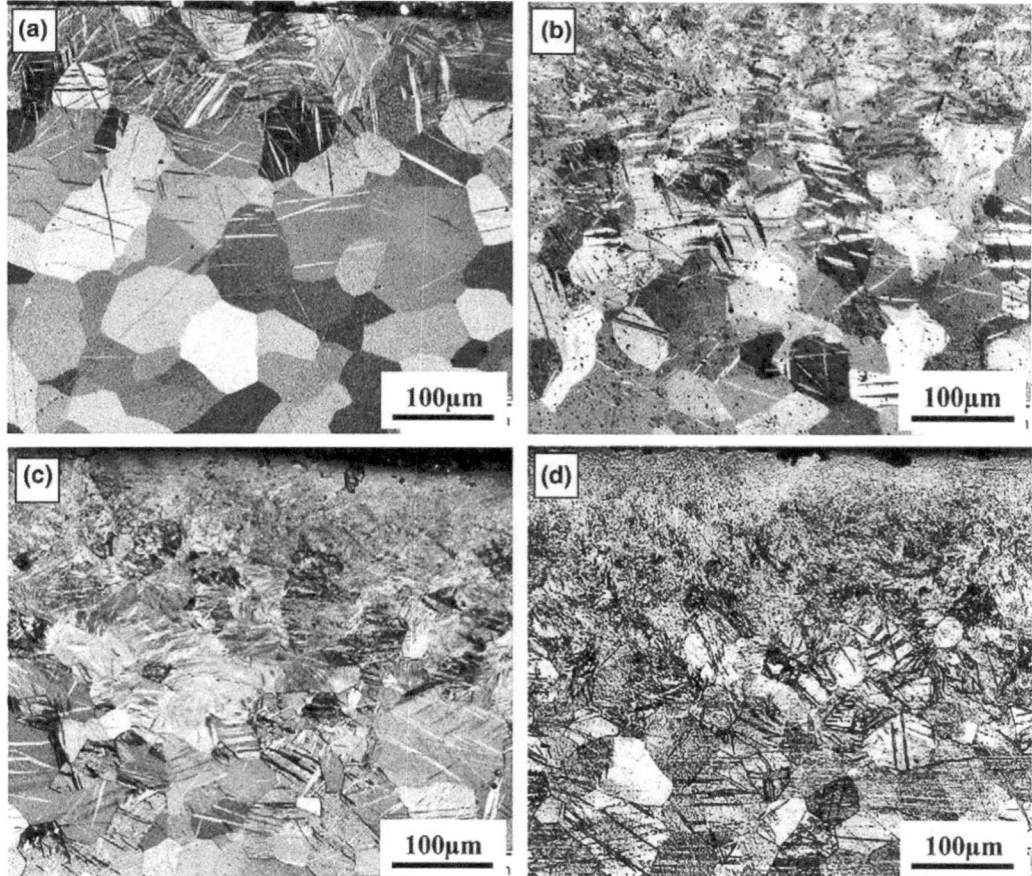

FIGURE 17.1 Cross-sectional images of SMATed pure Ti using 3 mm Ø stainless steel balls for various SMAT time: (a) 15; (b) 16; (c) 30; and (d) 60 min. respectively.

Source: Permission obtained from Elsevier; Zhu et al. (2004).

dislocation cells to disoriented blocks and polygonal submicronic grains, and (d) following by the multiple breakdown of submicronic polygonal grains to randomly oriented nanograins. The grain refinement at the end stage is directed by rotational recrystallization process. Similarly, Jelliti et al. (2013) have proposed the mechanism for the generation of NC surface on Ti-6Al-4V after SMAT (conditions: 2 and 3 mm Ø 100Cr6 balls, 20 kHz, 15 and 20 min). The GR is produced by twinning activity that subparts the coarse grains into finer grains till it attains a nanoscale range (averagely 50 nm). Upon SMAT, the multidirectional impinging actions enhance the activation of large number of twin systems. The change in microstructural features observed at different depth after SMAT is due to strain rate experienced during SPD.

Hu et al. (2011a, 2011b) have studied the impact of SMAT on the microstructural characteristics and phase content of Ni-Ti alloy (conditions: 2 mm Ø SS balls, 20 kHz, 60 min, and room temperature). According to them, SMAT produces a gradient microstructure in which the top surface layer consists of B2 NC and amorphous phases (grain size: ~20 nm; austenite phase only). They have also shown that intermediate regions from the top surface consist of B19' (strain induced martensite-SIM) and B2 phases followed by B19' (SIM) phase and B2 phase at the bulk. The phase variation in the gradient microstructure is due to strain and strain rate experienced by the alloy during repeated impingement of SS balls. The strain induced GR is triggered due to the formation of dislocations with high density and dislocation tangles (DTs) followed by strain induced martensite transformation (B2 to B19'). It consists of dislocation lines (DLs) and dense dislocation walls (DDWs) in the martensite plates that have led to the subsection of martensite plates. With a further rise in strain and its rate, the reverse transformation of martensite (B19' to B2) and amorphization occurs simultaneously on the SMATed surface. The continuous subdivision and amorphization have led to the formation of NC surface layer (B2) on the Ni-Ti alloy. The previously mentioned research works suggest that surface NC by SMAT is feasible on Ti based and Ni-Ti materials.

On the contrary, solution treated β-Ti alloy (TLM) after SMAT fails to show surface nanostructuring due to an unrecoverable strain that hinders GR, whereas it induces dislocations with high density and deformation twins. Acharya et al. (2020) have also reported that β-low modulus Ti-Nb-Ta-O alloy fails to show surface nanocrystallization after SMAT (4.75 mm Ø balls (500 numbers), 25 Hz, and 1 h). The oxygen presence in β-phase has increased the stability and minimized the feasibility of generating nanocrystallized grains.

The XRD data of untreated Ti shows the sharp diffraction peaks pertaining to (100), (002), (101), (102), (110), (103), and (112) planes. The GR induced by plastic deformation and an increase in micro-strain (atomic level) has led to broadening in diffraction peaks of SMATed Ti. A similar inference is also observed in β-type biomedical TiNbZrFe alloy after SMAT. The accumulation of micro-strain (0.61 to 0.79%) has shifted the diffraction patterns after SMAT. Annealing of SMATed β-Ti-25Nb-3Mo-3Zr-2Sn alloy at 200 °C has decreased the lattice micro-strain and dislocation density in the β phase, which has not caused any change in the grain size (Huang and Han, 2013).

17.3.2 Changes in Surface Roughness

SMAT has increased the surface roughness of Ti based materials. The increase in surface roughness is mainly due to the generation of numerous dimples and craters (peaks and valleys) on the treated surface. Jamesh et al. (2013) have showed the increase in average surface roughness (R_a) from 0.15 ± 0.12 µm to 3.27 ± 0.98 µm after SMAT of Ti (8 mm Ø Al_2O_3 balls, 50 Hz, and 45 min). Similarly, Zhao et al. (2011) have noticed the increase in R_a of Ti6Al4V alloy from 0.19 ± 0.07 µm to 3.93 ± 0.26 µm after SMAT (8 mm Ø SS balls, 50 Hz, and 60 min). These results indicate that the R_a is likely to increase with SMAT time due to the continuous and random impingement of balls. Nevertheless, a slight decrease in R_a from 1.26 µm to 1.09 is observed in Ti6Al4V after SMAT (5 mm Ø SAE 52100 steel balls, 50 Hz, and from 30 to 60 min): it is due to the flattening of the surface after the continuous impact of balls. The lack of generation of new dents/

dimples indicates the establishment of a dynamic equilibrium, which reduces the surface roughness. The progression of surface roughness mainly depends on the working capacity of materials to be treated and conditions employed for SMAT (Balusamy et al., 2015). In case of β-Ti-Nb-Ta-O, the R_a exhibits only a marginal increase from 74 ± 7 nm to 95 ± 16 nm after SMAT (500 numbers of hardened steel balls (4.75 mm Ø), 25 Hz for 1 h) (Acharya et al., 2020). The osseointegration rate and biomechanical fixation strongly depends on the surface roughness of implants (Becker et al., 2000). A rough surface profile is preferred rather than a smoother surface for the fast fixation to achieve long-term mechanical stability of a prosthesis. However, a high surface roughness of implants leads to periimplantitis and increase in ionic leakage, whereas an intermediate surface roughness facilitates improvement of those two attributes. Theoretically, the implant surface should be like a hemispherical pit (~1.5 μm in depth and 4 μm in diameter) to provide strong mechanical interlocking at the interface of implant and bone (Becker et al., 2000). Hence, with the optimized ball size and SMAT time the desired surface topography can be achieved on Ti based materials for bioimplant uses.

17.3.3 Changes in Mechanical Properties

SMAT causes the compressive residual stress (CRS) on the surface and sub-surface of metallic materials. The level of change in stress for Ti and its alloys depends on the strain and its rate, which is evidenced by an increase in CRS of Ti6Al4V (from −679 ± 20 to −706 ± 14 MPa) with SMAT duration of 30 to 60 min (Anand Kumar et al., 2013). The formation of nanocrystallized surface, CRS, and work hardening after SMAT of Ti6Al4V for 30 min has enhanced its fatigue life. Unlike the untreated Ti6Al4V, the CRS produced at the SMATed surface could suppress the initiation and propagation of cracks and retard the crack nucleation. Nevertheless, the increase in average surface roughness with anincrease in SMAT duration from 30 to 60 min could decrease in fatigue life. The low ductility of Ti6Al4V leads to heavy surface damage with the increase in SMAT duration. Thus to attain a suitable material performance, the SMAT duration should be carefully optimized considering the ductility, the change in microstructure, etc.

SMAT increases the surface hardness. The hardness profile reveals a decrease in hardness value from top surface to matrix, validating the generation of a gradient microstructure. The high hardness observed at a depth of 50 μm is attributed to the high strain rate (10^2–10^3m/s) experienced near the surface (Huang et al., 2006). Jamesh et al. (2013) have also observed that SMAT of CP-Ti (8 mm Ø alumina balls, 50 Hz for 15 to 45 min) has enhanced the hardness near the surface to about 4.4 GPa as compared to an untreated one (3.1 GPa). Similarly, SMATed Ti-6Al-4V (3 mm Ø SS balls at 20 kHz for 20 min) has increased the hardness value to 4.4 GPa (Jelliti et al., 2013). It is mainly attributed to the dislocation hardening/strain hardening, and density of twin boundaries rather than NC grains (Huang et al., 2006; Zhao et al., 2012; Jamesh et al., 2013; Jelliti et al., 2013; Anand Kumar et al., 2013; Acharya et al., 2020). Similarly, the increase in surface hardness (351 ± 5 Hv) after SMAT (4.75 mm Ø balls (500 numbers) for 1 h at 25 Hz) of low modulus β-Ti-Nb-Ta-O is noticed as compared to its untreated one (Acharya et al., 2020).

SMAT of Ti (8 mm Ø SS balls, 50 Hz for 60 min) has led to an increase in yield strength (YS) to 920 ± 40 MPa and ultimate tensile strength (UTS) to 970 ± 30 MPa due to the work hardening at the surface and sub-surface layers. The SMAT is capable of increasing the YS and UTS of Ti, which is comparable with untreated Ti6Al4V (YS: 902 MPa and UTS: 975 MPa). It may be considered to use Ti in place of Ti6Al4V alloy for implant uses. The difference in the Young's modulus of Ti and Ti6Al4V (~110 GPa) with the human bone (~30 GPa) leads to a stress shielding effect, resulting in failure of the implant. The low modulus β-Ti-Nb-Ta-O has experienced only a trivial change in the elastic modulus (from 64 ± 4 GPa to 66 ± 3 GPa) after SMAT (Acharya et al., 2020), which reduces the stress shielding effect. The formation of a NC surface, dislocation with high density and microbands due to SMAT are considered as the major factors in tuning the mechanical properties of Ti based materials after SMAT.

17.3.4 VARIATION IN TRIBOLOGICAL PROPERTIES

The tribological properties in load bearing implants play a major role. In addition, fretting wear is an important factor wherein micro movements of < 200 µm could cause a significant failure of the implants, such as loosening of the hip joints. Surface nanocrystallization caused by SMAT could enhance the hardness and tribological properties of Ti based materials (Zhao et al., 2012; Anand Kumar et al., 2013; Alikhani Chamgordani et al., 2018). Zhao et al. (2012) have reported that the sliding wear behavior (under dry conditions) of Ti after SMAT (8 mm Ø SS balls, 50 Hz for 60 min) is improved in comparison to untreated Ti. The improved wear resistance is due to the NC surface and high hardness. Both abrasive and adhesive wear mechanisms are operative in untreated Ti (at 1–2 N) and SMATed Ti (at 2 N), while it is found with only abrasive wear mechanism at 5 N, in untreated Ti and at 1 and 5 N in case of SMATed Ti.

According to Alikhani Chamgordani et al. (2018), SMAT of Ti decreased the wear rate and coefficient of friction (COF) by 60% and 66%, respectively. Surface nanocrystallization (12 nm) and an increase in surface hardness (2.8 times) have attributed to the improvement of tribological properties in SMATed Ti as opposed to the untreated one. Contrarily, Rajabi et al. (2019) have reported that the wear rate of SMATed CP-Ti is increased after SMAT (3 mm Ø high carbon steel balls) with the SMAT duration from 0.5 to 2 h under various loads (1–5 N) at sliding velocities of 0.052 and 0.1046 m/s. In SMATed Ti, the abrasive wear (through ploughing) under low forces is dominant although adhesive wear is operative in a few conditions. Inversely at high loads, the ploughing, plastic deformation, and delamination are found to be the dominant wear mechanisms. In this case, although the SMATed Ti surface consists of NC structure and high hardness (2.6 times as compared to untreated Ti), the wear resistance is inferior as opposed to the untreated one. In addition to hardness value, the toughness of material plays a major role. These combined factors are important to achieve the improved wear resistance rather than only the hardness. Similarly, Anand Kumar et al. (2013) have reported that the fretting wear resistance of SMATed Ti-6Al-4V is inferior as compared to the untreated one. Besides, the increase in SMAT duration from 30 to 60 min has decreased the fretting wear resistance largely, which is mainly due to the reduction in toughness of SMATed Ti-6Al-4V for 60 min. The study results reveals that 30 min SMAT duration is identified to be the optimum (5 mm Ø SAE 52100 steel balls) to obtain the improved fretting wear resistance.

In addition to the surface NC structure and increase in hardness value, the surface roughness has an impact on the tribological properties of Ti based materials (Alikhani Chamgordani et al., 2018; Anand Kumar et al., 2013). SMAT has increased the surface roughness of CP-Ti by 6 times due to the formation numerous peaks and valleys. During the wear process, the abrasive pin has less accessibility (contact) in the valley areas of a SMATed surface leading to a reduction in wear track (Alikhani Chamgordani et al., 2018). Similarly, smooth surfaces have lower surface roughness, which leads to a higher friction and fretting damage due to high adhesion of fretting ball. Moreover, accumulation of fretting debris in the same contact area leads to more fretting damage. However, on surfaces with high surface roughness, the fretting debris easily escapes from the contact area into adjacent valley areas, which decreases the abrasive wear (less fretting damage). As per Anand Kumar et al. (2013), in spite of an increase in surface roughness, SMATed Ti-6Al-4V has showed superior fretting wear behavior under normal loads due to the variation in the type of ball used for the test. Hence, it clear that the SMAT duration plays a major role in the tribological and fatigue resistance of Ti based materials.

The tribological properties of Ti based materials in physiological solutions are important for implant applications. Acharya et al. (2020) have studied the fretting wear (6 mm Ø ZrO_2 balls) of SMATed and untreated low modulus Ti-Nb-Ta-O samples. The fretting test was carried out at 5 N, 10 Hz and displacement of 200 µm in presence of 20% fetal bovine serum solution (pH: 7.40) at 37.4 °C. The experiment was performed till 50,000 cycles. The SMATed Ti alloy has reduced the wear volume as it is evidenced by the reduction in wear scar size in comparison to untreated one. However, the variation in COF is the same in both the untreated and SMATed Ti alloy. An increase in hardness after SMAT is likely to reduce the wear rate. However, a combination of both hardened surface and ductility (high strength-ductility combination) is important to achieve the service life of implants.

17.3.5 CORROSION BEHAVIOR

Corrosion resistance of SMATed Ti and its alloys in simulated body fluid (SBF) as well as 3.5% NaCl is studied by many researchers (Fu et al., 2015; Jelliti et al., 2013; Huang and Han, 2013; Acharya et al., 2020). In SMATed Ti, the GR and high density of grain boundaries act as nucleation sites for the formation of an intact and stable passive oxide layer as compared to that of untreated Ti. Corrosion resistance of Ti6Al4V with and without SMAT samples (SMAT conditions: 2 and 3 mm $\emptyset$ 100Cr6 balls, for 15 and 20 min) in Ringer's solution (pH: 7.2, at 37 °C) is addressed by Jelliti et al. (2013). Both Ti6Al4V samples (untreated and SMATed) exhibit the formation of a passive oxide layer and similar breakdown potentials in SBF. However, the SMATed sample shows a positive shift in E_{corr} from 330 mV$_{(SCE)}$ (for untreated Ti) to 235 mV$_{(SCE)}$, a decrease in i_{corr} from 52 nA/cm^2 (for untreated Ti) to 22 nA/cm^2, and an increase in R_{ct} value from 10.56 kΩ/cm^2 (for untreated Ti) to 1700 kΩ/cm^2. SMAT Ni-Ti alloy (2 mm $\emptyset$ SS balls for 60 min, 50 Hz) has showed the improved corrosion resistance as compared to untreated alloy in 0.9% NaCl (Hu et al., 2011c). The improved corrosion protection of Ni-Ti alloy after SMAT depends on the microstructure, the passive oxide layer forming ability, and its chemical constituents. The corrosion protection is associated with the NC structure that promotes the n-type semiconducting oxide layer formation. It is evidenced by a positive shift in E_{corr} towards the noble side and decrease in i_{corr} (from 4×10^{-8} to 4×10^{-9} A/cm^2). However, after increasing the immersion time to 126 h, there was no appreciable change in corrosion resistance in both SMATed and untreated Ni-Ti alloy due to aggressive nature of the Cl$^-$ ions.

Huang and Han (2013) have studied the electrochemical behavior of solution treated β-Ti-25Nb-3Mo-3Zr-2Sn alloy (TLM alloy) after SMAT (2 mm $\emptyset$ zirconia balls, 50 Hz for 30 min) in 0.9% NaCl and SBF. SMAT has shifted E_{corr} towards positive side from -660 mV$_{(SCE)}$ (untreated one) to -410 mV$_{(SCE)}$, decreased the i_{corr} from 2.08×10^{-6} A/cm^2 (untreated one) to 1.28×10^{-7} A/cm^2, and decreased the i_{pass} from 1.86×10^{-4} A/cm^2 (untreated one) to 1.35×10^{-6} A/cm^2. It reveals the formation thick stable passive oxide layer on the SMATed surface. Post-annealing of SMATed TLM alloy at 200 °C (for 10 h) did not exhibit any reasonable change in corrosion protection. The enhanced corrosion protection of SMATed alloy before and after annealing (at 200 °C for 10 h) is mainly due to the combined actions of GR and dissolution of segregated alloying elements near grain boundaries, which facilitates the stable and passive oxide layer formation.

The corrosion behavior of Ti-Nb-Ta-O alloy (β and α + β phases) in SBF was investigated by Acharya et al. (2020). In the first set of the alloy samples, both consisted of β phase only before (STQ) and after SMAT (STQ-SMAT). Another set of alloy samples consisted of α + β phases in aged condition (500 °C for 4 h) before (Aged) and after SMAT (Aged-SMAT). STQ samples consist of equiaxed β-grains, while aged condition has α + β phases. The alloy sample consists of uniformly distributed precipitates of α-platelets near the grain boundaries and inside the β-grains in β-matrix. The corrosion behavior of STQ, Aged, and their SMATed samples (STQ-SMAT and Aged-SMAT) in SBF is shown in Figure 17.2. SMAT has decreased the i_{corr} of STQ from $21.0 \pm 3.0 \times 10^{-3}$ to $7.4 \pm 2.7 \times 10^{-3}$ μA/cm^2 (for STQ-SMAT). Similarly, SMAT has decreased the i_{corr} of aged sample from $152.0 \pm 9.0 \times 10^{-3}$ to $69.9 \pm 0.3 \times 10^{-3}$ μA/cm^2 (for aged-SMAT). The critical current for passivation (i_{pass}) also follows a similar trend as i_{corr} is shown in Figure 17.2 (marked as arrow marks). However, i_{pass} appears to be more prominent than i_{corr}. The low i_{pass} of SMATed samples in both condition (SMATed STQ- and SMATed Aged) suggests the formation of a stable passive oxide layer with higher thickness as compared to STQ and Aged condition.

However, aged and aged-SMAT samples show inferior corrosion performance in comparison to that of STQ samples. The corrosion protection order is given as SMATed STQ > STQ > SMATed Aged > Aged. The passivation of STQ and SMATed STQ samples are analyzed by impedance

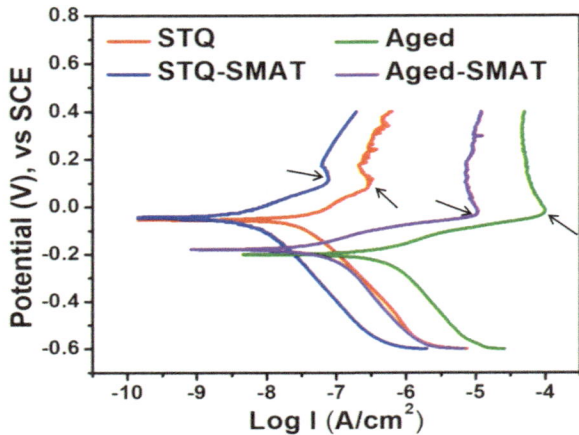

FIGURE 17.2   Tafel polarization curves show corrosion resistance of STQ, Aged, and their SMATed samples (STQ-SMAT and Aged-SMAT) in SBF at ~25 °C (pH:7.4).

*Source:*Reprinted by permission from Elsevier; Acharya et al. (2020).

method (EIS) and Mott-Scottky plots. A higher phase angle at low frequency and low charge carrier density conform the formation of stable oxide layer on SMATed STQ as compared to STQ. Hence, it is clear that SMAT on STQ and aged samples is beneficial in enhancing the corrosion performance in SBF solution.

17.3.6   BIOACTIVITY AND CELL RESPONSE

Ti and its alloys lack in bioactivity, due to the low amount of apatite formation on their surface even after long-term immersion in SBF (28 days). The low modulus β Ti alloys though capable of reducing the stress shielding effect and to avoid the release of toxic ions in SBF, they still lack in bioactivity. Zhao et al. (2011) have reported that the *in vitro* bioactivity of Ti6Al4V is increased after SMAT (8 mm Ø SS balls for 60 min) due to the nanostructured surface and high hydrophilicity. SMAT of β phase Ti alloys is a viable option to improve their bioactivity. SMAT of Ti (8 mm Ø alumina balls, 50 Hz, for 30 min) produces the desired surface topography such as high surface roughness (~2.0 μm), increased surface energy (~43 mJ.cm^{-2}), and decrease in contact angle (~54 °), which helps to promote the calcium apatites formation on Ti (in SBF for 28 days). However, SMAT of Ti (at the same condition for 45 min) limits the extent of apatite formation due to impregnation of fragmented alumina particles (Jamesh et al., 2013). Hence, SMAT with the right type of shots and treatment duration is very important to attain the improved bioactivity of Ti and its alloys.

Cell adhesion and proliferation involves adsorption of protein, interaction between cell and material, attachment of cells and is followed by spreading. These are highly influenced by the surface properties such as roughness, contact angle, surface energy, and hydrophilicity. The bio implant tends to absorb proteins that enhance the interactions between implant and cells. Consequently, the cell attachment and spreading occurs. SMAT of Ti and Ti6Al4V alloy improves the osteoblast cell adhesion and spreading as opposed to their untreated samples. It is due to the NC structure, high surface roughness and surface hydrophilicity, and improved surface energy after SMAT (Zhao et al., 2011, 2012; Lai et al., 2012; Azadmanjiri et al., 2016). Huang et al. (2013, 2014, 2019) have studied the osteoblast response (human fetal osteoblastic cells) of β-Ti-25Nb-3Mo-2Sn-3Zr (TLM) alloy with coarse grained (CG), nano grained (NG) and ultra-fine grained (UFG) surfaces. In comparison to CG surface, the extent of cell growth is much higher on NG and UFG surfaces after SMAT (steel balls for 30 min). This is mainly due to the decrease in contact angle (from 67.8 ± 2.4° to 50.4 ± 1.5°), an increase in surface energy and high degree of hydrophilicity. The number of cells is significantly increased on NG surface than the UFG and CG surface (Figure 17.3). After 24 h, cells grown on both CG and UFG surfaces exhibited spherical morphology, whereas a polygonal shaped morphology was observed on the NG surface. In addition, the amount of adsorption of fibronectin (Fn), vitronectin (Vn), and total protein on NG surface is increased considerably as opposed to the UFG and CG surfaces. SMATed Ti6Al4V alloy (8 mm Ø SS balls for 60 min) has exhibited a significant increase in bone mineral after 8 and 12 weeks of post operation as compared to the untreated one (Zhao et al., 2011). The study results indicate that SMAT can improve the osseointegration at bone-implant interface.

Acharya et al. (2020) have investigated the possibility of pre-osteoblasts cells (MC3T3-E1) on low modulus Ti-Nb-Ta-O alloy before and after SMAT. The cell evaluation depicts that SMAT did not influence the cell attachment. It may be due to the similar surface roughness (R_a from 74 to 95 nm) before and after SMAT (4.75 mm Ø hardened steel balls (500 numbers)). Besides, surface nanocrystallization is not fully realized in Ti-Nb-Ta-O alloy and only fragmentation of grains and twin boundaries are observed after SMAT. Consequently, it has limited the increase in surface roughness and less degree of hydrophilicity which restricts the extent of cell attachment.

A pictorial representation is shown in Figure 17.4 that explains the improvement in characteristic properties of Ti based materials after SMAT, such as mechanical properties, corrosion behavior, tribological properties, and cell attachment and growth. SMATed Ti based materials generate a gradient microstructure from the top surface (having NC layer) to the matrix (Figure 17.4(a)). The surface nanocrystallization accelerates the stable passive layer formation that enhances the corrosion protection of Ti based materials (Figure 17.4(b)). The CRS produced by SMAT helps to enhance the fatigue resistance of Ti based materials (Figure 17.4(c)). The NC surface facilitates apatite formation, and promotes the cell attachment and its proliferation, maturation, and mineralization (Figure 17.4(d)). The increase in hardness helps to improve tribocorrosion resistance (Figure 17.4(e)). The enhancement of these discussed characteristic properties indicates that SMAT is indeed a promising S^2PD method to engineer the metals/alloys for biomedical applications.

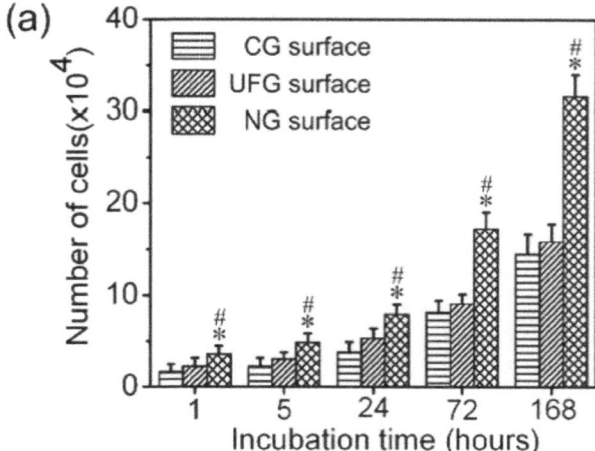

FIGURE 17.3 Cells count profile on CG, UFG, and NG β-TLM alloy surfaces after incubation for 1, 5, 24, 72, and 168 h. The data are mentioned as the mean ± SD, n = 4, *p < 0.01 as compared with CG surface and #p < 0.01 compared with UFG surface.

Source: Permission from Elsevier; Huang et al. (2013).

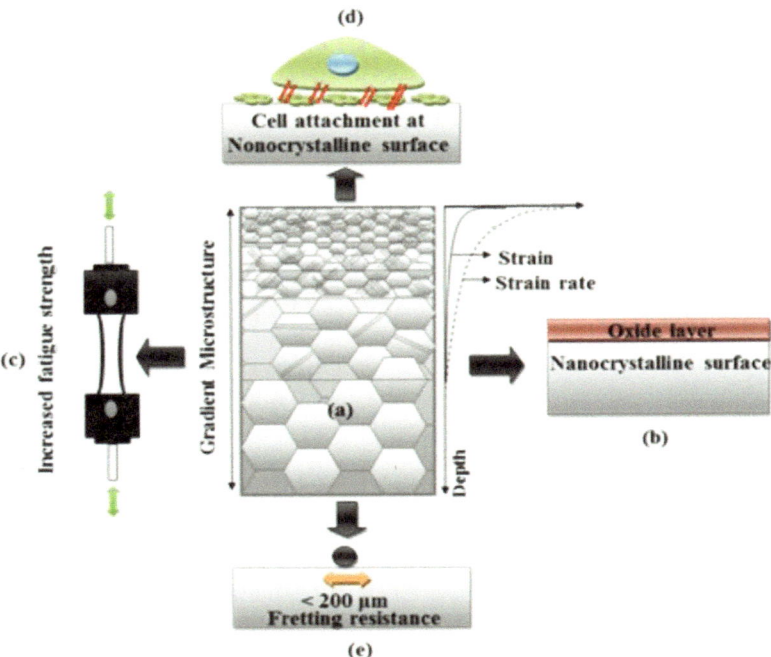

FIGURE 17.4 Pictorial representation of (a) gradient microstructure formation from the top surface to the matrix due to SMAT of Ti based materials, (b) stable passive oxide layer formation, (c) induced CRS which improves the fatigue strength, (d) facilitation of the adsorption of protein and interaction of osteoblasts cells leads to the attachment and spreading of cells, (e) the obtained high hardness would improve the fretting wear resistance.

17.3.7 DUPLEXES OR HYBRID TREATMENT/POST-TREATMENT

In a duplex and hybrid coating approach, SMAT can be utilized either as a pre-treatment or post-treatment. Anodizing is a renowned electrochemical surface modification approach to develop TiO_2 nanotubular arrays on Ti and its alloys. It would be of interest whether SMAT of Ti/Ti alloy can be used as pre-treatment for anodizing. Zhang and Han (2010) have studied anodizing of Ti after SMAT (5 mm Ø zirconia balls, 50 Hz, for 60 min). The results show that the pre-treatment of Ti by SMAT has no influence on the chemical nature and morphology of TiO_2 nanotubes which are formed in ethylene glycol solution (composed with 0.075 M NH_4F and 0.02 vol. % H_2O) at 60 V for 15 to 240 min at ambient temperature. However, pre-treating Ti by SMAT facilitated an increase in oxide layer thickness, which is confirmed with the observation of high current density in the current-time transients (CTT) as opposed to the untreated one. The GR and dislocations induced by SMAT act as fast diffusion channels for O^{2-} and promote the reaction between O^{2-} and Ti^{4+}, leading to the development of a thick oxide layer. In this way, Azadmanjiri et al. (2016) have demonstrated that SMATed Ti (3 mm Ø zirconia balls, 20 kHz, for 20–120 min) has enabled the development of TiO_2 nanotubes with a higher length (18 μm) and diameter (70 nm) than on the untreated one (10 μm and 40 nm, respectively). This is due to the increased reactivity of Ti after SMAT, wherein the NC surface and high density of dislocations facilitate diffusion of oxygen. These research findings reveal that SMAT could be employed as an efficient pre-treatment process to form the TiO_2 nanotubes with better aspect ratio which may explored for drug delivery applications.

The effectiveness of thermal oxidation can also be improved by SMAT process. The GR and dislocations with high density on the SMATed Ti assists to increase the diffusion of O, C, and N that enhances the hardness and lowers the COF as opposed to untreated Ti. The formation of TiO_2 oxide layer (rutile) with high crystallinity improves the biocompatibility of Ti, while the cell adhesion and spreading on Ti is enhanced due to the high surface energy (Wen et al., 2014). Similarly, the rate of thermochemical diffusion coatings like nitriding and boriding (Balusamy et al., 2012, 2013a, 2013b) and phosphate conversion coatings (Kavitha et al., 2013, 2014) could be enhanced by employing SMAT as a pre-treatment.

Wen et al. (2007) have attempted to develop nanoporous crystalline TiO_2 by chemical oxidation of Ti (in 30 wt. % H_2O_2) at 25 °C for 24 h and with other certain conditions. In pre-treated SMAT, the reactivity between Ti and H_2O_2 is increased, which promotes the formation of nanoporous crystalline TiO_2 under similar experimental conditions. Besides, the formed TiO_2 on SMATed Ti significantly enhances the bioactivity in SBF, as it is confirmed by the crack-free mesoporous anatase film formation (Wen et al., 2007).

SMAT can also be used as a post-treatment for Ti6Al4V alloy parts produced by selective laser melting (SLM) for biomedical applications. The high surface roughness, inherent defects, and tensile residual stress of SLM Ti6Al4V alloy parts have significantly limited their widespread acceptance for implant applications wherein long-term cyclic loading is expected (Eyzat et al., 2019; Yan et al., 2019). SMAT as a post-treatment could enhance the surface finishing and induce CRS, which could compensate the limitations induced during SLM of Ti6Al4V alloy parts. The reduction in surface roughness and change in residual stress from tensile to compressive is likely to delay the fatigue crack initiation and improve the fatigue strength (90%) in both low- and high-cycling regimes (LCF and HCF). The improved fatigue strength of SLM Ti6Al4V alloy parts after SMAT is much higher as compared to other post-treatments such as shot peening, electro-polishing, thermal treatment, and ultrasonic nanocrystal surface modification (UNSM).

17.4 SUMMARY

- SMAT of CP-Ti and Ti6Al4V alloy has facilitated the surface nanocrystallization that assisted the gradient microstructure formation consisting of NC microstructure at the top surface, submicron-sized grains at the sub-surface, followed by the coarse grains near the matrix.
- During SMAT, the repeated random multidirectional impingement of the balls generates many dimples and craters on the surface that enhance the surface roughness.
- Unlike the α-Ti alloy, the β-Ti alloy (TLM) is not highly suitable for surface nanocrystallization using SMAT due to the unrecoverable strain that hinders GR. The initial microstructure of alloy (α + β phase) has strong influence on producing the NC surface.
- SMAT induces CRS at the surface and sub-surface, and increases the surface hardness. The change in microstructure, deformation depth, magnitude of CRS, and extent of work hardening is a function of strain and its rate.
- The GR, dislocations with high density, work hardening, and CRS caused by SMAT have improved the hardness, mechanical properties, and tribological behavior of Ti and its alloys.
- GR and CRS caused by SMAT is likely to increase the fatigue life of Ti and its alloys. Nevertheless, overworking leads to surface and sub-surface damage (formation of microcracks, increased roughness), which could act as potential stress raisers promoting fatigue crack initiation. Besides, the decrease in ductility causes the severe surface damage when Ti and its alloys are overworked. The high surface roughness could exert a detrimental effect on fatigue life of SMATed Ti based materials and warrants for the suitable post-treatments.
- SMAT has enhanced the corrosion resistance of Ti based materials in SBF. In case of SMATed samples, the GR, high density of grain boundaries, and dilution of segregated alloying elements at grain boundaries promote the formation of a stable and thick passive oxide layer as opposed to untreated samples. The decrease in leaching of toxic metal ions

(Al and V) with the improved corrosion resistance could impart enhancement in biocompatibility of SMATed Ti6Al4V alloy.

- SMAT of low modulus β-Ti alloys has increased the corrosion resistance and fretting wear resistance (tribocorrosion) in SBF, which is required in load bearing applications. After SMAT, the formation of hard surface layer possesses both β and α + β phases which has improved the fretting wear resistance in SBF. The reduction in α + β interfaces after SMAT of aged samples offers an improved ductility, which helps to increase the wear resistance. A good combination of high corrosion resistance and improved tribological properties caused by SMAT on the low modulus β phase Ti alloy may be considered for the bioimplant uses.
- In case of SMATed Ti based materials, the surface nanostructuring, uniform surface profile, hydrophilicity, and high surface energy has enhanced the protein adsorption, cell-material interaction, cell attachment and spreading as compared to its coarse-grained counterparts.

17.5 FUTURE PERSPECTIVES

The rough surface caused by SMAT seems to be a deleterious factor on the corrosion resistance and tribological behavior, although it decreases the fatigue life of Ti and its alloys, which are important for biomedical applications. The uniform rough surface profile formation would increase the degree of hydrophilicity and promotes good interactions between implant and cells. Adopting a suitable post-treatment such as electropolishing, large pulsed electron beam (LPEB) irradiation, etc., could enhance the surface finish of SMATed Ti and its alloys. However, a proper choice of experimental conditions should be made for LPEB, which would otherwise induce tensile residual stress and affect the fatigue life. The GR, high density of grain boundaries, surface cracks, and the development of a gradient microstructure generated after SMAT of Ti alloy could serve as fast diffusion channels for hydrogen transportation and activate the alloy for hydrogen storage. The hydrogen can be stored at the defect free sub-surface region and it can be reverted back when required.

Numerous hybrid treatment approaches have been studied using SMAT as a pre-treatment such as anodizing, chemical oxidation, and thermochemical treatments. In recent years, additive manufacturing has gained momentum due to its unique ability to fabricate near-net-shape metallic components with a complex geometry and high accuracy. However, the inherent process limitations like inhomogeneous microstructure, defects, and so on, decrease the fatigue life of Ti and its alloys. Exploring SMAT as a post-treatment for additively manufactured Ti based implants would improve the surface finish and impart compressive residual stress, the combination of which is likely to enhance the fatigue life.

β-Ti alloys possess good mechanical properties and biocompatibility which makes them as potential candidate materials for the next generation of metallic implants. The feasibility of β-Ti alloys for surface nanocrystallization by SMAT depends on their microstructure. Since overworking leads to decrease in the ductility of materials, this ultimately influences mechanical properties. It is imperative to optimize the SMAT process parameters, which is specific to the type of β-Ti alloy. In addition, *invivo* studies need to be performed using these materials to ascertain their suitability for biomedical uses. Treating complex shaped implants by SMAT is difficult and warrants design change of the existing machines. A large amount of work needs to be done to realize the potential of SMAT of Ti and its alloys for biomedical applications.

REFERENCES

Acharya, S., Panicker, A. G., Gopal, V. et al. 2020. Surface mechanical attrition treatment of low modulus Ti-Nb-Ta-O alloy for orthopedic applications. *Materials Science and Engineering C* 110:110729.
Alikhani Chamgordani, S., Miresmaeili, R. and Aliofkhazraei, M. 2018.Improvement in tribological behavior of commercial pure titanium (CP-Ti) by surface mechanical attrition treatment (SMAT). *Tribology International* 119:744–752.

Anand Kumar, S., Ganesh Sundara Raman, S., Sankara Narayanan, T. S. N. and Gnanamoorthy, R. 2013. Influence of counterbody material on fretting wear behavior of surface mechanical attrition treated Ti—6Al—4V. *Tribology International* 57:107–114.

Azadmanjiri, J., Berndt, C. C., Kapoor, A. and Wen, C. 2015. Development of surface nano-crystallization in alloys by Surface Mechanical Attrition Treatment (SMAT). *Critical Reviews in Solid State and Materials Sciences* 40:164–181.

Azadmanjiri, J., Wang, P.-Y., Pingle, H., Kingshott, P., Wang, J., Srivastava, V. K. et al. 2016. Enhanced attachment of human mesenchymal stem cells on nanograined titania surfaces. *RSC Advances* 6:55825–55833.

Balusamy, T., Kumar, S. and Sankara Narayanan, T. S. N. 2010. Effect of surface nanocrystallization on the corrosion behavior of AISI 409 stainless steel.*Corrosion Science* 52:3826–3834.

Balusamy, T., Sankara Narayanan, T. S. N. and Ravichandran, K. 2012. Effect of Surface Mechanical Attrition Treatment (SMAT) on boronizing of EN8 steel.*Surface and Coatings Technology* 213: 221–228.

Balusamy, T., Sankara Narayanan, T. S. N., Ravichandran, K., Min Ho Lee, Nishimura, T. 2015. Surface nano-crystallization of EN8 steel: Correlation of change in material characteristics with corrosion behavior. *Journal of The Electrochemical Society* 162:C285–C293.

Balusamy, T., Sankara Narayanan, T. S. N., Ravichandran, K., Park, I. S. and Lee, M. H. 2013a. Surface Mechanical Attrition Treatment (SMAT) on pack boronizing of AISI 304 stainless steel.*Surface and Coatings Technology* 232: 60–67.

Balusamy, T., Sankara Narayanan, T. S. N., Ravichandran, K., Park, I. S. and Lee, M. H. 2013b. Plasma nitriding of AISI 304 stainless steel: Role of surface mechanical attrition treatment. *Materials Characterization* 85: 38–47.

Becker, W., Becker, B. E., Ricci, A. et al. 2000. A prospective multicenter clinical trial comparing one- and two-stage titanium screw-shaped fixtures with one-stage plasma-sprayed solid-screw fixtures.*Clinical Implant Dentistry and Related Research* 2:159–165.

Du, H.-Y., An, Y.-L., Wei, Y-H. et al. 2019. Experimental and numerical studies on strength and ductility of gradient-structured iron plate obtained by surface mechanical-attrition treatment.*Materials Science and Engineering A* 744: 471–480.

Duan, M., Luo, L. and Liu, Y. 2020. Microstructural evolution of AZ31 Mg alloy with surface mechanical attrition treatment: Grain and texture gradient. *Journal of Alloys and Compounds* 823:153691.

Eyzat, Y., Chemkhi, M., Portella, Q., Gardan, J., Remond, J. and Retraint, D. 2019. Characterization and mechanical properties of As-Built SLM Ti-6Al-4V subjected to surface mechanical post-treatment. *Procedia CIRP* 81: 1225–1229.

Fu, T., Zhan, Z., Zhang, L., Yang, Y., Liu, Z., Liu, J. et al. 2015. Effect of surface mechanical attrition treatment on corrosion resistance of commercial pure titanium.*Surface and Coatings Technology* 280: 129–135.

Gao, T., Sun, Z., Xue, H. and Retraint, D. 2020. Effect of surface mechanical attrition treatment on high cycle and very high cycle fatigue of a 7075-T6 aluminium alloy.*International Journal of Fatigue* 139: 105798.

Grosdidier, T. and Novelli, M. 2019. Recent developments in the application of surface mechanical attrition treatments for improved gradient structures: Processing parameters and surface reactivity. *Materials Transactions* 60: 1344–1355.

Hu, T., Chu, C. L., Wu, S. L., Xu, R. Z., Sun, G. Y., Hung, T. F. et al. 2011a. Microstructural evolution in NiTi alloy subjected to surface mechanical attrition treatment and mechanism. *Intermetallics* 19: 1136–1145.

Hu, T., Chen, L., Wu, S. L., Chu, C. L., Wang, L. M., Yeung, K. W. K. et al. 2011b. Graded phase structure in the surface layer of NiTi alloy processed by surface severe plastic deformation. *ScriptaMaterialia* 64: 1011–1014.

Hu, T., Xin, Y. C., Wu, S. L., Chu, C. L., Lu, J., Guan, L. et al. 2011c. Corrosion behavior on orthopedic NiTi alloy with NC/amorphous surface.*Materials Chemistry and Physics* 126: 102–107.

Huang, L., Lu, J. and Troyon, M. 2006. Nanomechanical properties of nanostructured titanium prepared by SMAT.*Surface and Coatings Technology* 201: 208–213.

Huang, R. and Han, Y. 2013. The effect of SMAT-induced GR and dislocations on the corrosion behavior of Ti-25Nb-3Mo-3Zr-2Sn alloy.*Materials Science and EngineeringC* 33: 2353–2359.

Huang, R., Lu, S. and Han, Y. 2013. Role of grain size in the regulation of osteoblast response to Ti-25Nb-3Mo-3Zr-2Sn alloy.*Colloids and Surfaces B Biointerfaces* 111: 232–241.

Huang, R., Zhuang, H. and Han, Y. 2014. Second-phase-dependent GR in Ti—25Nb—3Mo—3Zr—2Sn alloy and its enhanced osteoblast response. *Materials Science and Engineering C* 35: 144–152.

Huang, R., Zhang, L., Huang, L. and Zhu, J. 2019. Enhanced in-vitro osteoblastic functions on β-type titanium alloy using surface mechanical attrition treatment.*Materials Science and Engineering C* 97: 688–697.

Jamesh, M., Sankara Narayanan, T. S. N., Chu, P. K., Park, I. S. and Lee, M. H. 2013. Effect of surface mechanical attrition treatment of titanium using alumina balls: Surface roughness, contact angle and apatite forming ability. *Frontiers of Materials Science* 7: 285–294.

Jelliti, S., Richard, C., Retraint, D., Roland, T., Chemkhi, M. and Demangel, C. 2013. Effect of surface nano-crystallization on the corrosion behavior of Ti-6Al-4V titanium alloy.*Surface and Coatings Technology* 224: 82–87.

Kavitha, C., Ravichandran, K. and Sankara Narayanan, T. S. N. 2013. Effect of Surface Mechanical Attrition Treatment (SMAT) on zinc phosphating of steel.*Transactions of the IMF* 92: 161–168.

Kavitha, C., Sankara Narayanan, T. S. N., Ravichandran, K., and Lee, M. H. 2014. Improving the reactivity and receptivity of alloy and tool steels for phosphate conversion coatings: Role of surface mechanical attrition treatment.*Industrial and Engineering Chemistry Research* 53: 20124–20138.

Lai, M., Cai, K., Hu, Y., Yang, X., and Liu, Q. 2012. Regulation of the behaviors of mesenchymal stem cells by surface nanostructured titanium.*Colloids and Surfaces B Biointerfaces*97: 211–220.

Li, N. and Wang, N. 2018. The effect of duplex surface mechanical attrition and nitriding treatment on corrosion resistance of stainless steel 316L.*Scientific Reports* 8: 8454.

Lu, K. and Lu, J. 1999. Surface Nanocrystallization (SNC) of metallic: Materials-presentation of the concept behind a new approach.*Journal of Materials Science and Technology* 15: 193–197.

Lu, K. and Lu, J. 2004. Nanostructured surface layer on metallic materials induced by surface mechanical attrition treatment.*Materials Science and Engineering A* 375–377: 38–45.

Mani PrabuChandra, S. S., Perugu, S., Jangde, A., Madhu, H. C., Manikandan, M., Joshi, M. D. et al. 2020. Investigations on the influence of surface mechanical attrition treatment on the corrosion behavior of friction stir welded NiTi shape memory alloy. *Surface and Coatings Technology* 402: 126495.

Olugbade, T. O. and Lu, J. 2020. Literature review on the mechanical properties of materials after Surface Mechanical Attrition Treatment (SMAT).*Nano Materials Science* 2: 3–31.

Rajabi, M., Miresmaeili, R. and Aliofkhazraei, M. 2019. Hardness and wear behavior of surface mechanical attrition treated titanium. *Materials Research Express* 6: 065003.

Ratna Sunil, B., Thirugnanam, A., Chakkingal, U. and Sampath Kumar, T. S. 2016. Nano and ultra fine grained metallic biomaterials by severe plastic deformation techniques. *Materials Technology* 31: 743–755.

Singh, Swarnima, Pandey, Krishna Kant, Bose, Siva Kumar and Keshri, Anup Kumar. 2020. Role of surface nanocrystallization on corrosion properties of low carbon steel during surface mechanical attrition treatment. *Surface and Coatings Technology* 396: 125964.

Wei, H., Cui, Y., Cui, H., Zhou, C., Hou, L. and Wei, Y. H. 2018. Evolution of grain refinement mechanism in Cu-4wt.%Ti alloy during surface mechanical attrition treatment. *Journal of Alloys and Compounds* 763: 835–843.

Wen, M., Gu, J.-F., Liu, G., Wang, Z.-B. and Lu, J. 2007. Formation of nanoporoustitania on bulk titanium by hybrid surface mechanical attrition treatment. *Surface and Coatings Technology* 201: 6285–6289.

Wen, M., Wen, C., Hodgson, P. and Li, Y. 2014.Improvement of the biomedical properties of titanium using SMAT and thermal oxidation.*Colloids and Surfaces B Biointerfaces* 116: 658–665.

Wu, Y., Sun, Z., Brisset, F., Baudin, T., Helbert, A. L. and Retraint D. 2020.In-situ EBSD investigation of thermal stability of a 316L stainless steel nanocrystallized by surface mechanical attrition treatment.*Materials Letters* 263: 127249.

Yan, X., Yin, S., Chen, C., Jenkins, R., Lupoi, R., Bolot, R. et al. 2019. Fatigue strength improvement of selective laser melted Ti6Al4V using ultrasonic surface mechanical attrition. *Materials Research Letters* 7: 327–333.

Zhang, L. and Han, Y. 2010. Effect of nanostructured titanium on anodization growth of self-organized TiO_2 nanotubes.*Nanotechnology* 21: 055602.

Zhang, L.-C.and Chen, L.-Y. 2019. A review on biomedical titanium alloys: Recent progress and prospect. *Advanced Engineering Materials* 21: 1–29.

Zhao, C., Han, P., Ji, W. and Zhang, X. 2012. Enhanced mechanical properties and in vitro cell response of surface mechanical attrition treated pure titanium. *Journal of Biomaterials Applications* 27: 113–118.

Zhao, C., Ji, W., Han, P., Zhang, J., Jiang, Y. and Zhang, X. 2011. In vitro and in vivo mineralization and osseo-integration of nanostructured Ti6Al4V.*Journal of Nanoparticle Research* 13: 645–654.

Zhu, K. Y., Vassel, A., Brisset, F., Lu, K. and Lu, J. 2004. Nanostructure formation mechanism of α-titanium using SMAT.*ActaMaterialia* 52: 4101–4110.

18 Surface Modification by Electrospinning Technique on Mg Alloys for Biomedical Applications

Karthega Mani, Silpa Cheriyan and Nagarajan Srinivasan

CONTENTS

18.1 INTRODUCTION

Electrospinning is identified as a unique way of polymer deposition over the substrate that is most widely used in nanotechnology and considered as an efficient way of producing one-dimensional polymeric Nano fiber. It is accepted as the most powerful, simple, straightforward, and versatile technique capable of producing continuous, ultrafine, indefinitely long Nano fibers from polymeric solution. The diameter of these Nano fibers ranges from less than 100 nm to 500 nm can be electrospun from various materials which include polymer, composite, and ceramics [1]. The ability to spin Nano fibers of various diameters, large surface area to volume ratio, high aspect ratio; its easy fabrication, cost effectiveness, high reproducibility, and improved mechanical properties make this electrospinning technique to be mostly preferred for its many potential applications [2]. The successful use of Nano fibers in many engineering applications includes energy storage, energy conversion, super-capacitors, military protective clothing, and biomedical applications such as wound dressings, tissue engineering scaffolds, artificial blood vessels, pharmacy, drug delivery, etc. [3]. Further, based on the application, the fibers can be functionalized with a number of molecules such as nanoparticles, inorganic precursors, drugs, antioxidants, and antibacterial agents. However, the formation of nonwoven mats composed of Nano fiber bounds the applications of the electrospinning process in various fields [4].

18.1.1 ELECTROSPINNING PROCESS

In the electrospinning process, the polymeric solution is subjected to a static electric field. It is then transformed into fibers with an average diameter varying from a few microns to a nanometer scale,

DOI: 10.1201/9781003173533-18

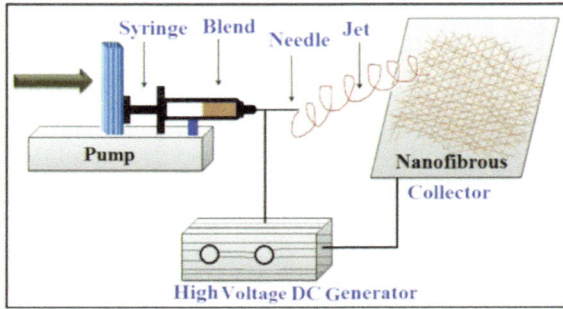

FIGURE 18.1 Basic electrospin setup.

and the fiber will be of the order of a few hundred micrometers in length. The basic electrospinning setup is shown in Figure 18.1.

A simple electrospinning instrument consists of three main components:

1. High voltage power supply (kV range)
2. Metallic needle attached to syringe, and
3. Solid substrate or liquid media as collector (which is been grounded).

This technique uses a very high intense voltage to produce an electric field between the charges induced in the needle tip and the collector using a power supply. The voltage applied to the polymeric solution is generally varied from 20 kV to 50 kV and it depends on various parameters like molecular weight of the polymer, viscosity, concentration, etc. A peristaltic pump with a controlled flow rate pushes the polymer solution present in the syringe through a metallic needle. The solution emerges out from the tip and appears as a semi-spherical droplet. As the voltage increases, the droplet is charged and it expands to form a cone-like shape at the needle tip which is named as Taylor cone. Further with increase in time, the charges get accumulated on the polymer droplet's surface till the voltage attains critical value. Once the critical voltage is reached, the surface tension of the polymer droplet dominates the electrostatic repulsive force between the charges, and forms a polymer jet. The polymer jet then moves towards the collector in air, evaporating the solvent present in the polymer jet. The charged polymeric fiber is then randomly collected over the surface of the collector plate as a thick solid mat of Nano fibers [5].

18.1.2 Fundamental Aspects of Electrospinning Technique

There are many theories, and simulated models are available to discuss the formation of Nano fibers in the electrospinning process. Generally, it involves four consecutive steps [2]:

1. Charging of polymer droplet results in development of cone-shaped jet known as Taylor cone
2. The charged polymer jet or the Taylor cone elongates in a straight line
3. As it elongates, the jet becomes thin and the formation of electrical bending is observed
4. Finally, the jet is deposited as a dry mat of solid fiber on the collector plate which has been grounded.

When the surface charge increases on the droplet or the Taylor cone, it becomes unstable, leading to the drawing of a single fluid jet and gets expelled from the tip of the needle (Figure 18.2). Later, the jet is accelerated by the electrically driven bending instability or whipping instability as it moves

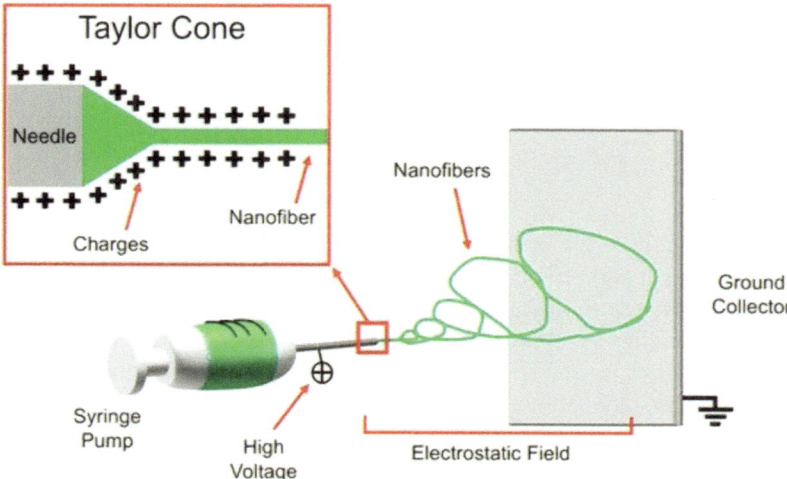

FIGURE 18.2 Taylor cone formation.

due to the applied field. To minimize the electrically driven instability, the polymer jet initially splits into two and then into more jets, and tries to accelerate along a straight line. But the presence of surface tension and viscoelastic force prevents forward motion of the jet and slowly attenuates it. Hence, the jet diameter reduces as it moves further away from the needle tip. In other words, the jet elongates continuously with an increase in distance from the needle tip. When it elongates, the solvent gets evaporated and it forms a solid dry fiber. Moreover, the polymer jet constantly experiences a force because of the electric field, which leads to continuous fiber formation. The formed fibers move towards the collector and then get deposited on the collector plate. The collector plate can be either a flat collector or drum collector based on which nonwoven or woven fibers can be obtained. The entire electrospinning process is carried out in room temperature with atmospheric conditions [13].

When the voltage reaches the critical value, the suspended polymer droplet flowing out of the needle tip overcomes the surface tension, and hence the polymer jet is produced. The surface of the droplet experiences electrostatic pressure (P_e) due to the external applied electric field which can be deduced from the relation

$$P_e = \frac{E^2 \varepsilon}{2}$$

Where, ε is the dielectric constant of the medium surrounding the droplet and E is the strength of the applied electric field. Moreover, the surface tension creates the capillary pressure (P_c) in the liquid and it is defined by the Young–Laplace equation given by

$$P_c = \frac{2\gamma}{r}$$

Where, γ is the surface tension and r is the mean radius of curvature of the surface, which can be given by the spinneret inner radius. The high repulsive force prevailing between the charges present in the solution will be sufficient enough to overcome the surface tension. Hence, the shape of spherical droplet changes into a conical shape [6]. The value of critical voltage (V_c) is defined by the following equation;

$$V_c = \frac{4H^2}{h^2}\left(In\left(\frac{2h}{R}\right)-1.5\right)(1.3\pi R\gamma)(0.09)$$

Where, H is the needle tip to the collector distance, h is the length of the spinneret, and R is the outer radius of the spinneret [1].The electrically charged jet will become thinner due to the influence of electro hydrodynamic forces. In the presence of proper operating conditions, the jet experiences electrically induced bending instabilities as it moves forward towards the collector plate resulting in extensive stretching. During the stretching process of the jet, the solvent present in the jet gets evaporated, resulting in a reduction in the jet diameter [7]. For the stretching of the jet the critical length (L) of the straight segment of the charged jet can be determined using the following equation:

$$L = \frac{4kQ^3}{\pi\rho 2I^2}\left(\frac{1}{R_0^2}-\frac{1}{r_0^2}\right)$$

$$R_0 = \sqrt[3]{\frac{2\sigma Q}{\pi k\rho E}}$$

Where σ is the surface charge density, Q is the rate of flow, k is the electrical conductivity of the fluid, ρ is the fluid density, E is the strength of the electric field, I is the current passing through the jet, and R_0 is the initial radius of the jet [2].

18.1.3 PARAMETERS INFLUENCING THE ELECTROSPINNING PROCESS

The structure, morphology, and formation of the Nano fibers in the process of electrospinning are influenced by various parameters. These parameters can be classified as intrinsic and extrinsic parameters.

18.1.3.1 Intrinsic Parameters

Solution Viscosity: To obtain a uniform and beadles fiber, the viscosity of solution plays a very crucial role. The solution viscosity can be modified by varying the solution concentration. At very low viscosity, the solution will be ejected out from the needle as a droplet without any fiber formation. At high viscosity, ejection of solution from the needle tip becomes difficult and leads to failure of Taylor cone formation. As a result, the solution fails to extend as a fiber. Hence, optimal viscosity is necessary for the electrospinning process. The spin ability of different polymers is achieved with specific viscosity varying from 1 to 215 poise. In general, the factors such as viscosity of the polymer solution, the concentration of the polymer, and molecular weight are connected to each other [4].

Solution conductivity and surface charge density: Generally, the solution used in the process of electrospinning must be highly conducting. Hence, more frequently conducting polymers are used in this process. The conductivity of the polymer depends on the type of polymer and solvent used. The surface charge density and the conductivity of the polymeric solution can be increased with the addition of ionizable salts to the solution. The increase in solution electrical conductivity results in decreased fiber diameter. Bending instability and larger diameter distribution are observed in highly conducting solution [2, 8].

Solution molecular weight distribution: Another most important factor that influences the fiber morphology is the molecular weight, as it affects viscosity, surface tension, and conductivity. Fibers with beads were observed for low molecular weight polymers, whereas, larger molecular weight polymers give a larger fiber diameter. The number of entanglements of polymer chains depends on molecular weight, which depicts the viscosity of the polymer. Even with low concentrations of high molecular weight polymer, the entanglement is maintained to ensure polymer viscosity to produce a uniform continuous fiber during electrospinning. If the solution contains sufficient intermolecular interaction to provide the interchain connectivity, even with low molecular weight it is considered to be ideal for electrospinning [7–9].

Solution Surface tension: Various surface tensions are obtained by using different solvents. Beadles and uniform fibers can be formed by reducing the polymer surface tension. A low electric field must be applied to lower the surface tension of the polymer.

18.1.3.2 Extrinsic Parameters

Applied Voltage: The cause for the jet formation followed by fiber deposition on the collector due to the applied voltage is considered to be a necessary parameter in the electrospinning technique. The voltage applied must exceed the critical value, so that the force of repulsion between the charges present in the solution can overcome the surface tension, resulting in the fiber formation. Studies reveal that more polymer solution is expelled out of the needle due to applied higher voltage, resulting in fiber with a larger diameter. Moreover, the increase in electrostatic repulsive force among the charges due to increase in the applied electric field results in the fiber formation with a lesser diameter. Higher voltage with a higher electric field leads to fiber with a lesser diameter and the solvent is evaporated rapidly from the fibers as it moves towards the collector. It is observed that as the intensity of the electric field is doubled, fibers formed will be half the original diameter. The influence of voltage on the diameter of the fiber depends on the polymer concentration and the tip of the needle to collector distance [4, 5].

Solution Flow rate: The solution flow rate is also an important parameter as it influences the fiber shape. A low flow rate will have sufficient time for solvent evaporation, accelerating the material transfer rate. In the presence of a very low flow rate, it is impossible to maintain the Taylor cone shape. Studies revealed that at a constant voltage, the fiber diameter increases as flow rate of the solution increases. Beaded fibers are obtained at high flow rates because of improper drying of the fibers as they move forward with respect to the collector plate [10].

Types of collector: Aluminum foil or metallic substrate is generally used for collecting random fibers on the stationary collector. It serves as a conductive substrate to collect Nano fibers. The drum collector with constant speed provides aligned fibers [5].

Distance between needle and collector: The needle and collector distance greatly affects the size and morphology of the fibers. An appropriate distance should be maintained for the solvent evaporation to take place before the fiber reaches the collector. Beaded fibers are obtained if the needle and the collector distance are too close or far away. Studies have revealed that the fiber diameter can be varied by adjusting the tip of needle and the collector distance [4, 5]. Other than solution and processing parameters, humidity and temperature are also very essential factors for the electrospinning process, because they affect the fiber diameter and morphology. Studies have proved that viscosity and temperature are inversely related to each other. At low humidity, the solvent evaporation is faster which leads to the drying of polymer in the needle itself. High humidity enhances the fibers discharge and results in formation of small pores on the surface of the fiber [4].

18.1.4 Methods of Electrospinning

Based on the formation of the jet and the way of using needles, there are plenty of methods to fabricate electrospun Nano fibers. Among that, syringes with needles are the basic electrospinning setup with two standard arrangements: vertical and horizontal, which are achieved by varying the position or angle of the basic components. There are two categories of electrospinning instruments: with needles and without needles. The electrospinning apparatus having needles is generally available in three configurations with different functionalities.

Single-needle electrospinning: Single-needle configuration is a basic instrumental setup of electrospinning which consists of three major parts: a single needle syringe, a stationary or rotating collector plate, and a high voltage source to create an electrical potential difference between the needle and the collector. This method is usually used to produce Nano fibers in small volumes in laboratory settings and is not suitable for industrial/wide-range applications.

Multi-jet electrospinning: single-needle spinnerets are the commonly used fiber fabrication method and helps in mass production within short interval of time. Because of the limited productivity for a wide-scale application of single needle electrospinning, multi-jet electrospinning was developed to overcome low productivity by generating multiple jets from a single needle. Even if it's an appropriate method in laboratories, the generation of numerous jets brought on other problems like evolution of jet repulsion, insufficient process control, lack of fiber quality, etc.

Multi-needle electrospinning: Multi-needle electrospinning has advanced to increase productiveness by adding more needles. In this method a number of needles can be used as spinnerets, which can contain one or more variant polymer solutions. So multi-needle electrospinning makes the process easier to combine different polymers in a required ratio. However, a high voltage is essential in this method for continuous electrospinning because of the huge amount of spinning solution. The effort in cleaning the number of needles, fluctuation in electric field strength, and imprudent fiber size distribution are also the drawbacks yet to resolve.

Needleless electrospinning has also gained a lot of attention recently and it was developed to propagate multiple jets of Nano fibers out of a polymer solution. The extensive range of production makes the technique more relevant for industrial applications. The needleless electrospinning has great control over the fiber diameter and amount of production. Researchers have opened new doors to the new techniques and altered the arrangements to produce more efficient Nano fibers to beat their requirements. There are several methods available which are suitable for needleless electrospinning such as bubble electrospinning, two layer fluid electrospinning, melt differential electrospinning, rotary cone electrospinning, etc.

18.2   BIOMEDICAL APPLICATIONS OF NANO FIBERS

The properties acquired by Nano fibers produced during electrospinning find a variety of applications in biomedical research, especially for fabricating bio-membranes, biosensors, wound healing materials, drug delivery, tissue engineering, etc.

In general, the fibers are designed and fabricated for the development of various biomedical devices. This is carried out by modifying the composition, structure, and properties of fibers, such as diameter, porosity, uniformity, biodegradability, biocompatibility, and other mechanical and biological properties. It helps to control cell repair or growth or various tissue regeneration and its interfaces. The structure, morphology, and function of the Nano fibers can be adapted to the natural extracellular matrix as a substrate to restore and enhance human tissue function [11]. Over last few years many reports have been published on the processing of polymers into fibers for various applications based on the electrospinning technique. Hence, a recent strategy is directed towards the usage of polymer composite coatings on the metallic substrate to deaccelerate the corrosion rate, improve the functional, mechanical, and biological performance of metallic implants. Many polymers are used to electrospin Nano fibers for biomedical applications because of their superior properties such as conductivity, ductility, biocompatibility, and biodegradability in nature [8]. In general, polymers are grouped as synthetic and natural polymers [12].

Synthetic polymers, which include polylactic acid, polycaprolactone, polyvinyl alcohol, polylactic-co-glycolic acid, etc., are employed for the formation of Nano fibers in controlled conditions and have been well studied [13]. These polymers are used widely owing to their better mechanical properties and tunable degradation kinetics. However, scaffolds fabricated from synthetic polymer fail to recognize the cells without the presence of proteins and peptides and hence prevent cell growth and attachment, proliferation, and differentiation. Synthetic polymers undergo sustaining challenges when placed inside our bodies. Even though these polymers are biocompatible and bio-degradable, the implantation of new materials in the body may reduce the amount of tissue integration, hence resulting in implant failure [14].

Natural polymers help in inducing cell growth, cell proliferation, and differentiation, by reducing implant interaction in the physiological medium. There are many natural polymers available such as collagen, glycosaminoglycan, chitosan, and alginates, etc. [15]. The natural polymers are gaining special attention in biomedical research, owing to their low toxicity and low chronic inflammatory responses. But these polymers have a complexity that consists of functional peptides, growth, and bioactive factors which are already present in the body [6]. Therefore, biologically derived materials can provide a signal for tissue engineering and scaffold applications, without the addition of growth factors. Research on bioactive coating materials indicates that biopolymers can be a better material for the construction of biodegradable and biocompatible devices [5]. There are many biopolymers, modified biopolymers, and blends of biopolymers with synthetic polymers that have been used for electrospinning and surface coating. Biopolymers lead in the biomedical industry as they possess non-toxic, antibacterial, biodegradable, and biocompatible properties [16]. Bio-fibers exhibit a large variation in their properties due to their complex structure; for example, natural bone is an inorganic-organic composite of Nano-hydroxyapatite and collagen fibers. Collagen is the most abundant extracellular matrix (ECM) protein and the major organic component of natural bone, which has a promising influence on attachment, growth activity, and function of osteoblast cell lines [7].

Metallic materials are widely used as implantable devices for orthopedic, dental, and cardiovascular applications. Non-biodegradable and non-porous metals like stainless steel, titanium and its alloys, cobalt-chromium alloys, and platinum have been used as load-bearing orthopedic implants. The surface composition of these implants has a huge influence on the tissue integration process and lifespan of an implant. Further, these metallic implants lack mechanical integration between the various implant material and their surrounding tissues. This leads to stress shielding between the implanted material and its nearby tissues, ultimately leading to bone restoration [12]. Moreover, the release of metal ions from the implant material due to mechanical wear and degradation of the implant material leads to toxicity and inflammation [13]. As a consequence, the patient has to undergo a secondary surgery with increased possibility of risk and various other complications.

18.2.1 Electrochemical Studies on Mg Alloys

Biodegradable materials such as magnesium (Mg) and its alloys are considered as one of the hopeful next-generation orthopedic materials due to their biodegradability, biocompatibility, and improved bone regeneration. Mg is a lightweight material with excellent properties to be used in orthopedic applications [14]. Magnesium ions are the fourth most abundant cation in our human body and also are present as natural components in bone tissue. Mg-based alloys have density nearly 1.7–2.0 g/cm^3 whereas for bone, the density value ranges around 1.8–2.1 g/cm^3. Mg-based alloys have elastic modulus (~ 45 GPa) closer to that of natural bone (3–20 GPa). Also, Mg alloys' Young's modulus is similar to bone, so it can be concluded that it has considerable potential to become desirable bio-degradable material for clinical applications, especially for orthopedic implants and cardiovascular stents [14–16]. Since it is a biodegradable material, the necessity for revision of surgery to take out the implanted material from the body after bone regeneration can be avoided. This helps to avoid the problems connected with the longer duration of implants in the human body. Furthermore, the corrosion products formed during the Mg alloys degradation are harmless, non-toxic, and get excreted from our human body through urine. But, the major problem encountered by Mg alloys is their poor resistance to corrosion in aqueous environments. The mechanism behind the corrosion of

Mg alloy in simulated body fluid solution (SBF) is discussed later. The Mg alloy degradation due to the overall redox reactions that take place in an electrolyte is as follows [17]:

Cathode Reactions

$$2H_2O + 2e- \rightarrow 2OH- + H_2 \uparrow \text{ (Eq. 1)}$$
$$O_2 + 2H_2O + 4e- \rightarrow 4OH- \text{ (Eq. 2)}$$

Anodic Reaction

$$Mg \rightarrow Mg^{2+} + 2e- \text{ (Eq. 3)}$$
Overall reaction involves $Mg + 2H_2O \rightarrow Mg(OH)_2 + H_2$ (Eq. 4)

In Eq. 4, the presence of high electronegative potential makes the Mg alloys undergo corrosion in an aqueous electrolyte via an electrochemical reaction. The major corrosion products formed during the degradation of Mg alloys are hydrogen gas evolution and the formation of magnesium hydroxide. Moreover, hydroxides' formation because of the presence of alloying elements and hydrated oxides is due to the variation in the local environment.

At times the physiological environment may also be enriched with chloride concentration; in such case Mg alloys will undergo a faster degradation rate because Mg hydroxide gets converted into soluble Mg chloride. However, fast degradation in the physiological environment hinders their further clinical applications. In order to overcome this drawback, either a newly developed magnesium alloy or surface modification technique needs to be employed for bone fracture fixation, so that the Mg alloy slowly degrades with healing and at the same time enhances the bone growth formation.

Alloying of Mg-based materials with elements such as Al, Ca, Zn, Mn, rare earth elements, etc. at times provides an unfavorable environment for the bone healing to take place, which is considered as a huge task in modifying the surface of Mg-based alloys. Hence, it's necessary to understand the biodegradability along with the biocompatibility of the alloying elements during the fabrication of orthopedic Mg-based implants. Also, these alloying elements form intermetallic phases on interaction with Mg; this in-turn influences the various properties of Mg-based alloys.

Hence, in past decades, various surface modification methods which include chemical conversion, micro-arc oxidation, sol-gel process, hydrothermal treatments, plasma electrolytic oxidation (PEO), and ion implantation acted as a protective coating for corrosion protection and also improved the biocompatibility of implants. However, at times these coatings don't provide protection for the implants for a longer duration of time. Thus the implants' surface modification by formation of a dense layer of polymeric coating and its influence on the corrosion behavior of the Mg alloys in SBF solution has been widely studied. But the experimental evidence indicates that these coatings cannot be effective in protecting the Mg alloy due to the formation of hydrogen gas and release of Mg ions beneath these coatings. Moreover, the creation of internal stress in thick, non-porous polymeric coating formed during the coating procedure may result in the existence of micro or Nano cracks. This defective layer provides a pathway for the solution or water molecules to diffuse through the coating and reach the substrate surface. This leads to the reduction in the adhesiveness between the coating and the material surface, delaminating the protective layer and subsequently increasing the degradation rate [17–19].

Hence, the recent research is focused on polymeric Nano fibrous coating on the metallic implant surface using the electrospinning technique. Polymers such as Polycaprolactone (PCL), polyvinyl alcohol (PVA), poly (3-hydroxybutaric acid-co-3-hydrovaleric acid) (PHBV), and polylactic acid (PLLA) were successfully electrospun into Nano fibers due to their biodegradability and biocompatibility [18]. The report shows that electrospun-based coatings are efficient in reducing the rate of corrosion of Mg alloys [19, 20]. In general, the corrosion resistance of any coated material depends on the bonding/adhesive strength at the interface of the coated layer and the metallic substrate. Low adhesive strength leads to easy penetration of electrolytes and accumulation of ions from the electrolytes at the interface which increases the corrosion process in the bonding region. This results in the delamination of coating and also hinders the application of polymeric coated metallic implants for the long term. Hence, the bonding between the substrate and coating layer at the interface should be through chemical interactions, rather than a physical deposition. For instance, a thick, dense, and porous layer of PLLA exhibited better adhesive property than PCL coating over Mg alloy. The oxygen present (C=O and C–O) in the molecular structure of PLLA, produces attractive force which helps in inducing the electrostatic interaction between the coated layer and Mg substrate. Hamid Reza et al. [21] examined the adhesive property of the PLLA and PLLA incorporated with akermanite (AKT) Nano fibers and observed that PLLA—AKT Nano fibers exhibited higher adhesive strength around 5.86 MPa which is higher than PLLA (4.12 MPa) Nano fibers. This might be because of the higher interfacial bonding between the PLLA-AKT Nano fibers and Mg substrate.

A study on the electrochemical behavior of electrospun PHBV Nano fibers coated over Mg alloy for orthopedic applications found these fibers had higher corrosion resistance than uncoated metal [20]. The experimental studies on the surface treatment and electrospun PLLA Nano fibers on Mg (AZ91) alloy indicated that the coated specimen exhibited a reduction in corrosion behavior in Hank's solution assessed by weight loss and hydrogen evolution method. Polymeric Nano composite fibers were also produced by incorporating different concentrations of TiO_2 nanoparticles into PCL solution and were deposited on AM50 alloy. A hollow uniform fibers without any beads were observed for PCL electrospun Nano fibers due to the use of acetone as a solvent. Whereas, the morphology of these uniform fibers was distorted with higher TiO_2 nanoparticle concentration. Most of the TiO_2 nanoparticles were observed to be uniformly dispersed within the fibers and few were observed on the surface of the fiber forming agglomeration. On immersion of these fibers in SBF solution, the fiber morphology was retained for PCL coated AM50 alloy [22]. But surprisingly, some discontinuity or break in the fiber structure was observed for PCL/TiO_2 Nano fiber deposited over AM50 alloy.

This shows the degradation of the PCL fiber in the SBF solution. The PCL/TiO_2 fiber degradation is mainly observed only at certain areas which possess some defects due to the irregularity in thickness due to TiO_2 nanoparticle agglomeration on the surface of the fiber, by which SBF solution would have penetrated into the fiber morphology, leading to breakage in the fiber morphology. As a result, the corrosion rate was observed to be higher for PCL/TiO_2 compared to PCL coated AM50 alloy. This kind of coating delamination and decrease in corrosion resistance due to increased intake of electrolytes through porous layers formed by the sol-gel layer doped with ceramic particles was reported earlier [23].

18.2.2 Biocompatibility Studies on Mg Alloy

The main principle behind any surface coating is to prevent the electrolytes from reaching the substrate surface. The bond developed between the magnesium surface and polymeric coating are highly stable, and that suppresses the acceleration of corrosion at the initial stages of immersion. The formation of coating matrix by mixing of PCL polymer and TiO_2 nanoparticle in acetone results in a nucleophilic substitution reaction, resulting in the hydrogen bond (H^+) formation between the carboxylic group of the PCL polymer and the hydroxyl group (OH^-) of the TiO_2 matrix. This coating matrix deposited over the surface of AM50 alloy is immersed in SBF solution [17]. The higher concentration of TiO_2 in PCL/6 wt. % TiO_2 Nano fiber leads to the formation of Ti-OH groups on the fiber surface, which stimulates the nucleation of apatite particles. As the apatite nucleation is initiated, the growth of it occurs rapidly, utilizing the positive calcium (Ca^+) ions and negative phosphate (PO^{4-}) ions in the SBF solution, resulting in hydroxyapatite formation as shown in Figure 18.3.

At the same time, the development of pores on the PCL/6 wt. % TiO_2 Nano fiber helps the SBF solution to diffuse easily through pores present in the fibers and reach the sample surface to undergo degradation. Kim et al. [24] characterized the formation and behavior of PCL/ZnO composite Nano

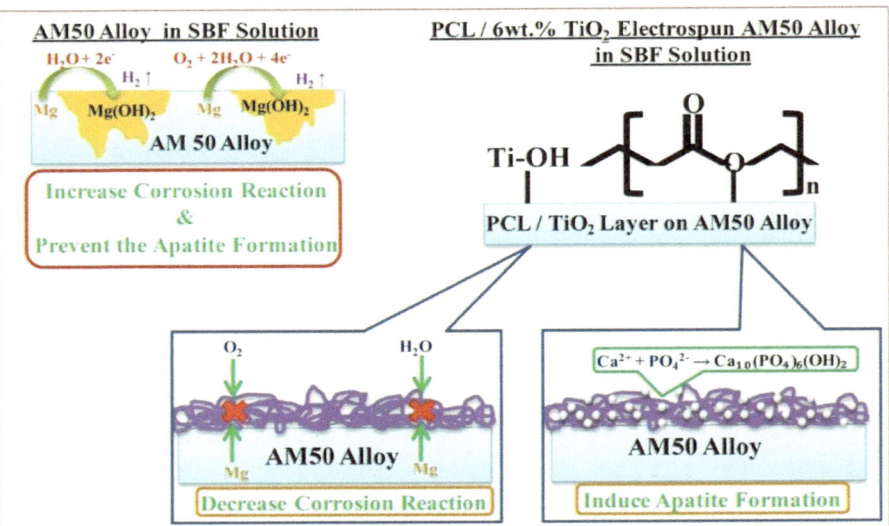

FIGURE 18.3 Corrosion mechanism for PCL/6wt% TiO$_2$ electrospun AM50 alloy on immersion in SBF solution.

fibers (1 and 3 wt. %) coated on the Mg alloy surface using the electrospinning technique. The fibers deposited on the surface were observed to be highly porous and became rougher than PCL fibers. The PCL/ZnO 3 wt. % composite coating on the Mg specimen exhibited strong adhesive property between the specimen surface and fibers in comparison with other coated specimens. This indicates that the incorporation of ZnO nanoparticles in PCL solution influenced the morphology and composition at the interface between the metal surface and polymeric fibers. Electrochemical studies revealed that the electrospun composite coated specimen exhibited improved corrosion resistance than that of bare metal and the samples coated only with the polymer.

This discussion indicates that molecular structure of the polymer along with its electrostatic force of attraction between the Mg substrate and the Nano fibrous coating and the adhesive strength of the polymer coating plays a very crucial role in controlling the corrosion—it exhibited higher corrosion resistance than that of the bare rate of Mg alloy. In other words, poor coating adhesion not only led to mechanical failure, but also created an inflammatory reaction in the surrounding tissue. The integration of Nano fibers with inorganic nanoparticles in a controlled amount enhances the barrier properties and specific functionalities over the surface.

18.2.3 Cell Proliferation and Antibacterial Studies on Mg Alloys

The cell proliferation and adhesion studies on bare Mg alloy in MTT assay have shown that the cell growth and viability is strongly influenced by the corrosion product formed on the materials surface. The magnesium hydroxide formation as well as the pH raise of the solution above 8 during the corrosion process leads to the removal of the corrosion layer. This results in the poor adhesion of cells on the surface of bare Mg alloy. Hence, any biological modifications such as cell growth and cell differentiation mainly depends on the substrate surface properties. The nature of coating, stability, adhesiveness, roughness, coating topographies which include micro and Nano scale at the interface plays a vital role for the interaction of the coating or bioactive surface for cell attachment, cell differentiation, and proliferation. Thus the surface modification, which in turn reduces the corrosion rate, is anticipated to promote the cell behavior over the surface. The nanostructured coating through the electrospinning process provides the surface with high surface area to volume ratio which increases the chemical reactivity for bone anchorage, surface roughness, mechanical interlocking, and regeneration of new cells. Also, incorporation of ceramic particles such as TiO_2, ZnO, Hap, and ZrO_2 into the Nano fibers promotes osteo-inductive and osteo-conductive properties. Hamid Esfahania and Mahsa Darvishghanbar [25] reported that the porous nanostructured surface of electrospun polyamide/hydroxyapatite coating on zirconia-toughened alumina (ZTA) coating provided the suitable environment for the MG63 osteosarcoma cells to differentiate and grow along the fibrous mat. The roughness formed by the Nano fibrous network also induced the fixation and growth of MG63 cells over the coating. The composite fiber (PCL/ZnO) coated over the Mg alloy surface also exhibited better cell attachment and proliferation compared to those of bare Mg surfaces [24].

The anticorrosion and antibacterial effect on Ta_2O_5-PCL/MgO-Ag coatings on the Mg-Ca alloy was studied by Hamid Reza Bakhsheshi-Rad et al. [21]. Electrospinning of PCL blended with different concentrations of MgO-Ag was coated on the Mg alloy surface, which resulted in smooth, porous, and plain Nano fibers with an average diameter ranging from 200–360 nm. The combined electrochemical studies and in-vitro tests revealed that PCL/MgO-Ag Nano fiber coatings resulted in better corrosion resistance properties and apatite forming ability in SBF. The results of antibacterial activity also demonstrated that the PCL/MgO-Ag and Ta_2O_5-coated Nano fiber coating contributed to decrease in the growth of bacteria than bare Mg alloy and Ta_2O_5-coated samples. These investigations highlight that the 3D structure with high specific surface area on the coated substrate surface is the predominant factor in the multiplication of cell adhesion and proliferation than non-coated Mg alloy.

18.3   DISCUSSION AND LIMITATIONS

In general, advancement of surface coating on any material to tune up the surface properties with respect to morphology, composition, and function of material is continuously a challenge for today's researcher. In this chapter, electrospinning of polymers over a metallic surface is considered to be one of the effective recent approaches in the biomedical field, especially for orthopedic applications. Wherein, a wide range of conducting biodegradable natural and synthetic biopolymers can be electrospun into fibers with specific alignment and structural integrity. These Nano fibers were deposited on the surface of various metallic implants such as titanium, stainless steel, and magnesium alloys which acted as a collector plate. The incorporation of polymeric Nano fibers with hydroxyapatite, collagen, etc., improves the biomineralization and mechanical integration between the coated layer and specimen surface.

Hybrid multifunctional composite coating blended with bone growth stimulating factors is found to have a better functional properties to overcome the disadvantages of Mg and its alloys for orthopedic applications. For instance, hydroxyapatite nanoparticles along with simvastatin incorporated into PCL Nano fibers acted as a dual functional coating which helped in bone tissue regeneration and controlling of the healing process with the release of drugs..

Advances in electrospinning technology and research based on the bioactive surface coating on biodegradable materials to enhance properties such as corrosion resistance, mechanical properties, and biocompatibility for orthopedic devices are very much needed for clinical applications. Still, several drawbacks need to be considered and solved accordingly to meet the practical criteria. Though there are a lot of reports available related to the applications of electrospinning technology in the biomedical field, these are still in the infant stage. All these reports are mainly performed in-vitro. Hence, attention has to be paid to real-time in-vivo and clinical research for an Mg alloy to be recognized as one of the potential material for applications. Further, the synthesis of Nano fibers, composite fibers, and hybrid multifunctional composite coating using natural as well as synthetic polymers produces hollow, core-shell, and single-multiwall Nano fibers. These Nano fibers help in improving material properties and hence can be used for a broad array of applications. For instance, coaxial set-up in the electrospinning process helps in studying the drug delivery system, where drugs can be incorporated into hollow fibers and can be delivered at the required sites. Though polymers, in general, possess biocompatibility, they result in poor antibacterial performance and hence lead to infections around the implants. Therefore it is necessary to incorporate antibiotics into polymer Nano fibers to lessen the viability, adhesion, and microbial growth on the implant surface.

Though the preparation of new and modified polymers can offer improved corrosion resistance and biocompatibility, even then, many challenges have to be faced by coating on Mg and its alloys. For example, the quality and stability of coatings on the substrate depend on the interfacial bonding strength difference between the polymeric coated layer and the Mg alloy surface such as adhesion, delamination, and coating efficiency. Also, other parameters such as pH, solvent, temperature, pressure, etc. are all of prime importance in understanding the formation of coating/film on the Mg-based alloys surface.

18.4 CONCLUSION AND FUTURE SCOPE

The surface modification by electrospinning technique is a prominent method to enhance the performance of metallic implants, especially in orthopedics for long-term application. The electrospun fibers act as a barrier coating on the metallic surface, help to lower the rate of degradation, and enhance the biocompatibility of metal. Hence, the properties of Nano fibers make them a good candidate for various in-vitro and in-vivo studies. The electrospun technique has been actively explored as a polymer surface coating on metallic implants and it can be directly deposited on it based on different clinical needs.

The improvement of materials and coating performance using the electrospinning technique for orthopedic applications has been growing tremendously. A multifunctional coating using the electrospinning process over Mg alloys surface provides the pathway for the development of smart surfaces in near future and are yet to be explored. The reports related to animal in-vivo studies are limited. There is a need for collaboration between material scientists and biologists to understand the behavior of various electrospun Mg alloys under physiological conditions; the incorporation of drugs/nanoparticles into the fiber and the release of it at the required site; the mixing of antibiotics with polymer solutions for early healing of wounds; the bioactivity and degradation rate of various polymer coated Mg alloys; etc. It's the right time to understand the electrospinning on Mg alloys for orthopedic applications and this basic research needs to be transferred into clinical use. Hence, it's well known that these electrospun materials are going to play a vital role for biomedical research in the future.

REFERENCES

1. Katayoon Kalantaria, Amalina M. Afifia, Hossein Jahangirianb and Thomas J. Webster. 2019. Biomedical applications of chitosan electrospun nano fibers as a green polymer—review. *Carbohydrate Polymers* 207: 588–600.
2. Jiajia Xue, Tong Wu, Yunqian Dai and Younan Xia. 2019. Electrospinning and electrospun nanofibers: Methods, materials and applications. *Chemical Reviews* 119: 5298–5415.
3. Duque Sanchez Lina, Brack Narelle, Postma Almar, J. Pigram Paul and Meagher Laurence. 2016. Surface modification of electrospun fibres for biomedical applications: A focus on radical polymerization methods. *Biomaterials* 106: 24–45.
4. Seema Agarwal, Joachim H. Wendorff and Andreas Greiner. 2008. Use of electrospinning technique for biomedical applications. *Polymer* 49: 5603–5621.
5. David J. Lockwood. 2014. Nano medicine. In *Nanostructure science and technology*. Edited by Yi Ge, Songjun Li, Shenqi Wang Richard Moore. Springer, New York, Heidelberg, Dordrecht, London. https://link.springer.com/content/pdf/bfm%3A978-1-4614-2140-5%2F1.pdf.
6. Gyeong-Man Kim, Seung-Mo Lee, Mato Knez and Paul Simon. 2014. Single phase ZnO submicrotubes as a replica of electrospun polymer fiber template by atomic layer deposition. *Thin Solid Films* 562: 291–298.

7. Alberto Sensini and Luca Cristofolini. 2018. Biofabrication of electrospun scaffolds for the regeneration of tendons and ligaments. *Materials 11*: 1963–2006.

8. Nandana Bhardwaj and Subhas C. Kundu. 2010. Electrospinning: A fascinating fiber fabrication technique: Biotechnology advances. *Biotechnology Advances* 28: 325–347.

9. Adnan Haider, Sajjad Haider and Inn-Kyu Kang. 2015. A comprehensive review summarizing the effect of electrospinning parameters and potential applications of nanofibers in biomedical and biotechnology. *Arabian Journal of Chemistry* 11: 1165–1185.

10. Hao Shao, Jian Fang, Hongxia Wang and Tong Lin. 2015. Effect of electrospinning parameters and polymer concentrations on mechanical-to-electrical energy conversion of randomly-oriented electrospun poly (vinylidene fluoride) nanofiber mats. *RSC Advances* 5: 14345–14350.

11. O. K. Pereao, C. Bode-Aluko, G. Ndayambaje, O. Fatoba and L. F. Petrik. 2016. Electrospinning: Polymer nanofibre adsorbent applications for metal ion removal. *Journal of Environmental Polymer Degradation* 25: 1175–1189.

12. Hanuma Reddy Tiyyagura, Tamilselvan Mohan, Snehashis Pal and Mantravadi Krishna Mohan. 2018. Surface modification of Magnesium and its alloy as orthopedic biomaterials with biopolymers. In *Fundamental Biomaterials: Metals*, 197–210. DOI 10.1016/B978-0-08-102205-4.00009-X

13. J. G. Acheson, S. McKillop, P. Lemoine, A. R. Boyd and B. J. Meenan. 2019. Control of magnesium alloy corrosion by bioactive calcium phosphate coating: Implications for resorbable orthopaedic implants. *Materialia* 6: 100–291.

14. J. Wang, J. Tang, P. Zhang, Y. Li, J. Wang, Y. Lai and L. Qin. 2012. Surface modification of magnesium alloys developed for bioabsorbable orthopedic implants: A general review. *Journal of Biomedical Material Research Part B Applied Biomaterial* 6: 1691–1701.

15. Yixuan Li, Jing Gao Liyun Yang, Jian Shen, Qing Jiang, Chi Wu, Dan Zhu and Yifeng Zhang. 2019. Biodegradable and bioactive orthopedic magnesium implants with multilayered protective coating. *ACS Applied Bio Materials* 8: 3290–3299.

16. Lili Tan, Xiaoming Yu, Peng Wan and Ke Yang. 2013. Biodegradable materials for bone repairs: A review. *Journal of Materials Science & Technology* 29: 503–513.

17. M. Karthega, Mogan Pranesh, Chockalingam Poongothai and Nagarajan Srinivasan. 2020. Poly caprolactone/titanium dioxide nanofiber coating on AM50 alloy for biomedical application. *Journal of Magnesium and Alloys* 9: 532–537.

18. Jian Tang, Jiali Wang, Xinhui Xie, Peng Zhang, Yuxiao Lai, Yangde Li and Ling Qin. 2013. Surface coating reduces degradation rate of magnesium alloy developed for orthopaedic applications. *Journal of Orthopaedic Translation* 1: 41–48.

19. S. Agarwal, J. Curtin, B. Duffy and S. Jaiswal. 2016. Biodegradable magnesium alloys for orthopaedic applications: A review on corrosion, biocompatibility and surface modifications. *Materials Science and Engineering* C 68: 948–963.

20. M. Licciardello, G. Ciardelli and C. Tonda-Turo. 2021. Biocompatible electrospun polycaprolactone-polyaniline scaffold treated with atmospheric plasma to improve hydrophilicity. *Bioengineering* 8: 24.

21. Hamid Reza Bakhsheshi-Rad, Ahmad Fauzi Ismail, Madzlan Aziz, Zhina Hadisi, Mahdi Omidi and Xiongbiao Chen. 2019. Antibacterial activity and corrosion resistance of $Ta2O_5$ thin film and electrospun PCL/MgO-Ag nanofiber coatings on biodegradable Mg alloy implants. *Ceramics International* 45: 11883–1189.

22. Pedro J. Rivero, Deyo Maeztu Redin and Rafael J. Rodríguez. 2020. Electrospinning: A powerful tool to improve the corrosion resistance of metallic surfaces using nanofibrous coatings. *Metals* 10: 350.

23. J. Astro, K. Gokula Krishnan, S. Jamaludeen, et al. 2017. Degradation and corrosion behavior of electrospun PHBV coated AZ-31 magnesium alloy for biodegradable implant applications. *Journal of Bio and Tribo Corrosion* 3: 52–24.

24. J. Kim, H. M. Mousa, C. H. Park and C. S. Kim. 2017. Enhanced corrosion resistance and biocompatibility of AZ31 Mg alloy using PCL/ZnO NPs via electrospinning. *Applied Surface Science* 396: 249–258.

25. Hamid Esfahania and Mahsa Darvishghanbar. 2018.Enhanced bone regeneration of zirconia-toughened alumina nanocomposites using PA6/HA nanofiber coating via electrospinning. *Journal of Material Research* 33: 4287–4296.

19 Advanced Biomedical Devices
Technology for Health Care Applications

Arul Kashmir Arulraj

CONTENTS

19.1 INTRODUCTION

The advanced biomedical device can be demarcated as an autonomous device located noninvasively to perform specific sensory functions over a prolonged period. The term "wearable" represents the support provided by placing it on the human body or externally as a piece of clothing on the human body. Generally, widely used medical equipment such as splints, bandages, gloves, bands, eyeglasses, and contact lenses are referred to as wearables. Anything mounted on such a wearable platform to monitor the health conditions is called a "wearable biomedical device". In this essence, we come across several advanced and sophisticated devices/sensors that integrate the sensory functions with the aid of microelectronics and computing, with a degree of intelligence in them. In general, this is called "human-machine intelligence" for biomedical health technology. The functions currently performed by such intelligent medical devices generally cover the applications such as medical monitoring, personalized rehabilitation assistance, and long-term controlled clinical aid. Wearable biomedical devices assist in both the monitoring and treating phases of health

DOI: 10.1201/9781003173533-19

care applications. For instance, biomedical devices have been used for monitoring and treating chronic diseases such as heart diseases, asthma, and diabetes (Fleming et al., 2020; "Physicians' Knowledge of Inhaler Devices and Inhalation Techniques Remains Poor in Spain" 2012). In addition to this, analysis-based medical devices such as sensors for heart rate, sweat sensing, blood oxygen level, respiration, and body fat were also developed (Bui et al., 2020; Prawiro, Yeh, Chou, Lee, & Lin, 2016; Shin et al., 2019; Vanegas, Igual, & Plaza, 2019). These functionalities are generally attained noninvasively via reliable design and communication capabilities. However, most medical aid devices available on the market usually require user intervention to record and process the data offline. It is emphasizing that human interventions are found essential to report the final results accurately. In heart rate monitors, any abnormality in heart activity is correlated with the patients with ideal position. However, the next-generation wearable devices drive and function beyond the recording, which means it provides intelligent patient monitoring that involves decision making. These intelligence devices acquire the data, process it with standards, and display the output decisions instantly. Also, these devices can often transmit information related to the user's health condition to caretakers and medical practitioners. In general, the processing unit deals with feedback generated by using measured sensor data to the users. Some devices corresponding with wireless communication capabilities can instantly provide health care alerts to health care workers, which provide ambulant health monitoring. Another class of wearables that is widely used in the medical field is sleep apnea monitors (Krishnaswamy, Aneja, Kumar, & Kumar, 2015; Mendonça, Mostafa, Ravelo-García, Morgado-Dias, & Penzel, 2018).

These devices monitor patients with high risks, such as infants susceptible to infant death syndrome. The infant wears the respiratory belt while sleeping, and these belts produce instantaneous alerts to parents or doctors if any breathing pattern abnormalities are found. The miniaturized wearable biomedical devices can deliver clinical screening beyond the hospital (i.e., at home or during routine daily activities). Another evolving concept in biomedical technology is multi-signal monitoring which incorporates multimodal sensor systems compounded in a single platform by intelligence to improve health care diagnostics (Diaz & Payandeh, 2017; Futagawa, Iwasaki, Murata, Ishida, & Sawada, 2012). Advanced biomedical products with these capabilities already exist to target the health care market, researchers, and health-conscious individuals. In such devices, users can regularly monitor several vital signs of the body such as electrocardiography (ECG) (Rashkovska, Depolli, Tomašić, Avbelj, & Trobec, 2020; Serhani, El Kassabi, Ismail, & Nujum Navaz, 2020), respiration (Chu et al., 2019), skin conductance (Gatti, Calzolari, Maggioni, & Obrist, 2018), blood pressure/pump rate (Al-Qatatsheh et al., 2020), or blood oxygen saturation (Das, Aggarwal, & Aggarwal, 2010) as well as behavioral, sensorial, emotional, and cognitive reactivity (Wilhelm et al., 2020). Moreover, devices intended for athletes can measure parameters such as heart rate, pace, acceleration, speed, and calorie burn, with the constant acquiring of information during exercise for the long term. Interestingly, the more advanced systems bearing the data processing capabilities such as speech/voice recognition, face recognition, and locomotor alerts or feedback are interesting current medical scenarios (Ache, Haupt, & Dürr, 2015; Kortli, Jridi, Al Falou, & Atri, 2020; Wolff, Socolinsky, & Eveland, 2003; Wolff, Socolinsky, & Eveland, 2003). In general, such devices include a range of global positioning systems (GPS)-based navigation to assist the visually impaired towards way finding, obstacle disturbance, and location detecting navigation. Another important wearable biomedical device category relies on rehabilitation assistance, which combines functions such as monitoring and aiding the patients in rehabilitation. For example, the wearable that provides advanced alerts towards the risk of heart attack, and the wearable that provides medical dosage are commonly proposed under this category (Chowdhury et al., 2019; Wilson, 1999). From the technical point of view, wearable biomedical devices consist of two main compartments: a wearable sensor as a data generator/input component and a data analyzer/output component (Figure 19.1). The sensor refers to the element/component that involves collecting/sensing information automatically. These components are placed on the wearer to collect/enter/sense the health information into a system. The wearable data analyzer is the heart of the system that involves data processing, data storage, and communication. The individual-centric functions of components of the advanced bio-medical device are as follows.

19.1.1 Components of Advanced Biomedical Devices

The advanced biomedical devices come with the capabilities of a signal sensing probe, a data col-lection component, an analytics-enabled decision-making component, and communication compo-nents in a single entity (Figure 19.1). The sophisticated functionalities with structural complications are involved in the design and fabrications of advanced biomedical devices. The overall coordina-tion of these individual components is essential for reliability in health care applications.

19.1.2 Sensor Materials

The data generators are sensing probes that are impeccably connected to the human subject to collect clinical information. This component has been considered as the heart of the biomedical devices where the chemical/electrochemical environment of the wearer has been monitored. The sensor data then further is converted into electrical signal input to the data processors for the data handling, decision-making, and communicating components of the biomedical device. In general, sensor probes commonly consist of semiconducting, conducting, and stimuli-responsive materials that render decent biocompatibility. The sensor elements could detect the electrical or electrochemi-cal data from the analytes and subjects that are further processed with the microprocessor. The range of sensor materials used in developing biomedical devices include metals, polymers, polymer composites, and allotropes of carbon. The details of materials used and types of devices developed for advanced human interactive biomedical applications are given in Table 19.1.

TABLE 19.1
The Types of Devices and Materials Used for Advanced Biomedical Sensing.

Device Type	Materials Used	Applications	Ref.
Tactile Sensors	PDMS	Artificial skin, e-skin, pressure sensors	(Mannsfeld et al., 2010)
	Micelles mediated polymer	Piezo sensors, human-machine intelligence	(Jung et al., 2014)
	Semiconductor Nanowire	Artificial skin	(Takei et al., 2010)
	Metal oxide-based triboelectric	Triboelectric sensor, pressure sensor, sensor arrays	(X. Wang, Zhang, et al., 2016)
Motion Sensors	Graphene	Locomotion sensors, human-machine interaction	(Lim et al., 2015)
	PDMS-PEIE	Finger motion, strain	(S. H. Jeong, Zhang, Hjort, Hilborn, & Wu, 2016)
	Carbon black-PDMS	Hand gesture, motion	(Lu, Lu, Yang, & Rogers, 2012)
	Alloy/PDMS	E-skin, strain	(S. H. Jeong, Hjort, & Wu, 2015)
Electrophysiology Sensors	Au	Human-machine interfacing, electromyography sensors	(J.-W. Jeong et al., 2013)
	Metal wires	Epidermal electronics, human-machine control, eyelid movement	(Kim et al., 2011)
	Stimuli measuring materials	Head posture, image processing, physiological movement	(Barnett-Cowan & Harris, 2011)
	Metal patches	Eye motion, electrooculography, image processing	(Guo et al., 2016)

19.1.3 Data Handling

This involves handling the type of data acquired from the human object and logistics on storage that is further commuted/transmitted to the next component connected with it. This component formally acts as a regulator of the wearable device where it regularizes and prioritizes the clinical data either as shorted or stimulated as per the proposed applications. In most cases, data handling also comes with the data storage components where the amount of data storage in the wearable biomedical devices is decided.

19.1.4 Decision Making

This component involves the analysis of medical data recorded by the biosensors to prepare the decision report as a legible and precise output/decision. In general, it is a custom-developed algorithm that consists of data handler and arithmetic operations. The arithmetic algorithms such as neural network models, human-machine intelligence, image processing, and artificial intelligence-based algorithms are commonly employed. This component in the biomedical devices will act as feedback provided to the patient and caretakers. In most cases, the decision-making process involves the comparison of preloaded data along with the measured data.

19.1.5 Device Communication

For any measured clinical data, the advancement comes with how the devices communicate to the users and caretakers. This part of the advanced biomedical system is denoted as a linkage medium between the sensors/devices and health providers. It can be either via wired or wireless transmission. The efficiency of communication is generally modulated with distance, speed, and data logging. By the systematic choice of communication medium, the end user can communicate their health status via smart phones, computers, and through any other electronic medium to the practitioners.

19.1.6 Design and Structure

The fabrication of medical devices involves mounting the array of sensors to the processor with the decision-making algorithm having a potential communication system. The appropriate placements of all these components with an effectively controlling algorithm make the system serve its purpose. Most of the time, the end users are human objects, which dictates the developed system must be compatible and compliable with the human body. The design must be fabricated in a way to attain high performance at an affordable cost. The cost of medical devices plays a prominent role in achieving affordable and cost-effective health care. Thus, it is imperative to analyze the market size and status of this biomedical device technology.

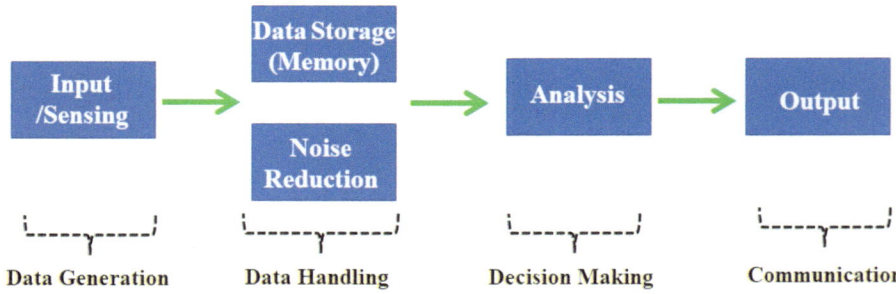

FIGURE 19.1 The schematic diagram of wearable sensor devices.

19.2 MARKET SIZE AND MARKET POTENTIAL

The consumer market status for medical devices has dramatically increased in recent years since 1992, when it was $1.6 billion. Interestingly, in 2001, the market for advanced biomedical devices alone reached $5.2 billion and is expected to bypass this in the future (www.businesswire.com/news/home/20200316005453/en/Insights-into-the-Global-Wearable-Medical-Device-Market-to-2027—Featuring-Sotera-Wireless-Zephyr-Technology-Corporation-Omron-Corporation-Among-Others—ResearchAndMarkets.com). However, the growth of advanced, miniaturized biomedical technology has not seen a corresponding decrease in hospital/health care utilization. This lag could be attributed to issues associated with the difficulties persisting with the development of advanced medical products, the design and integration of vast components in single modules, devices' vulnerability in clinical testing, and the strict rules and regulations to be followed for the bulk manufacturing and testing clinical products. The most recent and advanced biomedical device technology, such as artificial intelligence, human-machine interaction, image processing, have been considered in biomedical device applications. These device technologies could overcome the existing imbalances of developing the user interfaced wearable biomedical systems and advanced health care units. Thus, this chapter will be mainly focusing on the recent advanced biomedical technologies such as

1. Artificial Intelligence-based Biomedical Device Technology
2. Human-Machine Intelligence-Based Biomedical Device Technology
3. Image Processing Based Biomedical Device Technology.

19.3 ARTIFICIAL INTELLIGENCE-BASED BIOMEDICAL DEVICE TECHNOLOGY

Artificial intelligence (AI) and its related technologies such as neural networks and machine learning are interestingly prevalent in modern science. The AI-based technology is at its early stages in health care applications. These technologies have been found promising in transforming several aspects of patient care, like rehabilitation, regenerative medicines, and physical monitoring into the next phase. It has been evident that AI can be a viable alternative to human interventions in health care technology in diagnosing and monitoring health status. In this section, we deliberate on how AI promotes automation in health care diagnostics and decision making.

The AI-based technology associated with health care monitoring is currently going through a transformation where the diagnosis of people/patients is made possible without prolonged hospitalization. The recent advancement of wearable biomedical sensing technologies helps make it convenient to develop smart wearable systems to regularly monitor human behaviors (Uddin, 2019) (Figures 19.2a, 19.2b, and 19.2c). This wearable biomedical system predicts the physical activities of patients using a Recurrent Neural Network (RNN). The results are communicated through electronic interfaces like computers and smart phones efficiently. The clinical data are obtained from wearable sensors distributed on various parts of the human body like electrocardiography (ECG) meters (Fu, Hong, Zhang, & Du, 2021), magnetometers, accelerometers, and gyroscope sensors. Then, the RNN is used for predicting patient activities in real time. The Graphics Processing Unit (GPU) in the module helps fast computation of experimental data to the interface. The system results are compared against the conventional data available as a standard dataset. In RNN networks, the human activities are stored as time-sequential changes and they have been encoded by a machine learning model to make them readable signals. Besides, RNN has a recurrent connection in which stored past data is compared internally to the present real-time data to provide feedback to the medical aiders. The clinical signals are stored in the Long Short Term Memory (LSTM) of the device. The fabricated intelligence devices can predict human activity in unknown sensor data with a mean prediction of 99.69% compared to the conventional approaches (i.e., HMM and DBN of 92.01%).

One of the classic AI-based devices fabricated for special needs like hearing and visually impaired people used for autonomous vehicles has been developed (Son, Jeong, & Lee, 2019) (Figure 19.2d).

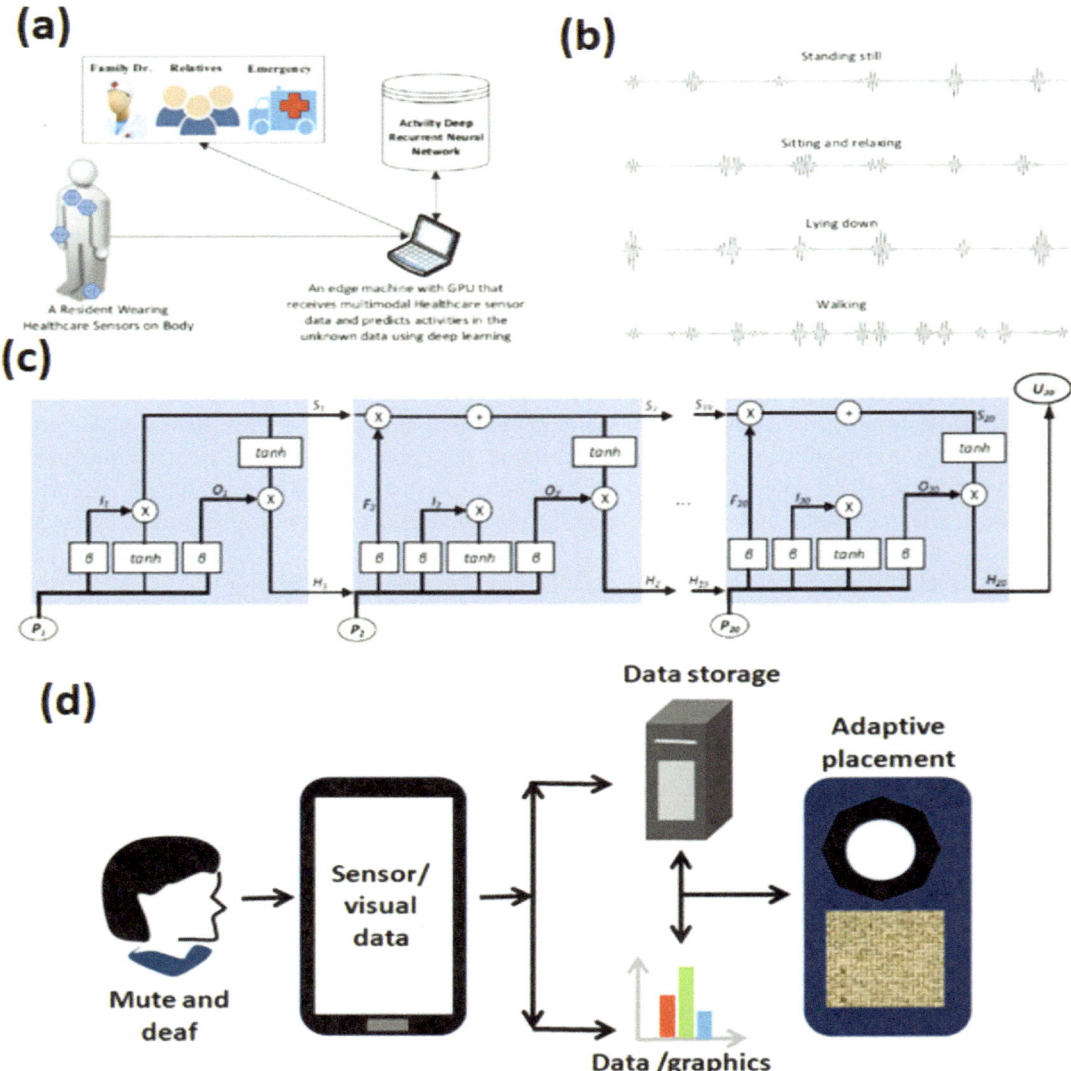

FIGURE 19.2 The schematic setup of the wearable biomedical sensor for human activity prediction system (a); the sample ECG signal from MHEALTH Dataset (b); the basic structure of RNN used in module fabrication (c) (adopted from Uddin, (2019)); the structure of data collection and management model and visualization model (DVM) (d).

These biomedical devices delivered an accurate visualization screen for people who are disabled with hearing problems, and a perfect audifying system with speech-to-text (STT) and text-to-speech (TTS) functions for the visually impaired populations. The audification and visualization system (AVS) of visually and hearing-impaired people can store the input as graphic data through cameras fixed in vehicles and systematically select the required data set using a touch interface and speech recognizer sensors. The AVS system mainly consists of a Data Collection and Management Module (DCMM), Audification Conversion Module (ACM), and Data Visualization Module (DVM). The DCMM stores and manages the data collected from the vehicle, the ACM has a speech-to-text sub module (STS) that recognizes the input speech and then converts it to text, and into wave sub module (TWS) that converts the input text to voice. The DVM components generally receive sensor data, diagnose and study the data through the touch interface. The performance on the learning ability, control over the learning times of HMM, RNN, LSTM, are controlled by systematically chosen processing units. The computational time of Tacotron, Deep Voice, and Deep Mind of the sensors are optimized in DVM. The fabricated AI base delivering the Hidden Markov Model (HMM) adopted in these devices was about 2.5 times faster than the cloud, and the HMM used in STS was about 25% faster than the RNN and ~42% faster than the LSTM. More interestingly, TWS converted text to voice 20 ms and 50 ms speedier than Deep Voice and Deep Mind, respectively. Finally, the DVM visualized the data 2 ms faster than existing instrument clusters.

The prosthetics that mimic the human legs made of an Ankle-Foot Exoskeleton (AFE) with a muscle-tendon unit have been reported (Wang, Guo, Qu, & Song, 2019). These devices are capable of storing energy and assistance during human movement. The sensor controller-based integrated system or a clutch has been designed and incorporated to detect the gait phases mechanically. There are two sensing components acting as an input to the data controller between the shoe sole and the ground that operates independently. Upon mechanical movement of a human, the energy storage and release process concurrently inferred on the clutch that is suspended subsequently behind the calf muscles is engaged. The total energy stored during flexing and its corresponding release during the push-off amply provides walking assistance via the integrated sensor platforms with microcontroller systems. Upon unclutching, the spring gets disengaged, and a reverse action occurs to the system without impeding the free rotation. The same prosthetic model has been simulated with the force and power of plantar muscles under various stiffness conditions. The biomedical prosthetic was found entirely passive and user-friendly with good reliability that could resist the disturbance of shock during human walking. Interestingly, the biomedical device and its detection of gait on the contact state of the feet and the ground, so the clutch is made universal and does not require customization for individuals.

Among the health care technologies, wearables for heart rate monitoring, heart disease sensing are highlighted as these diseases are a prolonged threat to society. There is a portable AI-based automatic system to improve cardiovascular health management that has been fabricated (Fu et al., 2021). Relatively, a tiny hardware module has been designed to acquire electrocardiogram (ECG) data from the wearer. A novel AI sub technique like deep-learning technology has been employed to achieve accurate cardiovascular disease detection. The developed devices could support about 20 types of diagnostic systems including sinus rhythm, tachyarrhythmia, and bradyarrhythmia. The experimental results show the hardware device can provide high-quality ECG data by systematically removing high-/low-frequency distortion/noises with the lead detection with 0.9011 Area Under the Receiver Operating Characteristic Curve (ROC-AUC) score. The fabricated devices inclusive of deep-learning-based cloud service supports about 20 types of diagnostic items, among which 17 of them have more than 0.98 ROC-AUC score. Surprisingly, the cardiac health care management through a lightweight mobile application on the WeChat Mini Program platform has been demonstrated over 20,000 user interfaces.

19.4 HUMAN-MACHINE INTELLIGENCE-BASED BIOMEDICAL DEVICES

Human-Machine Intelligence (Ajeev et al.) is a modern technique that seamlessly synergizes human and machine computing, enabling them as a single and collective to tackle challenging tasks in a very specialized domain that is impossible to execute individually. The functions of HMI are extracting the data/signals, understanding them, and finally utilizing human expertise for analyzing and feedback generation. Furthermore, the conventional human decision-making process is a multimodal step involving either bottom-up, where the source data will drive out to make a decision (data-driven), or top-down, where the existing knowledge towards certain things will identify the plausible feedback (knowledge-driven). But, if the data collected is unique and novel, a combination of approaches is required to appropriately elicit and extract the data representing the inherited knowledge, in some form, from human experts. This is called Human-Machine Intelligence. The HMI modeling outcome will guide the systematic fusion of the human knowledge/expertise data with the measured/sensed data, which ultimately a human-machine intelligence integration.

One of the remarkable improvements made in HMI-based biomedical devices is minimally invasive surgery (MIS) (López-Casado, Bauzano, Rivas-Blanco, Pérez-del-Pulgar, & Muñoz, 2019) (Figure 19.3(a)–19.3(d)). It has become one of the unique surgical methods as it reduces postoperative recuperation for medical patients. However, the restrictions on the freedom of surgeons' movements in MIS are high and cumbersome. Mostly the surgeons found difficulties in visual depth perception, limited tactile sensation, and the reversed motions of the laparoscopic tools during operations. Thus, the development of the HMI interfaced MIS tool must overcome the existing restrictions involved during MIS. The most suitable solution for this imbalance is to introduce the robotic assistants via HMI-MIS techniques. Robotics aims to improve both patient outcomes and surgeon movements and perceptions. This robotic system comes with an intuitive human-machine interface (Ajeev et al.) to assist the surgeon during any surgical practice. The conventional input method (i.e., pushing buttons and vocal commands) to operate the robotic elements have been considered as interruptions during surgery. Thus, deployment of advanced inputs like tracking health providers' eye movements or hand gesture movements via HMI for controlling the endoscope is most appreciated. The Human Machine Interfacing enabled Hand Gesture Recognition (HMI-HGR) system demonstrated the on-line reinforcement process to boost the device performance in terms of increasing accurate recognized gestures. The HGR device compares the gestures of a surgeon using the Conditional Random Field (CRF) algorithm. Although the gesture recognition was found identical to that of the conventional algorithms, the confidence index (Mendonça et al.) of false predictions has been found high with the Hidden Markov Model (HMM). Thus, incorporating the reinforcement learning (RL) algorithm with the HMM model is advised to improve the mean confidence index. The RL-HMM is suggested as more suitable for surgical applications. The accuracy obtained using RL-HMM in gesture recognition was found to agree with the values obtained from different methodologies, such as recurring neural networks (91.9%) or dynamic time warping (91.4%). It is noteworthy that the use of the RL algorithm notably improves both the accuracy and the CI. However, these biomedical devices have a limitation on the surgeon-dependency—that is, the gestures must be preloaded and must get trained by the same surgeon to perform the surgical practice. Also, the stored gestures will be made available on the single library that benefits to relocate the previous work and collaborate with new data. Interestingly, when the gesture recognition system gets stuck or fails, the feedback will be provided as a voice command. Finally, during the experimental stage on surgical operations, three gestures were trained and evaluated. Notably, these gestures were considered sufficient in the surgical scenario. These devices could be extended further to be implemented in multimodal surgical operations.

One of the traits of human-machine intelligence-based wearable biomedical devices that is highly useful in routine life is physiological signals monitoring for drowsiness (Kundinger, Sofra, & Riener, 2020). Drowsy driving especially imposes a high risk to human life. The current systems often use/detect the driving behavior parameters for detecting the driver drowsiness. Developing wearable devices that include physiological signal monitoring seems to be a promising alternative for such potential applications. However, in a dynamic driving environment, incorporating the non-intrusive or minimal instructive method is appreciable as it will not interfere with the driving sequences. Besides, the intended vibrations from the roadbed may also to degrade the sensor ability. Thus, developing an impeccable wrist band on the human wrist could be a potential device to detect driver drowsiness by using an advanced machine learning approach.

In this approach, the wrist-worn wearable device measures the heart rate and drowsiness during driving by monitoring the eyelid blinks. The various User Dependent Tests (UDT) were performed and found that accuracy of about 92.13% has been achieved using the wrist bands. These results were comparable with the conventional ECG ~97.37% in detecting drowsiness. The device performance has been compared with reference data from a medical-grade ECG device to check its accuracy and feasibility using 30 participants with the driving simulator. These wearable wrist bands have shown excellent compatibility and accuracy (> 92%) similar to medical-grade devices.

Another class of wearable HMI-based intelligence devices was reported to be used in nonverbal communication for the defense application (Ajeev et al., 2020) (Figure 19.3). The prototype module has been demonstrated in the use of human-machine intelligence sensing for gesture-mediated nonverbal communication and for detecting the heart rate and voice recognition of the wearer. The wearable strain sensor modules are sensitive in transducing high-strain human body actions such as hand gestures/neck postures into readable text and commands using HMI interfacing. Interestingly, the same modules are deliberate in sensing the ultra-low strain associated with the human body such as heart pulse rate and voice recognition applications. The successful demonstration has been done for the real-time application of these devices in nonverbal communication and biomedical

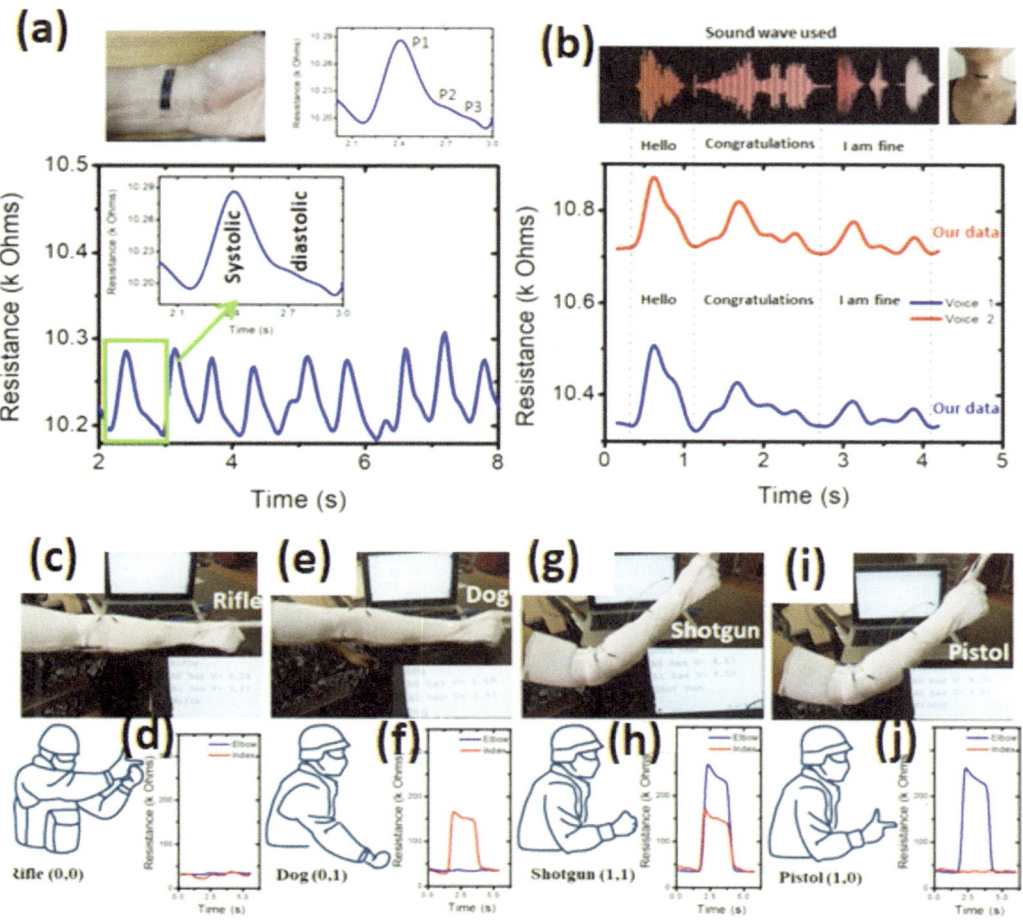

FIGURE 19.3 The LR sensors attached to the wrist to monitor cardiac activity (a); the LR bands attached on the speaker to monitor the noise/sound recognition of the words "Hello", "Congratulations", and "I am fine" with the sound wave profile used for the study (b). The strain sensors module made of LR bands mounted on the index finger and elbow to convert the gestures used in military commands like a rifle (c–d), dog (e–f), shotgun (g–h), and pistol (i–j) with its resistance response.

Source: Adapted from Ajeev et al. (2020).

applications using HMI. The wearable nonverbal communication sensor module consists of a glove attached with resistive type strain sensors, an Arduino circuit reader as a signal processor, and a display for feedback. The microcontroller is programmed to determine exact gestures/commands by generating a two-digit binary concerning on/off states strain bands attached to the wearer's elbow and index finger of the glove. The flexible, conductive strain bands are placed on the wearable glove, and the whole unit is clutched on the wearer to perform commands/gestures such as "Rifle", "Dog", "Shotgun", and "Pistol". Upon performing hand actions, the variation in the electrical output has been observed concerning the strain. This variation has been commuted with a microprocessor which converts the gestures into readable comments. By feeding this serial voltage output to analog pins A0 and A1 of the processor, the binary key generated as (0, 0), (0, 1), (1, 1), and (1, 0) congruently displays the command as Rifle, Dog, Shotgun, and Pistol, respectively. By this, the human hand gestures have been converted into translating the sign languages into readable commands using human-machine intelligence. Besides, these same sensor devices are capable of recognizing the same commands via voice processing as well. This broad range of sensitivity achieved by these HMI-based wearable devices could well be suitable for the multipurpose sensing applications associated with health care, robotics, and human-machine intelligence.

19.5 IMAGE PROCESSING BASED BIOMEDICAL DEVICE TECHNOLOGY

Among several advanced biomedical technologies that evolved to ease the diagnosis and treatment in the health care sector, image processing-based biomedical devices have considerably attained interest due to their use in cardiac activity monitoring and decision making. Compared to other advanced biomedical technologies, image processing comes with a fully computerized and automated model for detecting heart functioning that never requires any human interventions. The conventional image processing technique with suitable AI algorithms could produce faster and more reliable detection of heart rate variation. A few of such exciting devices and their technology have been deliberated in this section.

The region of interest (ROI)-based image processing methodology has been demonstrated for heart rate monitoring. Elucidating the heart rate using the face recognition process consist of three steps, namely,

1. Detecting facial region
2. Predicting the desired Region of Interest (ROI), and
3. Extracting the plethysmographic signals from the data extracted.

(Kaushal Kanakia, 2018)

Once the facial expression is captured by the image processing input, it commutes as boxes using the Haar cascade classifiers. Then it is narrowed into the single bounded box on the face of the subject. Once the ROI is identified, it converts it into a digital image, and the analysis of peak-to-peak variation is carried out via suitable image processing software. The various ROI selection options include a far-narrower box, with the eye portion and forehead removed, which can be chosen as per the user-defined ROI prediction. Among these, the far-narrower bounding box has been the most commonly used skin pixel for frontal imaging. For any tilt, straight face reimplementation is automatically adjusted by the GrabCut. After optimizing the facial recognition and ROI, the algorithm commutes to find the heart rate from the selected ROI pixels. The algorithm includes the commands to find the mean RGB pixel values within the ROI and normalize across a 30 seconds window that will be extracting the independent source signals to find prominent frequencies.

The optical fiber-based image processing technology has been developed for the accurate sensing of cardiac activity of patients with affordable technology. For this, a new miniaturized three-axis-force sensor integrated with a cardiac catheter has been studied (Noh et al., 2016). The tip of the catheter acts as a contact force with the subject to acquire the signals. The biomedical sensors are fabricated using 3D printed deformable structures with a 4-core fiber optics bundle. The light intensity and brightness have been modulated by the USB bounded camera to ascertain the three axial forces between the catheter tip and the cardiac tissues. As these fiber optic-based sensor catheters are free of applied current, the sensor data is not much influenced by the external electromagnetic field that is generally observed in MRI or radio frequency-based catheters used commercially to date. Also, the fiber optics-based catheters are found suitable for cardiac monitoring using image processing as it satisfies the conditions as follows.

1. The sensor fabrication technology must be scalable for bulk monitoring and screening, and it must render diverse applications such as surgical robot tips, laparoscopic tips, image collecting tips in biomedical engineering
2. The cardiac sensors must be capable of measuring the ultra-low force involved in the myographic or cardiac forces with the subject (i.e., the forces Fx, Fy, and Fz with 0.2 to 0.3 N force within the acceptable error and detection limit
3. The high rate of frequency is generally observed in heart rate (60–100 times/min). Thus the sensor elements should be capable of detecting such high frequency (2Hz) operations in a real environment such as electric current, temperature, and MR compatibility.

Another class of biomedical devices that has huge potential in prosthetic and gesture analysis applications is force myography (FMG) sensors. In general, the FMG devices are used in biomedical wearable device fabrication such as prosthetic hands, legs, and other body organs in mechanical deformation induction (Figure 19.4). Mostly the rubber-based prosthetic wearables are non-conducting in nature.

Thus the prosthetics are always embedded with myoelectric skin sensors. The myoelectric sensors measure the electrical impulse from the active area of the human body (generally limbs) and control the robotic arms/legs-based prosthetics. Surprisingly, the skin electric signals are immeasurable by the conventional electric probe as it is too low to be detected and sometimes too noisy. If that tiny electric impulse from human skin could be measured, it can then be further used to

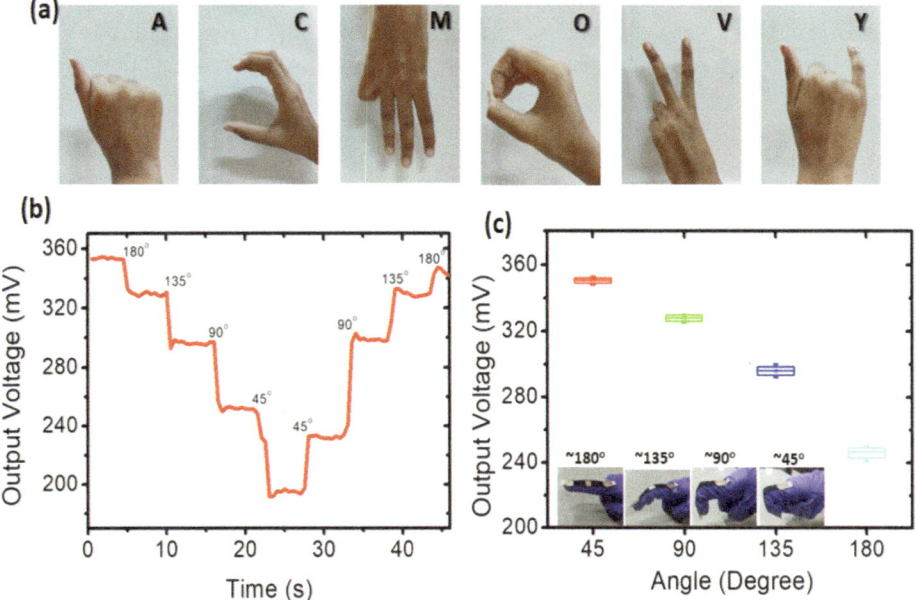

FIGURE 19.4 The postures of letters A, C, M, O, V, and Y to convert/sense its virtual reality (a) and its electrical output sensing data (b); and the figure shows the sensor modules placed on the human object and its sensing results (c).

trigger the functions and movements of the prosthetic wearables. There is an expedient attempt made towards this objective to remodel a typical optical fiber FMG sensor with completely portable components such as a Raspberry Pi-based microprocessor which ensures comfortability and fast communication (Wu, Gomes, da Silva, Lazari, & Fujiwara, 2020). The remodeled sensor could demonstrate the 2 input channels and 9 postures classification with an average precision and accuracy of ~99.5% and ~99.8%, respectively, using a feed forward artificial neural network (Wu et al., 2020).

Apart from developing the biomedical devices, validating and testing their compliance with the end-user community must be considered priorly. Though most of the lab-scale wearable biomedical devices render superior performance and functionality at the small scale testing environments, the main advantages could be accounted for by comparing its reliability in large scale and real testing environments and validation. Thus, the following analytics in biomedical device development could shed light on the most commercial aspects of the research into the welfare of life.

19.6 PREREQUISITES OF BIOMEDICAL DEVICES

19.6.1 USER ACCEPTABILITY

Medical health care has been considered a big consumer market dealing with fluctuating market size with nominal acceptability. Generally, user acceptability has been primarily determined by the cost, user-friendly nature, the efficiency of sense, and comfort of the wearables. Thus, any research in wearable and advanced biomedical technology must be focused on receiving the end-user acceptability, feedback, and its functional value over the time perceived by the user. Thus, a protocol must be regularized in this aspect to stimulate wearable and advanced biomedical device research into a commercial platform.

19.6.2 LEGAL AND ETHICAL SAFETY ISSUES

A device must be approved with a certain standard protocol to protect human life, safeguard the personal and health details of the wearer, and at the same time it must be approved and follow through with the standard ethical, operational risks involved, environmental conditions, and other clearances before being examined on human subjects.

19.6.3 VALIDATION PLAN

A set of validation trials are necessary to make a product launch into medical applications, such as a harsh testing environment, thermal shock test, power leakage test, skin compatibility, and irritation test, etc. Then it will undergo certain exacting condition testing such as high pressure and high altitude tests. If X-ray and other high energy radiations are involved, it must be made viable with the user interfacing and monitoring protocols.

19.6.4 INACCURACY IN DATA MANAGEMENT

The conventional data used in the experiment is generally collected in the laboratory by the health care workers and comes as real data while interpreting the outcomes. Whereas, advanced technologies such as real-time health, cardiac, motion, and effluents monitoring render an imbalance in test data with the data available with the open source. This unintentional imbalance invariably impacts on the test result and false prediction over health conditions of the end users. Also, the privacy of data, malfunctioning of wearables, and maintenance of the miniaturized machinery are still considered challenges that must be rectified at the laboratory level into commercial scale.

19.6.5   LIMITATIONS OF SENSOR PROBES

Advanced biomedical devices come with wearable probes that detect the electrical, chemical, electrochemical, and mechanical signals of the human body instantly. But, these impulses generally interfere with the external environment as vast as atmospheric and climatic changes, external and internal noise, signal overlapping, and durability. Also, the sensor probes used in advanced biomedical devices are position-specific where they are implanted. Therefore, it is essential to identify the methods which are independent of position and mode of the placement of implants. In some cases, the user intends to wear several probes with a large number of devices to measure single behavior (such as myoelectric sensors), which may cause discomfort to the wearer in addition to the powering issues associated with many probes at a given amount of time.

19.6.6   BROADNESS IN INDIVIDUAL ACTIVITY

The sensor elements used on advanced biomedical devices are mostly user-specific. The certain activity observed in one object/subject may not resemble others, which invariably affects the accuracy and precision of the sensitivity. Besides, several ethnic groups have different body functionalities, work lifestyle, food habits, socioeconomic status, leading to how broader consideration must be taken with the simplified device model which is tedious to attain. Thus, generating the classification model in design and sensitivity with a more widely applicable domain remains to be explored in advanced biomedical research.

19.6.7   LIMITATIONS IN POWER CONSUMPTION AND STORAGE OF WEARABLE MODULES

Most of the wearable biomedical devices come with a localized battery-based power supply unit which is insufficient for continuous health care monitoring for the long run. Also, in the total area of wearable devices, more than about 50% of the space is occupied by the power supply units, which must be replaced with lightweight and high power sources. Similarly, for a device to monitor health status continuously 24/7, the acquired data must be stored for comparative analysis. But, due to the limitations that persist with the size and shape of the wearable biomedical devices, the space for data storage is very limited. So, it is imperative to further work on the storage consumption model without losing precision and accuracy.

19.7   FUTURE PERSPECTIVES ON ADVANCED BIOMEDICAL TECHNOLOGY

Advanced health care biomedical technology, such as artificial intelligence, image processing, neural networks, and myographic devices, has played an important role in future biomedical electronics applications. The advanced topics such as high precision medicine and drug release are expected to boost the health care market in the near future. But, it is still under developing stage at laboratory scale. Although the enormous efforts in developing advanced health diagnostics and treatment are challenging, we anticipate that the advanced technology discussed in this chapter will ultimately master the domain in the future. Considering the rapid advancement in AI, image processing, and neural networks, it promises the most cardiology, radiology, and pathology data can be examined sometime soon by a machine. Besides, speech recognition, gesture sensing, myographic image processing devices have been already employed in real-time tasks such as in patient-caretaker communication and in nonverbal communication in the disabled population, and their usage is expected to increase drastically in a consumer market. The greatest challenges anticipated in these advanced biomedical technologies lie in ensuring they are useful in regular clinical practice. For this, widespread adaptation of these advanced technologies must take place with appropriate regulators, integrations, and standardizations. Besides, the products must be available through bulk manufacturing, the degree of similarity of products must be taught to the caretakers, and there must be investment by public-private sectors. But, these paradigm challenges will ultimately take a much longer time during which the advanced technologies will mature, with the efficient use of AI in regular clinical practice in the next 5 to 10 years. On the contrary, advancements in the medical field perhaps may end up in the reduction of job opportunities for health workers due to automation by machines. From a technical point of view, although the augmented advanced biomedical technology promises advanced utilities in health care diagnostics and monitoring, the methodologies still face many challenges that require further exploration before meeting the needs of end users.

19.8   CONCLUSION

The present trends dictate that future biomedical wearables are likely to store the clinical data and execute the analysis for reporting to-and-fro between users. It is evident that they are also compatible for integration with common digital gadgets such as mobile phones, electronic bands, computers to execute the applications such as speech recognition, audio, video, and virtual reality. To do so, the system must be developed to for human-machine interfacing, artificial intelligence, and image processing entities. Furthermore, the proposed device counterparts are viable to be assembled into clothing, skin, and other external parts of users. Considering the clinical point of view, the trend of biomedical devices now moves beyond monitoring to actively assist in the assessment of a health condition. The detailed technologies in biosensing discussed in this chapter are the key and foremost recent technologies that will be adopted in commercial-scale health care diagnostics in the near future.

REFERENCES

Ache, J. M., Haupt, S. S., & Dürr, V. (2015). A direct descending pathway informing locomotor networks about tactile sensor movement. *The Journal of Neuroscience: The Official Journal of the Society for Neuroscience, 35*(9), 4081–4091. doi:10.1523/JNEUROSCI.3350-14.2015

Ajeev, A., Javaregowda, B. H., Ali, A., Modak, M., Patil, S., Khatua, S., . . . Arulraj, A. K. (2020). Ultrahigh sensitive carbon-based conducting rubbers for flexible and wearable human—machine intelligence sensing. *Advanced Materials Technologies, 5*(12), 2000690. https://doi.org/10.1002/admt.202000690

Al-Qatatsheh, A., Morsi, Y., Zavabeti, A., Zolfagharian, A., Salim, N., Kouzani, A. Z., . . . Gharaie, S. (2020). Blood pressure sensors: Materials, fabrication methods, performance evaluations and future perspectives. *Sensors, 20*(16), 4484.

Barnett-Cowan, M., & Harris, L. R. (2011). Temporal processing of active and passive head movement. *Exp Brain Res, 214*(1), 27–35. doi:10.1007/s00221-011-2802-0

Bui, N., Nguyen, A., Nguyen, P., Truong, H., Ashok, A., Dinh, T., . . . Vu, T. (2020). Smartphone-based SpO_2 measurement by exploiting wavelengths separation and chromophore compensation. *ACM Trans. Sen. Netw., 16*(1), Article 9. doi:10.1145/3360725

Chowdhury, M. E. H., Alzoubi, K., Khandakar, A., Khallifa, R., Abouhasera, R., Koubaa, S., . . . Hasan, M. A. (2019). Wearable real-time heart attack detection and warning system to reduce road accidents. *Sensors (Basel, Switzerland), 19*(12), 2780. doi:10.3390/s19122780

Chu, M., Nguyen, T., Pandey, V., Zhou, Y., Pham, H. N., Bar-Yoseph, R., . . . Khine, M. (2019). Respiration rate and volume measurements using wearable strain sensors. *NPJ Digital Medicine, 2*(1), 8. doi:10.1038/s41746-019-0083-3

Das, J., Aggarwal, A., & Aggarwal, N. K. (2010). Pulse oximeter accuracy and precision at five different sensor locations in infants and children with cyanotic heart disease. *Indian Journal of Anaesthesia, 54*(6), 531–534. doi:10.4103/0019-5049.72642

Diaz, C., & Payandeh, S. (2017). Multimodal sensing interface for haptic interaction. *Journal of Sensors, 2017*, 2072951. doi:10.1155/2017/2072951

Fleming, G. A., Petrie, J. R., Bergenstal, R. M., Holl, R. W., Peters, A. L., & Heinemann, L. (2020). Diabetes digital app technology: Benefits, challenges, and recommendations: A consensus report by the European Association for the Study of Diabetes (EASD) and the American Diabetes Association (ADA) Diabetes Technology Working Group. *Diabetologia, 63*(2), 229–241. doi:10.1007/s00125-019-05034-1

Fu, Z., Hong, S., Zhang, R., & Du, S. (2021). Artificial-intelligence-enhanced mobile system for cardiovascular health management. *Sensors (Basel), 21*(3). doi:10.3390/s21030773

Futagawa, M., Iwasaki, T., Murata, H., Ishida, M., & Sawada, K. (2012). A miniature integrated multimodal sensor for measuring pH, EC and temperature for precision agriculture. *Sensors, 12*(6), 8338–8354.

Gatti, E., Calzolari, E., Maggioni, E., & Obrist, M. (2018). Emotional ratings and skin conductance response to visual, auditory and haptic stimuli. *Scientific Data, 5*(1), 180120. doi:10.1038/sdata.2018.120

Guo, X., Pei, W., Wang, Y., Chen, Y., Zhang, H., Wu, X., . . . Liu, R. (2016). A human-machine interface based on single channel EOG and patchable sensor. *Biomedical Signal Processing and Control, 30*, 98–105. https://doi.org/10.1016/j.bspc.2016.06.018

Jeong, J.-W., Yeo, W.-H., Akhtar, A., Norton, J. J. S., Kwack, Y.-J., Li, S., . . . Rogers, J. A. (2013). Materials and optimized designs for human-machine interfaces via epidermal electronics. *Advanced Materials, 25*(47), 6839–6846. https://doi.org/10.1002/adma.201301921

Jeong, S. H., Hjort, K., & Wu, Z. (2015). Tape transfer atomization patterning of liquid alloys for microfluidic stretchable wireless power transfer. *Scientific Reports, 5*(1), 8419. doi:10.1038/srep08419

Jeong, S. H., Zhang, S., Hjort, K., Hilborn, J., & Wu, Z. (2016). PDMS-based elastomer tuned soft, stretchable, and sticky for epidermal electronics. *Advanced Materials, 28*(28), 5830–5836. https://doi.org/10.1002/adma.201505372

Jung, S., Kim, J. H., Kim, J., Choi, S., Lee, J., Park, I., . . . Kim, D.-H. (2014). Reverse-micelle-induced porous pressure-sensitive rubber for wearable human—machine interfaces. *Advanced Materials, 26*(28), 4825–4830. https://doi.org/10.1002/adma.201401364

Kaushal Kanakia, S. P., Sabnis, S., & Shah, V. (2018). Emotion & heartbeat detection using image processing. *International Journal of Scientific & Engineering Research, 9*(3), 43–47.

Kim, D.-H., Lu, N., Ma, R., Kim, Y.-S., Kim, R.-H., Wang, S., . . . Rogers, J. A. (2011). Epidermal electronics. *Science, 333*(6044), 838–843. doi:10.1126/science.1206157

Kortli, Y., Jridi, M., Al Falou, A., & Atri, M. (2020). Face recognition systems: A survey. *Sensors, 20*(2), 342.

Krishnaswamy, U., Aneja, A., Kumar, R. M., & Kumar, T. P. (2015). Utility of portable monitoring in the diagnosis of obstructive sleep apnea. *Journal of Postgraduate Medicine, 61*(4), 223–229. doi:10.4103/0022-3859.166509

Kundinger, T., Sofra, N., & Riener, A. (2020). Assessment of the potential of wrist-worn wearable sensors for driver drowsiness detection. *Sensors, 20*(4), 1029.

Lim, S., Son, D., Kim, J., Lee, Y. B., Song, J.-K., Choi, S., . . . Kim, D.-H. (2015). Transparent and stretchable interactive human machine interface based on patterned graphene heterostructures. *Advanced Functional Materials, 25*(3), 375–383. https://doi.org/10.1002/adfm.201402987

López-Casado, C., Bauzano, E., Rivas-Blanco, I., Pérez-del-Pulgar, C. J., & Muñoz, V. F. (2019). A gesture recognition algorithm for hand-assisted laparoscopic surgery. *Sensors, 19*(23), 5182.

Lu, N., Lu, C., Yang, S., & Rogers, J. (2012). Highly sensitive skin-mountable strain gauges based entirely on elastomers. *Advanced Functional Materials, 22*(19), 4044–4050. https://doi.org/10.1002/adfm.201200498

Mannsfeld, S. C. B., Tee, B. C. K., Stoltenberg, R. M., Chen, C. V. H. H., Barman, S., Muir, B. V. O., . . . Bao, Z. (2010). Highly sensitive flexible pressure sensors with microstructured rubber dielectric layers. *Nature Materials, 9*(10), 859–864. doi:10.1038/nmat2834

Mendonça, F., Mostafa, S. S., Ravelo-García, A. G., Morgado-Dias, F., & Penzel, T. (2018). Devices for home detection of obstructive sleep apnea: A review. *Sleep Med Rev, 41*, 149–160. doi:10.1016/j.smrv.2018.02.004

Noh, Y., Liu, H., Sareh, S., Chathuranga, D. S., Würdemann, H., Rhode, K., & Althoefer, K. (2016). Image-based optical miniaturized three-axis force sensor for cardiac catheterization. *IEEE Sensors Journal, 16*(22), 7924–7932. doi:10.1109/JSEN.2016.2600671

Physicians' Knowledge of Inhaler Devices and Inhalation Techniques Remains Poor in Spain. (2012). *Journal of Aerosol Medicine and Pulmonary Drug Delivery, 25*(1), 16–22. doi:10.1089/jamp.2011.0895

Prawiro, E. A. P. J., Yeh, C.-I., Chou, N.-K., Lee, M.-W., & Lin, Y.-H. (2016). Integrated wearable system for monitoring heart rate and step during physical activity. *Mobile Information Systems, 2016*, 6850168. doi:10.1155/2016/6850168

Rashkovska, A., Depolli, M., Tomašić, I., Avbelj, V., & Trobec, R. (2020). Medical-grade ECG sensor for long-term monitoring. *Sensors, 20*(6), 1695.

Serhani, M. A., El Kassabi, H. T., Ismail, H., & Nujum Navaz, A. (2020). ECG monitoring systems: Review, architecture, processes, and key challenges. *Sensors (Basel, Switzerland), 20*(6), 1796. doi:10.3390/s20061796

Shin, S.-C., Lee, J., Choe, S., Yang, H. I., Min, J., Ahn, K.-Y., . . . Kang, H.-G. (2019). Dry electrode-based body fat estimation system with anthropometric data for use in a wearable device. *Sensors, 19*(9), 2177.

Son, S., Jeong, Y., & Lee, B. (2019). An Audification and Visualization System (AVS) of an autonomous vehicle for blind and deaf people based on deep learning. *Sensors (Basel), 19*(22). doi:10.3390/s19225035

Takei, K., Takahashi, T., Ho, J. C., Ko, H., Gillies, A. G., Leu, P. W., . . . Javey, A. (2010). Nanowire active-matrix circuitry for low-voltage macroscale artificial skin. *Nature Materials, 9*(10), 821–826. doi:10.1038/nmat2835

Uddin, M. Z. (2019). A wearable sensor-based activity prediction system to facilitate edge computing in smart healthcare system. *Journal of Parallel and Distributed Computing, 123*, 46–53. https://doi.org/10.1016/j.jpdc.2018.08.010

Vanegas, E., Igual, R., & Plaza, I. (2019). Piezoresistive breathing sensing system with 3D printed wearable casing. *Journal of Sensors, 2019*, 2431731. doi:10.1155/2019/2431731

Wang, X., Guo, S., Qu, H., & Song, M. (2019). Design of a purely mechanical sensor-controller integrated system for walking assistance on an ankle-foot exoskeleton. *Sensors (Basel, Switzerland), 19*(14), 3196. doi:10.3390/s19143196

Wang, X., Zhang, H., Dong, L., Han, X., Du, W., Zhai, J., . . . Wang, Z. L. (2016). Self-Powered high-resolution and pressure-sensitive triboelectric sensor matrix for real-time tactile mapping. *Advanced Materials, 28*(15), 2896–2903. https://doi.org/10.1002/adma.201503407

Wilhelm, S., Weingarden, H., Ladis, I., Braddick, V., Shin, J., & Jacobson, N. C. (2020). Cognitive-behavioral therapy in the digital age: Presidential address. *Behavior Therapy, 51*(1), 1–14. https://doi.org/10.1016/j.beth.2019.08.001

Wilson, C. B. (1999). Sensors 2010. *BMJ (Clinical Research ed.), 319*(7220), 1288–1288. doi:10.1136/bmj.319.7220.1288

Wolff, L., Socolinsky, D., & Eveland, C. (2003). *Using infrared sensor technology for face recognition and human identification* (Vol. 5074). SPIE.

Wu, Y. T., Gomes, M. K., da Silva, W. H., Lazari, P. M., & Fujiwara, E. (2020). Integrated optical fiber force myography sensor as pervasive predictor of hand postures. *Biomed Eng Comput Biol, 11*, 1179597220912825. doi:10.1177/1179597220912825

www.businesswire.com/news/home/20200316005453/en/Insights-into-the-Global-Wearable-Medical-Device-Market-to-2027-Featuring-Sotera-Wireless-Zephyr-Technology-Corporation-Omron-Corporation-Among-Others-ResearchAndMarkets.com.

20 Biosensors

Taner Daştan and Sevgi Durna Daştan

CONTENTS

20.1 INTRODUCTION

Living creatures struggling to survive in nature perceive all the changes that occur in the environment in a certain process by which they live. They have to keep up with these changes in order to survive. This sensing mechanism is carried out with the five basic sensory organs: the eyes, nose, skin, tongue and ears in living organisms [1]. These mechanisms are sampled by researchers in a laboratory environment. Biosensors discovered as a result of the transformation of a biological event into an electrical signal in a laboratory environment. Living things break down biological substances into basic parts, which allow the detection of stimulants that occur in the life process. These pieces and their relationships are then examined as a whole. Biosensor structures are formed by combining the systems obtained as a result of these conversion research studies of a biological event into an electrical signal [1]. Thanks to today's technological developments, biosensors are developing very rapidly. The inclusion of sensor technology in areas of study such as chemistry, biology, materials science and engineering, with new technologies updating these disciplines, has clearly contributed to the development of biosensors.

Biosensors are the result of combining biological systems such as immunological agents, microorganisms, enzymes, cells and chemical receptors, thanks to existing structures such as amperometrics, potentiometrics, piezoelectric, thermal, acoustic and optical sensors [1, 2]. Biosensors are defined as the conversion of a biological event depending on external factors into an electrical signal in organisms. Using the knowledge obtained through various disciplines such as chemistry, biochemistry, physics, genetics, biology and engineering, biosensors have developed as a result of combining advanced electronic processing capabilities, thanks to the elective properties of biological molecules or systems. Biosensors are bioanalytic devices developed for

DOI: 10.1201/9781003173533-20

use in many different industrial fields as well as medicine, engineering, ecology and pharmacy. Biosensors have become small-sized structures as a result of the development of electronic circuits over time. Biosensors are mounted on chips because of nano-sized studies [1, 3]. When biosensor technology is examined, it is seen that it is affected by nanotechnology. The biosensor dimensions are obtained in a way that is smaller, precise, long-lasting and inexpensive according to the development in nanotechnology.

20.2 BIOSENSORS

In order to survive, living things have to perceive the changes occurring in their environment and adapt as soon as possible. The organisms often use natural sensors with pure enzyme, tissue and microorganism properties to create protective shields against external hazards [1, 4]. Biosensors were obtained by many scientists sampling the sensing mechanism that exists in living things. When the enzyme electrode in the studies contacts the biological solution, glucose and oxygen pass through the enzymatic membrane. The electrode in the solution oxides glucose to gluconic acid using oxygen. In this way, the enzyme electrode measures the amount of decrease in the partial pressure of oxygen. Biological sensors take advantage of these changes to obtain physical data. Physical signals are measured by combining analyte-like substances or materials analyzed in biological or non-biological applications, biological sensors like tissues, nucleic acids, microorganisms, enzymes, antibodies, cells, artificial biological inducers and selectively interacting physiochemical converters like electrodes, transistors, thermistors, optical fibers, piezoelectric crystals. Electrical signals obtained from such materials are measured as physical input. Structures that exhibit work in this way are called biosensors [1]. Biosensors are pieces of equipment that are evaluated to target analyte materials in biological reactions and have sensing properties. It consists of two key structures of intertwined biochemical and electrochemical properties. The biochemical part interacts with the substance to be analyzed and ensures its recognition. During this process, a biochemical product may occur. In the electrochemical part of the biosensors, the data obtained as a result of the recognition event is transformed into a readable (quantitative) numerical value [1, 4]. Biosensors must have bioactive and biorecognition material when performing their function. These materials should be able to recognize analytes and be in close contact with the transducer. The biosensor structures have many superior characteristics such as long storage stability, short analysis time, authenticity of substances processed in reactions. Since repeatability is high in the analyses, it can be easily applied in all laboratory environments. There is not much time needed for this method to be applied. During the measurement procedures, the number of process steps is less. Therefore, high-priced devices and various chemical materials are not needed during the procedures.

20.3 STRUCTURE AND FUNCTION OF BIOSENSORS

Biosensors are basically built on the basis of the formation of a signal proportional to the amount of analyte on the transducer surface and the transmission of this signal to the measuring device as a result of the interaction of the substance to be analyzed with the biocomponent on the biosensor surface [5, 6]. Enzymes, microorganisms, botanical and animal tissues, receptors, antibodies and nucleic acids can be used as biosynthesis in biosensors. A suitable transducer should be selected in accordance with the molecule to be analyzed, which translates the electrochemical, optical or gravimetric signal formed as a result of the conversion of the analyte into an electrical signal. Transducers and biocomponents can be connected by appropriate physical or chemical method [5, 7]. Today, biosensors contain many different substances and systems as biocomponents and transducers. The most important of these are given in Table 20.1.

Biosensors allow a biological event to be converted into an electrical signal. In general, it is the result of combining the substance to be analyzed with a transmitting mechanism that transmits the

TABLE 20.1
Biosensor Component Contents Modified According to the Article by Akbayirli and Akyilmaz.

Analyte (Sample)	Biocomponent (Receptor)	Signal Transducing System
Metals		Electrochemical
Hormones	Enzymes	Electrochemical
Enzymes	Antibodies	Semiconductor
Co-enzymes	Cell	Optical
Substrate	Tissue slices	Photometric
Activator	Receptors	Fluorometric
Inhibitor	Microorganisms	Fluorometric
Antibody-Antigen	Nucleic acids	Piezoelectrical
Nucleic acid	Lipids	Quartz crystal microbalance
Microorganisms	Cellular organelles	Microquantiller
Glucose	Glucose oxidase	Amperometrical
L-amino acids	Horseradish peroxidase	Potantiometrical
Herbicide	Antibodies	Piezoelectrical quartz crystal

Source: [1, 7].

signal resulting from the interaction of the bioactive component that interacts on demand and is used in conjunction with the measuring mechanisms. Depending on the system characteristics, structures such as magnifiers, microprocessors and digital viewers can be incorporated later into the system when needed. These physical components, which are later incorporated into the system, convert them into physical signals depending on the ability to measure the biological functions of biocomponents. The appropriate physical size is selected taking into account the resulting biochemical reaction. Electrodes (conductive rods) used in the system as physical transducers are used in the measurement processes such as amperometric and potantiometric [8]. In addition to these measurement processes, electronic elements such as transistors and thermistors can also be used within the system as physical components. In biosensors, electronic devices such as electrochemical, optical, calorimetric and piezoelectric are used to measure and transmit the signal that presents an output of the interaction of the substance that determines the bioactive components. Thanks to these mechanisms used in the system, long-lasting tests are concluded in a shorter time. Biosensor components are given in Figure 20.1.

In biosensor applications, the substances and structures to be analyzed are defined as analytes. The elements that make up the compounds or mixtures in the biosensor structure are called bioreceptors. Bioreceptors convert analyte substances into a different form than they are. Analytes commonly used during these processes are enzymes and antibodies. The first step in the relationship between enzyme-substrate and antibody-antigen is the binding of analytes to protein molecules. Catalytic antibodies developed in recent years not only adopt antigens, but also allow a chemically reactive substance to react or change the rate of reaction. Large molecules are formed as a result of polymerization of small protein-structured substances [9]. Nucleic acids and carbohydrates are also used in areas such as gene ring analysis and cell surface characterization specific to a species. Thus, biosensor application areas are developed day to day [1]. Transducers are structures that measure biological reactions generated by receptors and convert them into a physical signal. After the recognition of analytes, physical and chemical change is detected, measured and converted into digital signals. Transducer selection is made taking into account biochemical reactions. Enzyme electrodes are a target substance used in amperometric and potentiometric measurement processes. Devices (transducers) that convert one form of energy into another form of energy in the biosensor

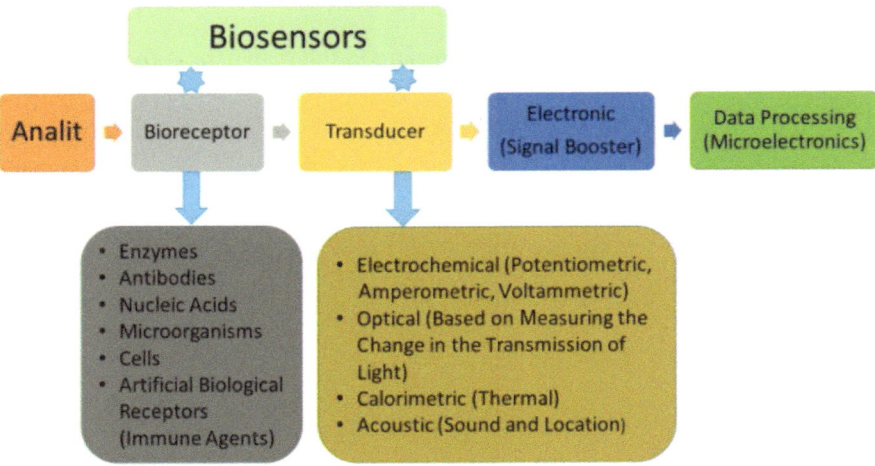

FIGURE 20.1   The components of biosensors [1].

structure possess electrochemical (amperometric, voltageametry, potentiometric, capacitive, conductive, impedance), optics (including surface plasmon resonance, absorption, chemical glow, biological glow, fluorescence, fiber optical ones), piezoelectric (quartz crystal microbalance, surface sound wave), calorimetric and magnetic properties [1]. Biosensors are mainly used for diagnostic purposes in biomedical areas today. The enzyme biosensor is the first sensor to be applied in the biomedical field.

20.4   THE CLASSIFICATION OF BIOSENSORS

20.4.1   BASED ON INTERACTION BETWEEN THE ANALYTE AND BIOACTIVE COMPONENT

Biosensors are divided into biocatalytic and bioaffinity-based sensors according to the substance-bioactive component relationship to be analyzed [5, 10].

1. Biocatalytically based biosensors using microorganisms and enzymes
2. Bioaffinity-based biosensors using interactions such as antibody-antigen and receptor-ligand.

20.4.2   BASED ON THE COMPONENTS OF BIOACTIVE SURFACE, TRANSDUCING AND MEASUREMENT SYSTEMS

Biosensors are classified into four different titles according to measurement principles and transducer types [11]. The classification for this is as shown in Table 20.2 [1]. Also wearable biosensors are very famous nowadays. Wearable biosensors can detect human body blood sugar, heart rate, bipolar disorder, etc. It is a digital device used in different outfit systems such as smart watches, smart shirts and tattoos.

20.4.3   BASED ON BIOCOMPONENTS IN THE BIOACTIVE SURFACE

Biosensors are classified into five different groups according to biocomponents in the bioactive surface as follows [5].

TABLE 20.2
Classification of Biosensors According to Transmission and Measurement Systems.

Biosensors	Classification
Electrochemical Based Biosensors	Amperometrical electrodes
	Potantiometrical electrodes
	Conductiometrical transistors
Optical Based Biosensors	Photometrical optic fibers
	Fluorometrical optic fibers
	Bioluminescent optic fibers
Calorimetrical Biosensors	Thermistors (e.g. isothermal, isoperibols)
Piezoelectrical Biosensors	Piezoelectrical crystals, quartz crystal microbalances, microcantilevers

Source: [1].

20.4.3.1   Enzyme Based Biosensors

Glucose oxidase enzyme electrodes for glucose determination, reported by Clark and Lyons in 1962 and also Updike and Hick in 1967, constitute the first examples in this regard [1, 5, 11]. The first examples in biosensor technology appeared especially in the form of enzyme electrodes based on amperometric and potentiometric measures. The first biosensor developed for commercial purposes is for the diabetes diagnosis, which is the glucose oxidase electrode for the defining of glucose in body liquids like blood and urine. These enzyme sensors are frequently used to track biological products such as glucose, urea and sugar in the body, to monitor cancers and to detect cases caused by microbes [12]. When the biosubstances used in biosensor technology are sorted according to the increasing quality of complexity, ionophores, antibodies, enzymes, liposomes, biomembrane fragments (e.g. receptor), cell organelles (e.g. mitochondria) tissue or all cells and organs (e.g. vision and smell) can be listed. In the most general sense, enzyme sensors, like other biosensors, consist of bioactive layer, transducers and measuring system [7]. The only difference from other biosensors is that enzymes are present as biomolecules in the bioactive layer. In contrast, as with other biosensors, membranes on the inner and outer surfaces of the bioactive layer signal amplifiers between the transducer and the measuring mechanism; microprocessors or recorders or computer systems connected to the measuring mechanism are elements added according to their requirements [13].

20.4.3.2   Tissue Based Biosensors

It is known that some animal and vegetable tissues and organelles are rich in certain enzymes. Instead of isolated preparations of these enzymes, these tissue fragments, where they are directly dense, are also used as biocomponents [10, 14]. It is especially advantageous to use the tissue containing these enzymes instead of using different enzymes in the multi-layered transformation of the target analyte especially if the enzyme is not commercially available. Due to the complexities of the tissues, optimization is required for each biosensor because the response characters are different for each new type of electrode. Since enzymes of this type of biosensor will be found in their natural environment, they are very advantageous in terms of catalytic stability. However, the response time is prolonged due to the increase of the diffusion barrier that the substrate must overcome in order to reach the enzyme [14]. In order to partially reduce this disadvantage, the tissues are homogenized and used. In addition, since the tissues contain many enzymes in their bodies, substances and conditions that will inhibit other enzyme systems can be used when preparing biosensors to prevent errors that will occur by converting the analyte with other enzymes other than the desired enzyme [15].

20.4.3.3 Deoxyribonucleic Acid Based Biosensors

Biosensor systems use single chain DNA oligomers as biocomponents. The emergence of DNA structure, hybridization, amplification and the development of recombinant DNA technologies led to the development of DNA sensors. Generally, DNA sensors are based on the principle of hybridization of a single-chain DNA probe fixed to the transducer surface and a single-chain piece of DNA that symbolizes a particular disease, an inherited character, or the pathogenicity of a bacteria or virus [1, 6, 7]. Thanks to the double-chain DNA generated by this hybridization, an electrochemical or optical signal is formed and the signal is made readable by an optical, piezoelectric or electrochemical transducer [16].

20.4.3.4 Antibody/Antigen Based Biosensors (Immunosensors)

Biosensors using antibodies as biocomponents are called antibody-based biosensors or immunosensors. Protein-structured substances produced by immune system cells against foreign organisms as antigenic particulates like viruses, bacteria and protozoa or their protein products are called antibodies. Since there is a specific interaction between antibodies and antigens, highly specific and sensitive analyses can be performed with immunosensors. By pairing antibodies with appropriate transducers, immunosensors can be developed for the determination of hormones, drugs, viruses, bacteria and other environmentally pollutant molecules, pesticides, biomedical substances [17].

20.4.3.5 Microbial Biosensors

A microbial biosensor is formed by the consolidation of living or non-living microbial cells (by immobilization) with a transducer [18]. Microorganisms can metabolize chemical compounds of great diversity. Microorganisms can effectively conform to severe conditions and can adapt their abilities and form new molecules over time. Microorganisms are also an economic source of intracellular enzymes for genetic modifications caused by mutation or recombinant DNA technology [1, 5].

20.5 IMMOBILIZATION METHODS IN BIOSENSORS

After the biocomponent and transducer is selected according to the sample intended to be analyzed, these two elements should be connected to each other, that is, fixed to the surface of the transducer. Since the life of the biosensor depends on how long the biocomponent can be held on the transducer surface through the stabilization process, it should prevent the biocomponent from leaving the surface for a long time. In the stabilization of enzymes, the enzyme should not be damaged during active central stabilization or there should be no decrease in enzyme activity due to steric inhibition [5, 19]. Basic methods used for stabilization and immobilization are adsorption (non-covalent binding), covalent binding, arresting and cross-linking.

20.6 DESIGN OF BIOSENSORS

Advanced technologies, samples, working conditions and results used as the basis for biosensor design are taken into account [20]. The analytes used consist of many known, unknown, organic and inorganic materials. Therefore, the concentration of analytes is important. Clarity is provided as a result of the separation of the target analytes. Thus, the concentration lower detection limit is improved and the desired value is reached. For this purpose, filtration, centrifuge and magnetic immuno-separation methods are usually used. Magnetic immuno-separation is preferred during biosensor studies as it is a fast, easy and automated application. In parsing method applications, magnetic nanoparticles and microparticles can be magnetically manipulated when using natural magnets or electro-magnets. Manipulated cells become selectable, parsable and sortable when they remain in a magnetic environment [21]. The properties considered in quality biosensors are sensitivity, electability, measuring range, measuring time, consistency, measurement limit, lifespan and stability. A quality measurement value takes into account changes in the sensitivity of the device over a period of time [5]. Certainness or selectivity in biosensors are the most important criteria used in performance evaluation. The biosensor's interest only in analyte matter is a sign of its specificity. Therefore, it does not react to other reagents and does not give incorrect results [22]. The ratio of the change in the biosensor result signal to the change in the target analyte amount determines the measure of its precision. The change in signal can be observed as a result of exposure of the biosensor to a standard solution containing target analyte at different concentrations. Depending on different biorecognition materials and conversion devices, there are many factors that affect the precision of a biosensor. The sensitivity of the biosensor is always expected to remain the same and give reliable numerical results. The accuracy of the results obtained must be high [23]. The detection time of the biosensor is the period from the sampling stage to the moment the results are displayed. Sometimes the response and reset time are included in this calculation. Response time is the time it takes for a biosensor measurement signal to be obtained within an equation. The reset time is the required preparation time of a biosensor until the start of use of the next instance. Repeatability and reproducibility are important in biosensor evaluations, as great variability is expected in biological samples and biorecognition materials. Biorecognition materials are affected by heat, humidity, pH and other factors over time. The lifespan of organic materials that deteriorate in these environments is another important factor to consider [24].

20.7   APPLICATION AREAS OF BIOSENSORS

Biosensors play a very important role in many areas like medicine, food, pharmacy, defense, metabolite measurement, disease screening and diagnosis, insulin therapy, drug improvement, clinical psychotherapy, military services and practices, agricultural and veterinary medicine applications, bacterial and viral diagnosis, industrial wastewater control, toxic gas analysis in mines and mining enterprises, environmental protection and ecological pollution control, and many other industrial activities, especially in automation, processing and monitoring, quality control, attitude controlling and energy storage. The glucose oxidase enzyme sensor used to diagnose diabetes is the biosensor that attracts the most attention [1, 5, 7]. The biosensors detect glucose in very little blood and urine very quickly. Biocompatibility becomes a very important parameter when it comes to use in the field of medicine. In the case of intra-body uses, fibrin accumulation and platelet formation may occur on the enzyme-containing membrane, or there may be inflammation of the body tissues [25]. To prevent these problems, the surfaces of the biosensor that will come into contact with the tissue are covered with special materials. Examples of other medical biosensors are urea and creatinine electrodes, cholesterol electrodes, acetylcholine electrodes. Thanks to these sensors, very important diseases can be diagnosed. For example, urea level follows kidney function, cholesterol level is monitored to prevent arteriosclerosis, acetylcholine electrode is used to evaluate muscle fatigue [26]. The immunological biosensors are widely used in drug active substance measurement. For example, the theophyline sensor, HCG hormone sensor for pregnancy test, alpha-fetoprotein sensor for cancer diagnosis, surface antigen sensor for hepatitis B, are examples of the immunological sensors used in medicine [27].

On the other hand, nanotechnology is a technology science that focuses on the design and production of new materials, devices and mechanisms by controlling matter in nano dimensions, that is, structures at the atomic, molecular and supramolecular levels. In the light of advances in technology, nanotechnology which solves many problems in the field of health, pharmacology and drug delivery systems, patient monitoring devices, regenerative sciences [28], in water decontamination and sterilization areas [29], is increasingly important in parallel with the increase in usage areas. The micro, nano or bionano electromechanical system biosensors (MEMS/NEMS/BioNEMS) produced using nanotechnology are expected to have more features in parallel with the developments achieved in the field of nanotechnology and biotechnology. The results are easier to monitor and evaluate, thanks to the new generation of biosensors developed using nanotechnology and chip technology, called modern-day biosensors. Because of this technology, the biochip will make it possible to track the variations that will occur in the individual without him or her going to the hospital. In addition, the rapid development of methods of replicating and imaging the basic structures of molecules, their evaluation together with nanotechnology, will lead to a further expansion of the uses of nucleotide-based biosensors [30, 31].

Also, with the development of silicon chip technologies and their applications, the use of biosensors in the pharmacology and toxicology fields has been mentioned, and it has increased day by day [32]. These silicon biosensors (cytosensors, silicon microphysiometers) can control the physiological changes of cultured cells and it can accurately determine which cells secrete products of acidic metabolism [32]. The cytosensor essentially consists of eight independent fluid assemblies. These parts of the silicon chip biosensor are reagent reservoir and include the components of computer controlled pump and valve, sensor chamber, and reference electrode. The porous polycarbonate membrane is used to compress cells in between them. In this membrane, both tied and non-adherent cells can be placed and can be grown directly. This prepared capsule is placed in a sensor chamber, and the piston creates a microvolume space containing cells and regulated nutrient flow, or an investigated effector agent is provided to the cells during the experiments for various pharmacological and toxicological analyses [32]. For drug discovery and effective screening in integrated systems, various biosensors can be utilized, which can be based on antibodies, enzymes, cells, synthetic membranes, etc. Particularly in the field of pharmacology and toxicology, concise determination of drugs is vital both in formulations and biological materials. Drug analysis in particular biosensors are classified according to the signal transduction through electrochemical, optical, thermal or acoustic transducing. Different biosubstances can be used for drug screening, which include enzymes, antibodies, cells and aptamers. With their high stability coupled with low-detection thresholds, electrochemical biosensors are favored. Optical biosensors also provide high sensitivity with high specificity in a lower expense and provide real-time and label-free detection [33].

On the other hand, biosensors may also be built to detect cancer biomarkers and determine targeted drug efficacy. Biosensor technologies have the potential for quick and accurate detection, reliable imaging, monitoring of different cancer processes like angiogenesis and metastasis, and defining the effective chemotherapy agents for anticancer activity [34]. Biomarkers are different molecules such as DNA, RNA, protein, and each of them can be obtained from various intracellular and extracellular environments such as blood, serum, plasma, saliva, urine, etc. Biomarkers for cancer are still insufficient in both specificity and sensitivity to be utilized in cancer treatment, which is a point of interest for the biosensor development. In this context, the advancements in the nanotechnology, in particular the quantum dots, can be utilized to improve cancerous cell diagnosis and labeling along with concise delivery of drugs into targeted locations, and would also contribute to the fine imagining. The advancements in the nanotechnology applications would certainly improve the fields of cancer diagnosis and therapy. Nanotechnology based biosensors would enable early and sufficient diagnosis of diseases despite the apparent complexity of cancerous diseases [34].

Cardiovascular diseases (CVDs) and strokes are accepted as two of the most prominent causes of death worldwide. It is vital to diagnose CVD and stroke as early and swiftly as possible; therefore various cardiac-specific biomarkers that include optical, fiberoptical/biooptrode, acoustic, electrochemical and magnetic biosensors are developed for this objective [35, 36]. Currently the classical diagnostic approach of laboratory-based yet rather slow acting biomarkers are in use, which stresses the demand for a fast-acting, real-time, decentralized and inexpensive diagnostic system for CVDs. For this context integration of artificial intelligence (AI) and large-scale data mining coupled with data processing with the use of machine-learning algorithms are in hot debate for use in CVD and stroke diagnosis and monitoring, which would lead into improved establishments in medical practices [35]. Large-scale data interpretation via AI managed data processing provided efficient feedback for diagnosis and clinical decisions in cases of abnormalities [35, 37]. These analysis components are in smartphone-based devices, which are both small and composite [35, 38]. It is also debated that AI integration might include other biomarkers, such as hematocrite, oxygen saturation, HbA1C, lipids, and other biomarkers associated with infections and inflammations. More recently, deep mind algorithms to interpret and process physiological data through touch-free or pseudo-touch-based biosensors is highly considered. Various disciplines now have high interest in developing stable therapeutic biosensors through novel and advanced devices that will provide higher sensitivity and specificity [35, 39].

Commercially available biosensors offered by companies enable effective and detailed understanding of applications in different fields, such as medical, bio-processing, and the ecology. Some of the companies make the parts of biosensors while others assemble the parts and launch in the markets and still some are manufacturers of parts and complete sets of biosensors. Hence, it is difficult to follow the criteria to give credit of development of commercial biosensors for all of them. Based on their centralization, commercially available biosensors are divided into laboratory-based and portable/field devices. An electrochemical-based biosensor, the blood glucose monitor is known as the most prominent portable biosensor [40, 41], which is manufactured and offered by many different companies. As for the laboratory-based devices, commercial offerings are mostly based on optical-based biosensor, such as optical microarray detectors that can be used in scanning genomic and proteomic data [40].

As the most frequently seen form of dementia, Alzheimer's Disease (AD) is of prime interest for developing specified molecular biomarkers for diagnosis and therapy. Towards this end, miR-137, which is a confirmed biomarker for the AD, serum quantification through ultrasensitive nanobiotechnology-based sensors is highly considered [42]. This considered sensor model is based on an electrochemical sensor basis yet in nano-scale which includes electrochemically reduced graphene oxide (ERGO) and gold nanowires (AuNWs) that modify screen-printed carbon electrodes (SPCE) through intercalated Doxorubucin (DOX) label. For the context of early diagnosis, preliminary results are showing promise. Specificity towards oligonucleotide discrimination from this sensor model on target and off-target miR-21 and miR-155 proved to be effective, which indicates solid use for clinical trials for early diagnosis of AD [42]. Apparent advantages of the nanobiosensor model strongly indicate its successful use in early-phase detection of AD [42].

20.8 BIOSENSORS IN DIAGNOSIS OF NOVEL CORONAVIRUS (SARS-COV-2) AND ITS ASSOCIATED DISEASE (COVID-19)

Coronavirus disease (COVID-19) is a highly contagious viral disease caused by the virus SARS-CoV-2. Various methods have been developed to diagnose the disease. On the other hand, significant investment is being made in technologies that will improve the process of detecting individuals infected with the virus. One of the technological methods planned to be utilized for this purpose is biosensors [43]. It is thought that biosensors can serve as a key for comprehensive coronavirus scans. For this reason, it is known that scientists are working on the production of chips that can detect coronavirus RNA, antibodies and antigens. By integrating the biosensors into the COVID-19 diagnosis process, the sample containing the blood taken from the finger or the swab from the nose can be read on the semiconductor chip, and it will be possible to find out whether there is antibody in the sample in just a few minutes [43]. In another biosensor research study for the diagnosis of COVID-19 disease, the laser-based biosensor developed by scientists scans the molecules in the sample and aims to determine the light signals of the molecules. Studies are also being carried out to show that the antibodies in the sample can be detected in a few minutes with the help of miniature chips, with metal-oxic semiconductors called nanophotonic biosensors. In this test, which can detect antibodies produced against the SARS-CoV-2 virus in a very short time, the graphene oxide used as the sensor platform provides conductivity and biocompatibility. In addition to these, in another biosensor study conducted to use in the diagnosis of COVID-19, it was reported that the disease can be diagnosed by recording the changes in wavelengths reflected from fiber optic biosensors. In the diagnosis of COVID-19, it is considered that biosensors with nano gold particles can separate viral cells using biological or synthetic molecules [43].

20.9 CONCLUSION

Biosensors, which are the product of the combination of technology and biological systems with impeccable functioning, are at a dazzling level today. Undoubtedly, the emergence of new knowledge in biology every day and the continued progress of technology are a precursor to the development of new biosensors very soon. Due to advances in biochip technology, the importance and weight of biosensors will increase gradually [44, 45]. Nanotechnology will greatly open the horizons of biosensors and further increase its weight in the economy and daily life [46]. Biosensors greatly shorten the time between taking analyte and obtaining results. Biosensors produced by nanotechnology require fewer analytics and also have a high level of sensitivity and specificity. The fact that the measuring devices are suitable and portable for the automation system allows them to be used in different sectors. On the other hand, our world is struggling with the COVID-19 disease that emerged in 2019. This pandemic has brought about major changes in the economy, health care sectors and lives of people. In the diagnosis and treatment of COVID-19 and the other microbial diseases, and even the other common illnesses, scientists are trying to produce solutions by using technology. In this sense, biosensor technologies are seen as very suitable methods for detecting viral or bacterial infectious agents and antibody-antigen particles with high sensitivity and in a short time. So it is thought that biosensors will play an active role in combating the pandemic, especially due to the low cost they promise. The future of biosensors depends on the cost of production and its active use. Converting biosensors into trades will make significant gains on behalf of manufacturers or countries, while purchasers will have to pay large amounts of money to producing countries.

REFERENCES

1. Tuylek Z. 2017. Biosensor an Nanotechnological Interaction. *BEU Journal of Science*, 6(2):71–80.
2. Blum L.J., Coulet P.R. 1991. *Biosensor Principles and Applications*, CRC Press, Boca Raton.
3. Ratner B.D., Hoffman A.S., Schoen F.J., Lemons J.E. 1996. *Biomaterials Science*, Academic Press, San Diego.
4. Coulet P.R. 1991. What is a Biosensor? Chapter 1; *Biosensor Principles and Applications*. Edited by L. J. Blum, P. R. Coulet, pp. 1–6. Marcel Dekker Inc., New York.
5. Kokbas U., Kayrın L., Tuli, A. 2013. Biosensors and Their Medical Applications. *Archives Medical Review Journal*, 22(4):499–513.
6. Dinçkaya E. 1999. Enzyme Sensors. In *Biosensors, Biochemistry Summer School Book*. Edited by A. Telefoncu, pp. 81–142. Ege University Press, Izmir.
7. Akbayirli P., Akyilmaz E. 2007. Activation-Based Catalase Enzyme Electrode and Its Usage for Glucose Determination. *Analytical Letters*, 40:3360–3372.
8. Zhou Y.L., Zhi J.F. 2006. Development of an Amperometric Biosensor Based on Covalent Immobilization of Tyrosinase on a Boron-doped Diamond Electrode, *Electrochemistry Communications*, 8:1811–1816.
9. Kindschy L.M., Alocilja E.C. 2004. A Review of Molecularly Imprinted Polymers for Biosensor Development for Food and Agricultural Applications, Trans. *ASAE*, 47:1375–1382.
10. Akyilmaz E., Baysal S.H., Dinçkay E. 2007. Investigation of Metal Activation of a Partially Purified Polyphenol Oxidase Enzyme Electrode. *International Journal of Environmental Analytical Chemistry*, 87:755–761.
11. Akyilmaz E., Yorganci E. 2008. A Novel Biosensor Based on Activation Effect of Thiamine on the Activity of Pyruvate Oxidase. *Biosens Bioelectron*, 23:1874–1877.
12. Rainina E.I., Efremenco E.N., Varfolomeyev S.D., Simonian A.L. 1996. The Development of a New Biosensor Based on Recombinant *E. coli* for the Direct Detection of Organophosphorus Neurotoxins. *Biosens Bioelectron*, 11:991–1000.
13. Bartlett P.N. 2008. *Biosensors* (Ed. Cass, A.E.G.), 42. Oxford University Press, Oxford.
14. Garipcan B., Andac M., Uzun L., Denizli A. 2004. Methacryloylamidocysteine Functionalized Poly (2-Hydroxyethyl Methacrylate) Beads and Its Design as a Metal-Chelate Affinity Support for Human Serum Albumin Adsorption. *Reactive and Functional Polymers*, 59:119–128.

15. Beilen J.B., Li Z. 2002. Enzyme Technology: An Overview. *Current Opinion in Biotechnology*, 13:338–344.

16. Campàs M., Katakis I. 2004. DNA Biochip Arraying, Detection and Amplification Strategies. *Trends in Analytical Chemistry*, 23:49–62.

17. Dong S., Chen X. 2002. Some New Aspects in Biosensors. *Journal of Biotechnology*, 82:303–323.

18. Hemachandera C., Bosea N., Puvanakrishnan R. 2001. Whole Cell Immobilization of *Ralstonia pickettii* for Lipase Production. *Process Biochemistry*, 629:633–637.

19. Cooper J.C., Hall E.A. 1988. The Nature of Biosensor Technology. *Journal of Biomedical Engineering*, 10:210–219.

20. Ferrari M., Bashir R., Wereley S. 2007. *BioMEMS and Biomedical Nanotechnology*, Springer, USA.

21. Tibbe A.G., de Grooth B.G., Greve J., Dolan G.J., Rao C., Terstappen L.W. 2002. Magnetic Field Design for Selecting and Aligning Immunomagnetic Labeled Cells. *Cytometry*, 47:163–172.

22. Singh N., Manshian B., Jenkins G.J., Griffiths S.M., Williams P.M., Maffeis T.G., Wright C.J., Doak S.H. 2009. NanoGenotoxicology: The DNA Damaging Potential of Engineered Nanomaterials. *Biomaterials*, 30(23–24):3891–3914.

23. Ng H.T., Li J., Smith M.K., Nguyen P., Cassell A., Han J., Meyyappan M. 2003. Growth of Epitaxial Nanowires at the Junctions of Nanowalls. *Science*, 300(5623):1249.

24. Rasooly A. 2005. Biosensor Technologies. *Methods*, 37(1):1–3.

25. D'Souza S.F. 2001. Immobilization and Stabilization of Biomaterials for Biosensor Applications. *Applied Biochemistry and Biotechnology*, 96:225–238.

26. Rainina E.I., Efremenco E.N., Varfolomeyev S.D., Simonian A.L. 1996. The Development of a New Biosensor Based on Recombinant *E. coli* for the Direct Detection of Organophosphorus Neurotoxins. *Biosensors and Bioelectronics*, 11:991–1000.

27. McGlennen R.C. 2001. Miniaturization Technologies for Molecular Diagnostics. *Clinical Chemistry*, 47:393–402.

28. D'Souza S.F. 2001. Immobilization and Stabilization of Biomaterials for Biosensor Applications. *Applied Biochemistry and Biotechnology*, 96:225–238.

29. Snejdarkova M., Svobodova L., Gajdos V., Hianik T. 2001. Glucose Biosensors Based on Dendrimer Monolayers. *Journal of Materials Science: Materials in Medicine*, 12:1079–1082.

30. Arlett J.L., Myers E.B., Roukes M.L. 2011. Comparative Advantages of Mechanical Biosensors. *Nature Nanotechnology*, 6:203–215.

31. Pan Y., Sonn G.A., Sin M.L., Mach K.E., Shih M.C., Gau V., Wong P.K., Liao J.C. 2010. Electrochemical Immunosensor Detection of Urinary Lactoferrin in Clinical Samples for Urinary Tract Infection Diagnosis. *Biosens Bioelectron*, 26:649–654.

32. Catroux P., Cottin M., Rougier A., Leclaire J. 1995. Biosensors in Pharmacology and Toxicology in Vitro. In Book: *Modulation of Cellular Responses in Toxicity*. Edited by C. L. Galli, A. M. Goldberg, M. Marinovich, pp. 145–156. Springer-Verlag, Berlin, Heidelberg.

33. Aydin E.B., Aydin M., Sezginturk M.K. 2019. Biosensors in Drug Discovery and Drug Analysis. *Current Analytical Chemistry*, 15:467–484.

34. Bohunicky B., Mousal S.A. 2011. Biosensors: The New Wave in Cancer Diagnosis. *Nanotechnology, Science and Applications*, 4:1–10.

35. Vashistha R., Dangi A.K., Kumar A., Chhabra D., Shukla P. 2018. Futuristic Biosensors for Cardiac Health Care: An Artificial Intelligence Approach. *BioTech*, 8(8):358.

36. Qureshi A., Gurbuz Y., Niazi J.H. 2012. Biosensors for Cardiac Biomarkers Detection: A Review. *Sens Actuator B Chem.*, 171:62–76.

37. Wu Y., Yao X., Vespasiani G., Nicolucci A., Dong Y., Kwong J., Li L., Sun X., Tian H., Li S. 2017. Mobile App-Based Interventions to Support Diabetes Self-Management: A Systematic Review of Randomized Controlled Trials to Identify Functions Associated with Glycemic Efficacy. *Journal of Medical Internet Research mHealth uHealth*, 5(3):e35.

38. Satija U., Ramkumar B., Manikandan M.S. 2017. Real-Time Signal Quality-Aware ECG Telemetry System for IoT-Based Health Care Monitoring. *IEEE Internet of Things Journal*, 4(3):815–823.

39. Bandodkar A.J., Jeerapan I., Wang J. 2016. Wearable Chemical Sensors: Present Challenges and Future Prospects. *ACS Sensors,* 1(5):464–482.

40. Mongra A.C. 2012. Commercial Biosensors: An Outlook. *Journal of Academia and Industrial Research*, 1(6):310–313.

41. D'Orazio P. 2003. Biosensors in Clinical Chemistry. *Clinica Chimica Acta*, 334:41–69.

42. Azimzadeh M., Nasirizadeh N., Rahaie M., Naderi-Manesh H. 2017. Early Detection of Alzheimer's Disease Using a Biosensor Based on Electrochemically-Reduced Graphene Oxide and Gold Nanowires for the Quantification of Serum microRNA-137. *RSC Advances*, 7:55709–55719.

43. STM ThinkTech 2020. Biosensors in Coronavirus Diagnosis. STM Thinktech Future Technology Institute. https://thinktech.stm.com.tr/detay.aspx?id=376.

44. Bhushan B. 2007. Nanotribology and Nanomechanics of MEMS/NEMS and BioMEMS/BioNEMS Materials and Devices. *Microelectronic Engineering*, 84:387–412.

45. Okandan M., Galambos P., Mani S., Jakubczak J. 2001. Development of Surface Micromachining Technologies for Microfluidics and BioMEMS. Presented at SPIE Micromachining and Microfabrication Conference, San Francisco, CA.

46. Jianrong C., Yuqing M., Nongyue H., Xiaohua W., Sijiao L. 2004. Nanotechnology and Biosensors. *Biotechnology Advance*, 22:505–518.

21 Future Trends in Biomedical Applications

Somasundaram Ambiga, Raja Suja Pandian, Abdul Bakrudeen Ali Ahmed, Raju Ramasubbu, Ramu Arun Kumar, Lazarus Vijune Lawrence, Arjun Pandian, Sasikumar Yesudass and Sivakumar Bose

CONTENTS

21.1 INTRODUCTION

A biomaterial is basically a substance, whether natural or manmade, that constitutes an entire or a portion of a living structure or biomedical tool that executes, performs and enhances a natural feature, which can be used for a medical application and modified for it. Biomaterials have a benign role, including to be used for a heart regulator, which can be bioactive, similar to hydroxyapatite covered hip implants, designed for a much more active purpose. In surgery, dental applications and medicine delivery, biomaterials are still being used every day (Isa et al. 2000).

The synthetic or natural materials such as ceramic, metals, polymerics and their composite materials can efficiently integrate into the body and are known as biomaterials, being biocompatible with living tissues (Ozkizilcik and Tuzlakoglu 2014). Furthermore, biomaterials are bioactive materials, while the biomimetic materials and biodegradable materials are biologically inactive materials (Li et al. 2007). The materials which interact with body parts and body fluids are important in hand bones, tendons, nerves, heart valves, blood vessel prosthesis and cochlear replacements (Cicco et al. 2015).

DOI: 10.1201/9781003173533-21

The natural polymers are measured as the first eco-friendly (biodegradable) biomaterials used in the human experimental circumstances (Ueno et al. 2001). These natural materials, owing to their bioactive properties, are inclined to contain superior natural interactions through the cells, which allocate them to contain enhanced presentation in the natural system. Natural polymers can be classified as proteins (silk, collagen, fibrinogen, elastin, cheratin, actin and myosin) and polysaccharides (cellulose, amylose, dextran, chitin and glycosaminoglycans) or polynucleotides like DNA and RNA (Raghavendra et al. 2013; Steven et al. 2015).

In biological systems and medical equipment, biomaterials are also used, certain specific bioactive glass (Pillai and Sharma 2010), ceramic materials, polymeric materials and their composites, few metal/alloy based materials, synthesized resources of mammals, biopolymers and their composites (Chattopadhyay and Raines 2014). In assessments to certain supplementary conventional resources, the application of biomaterials has been growing substantially (Chang et al. 2003).

Several advanced methods are used to synthesize different types of biomaterials according to the required properties: a few methods like surfactant models, hydrothermal or thermal synthesis can engineer the bionanomaterials in a sensible way. Amongst various methodologies used for the biological nanomaterial synthesis, the most commonly used method is hydrothermal carbonization, which is performed in a water medium at a moderate heat and pressure (Young and Engler 2011). Resorbable biomaterials suggest that after implant in the human body, the biomaterial is degraded and absorbed through human body tissues (Huh et al. 2011). In order to enhance the functioning of raw materials, the application of biomaterials, design, selection and properties, such as biological, chemical, physical and mechanical, must be emphasized (Figure 21.1; Table 21.1).

In the recent past, numerous research works and review articles have focused on the different types of synthesis of biomaterials (from nano to macro scale), types of applications, and advancement in methodologies/developments for diagnosing/treatments. In the present chapter, we have also tried to elicit the brief of biomaterials, its applications, advancements and recent strategy for the benefit of human beings, to a certain extent.

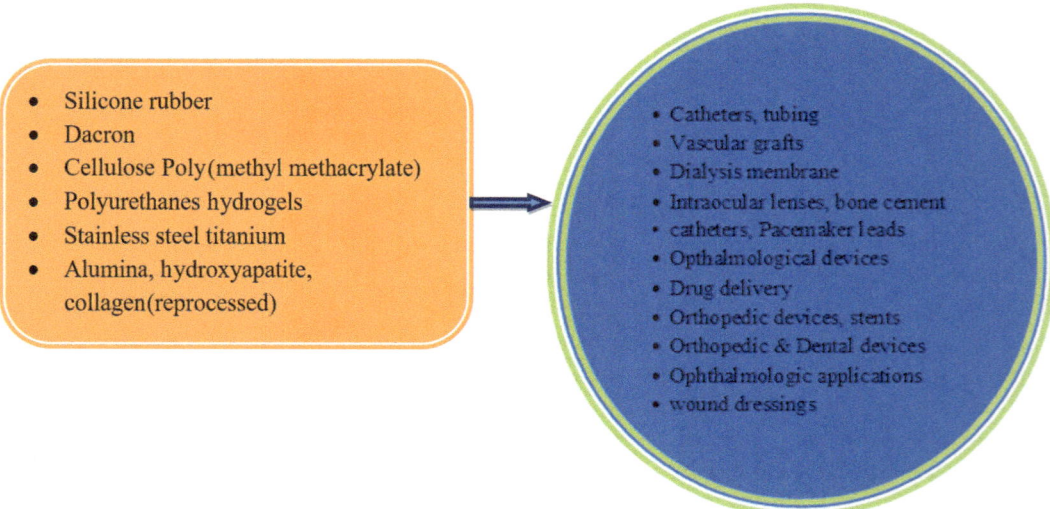

FIGURE 21.1   Commonly used biomaterial and their applications.

TABLE 21.1
Biomedical Applications of Enzyme-Like Catalytic Nanomaterials.

Enzyme Mimicked	Nanozyme	Different Types of Applications
Peroxidase	Iron oxide	Glucose detection
Peroxidase	Iron oxide	HepB ELISA
Peroxidase	Iron oxide	Tumor staining in histology
Peroxidase	Silver halides	Cancer cell detection
Peroxidase	Iron oxide	Ebola virus detection
Haloperoxidases	Vanadium oxide	Anti-biofilm
DNase	Gold-cerium	Anti-biofilm
Peroxidase	Gold-graphitic carbon nitride	Anti-biofilm/wound healing
Peroxidase	Graphene oxide	Wound healing
Peroxidase	Iron oxide	Anti-tumor
SOD	Cerium oxide	Retinal degeneration
SOD	Cerium oxide	Stent coating
SOD	Molybdenum sulfide	Radiation protection
Peroxidase	Iron oxide	Anti-ischemia
SOD and catalase	Cerium oxide	Anti-ischemia

21.2 BIOMATERIAL—PROTEIN SURFACE INTERACTIONS

Protein-surface interactions are primarily essential for medical devices biocompatibility. The action of proteins on the layer of a biomaterial performs an important function in deciding its nature of the interface of tissue-implant (Helmus et al. 2008).

21.3 FUTURE APPLICATION IN BIOACTIVE SPECIFIC BIOMATERIALS—BONE TISSUE ENGINEERING

Biomaterials should have distinctive features: biocompatibility or the capability of the material to work without an immune response from the host, biodegradability or the ability of the biomaterial to disintegrate when new bones are developed, basic structural features, sufficient osteoinductive and osteoconductive characteristics to promote cell growth and osteogenic multiplication in the healing place (Hubbell 2003). In addition, there are three main bone repair techniques in tissue engineering depending on the severity of the trauma: direct implantation of biomaterial, stem cell separation from patients and biomaterial planting as freshly collected bone marrow cells (Carvalho et al. 2013).

As bone graft replacements, synthetic biomaterials are presently being used, biomaterials in bone tissue engineering should be capable of promoting the multiplication of progenitor cells into osteoblasts, promoting bone formation, promoting the development of the neighboring bone, and incorporating into developed bone cells (Billiet et al. 2014). Depending on their biomechanical behavior, these biomaterials were primarily identified for structural reconstruction. Later scaffolds were designed to be bioactive, in order to improve tissue regeneration (Chen et al. 2008). Scaffolds are now planned to induce the development and vascularization of bones. All bone tissue engineering biomaterials must be bioresorbable and substituted in the body with recently regenerated biological tissue. Usually, these scaffolds are porous, biodegradable substances which carry various growth factors, medicines, stem cells (de La Puente et al. 2013).

The creation and formulation of 3D porous scaffolds of the same structure to the bone are one of the most applied techniques in bone tissue engineering (Kolesky et al. 2014). Bioceramic scaffolds, primarily hydroxyapatite (HA), have been frequently used because their chemical composition is identical to the mineral part of normal bone, and their biocompatibility and bioactivity facilitates the development of new bones *in vivo*. A biological action of the body, including such bonding to tissue is triggered by bioactive materials (Highley et al. 2015). The design and development of biocompatible scaffolds for bone tissue engineering needs a coordination of the osteoinductive cell microenvironment, soluble factor diffusion, versatility and mechanical filling ideal for the anatomical location (Singh et al. 2019).

Proteins/enlargement factors such as vascular endothelial growth factor, bone morphogenetic protein, transform expansion factor-β, insulin resembling enlargement factor and fibroblast expansion factor are biomolecules incorporated through scaffolds (Tathe et al. 2010). By integrating and differentiating osteoprogenitor cells into particular lineages, these growth factors regulate osteogenesis, bone tissue development and ECM formation. The union of non-union type bone fractures can be strengthened by updating distinctive biomolecules. The successful integration of biomolecules and growth factors in scaffolds can decrease the time of wound repair but also assist in patient recovery (Anselme 2000).

It is a great potential to acquire sequential release of GF from the scaffold by integrating various molecules into sequential polymer layers, a technique called layer by layer. It is capable of delivering GF sequentially by integrating various biomolecules into different layers and taking account of matrix-degrading enzymes generated from cells. An important aspect is a polycaprolactone (PCL) LBL scaffold made by heparin and VEGF enriched layers designed for vascular graft implementations.

21.4 BIOMATERIALS IN MEDICAL 3D PRINTING TECHNIQUES

Today, in the clinical sector, advanced production technology has broad applications and is growing rapidly (Hunt et al. 1984). In the development of biocompatible devices for the replacement of damaged tissues, the sector of tissue engineering performs a crucial function (Pramanik et al. 2005). The perception of tissue engineering is that biocompatible substances, subsist cells and expansion factors are combined to engender implants which promotes the proper tissue development in the engineered part. *In vitro*, the tissue engineering applications have provided the way for medically useful, supplementary physiological models instead of tissue resembling animal and human mini environments to be built (MacGregora et al. 2001). Redesigning the prostheses and implants, constructing biomedical devices, medical aids customized to the enduring, living scaffolds, and bioprinting tissues for regenerative medicine, have revolutionized the systems of health care (Ulery et al. 2011).

A wide variety of biomaterials, such as polymers, metals, ceramics and composites are presently used in medical 3D printing. At present, the development of 3D printing technologies is highly significant in the manufacture of advanced devices, prostheses, medicine liberation systems and 3D-scaffolds intended for tissue engineering, including therapy for regeneration (Iftekhar 2004). Materials used for the biomedical parts are developed by the medical 3D printing technology, including inkjet, extrusion based bio-printing, polyjet printing techniques and fused deposition modelling. The perfect 3D printing biomaterial is biocompatible, conveniently printable with tuneable degradation rates and morphologically mimic living tissue.

Emerging medical applications using 3D printing technologies are in tissue manufacturing for organ transplantation, disease modelling, discovery of drugs, production of customizable prosthetics/implants and anatomical models and implantable biosensors. Several researchers are presently using 3D printing to generate scaffolds for tissue engineering applications and to produce specific and complex 3D structures that facilitate cell adhesion, multiplication and movement in 3D printed models (Awad et al. 2003).

A vascularized thyroid gland model is created by biomaterial extrusion bio-printing technique and a collagen based bioink (Holzl et al. 2016). Throughout bioprinting, the authors of the study were intelligent to build representation embryonic stem cell derived thyroid and epithelial spheroids, which permitted the epithelial cells to invade and vascularize the thyroid spheroids. Over the last few years, bioprinting has emerged as a capable strategy of developing massively complex organoids and lab animal models (Chawla et al. 2018). The most elastic biomaterials could make complex models to be created which can retain their shape, but limit cellular interactions and movement at the same time. Highly viscous biomaterials generate a liquid cell environment, but also are less capable of producing great models without collapsing with their individual weight (Kumar et al. 2004).

The 3D printing use in the medical field is increasing dramatically, due to its ability, like medicine customization, cost efficiency, speed and increased productivity. The use of only a few environmental polymers for 3D printing are currently available. Moreover, for either drug delivery or space-filling implantation purposes, all of these 3D printing biomaterials have been used. These polymers are often blended with conventional biomaterials (such as TCP, HA) and used as composites to deliver orthopedic applications with better printability, mechanical durability and stronger tissue integration (Briganti et al. 2010).

21.5 RAPID PROTOTYPING TECHNOLOGY

Rapid prototyping (RP) is a technique in which physical design can be directly developed with the assistance of magnetic resonance imaging (MRI), computer assisted modelling (CAD) design and computed tomography (CT) data or any reverse engineering techniques. Used in numerous applications such as manufacturing industries, clinical operations, textile industries, aerospace applications and automotive industries, RP machines are now being used every day. In most cases, patient-specific medical models are developed. In order to generate medical models with high precision with less time, CT and MRI images are really a popular source of input. In the case of manufacturing medical models, high precision and adequate selection of materials is necessary. A wide variety of materials, such as cobalt chromium alloys, stainless steel, PEEK, titanium alloys, etc., are available for fabrication of medical devices. In the near future, bio fabrication is projected to grow to where machines can print human organs that can be effectively inserted into a patient to replace the damaged portion. Bio fabrication is currently used in various medical fields, such as dental surgery, oral and maxillofacial surgery, cardiac surgery, reconstructive and orthopedic surgery, etc.

Rapid prototyping models are not simply utilized for implants in current conditions, but also as diagnostic manuals or instruments. With the aid of RP technology, these tools are very simple to fabricate. A 6-month-old baby had a serious breathing issue. Due to a block in respiratory system, it was not functioning correctly and the doctor lost hope. They made an RP model as a last attempt, a biocompatible valve type of thing that was inserted into the blocked area to keep it open.

21.6 NANOFIBERS CHITOSAN APPLICATION

Very recently, the development of new antibacterial drugs for the treatment of wounds infected with antibacterial resistant microorganisms has been a major focus. It has been identified that chitosan substances through quaternary groups of ammonium are highly effective alongside fungi and bacteria. Chitosan scaffolding is used for the treatment of patients with deep burns, injuries, etc. Photo cross connected electrospun mats with quaternized chitosan effectively prevent the enlargement of both gram (+) and (−) bacteria and also it is a favorable material for wound dressing application (Park et al. 2013). A nanofibrous poly blend membrane of chitosan/collagen has been known to stimulate cell proliferation and migration while helping to heal wounds. Because of the very large specific surface area, nanofibers possess higher water holding capacity and are also very soft, but the dressing will not scrape the wound. A wound dressing consists of a nanofiber based on chitosan as a potential platform for dressing (Ahmed et al. 2008).

21.7   DRUG RELEASE

Owing to their wide surface area and regulated degradation, nanofiber scaffolds have been described as specific drug delivery systems. Chitosan nanofibrous structures have recently identified their place as a specific process of drug release. Electrospinning-prepared composite membranes of PLGA and PEG-g-chitosan have shown the capabilities to be packed with an anti-inflammatory substance called ibuprofen. Recently, electrospun fibers have gained a lot of attention as antitumor drug carriers, but since it has been a successful strategy for locating the tumor tissue delivery of antitumor drugs, particularly in postoperative local chemotherapy. By adjustment of the porosity, morphology and composition of nanofiber, the release of drug patterns from these systems has been controlled. Chitosan and its byproduct have attracted considerable attention as carriers of antitumor drugs such as doxorubicin hydrochloride (DOX).

The d-glucosamine compounds of chitosan will react with the negatively charged sialic acid byproducts of the mucus-composing. Therefore, muco-adhesion is related directly to chitosan deacetylation degree, with a larger chitosan deacetylation degree leading to higher positive charges that strengthen its muco-adhesive properties. Chitosan may associate to the cell membrane's negative portion, improving the diffusion of an energetic mediator via the epithelium layer which includes stretched junctions. Chitosan is therefore an ideal excipient for preparing ocular, nasal, oral, vaginal and subcutaneous modes of delivery and is used as a vaccine adjuvant to improve the immunogenicity and bioavailability of antigens due to muco-adhesion and increased diffusion properties. Such nanofibers have shown significant antitumor effect, making such forms of nanofibrous substances ideal candidates for cervical tumor treatment.

21.8   CHITOSAN 3D-SCAFFOLDS FOR TISSUE ENGINEERING

Chitosan provided distinctive opportunities for the improvement of biomedical applications with several remarkable characteristics. Specialized consideration must be given to porosity, compatibility, mechanical characteristics, scaffold morphology and also healing, tissue replacement capability throughout the fabrication of implantable scaffolds.

Chitosan's hemostatic function may also be correlated to the attendance of positive (+) charges on a vertebral column of chitosan. In fact, the membranes of red blood cells (RBCs) are unenthusiastically charged and therefore can interact with chitosan, which is positively charged. In addition, chitin exhibits very lower hemostatic activity than chitosan. Chitosan can also react with the negative (−) part of cell membranes due to the positive charges that can show the reorganization and opening of tight junction proteins explaining the possessions of this polysaccharide that enhances permeation. As for muco-adhesion, when the levels of chitosan deacetylation degree (DD) increase, the capacity for permeation also improves. Because of its effective chelating or binding ability of protonable amino groups of different species, such as metal ions, chitosan may also be utilized in waste water treatment and beverage clarification. Similarly, for tissue engineering, the very valuable pharmaceutical and medical applications of chitosan are medicine delivery, injuries dressings, and biocomposite scaffolds. Chitosan biomaterial and its applications are shown in Figure 21.2.

21.9   ORGAN-ON-A-CHIP

A different forum for medicine development is Organ-on-a-chip (OOc). OOc is also called a microphysiological system. Organ-on-a-chip is a biomimetic system and tiny device (i.e. a physiological organ built on a microfluidic chip). Approximately one billion people across the world have struggled from neurodegenerative disorders, like Parkinson's disease and Alzheimer's (AD), according to the World Health Organization (WHO) (Batista et al. 2004). An OOc combines various disciplines of material, chemical and biological sciences. The aim for the invention is to simulate the human organs' physiological roles, activities and mechanics (Lutolf et al. 2003). Organ-on-a-chip

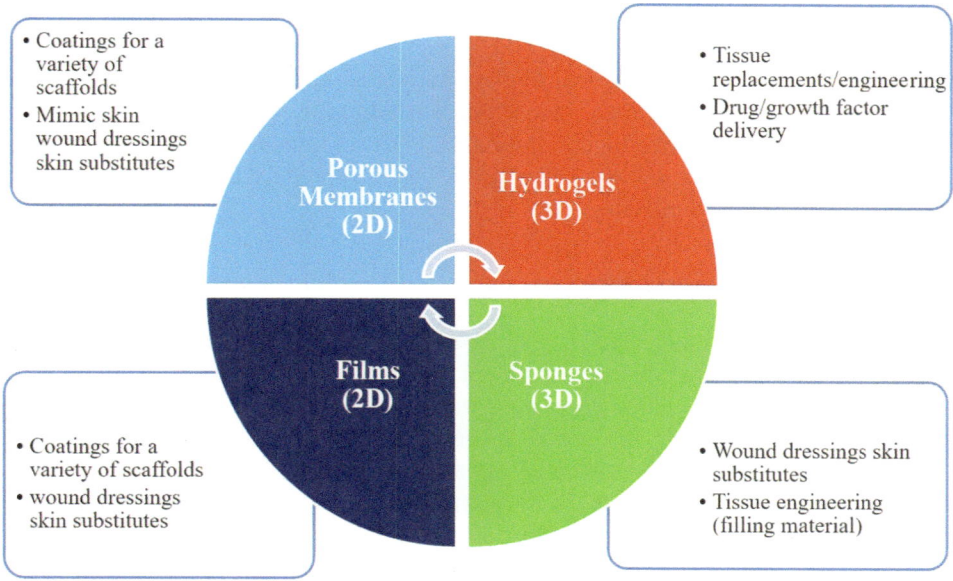

FIGURE 21.2   Chitosan biomaterial and its applications.

can mimic the human organs. It has the ability in the regulation of key parameters which includes tissue boundaries, cell patterning, shear force, concentration gradients, and tissue-organ interactions. These chips are designed for the purpose of controlling the physiological functions of the organs, and to control the fluid flow and viability of the cell. They are also important for the identification of any dysfunction of the body (ASTM 2003).

21.10   COMPONENTS OF THE ORGAN-ON-A-CHIP (OOC)

There are four main components in the chip. They are living-cell tissues, microfluidics, stimulation or medicine liberation and sensing. The microfluidics component is second-hand to transport targeted cells with the use of microfluidics to a pre-designed location. The livelihood cell tissue constituent is used to support a meticulous cell type spatially in the container of 3D or 2D systems. Hydrogels, a biocompatible material, are used to create the 3D arrangements and they prevent mechanical damage (Catanzano et al. 2015). So, with the drug delivery approaches, different signals are stimulated to the physiological functions. The sensing component is used to detect and to compile the data. Many studies have discussed setting a monitor to the cells in 3D microfluidics.

21.11   BRAIN-ON-A-CHIP

Compared to other 3D and 2D designs, microfluidics has differentiated brain-on-a-chip (brain chips) designs. Brain microfluidic chips have the ability to enhance the physiological functionality of the models to superior anatomical orders that in turn support designing models with superior order functionalities. Because the brain contains the central nervous system (CNS) along with the spinal cord, it is the most multifaceted organ in our human being body. The brain, as the higher backbone of processes of the CNS, incorporates and coordinates the sequence obtained, and subsequently makes decisions to systematize the activities of personality parts of the human body (Deng et al. 2014). The brain comprises frequent neurons that interact uni-directionally through synapses with each other, where one cell's axon terminal associates the dendrites of an additional in a particular direction. Nonneuronal cells, such as microglia, astrocytes and oligodendrocytes also interact with these neurons. By transmitting electrical or chemical signals between neurons via the synapses, the functions of the brain are regulated, and if this is not done properly, it may become a neurodegenerative disorder. One or more higher order anatomical characteristics, such as heterogeneous histotypic cell culture, cell orientation, 3-dimensionality, organotypic cell cultures, are periodically encouraged by brain chips (Sudheesh Kumar et al. 2010).

21.12   BRAIN CANCER CHIPS

More accurate dimensions and geometry give the advantage of brain-on-a-chip over other *in vitro* and *in vivo* models. These microfluidic chips assist the epithelium to check its movement of a physiological liquid which mimics blood circulation and helps better differentiation and maturation of the brain cells. Brain-on-a-chip has proved to be more exact than stationary cultures to assess blood-brain barrier permeability (Davis et al. 2008).

The tumor interacts strongly with its microenvironment, making a cancerous tissue unsuitable for mimicking spheroids, trans well cell cultures and organoids. The intestinal fluid pressure associated with the tumor and the mechanical/chemical transfer to the underlying cells and local tissue contribute to the complexity of the disease's pathophysiology. In developing new anticancer therapies as well as tracking the patient's progress in reaction to the treatment, a stable *in vitro* microfluidic model may aid significantly. Brain chips use microfluidic technology to provide a human-relevant microenvironment for tumor research by providing fluid flow regulation, biochemical and mechanical signals, etc. The following are examples of microfluidic brain cancer chips: tumor growth modelling, angiogenesis, metastasis, intra-vasation, and extra-vasation (Williams 1987).

21.13   THE FUTURE OF BRAIN-ON-A-CHIP

In addition to phenotypic, genotypic and circuit mapping, the characteristics of the brain chips are constructed perfectly for studying cell formation, migration, and signaling in several of the following fields, such as modelling of the blood brain barrier, simulation of brain cancer, and modelling of disease (Clark et al. 1982). While several brain chip models have been produced on these models, subsequent extra factors need to be considered to accurately replicate brain tissue structure and physiology, including cell-cell interactions, cell sources and interactions between cell and extracellular environment (Madri et al. 1988). In organ-on-chip engineering, re-circulation is a major problem, particularly when multiple organs were created by combining a multi-organ chip, also known as a human-on-chip (Dvir et al. 2011).

21.14   HEART-ON-A-CHIP

The chip in the heart is a microfluidic fragment that reproduces a heart's mechanisms to rapidly test medication and monitor the heart cells' reactions. The heart mechanisms in a synthetic structure, lined with live heart cells, are imitated with great care. Two revolutionizing events have hit the cardiac tissue engineering field; the introduction of microfluidic knowledge and induced pluripotent stem cell technology. Recent knowledge of cardiac tissue engineering researchers has been a powerful force in advancing research in this field and rapidly developing biologically important cardiac organ-on-a-chip termed as heart-on-a-chip (Rae 1986). The amalgamation of these two has empowered researchers to reconstruct cardiac tissues commencing patient-specific models for disease modelling and diagnosis as well as beneficial approaches in highly spatiotemporally regulated microenvironments (Gillette et al. 2010).

21.15   ELECTRIC STIMULATION

It is noted that cardiac muscle cells produce strong *in vivo* electrophysiological responses. In tissue growth, maintenance and regeneration, electrical signaling has an important function. Two goals are fulfilled by heart-on-a-chip systems employing electrodes. They moreover use the electrodes to stimulate the cells or exploit them to calculate the cells' electrical response to a specific stimulus as readout tools. Normally, electrical stimulation is carried out by electrodes that make contact with the cells. At single-cell resolution, the electrical indication may be applied to the bulk or to the cells. A pair of electrodes are typically used and connected to the unit to apply the electrical field to the bulk of the cells (Marijnissen et al. 2002).

21.16   NANOBOTS

Nanobots are very precise and 50–100 nm wide, and are robotic devices. For drug delivery, they can be very efficiently used (Sill and von Recum 2008). Nanorobots, frequently referred to as nanobots are nano-electromechanical devices that are deliberated to execute a vital function with nanoscale dimensional accuracy. The benefit of medicinal purposes lies in its size. The principle of using nanobots in nanomedicine comes from the fact that most biological structures are at the nanometer scale. At the molecular level, 1 nm is 10^{-9} of a meter or one billionth. For contrast, the usual distances of carbon-carbon bonds or the spacing in a molecule between these atoms are in the assortment of 0.12–0.15 nm, and a double-helix DNA has a diameter of concerning ~2 nm. There are chemical sensors in several nanobots which detect the target molecules. They will emit a power signal proportional to the observed quantity in the biological system as a response. This signal would enter a programmed microprocessor that governs the nanobot's direction and velocity (Melinda 2011).

However, as applications of nanotechnology develop, researchers around the world are seeking to exploit the advantages by equipping the nanorobots with methods that can easily photograph, diagnose, prescribe medication and conduct small surgical operations. Of course, nanorobots may be the future of nanomedicine and biotherapeutics may replace older models of drug delivery in the next few decades.

21.17   MICROBUBBLES

Not all bubbles that in the bloodstream are harmful, dissimilarity improved ultrasound has developed starting a predominantly investigational method to a schedule diagnostic procedure over the past decade. To define pathophysiology and to establish pioneering therapeutic strategies in the management of neoplastic and cardiovascular diseases, the capacity to non-invasively visualize molecular proceedings with beleaguered microbubbles is likely to be significant.

During echocardiography, microbubble dissimilarity agents and specialized imaging strategies intended to reduce tissue noise while enhancing contrast signals are presently used to enhance endocardial boundary description. During echocardiography, allowing enhanced visualization of myocardial segments has provided greater assurance and efficiency in partition motion analysis, as well as decreasing inter and intra spectra to run predictability. While echocardiography is used to identify acute coronary syndromes, throughout pressure echocardiography, correct measurement of wall motion is particularly necessary. In order to evaluate myocardial perfusion, to enumerate myocardial blood supply, microvascular honesty, and practicability, a comparison echocardiography should also be used. These applications use microbubble agents that inside the microvasculature are free-flowing (Pogorielov et al. 2017).

Gaseous microbubbles manufacture high backscatter ultrasound signals; they are 4–5 times supplementarily compressible than tissue or water. Microbubbles experience volumetric fluctuation in the auditory field throughout the irregular difficulty cycles of ultrasound, whereby they are compressed during the pressure peaks and extended during the troughs. Opportunity uses of contrast ultrasound have extended the usage of microbubbles beyond optimizing the left ventricular cavity, improving the description of cardiac and structural Doppler signals and also myocardial perfusion imaging. Microbubbles are currently used to determine molecular and cell pathology contained by the vascular space. The capability to angiogenic vessels, monitor inflammation and premature atheroma are only some few examples of how this technology can be applied. The diagnostic usefulness of ultrasound not just for heart disorders, but also for diseases involving other organs that can be imaged using ultrasound is expected to expand (Anderson 1993). In addition, microbubbles can be used for medicinal applications and can facilitate the delivery of medications, genes or other substances straight to the site of most need. Undoubtedly, even more testing, confirmation and preparation are required before these apparatuses are incorporated into experimental practice—but the future of contrast ultrasound holds enormous hope and enthusiasm (Ahmad and Faiyazuddin 2016).

21.18   TRANSDERMAL PATCHES

A transdermal piece is a medicated sticky square that is placed on the skin to distribute a particular dosage of medicine throughout the skin and through the bloodstream. This frequently promotes remediation of the wounded area of the body. The benefit of a transdermal medicine distribution path compared to other forms of drugs, such as topical, intravenous, dental, intramuscular, etc., is that the square allows managed discharge of the prescription to the user, generally moreover by a porous membrane that covers the medicine reservoir or through body heat that removes thin layers of the medication that are trapped in the patch (Tang et al. 2017). In attendance are five types of transdermal patches. They are single-single drug in adhesive, multilayer drug in matrix, adhesive, reservoir and vapour patch. They include a lining, the drug, adhesive; membrane, backing, permeable enhancer, matrix filter, stabilizer (antioxidant) and preservatives are also some of the components that are present in the transdermal patches (Wei 2011).

To enhance transdermal medicine transport, various methods have been used, including chemicals, electrical fields and ultrasound. These techniques have made transdermal delivery a viable method of systemic administration of drugs. In the last two decades, the scientific interest in this area has increased significantly. Various studies have been carried out to safely break the obstructive occupation of the skin and allow the therapeutic quantity of drugs to be administered. Studies conducted with solution or suspension formulations, however, have limited application potential (Wei 2011; Michael et al. 2015).

21.19   SUMMARY AND CONCLUSIONS

In conclusion, with the development of new biomedical materials (including the materials in nano scale), new treatment methodologies and diagnoses, the research scope is always opening up in the field of biomedical applications. The role of nanomaterials is highly promising and they are used as fluorescent markers in chemotherapy, radioactive materials (a nuclear imaging tool) biodetection assays and transplant materials. In particular, the novel nanomaterials having radioactivity that can be used as radioactive tracers, are highly helpful in earlier diagnosis of diseases (like thyroid, gall bladder, heart conditions, cancer, etc.) using nuclear imaging. A recent development of Amyloid PET imaging has helped to predict Alzheimer's progression. Graphene based enabling biotechnology is leading to the production of biofuel cells, label-free DNA biodetection and electrocatalytic devices. In view of medical devices development, a research group from the School of Medicine, New York University has come up with a wearable MRI Glove that can map the hand's anatomy with high accuracy, which aids from surgery to prosthetic design. Similarly, the bioprinting method has paved the way for the reconstruction of complex craniomaxillofacial defects which is a challenging task so far due to the organized layering of multiple tissue types. A study using a rat model has demonstrated ~80% of skin closure in 10 days and 50% of bone coverage in 6 weeks using intra-operative bioprinting approach. Personalized medicine, material innovations, miniaturization and additive manufacturing are key engineering trends that biomedical researchers are enthusiastic to integrate into their designs. Hence, this research works to emphasize the development of novel nano materials and new methodologies which will have a significant impact on the real biomedical applications for the benefit of humankind.

REFERENCES

Ahmad, U., and M. D. Faiyazuddin. 2016. Smart nanobots: The future in nanomedicine and biotherapeutics. *J Nanomedicine Biotherapeutic Discov.* 6(1):1000e140.

Ahmed, T. A. E., E. V. Dare, and M. Hincke. 2008; Fibrin: A versatile scaffold for tissue engineering applications. *Tissue Eng Part B Rev.* 14(2):199–215.

Anderson, J. M. 1993. Mechanism of inflammation and infection with implanted devices. *Cardiovasc Pathol* 2:33S–41S.

Anselme, K. 2000. Osteoblast adhesion on biomaterials, review. *Biomaterials.* 21:667–681.

ASTM. 2003. ASTM F 138: Standard Specification for Wrought 18chromium-14nickel-2.5 molybdenum Stainless Steel Bar and Wire for Surgical Implants (UNS S31673), West Conshohocken, ASTM International.

Awad, H. A., G. P. Boivin, M. R. Dressler, F. N. L. Smith, R. G. Young, and D. L. Butler. 2003. Repair of patellar tendon injuries using a cell—collagen composite. *J Orthop Res.* 21(3):420–431.

Batista, G., M. Ibarra, J. Ortiz, and M. Villegas. 2004. *Engineering biomechanics of knee replacement, applications of engineering mechanics in medicine.* Mayaguez: GED-University of Puerto Rico, pp. 1–12.

Billiet, T., E. Gevaert, T. de Schryver, M. Cornelissen, and P. Dubruel. 2014. The 3D printing of gelatin methacrylamide cell-laden tissue-engineered constructs with high cell viability. *Biomaterials.* 35(1):49–62.

Briganti, E., D. Spiller, and C. Mirtelli, 2010. A composite fibrin-based scaffold for controlled delivery of bioactive pro-angiogenetic growth factors. *J Control Release.* 142(1):14–21.

Carvalho P. P., I. B. Leonor, and J. Smith Brenda. 2013. Undifferentiated human adipose-derived stromal/stem cells loaded onto wet-spun starch—polycaprolactone scaffolds enhance bone regeneration: Nude mice calvarial defect in vivo study. *J Biomed Mater Res A.* 102(9):3102–3111.

Catanzano, O., V. DEsposito, and S. Acierno. 2015. Alginate—hyaluronan composite hydrogels accelerate wound healing process. *Carbohydrate Polymers.* 131:407–414.

Chang, S. C. N., G. Tobias, A. K. Roy, C. A. Vacanti, and L. J. Bonassar. 2003. Tissue engineering of autologous cartilage for craniofacial reconstruction by injection molding. *Plast Reconstr Surg.* 112(3):793–799.

Chattopadhyay, S., and R. T. Raines. 2014. Review collagen-based biomaterials for wound healing. *Biopolymers.* 101(8):821–833.

Chawla, S., S. Midha, A. Sharma, and S. Ghosh. 2018. Silk-Based Bioinks for 3D Bioprinting. *Adv Healthcare Mater.* 7(8):1701204.

Chen, Q., J. A. Roether, and A. R. Boccaccini. 2008. Tissue engineering scaffolds from bioactive glass and composite materials. In: N. Ashammakhi, R. Reis, F. Chiellini, editors. *Tissue engineering*, vol. 4.

Cicco, S. R., D. Vona, and E. deGiglio. 2015. Chemically modified diatoms biosilica for bone cell growth with combined drug-delivery and antioxidant properties. *Chempluschem.* 80(7):1104–1112.

Clark, R. A., J. M. Lanigan, P. DellePelle, E. Manseau, H. F. Dvorak, and R. B. Colvin. 1982. Fibronectin and fibrin provide a provisional matrix for epidermal cell migration during wound reepithelialization. *J Invest Dermatol.* 79:264–269.

Davis, M. E., Z. G. Chen, and D. M. Shin. 2008. Nanoparticle therapeutics: An emerging treatment modality for cancer. *Nat Rev Drug Discov.* 7(9):771–782.

de La Puente, P., D. Ludena, M. Lopez, J. Ramos, and J. Iglesias. 2013. Differentiation within autologous fibrin scaffolds of porcine dermal cells with the mesenchymal stem cell phenotype. *Exp Cell Res.* 319(3):144–152.

Deng, B., L. Shen, and Y. Wu. 2014. Delivery of alginate-chitosan hydrogel promotes endogenous repair and preserves cardiac function in rats with myocardial infarction. *J Biomed Mater Res A.* 103(3):907–918.

Dvir, T., B. P. Timko, D. S. Kohane, and R. Langer. 2011. Nanotechnological strategies for engineering complex tissues. *Nat Nanotechnol.* 6(1):13–22.

Gillette, B. M., J. A. Jensen, M. Wang, J. Tchao, and S. K. Sia. 2010. Dynamic hydrogels: Switching of 3D microenvironments using two component naturally derived extracellular matrices. *Adv Mater.* 22(6):686–691.

Helmus, M. N., D. F. Gibbons, and D. Cebon. 2008. Biocompatibility: Meeting a key functional requirement of next-generation medical devices. *Toxicol Pathol.* 36(1):70–80.

Highley, C. B., C. B. Rodell, and J. A. Burdick. 2015. Direct 3D printing of shear-thinning hydrogels into self-healing hydrogels. *Adv Mater.* 27(34):5075–5079.

Holzl, K., S. Lin, L. Tytgat, S. van Vlierberghe, L. Gu, A. Ovsianikov. 2016. Bioink properties before, during and after 3D bioprinting. *Biofabrication.* 8(3):032002.

Hubbell, J.A. 2003. Materials as morphogenetic guides in tissue engineering. *Curr Opin Biotechnol.* 14:551–558.

Huh, D., G. A. Hamilton, and D. E. Ingber. 2011. From 3D cell culture to organs-on-chips. *Trends Cell Biol.* 21(12):745–754.

Hunt, T. K., R. B. Heppenstall, E. Pines, and D. Rovee. 1984. *Soft and hard tissue repair: Biological and clinical aspects,* vol. 2. New York, NY: Praeger Scientific; pp. 283–292.

Iftekhar, A. 2004. Biomedical composites. In: *Standard handbook of biomedical engineering and design.* McGraw-Hill Companies. Inc.US.

Isa, Z. M., and I. A. Hobkir. 2000. Dental implants: Biomaterial, biomechanical and biological considerations. *Annal Dent Univ Malaya.* 7:27–35.

Kolesky, D. B., R. L. Truby, A. S. Gladman, T. A. Busbee, K. A. Homan, and J. A. Lewis. 2014. 3D bioprinting of vascularized, heterogeneous cell-laden tissue constructs. *Adv Mater.* 26(19):3124–3130.

Kumar, M. N. V. R., R. A. A. Muzzarelli, C. Muzzarelli, H. Sashiwa, and A. J. Domb. 2004. Chitosan chemistry and pharmaceutical perspectives. *Chem Rev.* 104(12):6017–6084.

Li, W. J., R. L. Mauck, J. A. Cooper, X. Yuan, and R. S. Tuan. 2007. Engineering controllable anisotropy in electrospun biodegradable nanofibrous scaffolds for musculoskeletal tissue engineering. *J Biomech.* 40(8):1686–1693.

Lutolf, M.P., F. E. Weber, H. G. Schmoekel, J. C. Schense, T. Kohler, R. Müller, and J. A. Hubbell. 2003. Repair of bone defects using synthetic mimetics of collagenous extracellular matrices. *Nat Biotechnol.* 21(5):513–518.

MacGregora, J. T., J. M. Collinsa, and Y. Sugiyamab. 2001. *In vitro* human tissue models in risk assessment: Report of a consensus-building workshop. *Toxicol Sci.* 59(1):17–36.

Madri, J. A., B. M. Pratt, and A. M. Tucker. 1988. Phenotypic modulation of endothelial cells by transforming growth factor-beta depends upon the composition and organization of the extracellular matrix. *J Cell Biol.* 106:1375–1384.

Marijnissen, W. J. C. M., G. J. V. M. van Osch, and J. Aigner. 2002. Alginate as a chondrocyte-delivery substance in combination with a non-woven scaffold for cartilage tissue engineering. *Biomaterials.* 23(6):1511–1517.

Melinda, W. M. 2011. Organs-on-a-chip for faster drug development. *Scientific American.* www.scientificamerican.com/article/organs-on-a-chip/ (Accessed September 2020).

Michael, N. P., N. K. Yogeshvar, H. Michael, and S.R. Michael. 2015. Transdermal patches: history, development and pharmacology. *British Journal of Pharmacology.* 172:2179–2209.

Ozkizilcik, A., and K. Tuzlakoglu. 2014. A new method for the production of gelatin microparticles for controlled protein release from porous polymeric scaffolds. *J Tissue Eng Regen Med.* 8(3):242–247.

Park, H., B. Choi, J. Hu, and M. Lee. 2013. Injectable chitosan hyaluronic acid hydrogels for cartilage tissue engineering. *Acta Biomater.* 9(1):4779–4786.

Pillai, C. K., and C. P. Sharma. 2010. Review paper: Absorbable polymeric surgical sutures: Chemistry, production, properties, biodegradability, and performance. *J Biomater Appl.* 25(4):291–366.

Pogorielov, M., O. Oleshko, and A. Hapchenko. 2017. Tissue engineering: Challenges and selected application. *Adv Tissue Eng Regen Med Open Access.* 3(2):330–334.

Pramanik, S., A. K. Agarwal, and K. N. Rai. 2005. Chronology of total hip joint replacement and materials development. *Trends Biomater Artif Organs.* 19(1):15–26.

Rae, T. 1986. The macrophage response to implant materials. *Crit Rev Biocompat.* 2:97–126.

Raghavendra, G. M., T. Jayaramudu, K. Varaprasad, R. Sadiku, S. S. Ray, and K. M. Raju. 2013. Cellulose polymer Ag nanocomposite fibers for antibacterial fabrics/skin scaffolds. *Carbohydr Polym.* 93(2):553–560.

Sill, T. J., and H. A. von Recum. 2008. Electrospinning: Applications in drug delivery and tissue engineering. *Biomaterials.* 29(13):1989–2006.

Singh, A., P. A. Shiekh, M. Das, J. Seppala, and A. Kumar. 2019, Aligned chitosan-gelatin cryogel-filled polyurethane nerve guidance channel for neural tissue engineering: Fabrication, characterization, and *In Vitro* evaluation. *Biomacromolecules.* 20(2):662–673.

Steven R. C., W. W. Daniel, K. G. William, M. Ziad, D. B. Marni, and A. C. H. Brendan. 2015. Collagen scaffolds incorporating coincident gradations of instructive structural and biochemical cues for osteotendinous junction engineering. *AdvHealthc Mater.* 4(6): 831–837.

Sudheesh Kumar, P. T., S. Abhilash, K. Manzoor, S. V. Nair, H. Tamura, and R. Jayakumar. 2010. Preparation and characterization of novel β-chitin/nanosilver composite scaffolds for wound dressing applications. *Carbohydr Polym.* 80(3):761–767.

Tang, L., Y. Zeng, H. Du, M. Gong, J. Peng, B. Zhang, M. Lei, F. Zhao, W. Wang, X. Li, and J. Liu. 2017. CRISPR/Cas9-mediated gene editing in human zygotes using Cas9 protein. *Mol Genet Genomics.* 292(3):525–533.

Tathe, A., M. Ghodke, and A. P. Nikalje. 2010. A brief review: Biomaterials and their application. *Int J Pharm Sci.* 2(4):19–23.

Ueno, H., T. Mori, and T. Fujinaga. 2001. Topical formulations and wound healing applications of chitosan. *Advanced Drug Delivery Reviews.* 52(2):105–115.

Ulery, B. D., L. S. Nair, and C. T. Laurencin. 2011. Biomedical applications of biodegradable polymers. *J Polym Sci, B, Polym Phys.* 49(12):832–864.

Wei, K. 2011. Future applications of contrast ultrasound. *J Cardiovasc Ultrasound.* 19(3):107–114.

Williams, D. F. 1987. Review: Tissue biomaterial interactions. *J Mat Sci.* 22(10):3421–3445.

Young, J. L., and A. J. Engler. 2011. Hydrogels with time-dependent material properties enhance cardiomyocyte differentiation *in vitro*. *Biomaterials.* 32(4):1002–1009.

Index

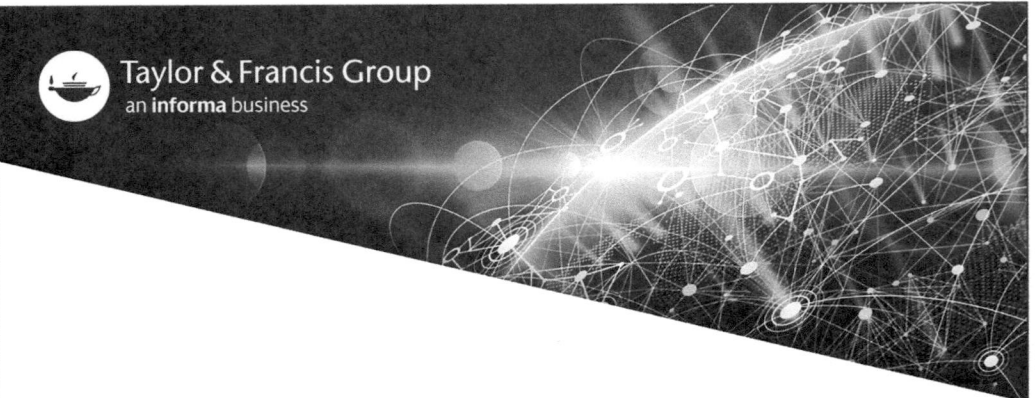

T - #0046 - 250924 - C85 - 254/178/16 - PB - 9781032002903 - Gloss Lamination